INTRODUCTION TO SPECTROSCOPY:
A Guide for Students of Organic Chemistry

DONALD L. PAVIA

GARY M. LAMPMAN

GEORGE S. KRIZ, Jr.

Department of Chemistry
Western Washington University, Bellingham, Washington

SAUNDERS GOLDEN SUNBURST SERIES

1979 W. B. SAUNDERS COMPANY • Philadelphia • London • Toronto

W. B. Saunders Company: West Washington Square
Philadelphia, PA 19105

1 St. Anne's Road
Eastbourne, East Sussex BN21 3UN, England

1 Goldthorne Avenue
Toronto, Ontario M8Z 5T9, Canada

Front cover illustration: Infrared, nuclear magnetic resonance, ultraviolet, and mass spectra of *p-anisaldehyde*.

Introduction to Spectroscopy: A Guide for Students of Organic Chemistry ISBN 0-7216-7119-5

 Made in the United States of America. Press of W. B. Saunders Company. Library of Congress catalog card number 77-11348.

Last digit is the print number: 9 8 7 6 5 4 3 2 1

TO
Neva-Jean, Marian, and Dian

PREFACE

This book is intended to serve as a supplement to the typical organic chemistry lecture textbook, for those sections which deal with aspects of spectroscopy. In addition, this book can stand alone as a textbook for a course in spectroscopic methods of structure determination or qualitative analysis in organic chemistry. Our intent is to create a book which is of an intermediate level of difficulty. We have tried not to dwell excessively on difficult aspects of theory, but we still have attempted to cover the important aspects of each spectroscopic technique.

The discussions in this text have evolved from materials which we have developed over a period of years for use in our own classes. We have used these materials in our one-year sequence of lecture courses in organic chemistry, in qualitative organic analysis and spectroscopic methods classes, and in some graduate courses. We have found our approach to be very successful with our students, and we believe that this textbook should prove very useful with courses taught in other institutions.

With each chapter of this text, a large number of problems have been included to provide the student with a great deal of practice in applying the principles developed. In addition, the final chapter consists of combined problems which use information gathered by each of the spectroscopic methods to reach a structure determination. Our philosophy of using large numbers of drill problems to reinforce the discussions in the book is reflected in these problems. Answers to the problems are provided at the end of the book.

We wish to express our gratitude to the people who assisted us with the preparation of this manuscript. Some of the infrared spectra were determined by Neva-Jean Pavia. Robert Carter determined some of the ultraviolet spectra. Dian Kriz assisted with preliminary graphing of mass spectra and with proofreading of various drafts of the chapters.

We wish to acknowledge the cooperation of Varian Associates and the Aldrich Chemical Company for permission to use nuclear magnetic resonance spectra from their catalogs. Special thanks are due to Professor William Trager, who generously provided us with the mass spectrum of dopamine, and to Professors Arnold Krubsack and Thomas Cogdell, who reviewed an earlier version of some of the chapters.

We also wish to express our special thanks to Mr. Robert LaRiviere, who illustrated the text. We feel that his work does much to present our ideas in a clear pictorial form. Joy Dabney redrew the mass spectrum of dopamine.

Finally, we wish to thank our wives, Neva-Jean, Marian, and Dian, for their patience and understanding. Their support was most valuable as we struggled to translate our ideas into a final, polished manuscript.

DONALD L. PAVIA
GARY M. LAMPMAN
GEORGE S. KRIZ, JR.

CONTENTS

CHAPTER 1

MOLECULAR FORMULAS AND WHAT CAN BE LEARNED FROM THEM

Before attempting to deduce the structure of an unknown organic substance from an examination of its spectra, one can simplify the problem somewhat by examining the molecular formula of the substance. The purpose of this chapter is to describe how the molecular formula of a compound is determined and how structural information may be obtained from that formula.

1.1 ELEMENTAL ANALYSIS AND CALCULATIONS

The process of determining the molecular formula of a substance involves three steps. The first step is to perform a *qualitative elemental analysis*, in order to find out what kinds of atoms are present in the molecule. The second step is to perform a *quantitative elemental analysis*, in order to determine the relative numbers of the different kinds of atoms in the molecule. This leads to an *empirical formula*. The third step is a *molecular weight determination*, which, when combined with the empirical formula, shows the actual numbers of the different kinds of atoms. This, then, is the *molecular formula*.

Virtually all organic compounds contain carbon and hydrogen. In most cases it is not necessary to determine whether these elements are present – their presence is assumed. However, if it should be necessary to demonstrate that either carbon or hydrogen is present in a compound, that substance may be burned in the presence of oxygen. If combustion produces carbon dioxide, carbon must have been present in the unknown material; if water is produced in the combustion, hydrogen atoms must have been present in the unknown.

Nitrogen, chlorine, bromine, iodine, and sulfur may be identified by tests similar to the *sodium fusion test*. To obtain details of such a test, the student is urged to consult a textbook on qualitative organic analysis. Examples of suitable texts are included at the end of this chapter.

It is important to note that there is no suitable direct method for determining the presence of oxygen in a substance. For this reason, the presence of oxygen is not demonstrated in a qualitative analysis.

To determine the precise amounts of carbon and hydrogen present in an unknown substance, a quantitative analysis is required. In practice, commercial laboratories frequently perform these analyses, but it is worthwhile to describe the procedure here. The method of determining the amounts of carbon and hydrogen in a substance involves combustion to carbon dioxide and water. In a quantitative analysis, the carbon dioxide and water are collected and weighed. Figure 1–1 describes an apparatus which might be used for such a combustion analysis. The combustion occurs over copper oxide in a quartz tube which is heated to 600–800°. The combustion products pass through two U-tubes. The first U-tube is filled with a drying agent and the second with a strong base. The U-tubes are weighed before installation in the combustion apparatus. Water which is produced in the combustion is absorbed by the drying agent (usually magnesium perchlorate). The weight of the U-tube after the combustion is compared to its weight before the combustion to determine the number of grams of water produced. In the second tube, the strong base absorbs the carbon dioxide which is produced in the combustion. A typical strong base is the commercial product Ascarite, which is composed of sodium hydroxide dispersed on asbestos. Again, the weight increase of this U-tube is due to the number of grams of carbon dioxide produced in the combustion.

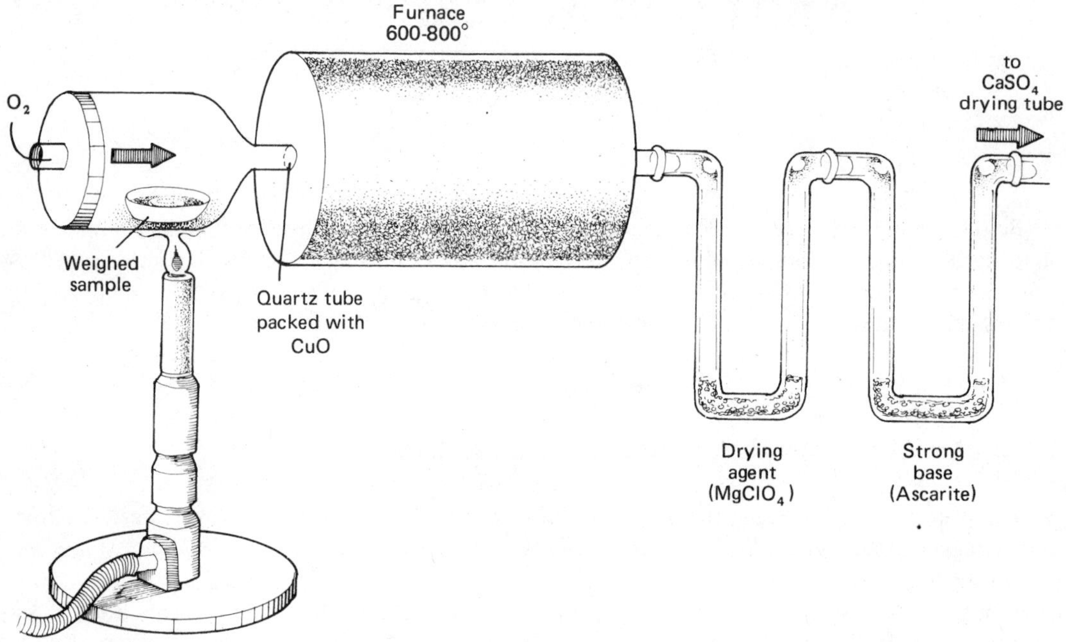

FIGURE 1–1 Combustion Apparatus for the Process:
C_xH_y + excess $O_2 \rightarrow x\ CO_2 + y/2\ H_2O$

Essentially the same type of procedure may be used to determine the number of milligrams of other elements which might be present in a molecule. For example, halogen may be determined by converting covalently bound halogen to ionic halide ions. This is achieved by treating the organic compound with hot, fuming nitric acid or sodium peroxide. This treatment oxidizes halogen to halide ions, which are precipitated with silver ion. The silver halide precipitate is weighed, and the number of moles of silver halide is equated to the number of moles of halogen in the original sample. Sulfur may be determined by the same oxidation process. Sulfur is oxidized to sulfate ion, which is precipitated with barium ions. The barium sulfate precipitate is weighed, and the number of moles of barium sulfate is equated to the number of moles of sulfur in the original sample. To determine the amount of nitrogen present, a combustion procedure is used which converts covalently bound nitrogen to nitrogen gas. The number of moles of nitrogen gas is obtained from the weight, pressure, volume, and temperature of the gas which has been collected. The ideal gas law is used to obtain the number of moles of gas at standard temperature and pressure. In this case, for every mole of nitrogen gas collected, there must be *two* moles of nitrogen atoms in the organic compound.

To illustrate how the combustion method may be used, let us examine a particular example. In this example, an unknown substance, which possessed the odor of bananas, and which acted as an alarm pheromone for honeybees, was subjected to a combustion analysis. The amount of unknown used in the analysis was 9.83 mg. The combustion produced 23.26 mg of carbon dioxide and 9.52 mg of water.

Taking the amount of carbon dioxide produced, we can calculate the weight of carbon contained in the sample. The molecular weight of carbon dioxide is 44.01 g/mole. The number of millimoles of carbon dioxide produced in the combustion is:

$$\frac{23.26 \text{ mg } CO_2}{44.01 \text{ mg/mmole}} = 0.5285 \text{ mmoles } CO_2$$

In the combustion process, one mole of carbon dioxide is produced for every mole of carbon in the sample, so the number of millimoles of carbon in the sample must be 0.5285 mmoles. This value can be converted to milligrams of carbon:

$$(0.5285 \text{ mmoles})(12.01 \text{ mg/mmole}) = 6.35 \text{ mg C in original sample}$$

Taking the amount of water produced, it is possible to proceed in the same manner to obtain the weight of hydrogen present in the sample. However, in this case, one must remember that one mole of water is produced for every *two* moles of hydrogen present in the sample. The number of millimoles of water must be multiplied by two in order to obtain the number of millimoles of hydrogen in the sample.

$$\frac{9.52 \text{ mg } H_2O}{18.02 \text{ mg/mmole}} = 0.528 \text{ mmoles } H_2O = 1.056 \text{ mmoles H}$$

$$(1.056 \text{ mmoles})(1.008 \text{ mg/mmole}) = 1.06 \text{ mg H in original sample}$$

The next step is to obtain the *percentage composition* of the unknown substance. The percentage of each atom in the sample is equal to weight of that atom, divided by the weight of the sample, and multiplied by 100.

$$\%C = \frac{6.35 \text{ mg C}}{9.83 \text{ mg sample}} \times 100 = 64.6\%$$

$$\%H = \frac{1.06 \text{ mg H}}{9.83 \text{ mg sample}} \times 100 = 10.8\%$$

At this point, one may notice that the percentages of carbon and hydrogen total 75.4%, rather than 100%. Not all of the atoms present in the molecule have been accounted for by the combustion analysis. We may assume, for the purposes of this example, that qualitative elemental analysis did not reveal the presence of elements other than carbon and hydrogen. However, it must be recalled that there is no simple direct chemical method of analysis for the presence of oxygen. Therefore, it must be assumed that the remaining atoms in the unknown substance are oxygen atoms. Subtracting from 100% gives the percentage oxygen as 24.6%. The percentage oxygen in samples generally is determined by difference, as has been illustrated here. The calculation of percentage composition is summarized in Table 1–1.

It is fairly rare to find a laboratory in which elemental analyses are performed as has been described here. Either samples are sent to a commercial analytical laboratory which specializes in quantitative elemental analyses, or modern instrumentation is used. In either case, the principles outlined in this section still apply, but the procedures tend to be too time-consuming to be conducted in every organic chemistry laboratory.

Commercially available elemental analyzers are instruments which are capable of determining the percentage of carbon, hydrogen, and nitrogen in a compound. These determinations are performed simultaneously. In these instruments, the sample is combusted in a stream of oxygen. The gaseous products are converted to carbon dioxide, water, and nitrogen. Each gas is detected independently by means of thermal conductivity detectors, similar to those used in gas chromatography equipment. Some instruments are capable of performing oxygen analyses.

TABLE 1–1 Calculation Of Percentage Composition From Combustion Data

$$C_xH_yO_z + \text{excess } O_2 \rightarrow x\,CO_2 + y/2\,H_2O$$

9.83 mg 23.26 mg 9.52 mg

$$\text{millimoles } CO_2 = \frac{23.26 \text{ mg } CO_2}{44.01 \text{ mg/mmole}} = 0.5285 \text{ mmoles } CO_2$$

$$\text{mmoles } CO_2 = \text{mmoles C in original sample}$$

$$(0.5285 \text{ mmoles C})(12.01 \text{ mg/mmole C}) = 6.35 \text{ mg C in original sample}$$

$$\text{millimoles } H_2O = \frac{9.52 \text{ mg } H_2O}{18.02 \text{ mg/mmole}} = 0.528 \text{ mmoles } H_2O$$

$$(0.528 \text{ mmoles } H_2O)\left(\frac{2 \text{ mmoles H}}{1 \text{ mmole } H_2O}\right) = 1.056 \text{ mmoles H in original sample}$$

$$(1.056 \text{ mmoles H})(1.008 \text{ mg/mmole H}) = 1.06 \text{ mg H in original sample}$$

$$\%C = \frac{6.35 \text{ mg C}}{9.83 \text{ mg sample}} \times 100 = 64.6\%$$

$$\%H = \frac{1.06 \text{ mg H}}{9.83 \text{ mg sample}} \times 100 = 10.8\%$$

$$\%O = 100 - (64.6 + 10.8) = 24.6\%$$

The percentage composition result may be used to calculate the empirical formula of the substance being studied. In order to illustrate how this calculation is accomplished, the unknown substance for which percentage composition values were calculated will be used as an example. If one had 100 grams of this substance, it would contain 64.6 g carbon, 10.8 g hydrogen, and 24.6 g oxygen. Converting these weights to moles of the elements, one obtains:

$$\text{moles C} = \frac{64.6 \text{ g}}{12.01 \text{ g/mole}} = 5.38 \text{ moles}$$

$$\text{moles H} = \frac{10.8 \text{ g}}{1.008 \text{ g/mole}} = 10.7 \text{ moles}$$

$$\text{moles O} = \frac{24.6 \text{ g}}{16.0 \text{ g/mole}} = 1.54 \text{ moles}$$

From these results, a formula

$$C_{5.38}H_{10.7}O_{1.54}$$

might be written. However, this formula is not in the form that one is accustomed to seeing. It is much better to convert these decimal fractions to the *simplest, whole number ratios*. This is accomplished by dividing each of the figures by the smallest of them:

$$\frac{5.38}{1.54} = 3.49 \approx 3.50 \text{ (carbon)}; \quad \frac{10.7}{1.54} = 6.95 \approx 7.00 \text{ (hydrogen)}$$

$$\frac{1.54}{1.54} = 1.00 \text{ (oxygen)}$$

In this case, to obtain a whole number ratio, it is necessary to multiply each of these ratios by two. Doing so leads to the *empirical formula* of the unknown. The empirical formula is the simplest formula which may be written for a substance with each of the elements indicated in the correct ratio. The empirical formula may not be the true, or molecular, formula of the substance being examined. The molecular formula may be some multiple of the empirical formula. In the example with which we have been dealing, the empirical formula is

$$C_7H_{14}O_2$$

The calculation of the empirical formula is summarized in Table 1–2.

TABLE 1–2 Calculation Of Empirical Formula

Using a 100 g sample:

$$\begin{aligned} 64.6\% \text{ of C} &= 64.6 \text{ g} \\ 10.8\% \text{ of H} &= 10.8 \text{ g} \\ 24.6\% \text{ of O} &= \underline{24.6 \text{ g}} \\ & \quad 100.0 \text{ g} \end{aligned}$$

$$\text{moles C} = \frac{64.6 \text{ g}}{12.01 \text{ g/mole}} = 5.38 \text{ moles C}$$

$$\text{moles H} = \frac{10.8 \text{ g}}{1.008 \text{ g/mole}} = 10.7 \text{ moles H}$$

$$\text{moles O} = \frac{24.6 \text{ g}}{16.0 \text{ g/mole}} = 1.54 \text{ moles O}$$

giving the result:

$$C_{5.38}H_{10.7}O_{1.54}$$

Converting to the simplest ratio:

$$C_{\frac{5.38}{1.54}}H_{\frac{10.7}{1.54}}O_{\frac{1.54}{1.54}} = C_{3.49}H_{6.95}O_{1.00}$$

which approximates

$$C_{3.50}H_{7.00}O_{1.00}$$

or

$$C_7H_{14}O_2$$

1.2 MOLECULAR WEIGHT DETERMINATION

The next step in determining the molecular formula of a substance is to determine the weight of one mole of that substance. This may be accomplished in a variety of ways. Without knowledge of the molecular weight of the unknown, one cannot tell whether the empirical formula, which is determined directly from elemental analysis, is the true formula of the substance or whether the empirical formula must be multiplied by some integral factor to obtain the molecular formula. In the example cited above, without knowledge of the molecular weight of the unknown, it is impossible to tell whether the molecular formula is $C_7H_{14}O_2$ or $C_{14}H_{28}O_4$.

The simplest method for determining the molecular weight of a substance is the *vapor density method*. In this method, a known volume of gas is weighed at a known temperature. After converting the volume of the gas to standard temperature and pressure, one may determine what fraction of a mole that volume represents. From this, the molecular weight of the substance may be determined easily.

The vapor density method may be demonstrated for the unknown substance which has been discussed in Section 1.1. Let us assume that a flask whose volume is 500 ml is filled with vapor of the unknown substance to a pressure of 750 mm Hg at 142°C. The weight of vapor is 1.886 g. The first step is to convert the pressure, volume, and temperature data to standard conditions, in order to determine what the volume of the vapor would be at standard temperature and pressure.

$$\text{volume at STP} = \frac{(750\text{ mm})(500\text{ ml})(273^\circ)}{(760\text{ mm})(273^\circ + 142^\circ)} = 324.6\text{ ml}$$

To determine the number of moles of vapor contained in this volume it is necessary to remember that one mole of any gas occupies 22,400 ml at standard temperature and pressure.

$$\frac{324.6\text{ ml}}{22{,}400\text{ ml/mole}} = 0.0145\text{ moles}$$

It is known that 0.0145 moles of vapor weighs 1.886 g. One mole of vapor weighs:

$$\frac{1.886\text{ g}}{0.0145\text{ moles}} = 130\text{ g/mole}$$

This is the molecular weight of the unknown substance.

Another method of determining the molecular weight of a substance is measuring the freezing point depression of a solvent brought about when a known quantity of test substance is added. This is known as a *cryoscopic method*. The equation which governs this method is:

$$\text{mol. wt.} = \frac{(K)(w)(1000)}{(\Delta t)(W)}$$

where K is the molal freezing point depression constant, w is the weight of the solute expressed in grams, Δt is the observed depression of the freezing (or melting) point of the solvent, and W is the weight of the solvent.

The value of K for water is 1.86° which means that a 1 molal solution of a substance in water will exhibit a freezing point of −1.86°. In this example, it is assumed that the solute does not dissociate into ions in the solvent.

The determination of molecular weights by a cryoscopic method requires rather painstaking procedures to determine the amount by which the freezing point of the solvent has been depressed. However, a much more convenient variation on this method is the *Rast method*. In the Rast method, the solvent used is *camphor* (m.p. 178°). Since the solvent is a solid, the camphor and the solute must be melted to form a homogeneous mixture. This mixture is cooled until it resolidifies, and then the melting point of the mixture is determined. The advantage of using camphor is that it has a molal freezing point depression constant of 40.0°. The variations in melting point caused by a solute are sufficiently large that the melting point of the mixture can be determined in a melting point tube with results of sufficient accuracy for most purposes.

If the unknown substance is a carboxylic acid, it may be titrated with sodium hydroxide of known normality. Using the following equation, a *neutralization equivalent* may be determined.

$$\text{neutralization equivalent} = \frac{\text{mg acid}}{(\text{normality of NaOH})(\text{ml of NaOH added})}$$

The neutralization equivalent is identical to the equivalent weight of the acid. If the acid has only one carboxyl group, the neutralization equivalent and the molecular weight are identical. If the acid has more than one carboxyl group, the neutralization equivalent is equal to the molecular weight of the acid divided by the number of carboxyl groups. Many phenols, especially those substituted by electron-withdrawing groups, are sufficiently acidic to be titrated by this same method, as are sulfonic acids.

Another method sometimes used to determine molecular weights is *vapor pressure osmometry*. In this method, the change in vapor pressure of a solvent when a test substance is dissolved in it is determined. Expressed in simple terms, when a solution containing a solute is evaporated, the vapor contains less of the solute than the liquid phase. This is a result of *Raoult's Law* and is quite analogous to the effect observed during distillation of a liquid mixture. When the vapor recondenses to the liquid phase, the liquid becomes warmer due to the liberation of the latent heat of vaporization of the solvent. In a vapor pressure osmometer, the equilibrium between liquid and vapor is established on the surface of a termistor, which is capable of measuring this temperature increase in the liquid. The quantitative relationship between the temperature of the liquid and the molality of the solution being studied is the *Clausius-Clapeyron equation*. Without treating the detailed derivation, it is possible to combine Raoult's Law and the Clausius-Clapeyron equation to obtain the significant equation for this method. The molecular weight of the test substance is obtained from the temperature change by the equation

$$\text{mol. wt.} = \frac{RT^2C}{(\Delta T)(\Delta h)(1000)}$$

where R is the gas constant, T is the temperature of the pure solvent, C is the weight in grams of the solute per 1000 grams of solvent, ΔT is the temperature change of the liquid phase of the solution, and Δh is the latent heat of vaporization *per gram* of the solvent. Where an instrument capable of performing these measurements is available, the method proves to be rapid and precise.

Finally, molecular weights can be determined by *mass spectrometry*. The details of this method and the means by which molecular weights can be determined may be found in Chapter 6, Section 6.2.

1.3 MOLECULAR FORMULAS

Once the molecular weight and the empirical formula are known, one may proceed directly to the molecular formula. Often the empirical formula weight and the molecular weight are the same. In such cases, the empirical formula is also the molecular formula. However, in many other cases, the empirical formula weight is less than the molecular weight. In these cases, it is necessary to determine how many times the empirical formula weight can be divided into the molecular weight. The factor which is determined in this manner is the factor by which the empirical formula must be multiplied in order to obtain the molecular formula. A simple example is found in the case of *ethane*. After quantitative elemental analysis, one determines that the empirical formula for ethane is CH_3. A molecular weight determination gives a result of 30 as the molecular weight of ethane. The empirical formula weight of ethane, 15, can be divided into the molecular weight, 30, two times. Therefore, the molecular formula of ethane must be $2(CH_3)$ or C_2H_6.

For the example initially introduced in Section 1.1, the empirical formula was found to be $C_7H_{14}O_2$. The formula weight is 130. In Section 1.2, the molecular weight of this substance was determined to be 130. Therefore, one must conclude that the empirical formula and the molecular formula are identical and that the molecular formula must be $C_7H_{14}O_2$.

1.4 THE INDEX OF HYDROGEN DEFICIENCY

Frequently, a great deal of information can be learned about an unknown substance simply from a knowledge of the molecular formula. This information is based on the following general molecular formulas:

alkane	C_nH_{2n+2}	
		difference of 2 hydrogens
cycloalkane or alkene	C_nH_{2n}	
		difference of 2 hydrogens
alkyne	C_nH_{2n-2}	

It should be noticed that each time a ring or a π-bond is introduced into a molecule, the number of hydrogens in the molecular formula is reduced by *two*. For every triple bond (2 π-bonds) introduced into a molecule, the number of hydrogens in the molecular formula is reduced by four.

When the molecular formula for a compound contains non-carbon or non-hydrogen elements, the ratio of carbon to hydrogen may change. Simple rules which may be used to predict how this ratio will change are:

1. To convert the formula of an open-chain, saturated hydrocarbon to a formula containing Group V elements (N, P, As, Sb, Bi), one additional hydrogen atom must be added to the molecular formula for each such Group V element present. In the following examples, each formula is correct for a two-carbon, acyclic, saturated compound:

$$C_2H_6, C_2H_7N, C_2H_8N_2, C_2H_9N_3$$

2. To convert the formula of an open-chain, saturated hydrocarbon to a formula containing Group VI elements (O, S, Se, Te), no change in the number of hydrogens is required. In the following examples, each formula is correct for a two-carbon, acyclic, saturated compound:

$$C_2H_6, C_2H_6O, C_2H_6O_2, C_2H_6O_3$$

3. To convert the formula of an open-chain, saturated hydrocarbon to a formula containing Group VII elements (F, Cl, Br, I), one hydrogen must be subtracted from the molecular formula for each such Group VII element present. In the following examples, each formula is correct for a two-carbon, acyclic, saturated compound:

$$C_2H_6, C_2H_5F, C_2H_4F_2, C_2H_3F_3$$

The *index of hydrogen deficiency* (sometimes called the *modes of unsaturation*) is a measure of the number of π-bonds and/or rings a molecule contains. The index of hydrogen deficiency is determined from an examination of the molecular formula of an unknown substance and from a comparison of that formula with a formula for a corresponding acyclic, saturated compound. The difference in the numbers of hydrogens between these formulas, when divided by two, gives the index of hydrogen deficiency.

The index of hydrogen deficiency can be quite useful in structure determination problems. A great deal of information can be obtained about a molecule before a single spectrum is examined. To give some examples, a compound with an index of one must have one double bond or one ring. A quick examination of the infrared spectrum could confirm the presence of a double bond. If there were no double bond, the substance would have to be cyclic and saturated. A compound with an index of two could have a triple bond; or it could have two double bonds, or two rings, or one of each. Knowing the index of hydrogen deficiency of a substance, the chemist is able to proceed directly to the appropriate regions of the spectra to confirm the presence or absence of π-bonds or rings. Benzene contains one ring and three "double bonds." This corresponds to an index of hydrogen deficiency of four. Any substance with an index of four or more can contain an aromatic ring; a substance with an index of less than four cannot contain such a ring.

In order to determine the index of hydrogen deficiency for a compound, the following steps are applied:

1. Determine the formula for the saturated, acyclic hydrocarbon containing the same number of carbon atoms as the unknown substance.

2. Correct this formula for the non-hydrocarbon elements present in the unknown. Add a hydrogen atom for each Group V element present and subtract a hydrogen atom for each Group VII element present.

3. Compare this formula with the molecular formula of the unknown. Determine the number of hydrogens by which these formulas differ.

4. Divide the difference in the number of hydrogens by two to obtain the index of hydrogen deficiency. This equals the number of π-bonds and/or rings in the unknown substance.

The following examples illustrate how the index of hydrogen deficiency is determined and how this information can be applied to the determination of a structure for an unknown substance:

I. The unknown substance introduced in Section 1.1 has a molecular formula of $C_7H_{14}O_2$.
   1. The formula for a seven-carbon, saturated, acyclic hydrocarbon is C_7H_{16}.
   2. Correcting for oxygens gives a formula, $C_7H_{16}O_2$.
   3. This latter formula differs from the original by two hydrogens.
   4. The index of hydrogen deficiency equals *one*. There must be one ring or one double bond in the unknown substance.

Having this information, the chemist can proceed immediately to the double bond regions of the infrared spectrum. There the chemist finds evidence for a carbon-oxygen double bond (carbonyl group). At this point the number of possible isomers which might include the unknown has been narrowed considerably. Further analysis of the spectral evidence leads to an identification of the unknown substance as *isopentyl acetate*.

$$CH_3-\overset{\overset{\displaystyle O}{\|}}{C}-O-CH_2-CH_2-\underset{\underset{\displaystyle CH_3}{|}}{CH}-CH_3$$

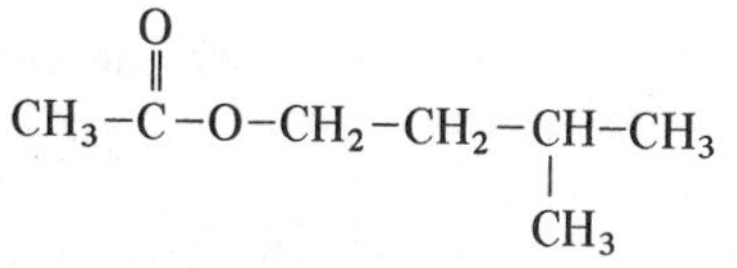

II. Nicotine has the molecular formula $C_{10}H_{14}N_2$.
   1. The formula for a ten-carbon, saturated, acyclic hydrocarbon is $C_{10}H_{22}$.
   2. Correcting for nitrogens gives a formula, $C_{10}H_{24}N_2$.
   3. This latter formula differs from the original by ten hydrogens.

4. The index of hydrogen deficiency equals *five*. There must be some combination of five π-bonds and/or rings in the molecule. Since the index is greater than four, an aromatic ring could be included in the molecule.

Analysis of the spectrum quickly shows that an aromatic ring is indeed present in nicotine. No other double bonds are indicated from the spectral results, suggesting that another ring, this one saturated, must be present in the molecule. More careful refinement of the spectral analysis leads to a structural formula for nicotine:

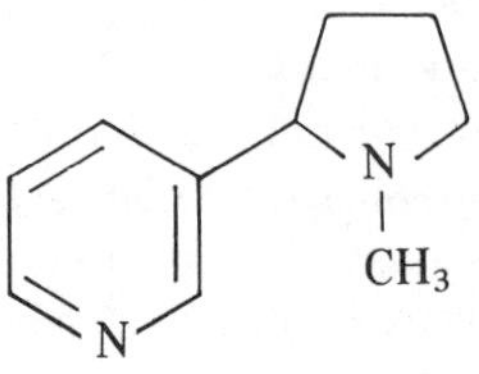

III. Chloral hydrate (knockout drops) is found to have a molecular formula of $C_2H_3Cl_3O_2$.
1. The formula for a two-carbon, saturated, acyclic hydrocarbon is C_2H_6.
2. Correcting for oxygens gives a formula, $C_2H_6O_2$.
3. Correcting for chlorines gives a formula, $C_2H_3Cl_3O_2$.
4. This formula and the formula of chloral hydrate correspond exactly.
5. The index of hydrogen deficiency equals *zero*. Chloral hydrate cannot contain rings or double bonds.

Examination of the spectral results is limited to regions which correspond to singly-bonded structural features. The structural formula which is obtained is:

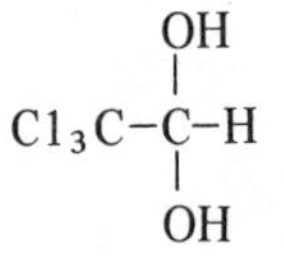

1.5 A QUICK LOOK AHEAD TO SIMPLE USES OF MASS SPECTRA

In Chapter 6 a discussion of the technique of *mass spectrometry* may be found. The details of that method will not be repeated here. However, the reader should be directed to Sections 6.1, 6.2, and 6.3 for applications of mass spectrometry to the problems of molecular formula determination.

Briefly, the mass spectrometer is an instrument which subjects molecules to a high energy beam of electrons. This beam of electrons converts the molecules into positive ions by removal of an electron. The stream of positively charged ions is accelerated along a curved path in a magnetic field. The radius of curvature of the path which the ions describe depends upon the ratio of the mass of the ion to its charge. The ions strike a detector at positions which are determined by the radius of curvature of their paths. The number of ions with a particular mass to charge ratio is plotted as a function of that ratio.

The particle with the largest mass to charge ratio, assuming that the charge is one, will be that particle which represents the intact molecule with only one electron removed. This particle is called the *molecular ion*. This molecular ion can be identified in the mass spectrum. From its position in the spectrum, the weight of that particle can be determined. The mass of the molecular ion is equal to the molecular weight of the original molecule. Thus, the mass spectrometer is an instrument capable of providing molecular weight information.

Virtually every element exists in nature in several isotopic forms. The natural abundance of each of these isotopes is known. Besides giving the mass of the molecular ion, when each atom in the molecule is the most common isotope, the mass spectrum also gives peaks which correspond to that same molecule with heavier isotopes. The ratio of the intensity of the molecular ion peak to the intensities of the peaks corresponding to the heavier isotopes is determined by the natural abundance of each isotope. Because each type of molecule has a unique combination of atoms, and because each type of atom and its isotopes exist in a unique ratio in nature, the ratio of the intensity of the molecular ion peak to the intensities of the isotopic peaks can provide information about the numbers of each type of atom present in the molecule. Thus, *isotope ratio studies* in mass spectrometry can be used to determine the molecular formula of a substance.

REFERENCES

D. L. Pavia, G. M. Lampman, and G. S. Kriz, Jr., "Introduction to Organic Laboratory Techniques," W. B. Saunders Co., Philadelphia, 1976. pp. 393 to 399 and p. 410.

D. J. Pasto and C. R. Johnson, "Organic Structure Determination," Prentice-Hall, Englewood Cliffs, N.J., 1969. pp. 73 to 80 and 316 to 326.

R. L. Shriner, R. C. Fuson, and D. Y. Curtin, "The Systematic Identification of Organic Compounds," 5th edition, John Wiley and Sons, Inc., New York, 1964. pp. 60 to 66.

J. B. Hendrickson, D. J. Cram, and G. S. Hammond, "Organic Chemisty," 3rd edition, McGraw-Hill Book Co., New York, 1970. pp. 15 to 19, 72 to 74, and 82 to 83.

PROBLEMS

1. A compound used as an antiknock additive in gasoline was analyzed by a combustion method. A sample of the compound weighing 9.394 mg yielded 31.154 mg of carbon dioxide and 7.977 mg of water in the combustion.

   a) Calculate the percentage composition of the compound.
   b) Determine its empirical formula.

2. The combustion of a sample of unknown substance weighing 8.23 mg gave 9.62 mg CO_2 and 3.94 mg H_2O. Another sample, weighing 5.32 mg, gave 13.49 mg AgCl in a halogen analysis. Determine the percentage composition and the empirical formula for this organic compound.

3. An important amino acid has the percentage composition, C 32.00%, H 6.71%, and N 18.66%. Calculate the empirical formula of this substance.

4. A compound known to be a pain reliever had the empirical formula $C_9H_8O_4$. When a mixture of 5.02 mg of the unknown and 50.37 mg of camphor was prepared, the melting point of a portion of this mixture was determined. The observed melting point of the mixture was 156°C. What is the molecular weight of this substance?

5. An unknown acid was titrated with 23.1 ml of 0.1 N sodium hydroxide. The weight of acid was 120.8 mg. What is the equivalent weight of the acid?

6. Determine the index of hydrogen deficiency for each of the following compounds:

   a) C_8H_7NO
   b) $C_3H_7NO_3$
   c) $C_4H_4BrNO_2$
   d) $C_5H_3ClN_4$
   e) $C_{21}H_{22}N_2O_2$

7. A substance has the molecular formula C_4H_9N. Is there any likelihood that this material contains a triple bond? Explain your reasoning.

8. a) An unknown solid, extracted from the bark of spruce trees, was analyzed to determine its percentage composition. A sample, containing 11.32 mg, was burned in a combustion apparatus. The carbon dioxide (24.87 mg) and water (5.82 mg) were collected and weighed. From the results of this analysis, calculate the percentage composition of the unknown solid.
   b) Determine the empirical formula of the unknown solid.
   c) From mass spectrometry, the molecular weight was found to be 420 g/mole. What is the molecular formula?
   d) How many aromatic rings could this compound contain?

CHAPTER 2

INFRARED SPECTROSCOPY

Almost any compound having covalent bonds, whether organic or inorganic, will be found to absorb various frequencies of electromagnetic radiation in the infrared region of the spectrum. The infrared region of the electromagnetic spectrum lies at wavelengths longer than those associated with visible light, which includes wavelengths from approximately 400 nm to 800 nm (1 nm = 10^{-9} m), but lies at wavelengths shorter than those associated with microwaves, which have wavelengths longer than 1 mm. For chemical purposes, we will be interested in the *vibrational* portion of the infrared region. This portion is defined as that including radiations with wavelengths (λ) between 2.5 μ and 15 μ (1 μ = 1 micron = 1 μm = 10^{-6} m). Although the more technically correct unit for wavelength in the infrared region of the spectrum is *micrometer* (μm), we shall follow common practice and use *micron* (μ) as the unit. The relationship of the infrared region to others included in the electromagnetic spectrum is illustrated in Figure 2–1.

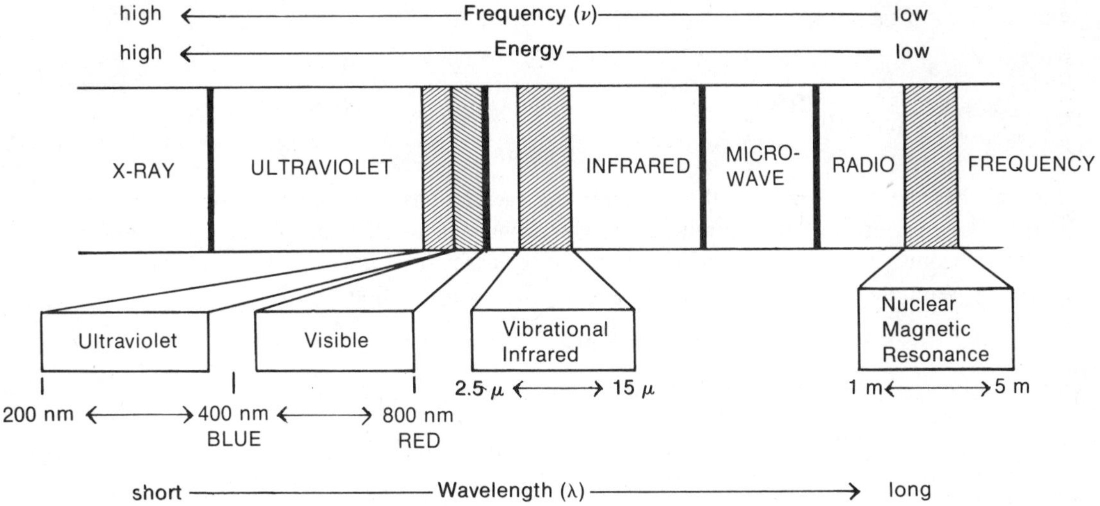

FIGURE 2–1 A Portion of the Electromagnetic Spectrum Showing the Relationship of the Vibrational Infrared to Other Types of Radiation.

In Figure 2–1, it is seen that the wavelength (λ) is inversely proportional to the frequency (ν) and is governed by the relationship $\nu = c/\lambda$, where c = speed of light. It should also be observed that the energy is directly proportional to the frequency: $E = h\nu$, where h = Planck's constant. From the latter equation, it can be seen qualitatively that the highest energy radiation corresponds to the X-ray region of the spectrum, where the energy may be great enough to break bonds in molecules. At the other end of the electromagnetic spectrum, radiofrequencies have very low energies, only enough to cause nuclear or electronic spin transitions (nmr or esr) within molecules.

The regions of the spectrum and the types of energy transitions observed there are summarized in Table 2–1. Several of these regions, including the infrared, give vital information about the structure of organic molecules. Nuclear magnetic resonance, which occurs in the radiofrequency part of the spectrum, is discussed in Chapters 3 and 4, while ultraviolet and visible spectroscopy are described in Chapter 5.

TABLE 2-1 Types of Energy Transitions in Each Region of the Electromagnetic Spectrum

Region of Spectrum	Energy Transitions
X-rays	Bond Breaking
Ultraviolet/Visible	Electronic
Infrared	Vibrational
Microwave	Rotational
Radiofrequencies	Nuclear Spin (Nuclear Magnetic Resonance) Electron Spin (Electron Spin Resonance)

Many chemists refer to the radiation in the vibrational infrared region of the electromagnetic spectrum in terms of a unit called *wavenumbers* ($\bar{\nu}$). Wavenumbers are expressed as cm^{-1} (reciprocal centimeters), and are easily computed by taking the reciprocal of the wavelength (λ) expressed in centimeters. They may be converted to a frequency (ν) by multiplying them by the speed of light (expressed in cm/sec).

$$\bar{\nu}\,(\text{cm}^{-1}) = \frac{1}{\lambda(\text{cm})} \qquad \nu(\text{Hz}) = \bar{\nu}c = \frac{c(\text{cm/sec})}{\lambda(\text{cm})}$$

This unit has the advantage, for those performing calculations, that it is directly proportional to energy. Thus, in terms of wavenumbers, the vibrational infrared extends from about 4000 to 650 cm^{-1}. Interconversions may be made between wavelengths (μ or μm) and wavenumbers (cm^{-1}) or between wavenumbers and wavelengths by using the following relationships:

$$\text{cm}^{-1} = \frac{1}{(\mu)} \times 10{,}000 \qquad \text{and} \qquad \mu = \frac{1}{(\text{cm}^{-1})} \times 10{,}000$$

2.1 THE INFRARED ABSORPTION PROCESS

As with other types of energy absorption, molecules are excited to a higher energy state when they absorb infrared radiation. The absorption of infrared radiation is, like other absorption processes, a quantized process. Only selected frequencies (energies) of infrared radiation will be absorbed by a molecule. The absorption of infrared radiation corresponds to energy changes on the order of from 2 to 10 kcal/mole. Radiation in this energy range corresponds to the range encompassing the stretching and bending vibrational frequencies of the bonds in most covalent molecules. In the absorption process, those frequencies of infrared radiation which match the natural vibrational frequencies of the molecule in question will be absorbed, and the energy absorbed will serve to increase the *amplitude* of the vibrational motions of the bonds in the molecule. It should be noted, however, that not all bonds in a molecule are capable of absorbing infrared energy, even if the frequency of the radiation exactly matches that of the bond motion. Only those bonds which have a *dipole moment* are capable of absorbing infrared radiation. Symmetric bonds, like those of H_2 or Cl_2, will not absorb infrared radiation. A bond must present an electrical dipole which is changing at the same frequency as the incoming radiation in order for energy to be transferred. The changing electrical dipole of the bond can then couple with the sinusoidally changing electromagnetic field of the incoming radiation. Thus, symmetric bonds which are symmetrically substituted will not absorb in the infrared. For the purposes of an organic chemist, the bonds most likely to be affected by this restraint are those of symmetric alkenes (C=C) and those of symmetric alkynes (C≡C).

2.2 USES OF THE INFRARED SPECTRUM

Since every different type of bond has a different natural frequency of vibration, and since the same type of bond in two different compounds is in a slightly different environment, no two molecules of different structure will have exactly the same infrared absorption pattern or *infrared spectrum*. Although some of the frequencies absorbed in the two cases might be the same, in no case of two different molecules will their infrared spectra (the patterns of absorption) be identical. Thus, the infrared spectrum can be used for molecules much as a fingerprint can be used for humans. By comparing the infrared spectra of two substances thought to be identical, one can establish whether or not they in fact are identical. If their infrared spectra coincide peak for peak (absorption for absorption), in most cases the two substances will be identical.

A second and more important use of the infrared spectrum is that it gives structural information about a molecule. The absorptions of each type of bond (N–H, C–H, O–H, C–X, C=O, C–O, C–C, C=C, C≡C, C≡N, etc.) are regularly found only in certain small portions of the vibrational infrared region. A small range of absorption can be defined for each type of bond. Outside this range, absorptions will normally be due to some other type of bond. Thus, for instance, any absorption in the range 3000 ± 150 cm^{-1} (around 3.33 μ) will almost always be due to the presence of a CH bond in the molecule; an absorption in the range 1700 ± 100 cm^{-1} (around 5.9 μ) will normally be due to the presence of a C=O bond (carbonyl group) in the molecule. The same type of range applies to each type of bond. The way these are spread out over the vibrational infrared is illustrated schematically in Figure 2–2. It is a good idea to try to fix this general scheme in one's mind for future convenience.

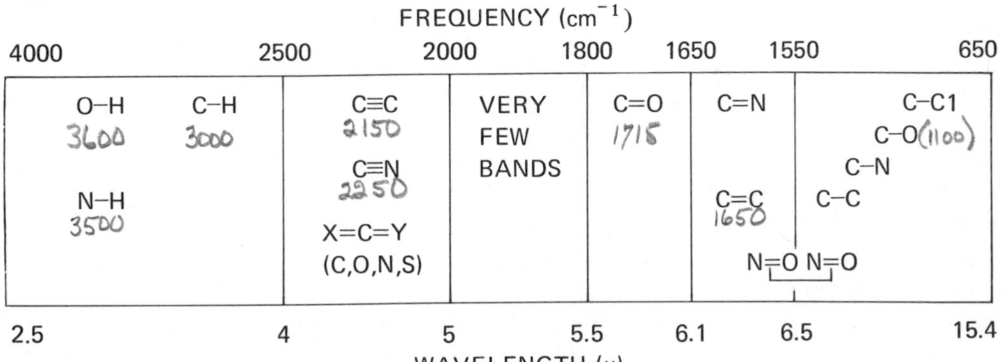

FIGURE 2–2. The Approximate Regions Where Various Common Types of Bonds Absorb (Stretching Vibrations Only; Bending, Twisting, and Other Types of Bond Vibrations Have Been Omitted for Clarity).

2.3 THE MODES OF VIBRATION AND BENDING

The simplest types, or *modes*, of vibrational motion in a molecule which are *infrared active*, that is, give rise to absorptions, are the stretching and bending modes.

However, other more complex types of stretching and bending are also active. In order to introduce several words of terminology, the normal modes of vibration for a methylene group are illustrated below. In general, asymmetric stretching vibrations occur at higher frequencies (lower wavelengths) than symmetric stretching vibrations; also, stretching vibrations occur at higher frequencies (lower wavelengths) than bending vibrations.

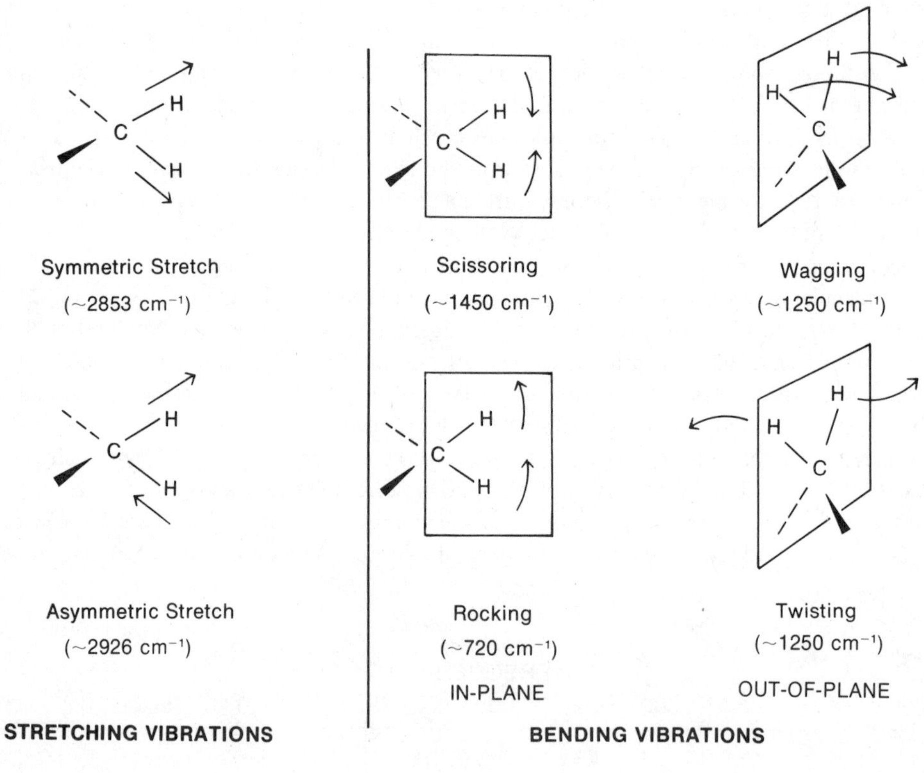

The terms scissoring, rocking, wagging, and twisting are commonly found in the literature to describe the origin of infrared bands.

In any group of three or more atoms, at least two of which are identical, there will be *two* modes of stretching and/or bending: the symmetric mode and the asymmetric mode. Examples of such groupings are $-CH_3$, $-CH_2-$ (shown above), $-NO_2$, $-NH_2$, and anhydrides. The methyl group gives rise to a symmetric stretching vibration at about 2872 cm^{-1} and an asymmetric stretch at about 2962 cm^{-1}. In the case of the anhydride, because of the asymmetric and symmetric modes of stretch, this functional group gives two absorptions in the C=O region. A similar phenomenon is seen for the amino group, where a primary amine usually has two absorptions in the NH stretch region, while a secondary amine (R_2NH) has only one absorption peak. Amides show similar bands. There are two strong N=O stretch peaks for a nitro group, with the symmetric stretch appearing at about 1350 cm^{-1} and the asymmetric stretch appearing at about 1550 cm^{-1}.

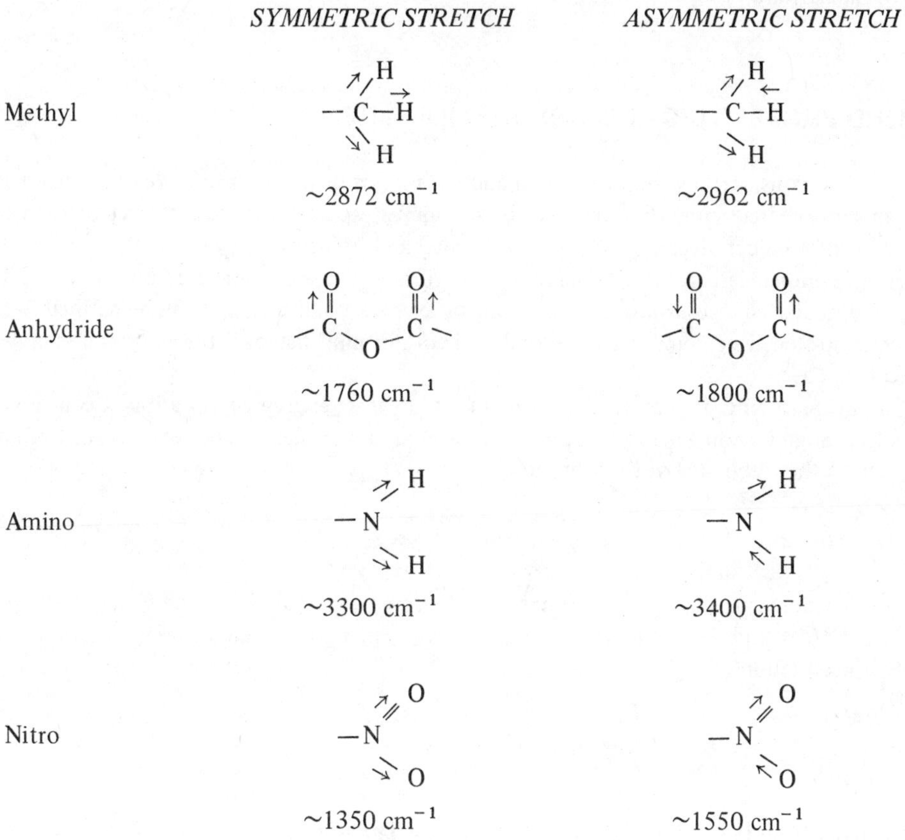

The vibrations described above are called *fundamental* absorptions. They arise from excitation from the ground state to the lowest energy excited state. Usually the spectrum is complicated because of the presence of weak *overtone*, *combination*, and *difference* bands. Overtones result from excitation from the ground state to higher energy states, which correspond to integral multiples of the frequency of the fundamental ($\bar{\nu}$). For example, one might observe weak overtone bands at $2\bar{\nu}$, $3\bar{\nu}$, ... Overtones are the same thing that one generates in any kind of physical vibration. If you pluck a string on a cello, the string will vibrate with a fundamental frequency. However, less intense vibrations will also be set up at several overtone frequencies. An absorption in the infrared at 500 cm^{-1} may well have an accompanying peak of lower intensity at 1000 cm^{-1} – an overtone.

When two vibrational frequencies ($\bar{\nu}_1$ and $\bar{\nu}_2$) in a molecule couple to give rise to a vibration of a new frequency within the molecule, and when such a vibration is infrared active, it is called a *combination band.* These frequencies are the sum of the two interacting bands ($\bar{\nu}_{comb} = \bar{\nu}_1 + \bar{\nu}_2$). Not all possible combinations occur. Rules exist that govern which combinations are allowed, but they are beyond the scope of our discussion here. *Difference bands* are similar to combination bands. The observed frequency, in this case, results from the difference between the two interacting bands ($\bar{\nu}_{diff} = \bar{\nu}_1 - \bar{\nu}_2$). Overtone, combination, and difference bands can be calculated by using direct manipulations of frequencies in wavenumbers by multiplication, addition, and subtraction, respectively. When a fundamental vibration couples with an overtone or combination band, the coupled vibration is called *Fermi resonance.* Again, only certain combinations are allowed. Fermi resonance is often observed in carbonyl compounds.

Although rotational frequencies of the whole molecule are not infrared active, they often couple with the stretching and bending vibrations in the molecule to give additional fine structure to these absorptions, thus further complicating the spectrum. One of the reasons that a band is broad rather than sharp in an infrared spectrum is because of rotational coupling which may lead to a considerable amount of fine structure.

2.4 BOND PROPERTIES AND ABSORPTION TRENDS

We will now consider how bond strength and the masses of the atoms bonded together affect the infrared absorption frequency. For the sake of simplicity, we will restrict the discussion to a simple heteronuclear diatomic molecule (two *different* atoms) and its stretching vibration.

A diatomic molecule can be considered as two vibrating masses connected by a spring. The bond distance continually changes, but an equilibrium or average bond distance can be defined. Whenever the spring is stretched or compressed beyond this equilibrium distance, the potential energy of the system will increase.

As for any harmonic oscillator, when a bond vibrates, its energy of vibration is continually and periodically changing from kinetic to potential energy and back again. The total amount of energy is proportional to the frequency of the vibration,

$$E_{\text{osc}} \propto h\nu_{\text{osc}}$$

which for a harmonic oscillator is determined by the force constant (K) of the spring, or its stiffness, and the masses (m_1 and m_2) of the two bonded atoms. The natural frequency of vibration of a bond is given by the equation

$$\bar{\nu} = \frac{1}{2\pi c}\sqrt{\frac{K}{\mu}}$$

which is derived from Hooke's law for vibrating springs. The term μ, or *reduced mass* of the system, is given by

$$\mu = \frac{m_1 m_2}{m_1 + m_2}$$

K is a constant that will vary from one bond to another. As a first approximation, the force constants for triple bonds are three times those of single bonds, while the force constants for double bonds are twice those of single bonds.

Two things should be noticeable immediately. One is that stronger bonds have a larger force constant K and will vibrate at higher frequencies than weaker bonds. The second is that bonds between atoms of higher masses (larger μ) will vibrate at lower frequencies than bonds between lighter atoms.

In general, triple bonds are stronger than double or single bonds between the same two atoms and have higher frequencies of vibration*:

C≡C	C=C	C–C
2150 cm^{-1}	1650 cm^{-1}	1200 cm^{-1}

← increasing K

*Recall that wavenumbers and frequency are proportional: $\nu = c\bar{\nu}$, where c is the speed of light.

The CH stretch occurs at about 3000 cm^{-1}. As the atom bonded to carbon increases in mass, the quantity μ increases, and the frequency of vibration goes down:

C–H	C–C	C–O	C–Cl	C–Br	C–I
3000 cm^{-1}	1200 cm^{-1}	1100 cm^{-1}	800 cm^{-1}	550 cm^{-1}	~500 cm^{-1}

$\longrightarrow$ increasing μ

Also, bending motions tend to be easier than stretching motions, and the force constant K is smaller.

CH stretching	CH bending
~3000 cm^{-1}	~1340 cm^{-1}

Hybridization affects the force constant K also. Bonds are stronger in the order $sp > sp^2 > sp^3$, and the observed frequencies of CH vibration illustrate this nicely.

sp	sp^2	sp^3
≡C–H	=C–H	–C–H
3300 cm^{-1}	3100 cm^{-1}	2900 cm^{-1}

Resonance will also affect the strength and length of a bond and, hence, its force constant K. Thus, while a normal ketone has its C=O stretch vibration at 1715 cm^{-1}, one which is conjugated with a double bond will absorb at a lower frequency near 1675–1680 cm^{-1}. This is because resonance lengthens the C=O bond distance and gives it more single bond character:

This has the effect of reducing the force constant K, and the absorption moves to a lower frequency.

The Hooke's law expression given above may be transformed in the following way to a very useful equation:

$$\bar{\nu} = \frac{1}{2\pi c}\sqrt{\frac{K}{\mu}}$$

$\bar{\nu}$ = frequency in cm^{-1}

c = velocity of light = 3×10^{10} cm/sec

K = force constant in dynes/cm

$$\mu = \frac{m_1 m_2}{m_1 + m_2}\text{; masses of atoms in grams}$$

$$\text{or } \frac{M_1 M_2}{(M_1 + M_2)(6.02 \times 10^{23})}\text{; masses of atoms in AMU}$$

Removing Avogadro's number (6.02×10^{23}) from the denominator of the reduced mass expression (μ) by taking its square root, one obtains the expression:

$$\bar{\nu} = \frac{7.76 \times 10^{11}}{2\pi c} \sqrt{\frac{K}{\mu}}$$

A new expression is obtained by inserting the actual values of π and c:

$$\bar{\nu}\,(\text{cm}^{-1}) = 4.12 \sqrt{\frac{K}{\mu}}$$

$$\mu = \frac{M_1 M_2}{M_1 + M_2}\text{; where } M_1 \text{ and } M_2 \text{ are atomic weights}$$

$$K = \text{force constant in dynes/cm}$$

This equation may be used to calculate the *approximate* position of a band in the infrared spectrum by assuming that K for single, double, and triple bonds are 5, 10, and 15×10^5 dynes/cm, respectively. Several examples are given in Table 2–2. One notices that excellent agreement can be obtained with the experimental values given below. It should be mentioned, however, that experimental and calculated values may vary considerably, owing to resonance, hybridization, and other effects which operate in organic molecules. Nevertheless, good qualitative values are obtained by such calculations.

TABLE 2-2 Calculation of Stretching Frequencies for Different Types of Bonds

C=C bond:

$$\bar{\nu} = 4.12\sqrt{\frac{K}{\mu}}$$

$$K = 10 \times 10^5 \text{ dynes/cm}$$

$$\mu = \frac{M_C M_C}{M_C + M_C} = \frac{(12)(12)}{12 + 12} = 6$$

$$\bar{\nu} = 4.12\sqrt{\frac{10 \times 10^5}{6}} = 1682 \text{ cm}^{-1} \text{ (calculated)}$$

$$\bar{\nu} = 1650 \text{ cm}^{-1} \text{ (experimental)}$$

C–H bond:

$$\bar{\nu} = 4.12\sqrt{\frac{K}{\mu}}$$

$$K = 5 \times 10^5 \text{ dynes/cm}$$

$$\mu = \frac{M_C M_H}{M_C + M_H} = \frac{(12)(1)}{12 + 1} = 0.923$$

$$\bar{\nu} = 4.12\sqrt{\frac{5 \times 10^5}{0.923}} = 3032 \text{ cm}^{-1} \text{ (calculated)}$$

$$\bar{\nu} = 3000 \text{ cm}^{-1} \text{ (experimental)}$$

C–D bond:

$$\bar{\nu} = 4.12\sqrt{\frac{K}{\mu}}$$

$$K = 5 \times 10^5 \text{ dynes/cm}$$

$$\mu = \frac{M_C M_D}{M_C + M_D} = \frac{(12)(2)}{12 + 2} = 1.71$$

$$\bar{\nu} = 4.12\sqrt{\frac{5 \times 10^5}{1.71}} = 2228 \text{ cm}^{-1} \text{ (calculated)}$$

$$\bar{\nu} = 2206 \text{ cm}^{-1} \text{ (experimental)}$$

2.5 WHAT TO LOOK FOR WHEN EXAMINING INFRARED SPECTRA

The instrument that determines the absorption spectrum for a compound is called an *infrared spectrophotometer*. The spectrophotometer determines the relative strengths and positions of all the absorptions in the infrared region and plots this on a piece of calibrated chart paper. This plot of absorption intensity versus wavenumber or wavelength is referred to as the *infrared spectrum* of the compound. A typical infrared spectrum, that of methyl isopropyl ketone, is shown in Figure 2–3.

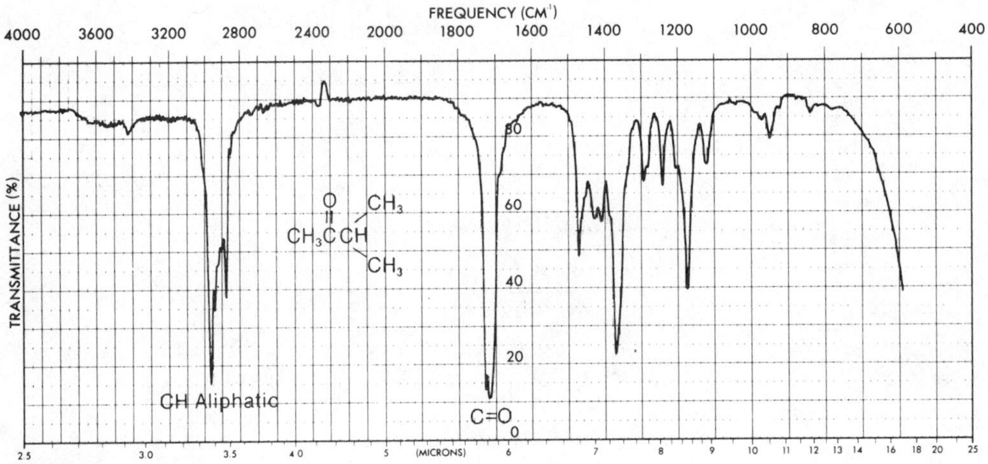

FIGURE 2–3 The Infrared Spectrum of Methyl Isopropyl Ketone (Neat Liquid, Salt Plates)

The strong absorption in the middle of the spectrum corresponds to C=O, the carbonyl group. Note that the C=O peak is quite intense. In addition to the characteristic position of absorption, the *shape* and *intensity* of this peak are also unique to the C=O bond. This is true for almost every type of absorption peak; both shape and intensity characteristics can be described, and these characteristics often allow one to distinguish the peak where the situation might be confusing. For instance, to some extent both C=O and C=C bonds absorb in the same region of the infrared spectrum,

C=O 1850–1630 cm^{-1} (5.40–6.13 μ)

C=C 1680–1620 cm^{-1} (5.95–6.17 μ)

However, the C=O bond is a strong absorber, whereas the C=C bond generally absorbs only weakly (Figure 2–4). Hence, a trained observer would not interpret a strong peak at 1670 cm^{-1} to be a carbon-carbon double bond, nor would he interpret a weak absorption at this frequency to be due to a carbonyl group.

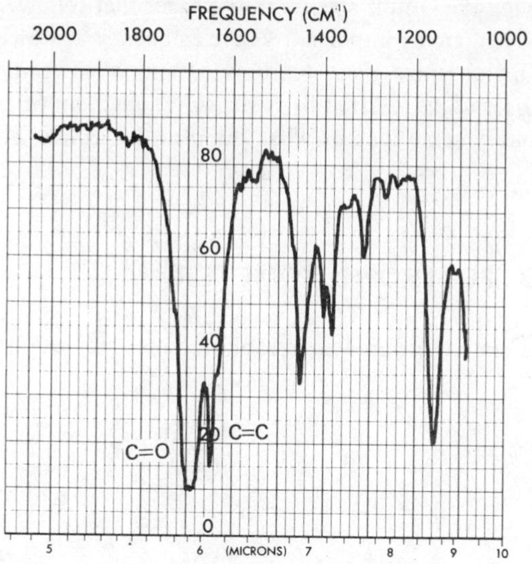

FIGURE 2–4 Comparison of the Intensities of the C=O and C=C Absorption Bands

The shape and fine structure of a peak will often give a clue to its identity as well. Thus, while the N–H and O–H regions overlap,

$$\text{O–H} \quad 3650\text{–}3200\ \text{cm}^{-1}\ (2.75\text{–}3.12\ \mu)$$
$$\text{N–H} \quad 3500\text{–}3300\ \text{cm}^{-1}\ (2.85\text{–}3.03\ \mu)$$

the N–H absorption usually has one or two *sharp* absorption bands of lower intensity while O–H, when it is in the N–H region, usually gives a *broad* absorption peak. Also, primary amines give *two* absorptions in this region, whereas alcohols as pure liquids give only one (Figure 2–5).

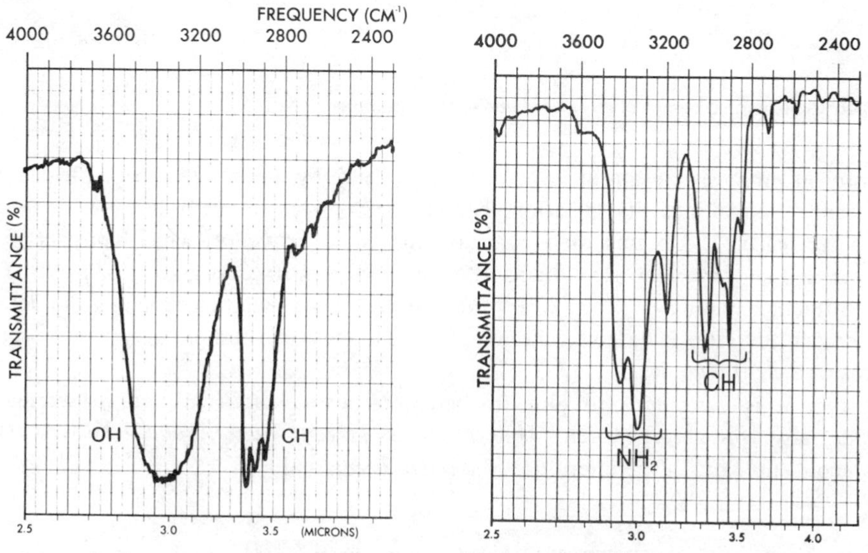

FIGURE 2–5 Comparison of the Shapes of the Absorption Bands for the O–H and N–H Groups

Therefore, while studying the sample spectra in the pages that follow, one should also take notice of shapes and intensities. They are as important as the frequency at which an absorption occurs, and the eye must be trained to recognize these features. Often, when reading the literature of organic chemistry, one will find absorptions referred to as strong (s), medium (m), weak (w), broad, or sharp. The author is trying to convey some idea of what the peak looks like without actually drawing the spectrum.

2.6 THE INFRARED SPECTROPHOTOMETER

The components of a simple infrared spectrophotometer are schematically illustrated in Figure 2–6.

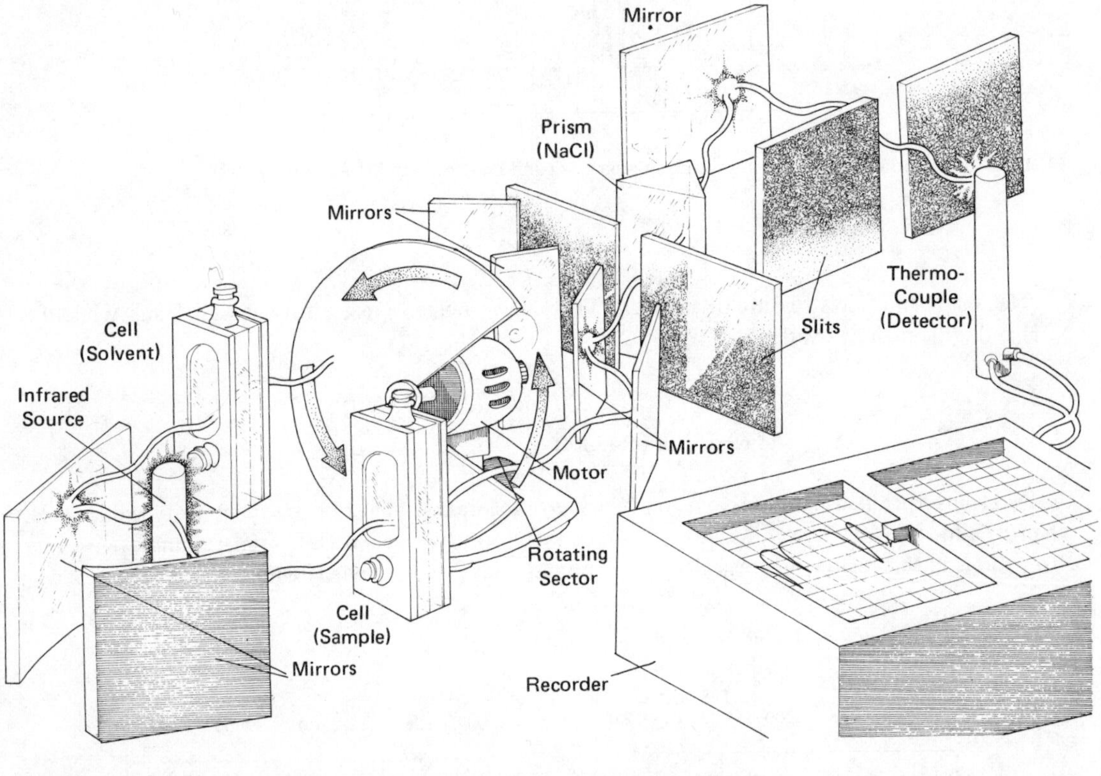

FIGURE 2–6 Schematic Diagram of an Infrared Spectrophotometer

The instrument produces a beam of infrared radiation and, by means of mirrors, divides it into two parallel beams of equal intensity radiation. The sample is placed in one beam, and the other beam is used as a reference. The beams then pass into the *monochromator*, which consists of a rapidly rotating sector that passes the two beams alternately to a diffraction grating or prism. The slowly rotating diffraction grating or prism varies the wavelength of radiation reaching the detector. The detector senses the ratio in intensity between the reference and sample beams and records these differences on a chart recorder.

Note that it is customary to plot frequency or wavelength (cm^{-1} or μ) versus the light *transmitted*, not light absorbed. This is recorded as *percent transmittance*. This is because the detector records the ratio of the intensities of the two beams and

$$\text{percent transmittance} = \left(\frac{I_s}{I_r}\right) \times 100$$

where I_s is the intensity of the sample beam and I_r is the intensity of the reference beam. In many parts of the spectrum the transmittance is nearly 100%, meaning that the sample is nearly transparent to radiation of that frequency (doesn't absorb). Maximum absorption is thus shown by a *minimum* on the chart. Even so, the absorption is traditionally called a *peak*.

Often the spectrum of a compound is obtained by dissolving it in a solvent. The solution is then placed in the *sample beam*. When doing this, pure solvent is placed in the *reference beam* in an identical sample cell, and the instrument automatically "subtracts" the spectrum of the solvent from that of the sample. This convenience feature is the reason most infrared spectrophotometers are double beam (sample + reference) instruments that measure intensity *ratios*; since the solvent absorbs in both beams, it is in both terms of the ratio I_s/I_r and cancels out.

2.7 CORRELATION CHARTS AND TABLES

To extract structural information from infrared spectra, one must be familiar with the frequencies or wavelengths at which various functional groups absorb. To facilitate this, tables exist, called infrared *correlation tables*, which have tabulated as much information as is known about where the various functional groups absorb. The books listed at the end of this chapter present extensive lists of correlation tables. Sometimes the absorption information is given in the form of a chart, called a *correlation chart*. A simplified correlation table is given in Table 2–3. A more detailed chart is given in Appendix One.

The mass of data in Table 2–3 seems as though it may be difficult to assimilate. However, it is really quite easy if one first starts simply and then slowly increases the familiarity with and the ability to interpret the finer details of an infrared spectrum. This is done most easily by first establishing the broad visual patterns of Figure 2–2 quite firmly in mind. Then as a second step, a "typical absorption value" can be memorized for each of the functional groups in this pattern. This value will be a single number that can be used as a pivotal value for the memory. For example, start with a simple aliphatic ketone as a model for all typical carbonyl compounds. The typical aliphatic ketone has a carbonyl absorption of about 1715 ± 10 cm^{-1}. Without worrying about the variation, memorize 1715 cm^{-1} as the base value for carbonyl absorption. Then, more slowly, familiarize yourself with the extent of the carbonyl range and the visual pattern of the way in which the different kinds of carbonyl groups are arranged throughout this region. See, for instance, Figure 2–32, which gives typical values for the various types of carbonyl compounds. Also learn how factors such as ring size (when the functional group is contained in a ring) and conjugation affect the base values (*i.e.*, which direction the values are shifted). Learn the trends—always keeping the memorized base value (1715 cm^{-1}) in mind. As a beginning, it might prove useful to memorize the base values for this approach given in Table 2–4. Notice that there are only eight of them.

TABLE 2–4 Base Values for Absorptions of Bonds

OH	3600 cm^{-1}	2.8 μ	C≡C	2150 cm^{-1}	4.6 μ
NH	3500	2.9	C=O	1715	5.8
CH	3000	3.3	C=C	1650	6.1
C≡N	2250	4.4	C–O	1100	9.1

TABLE 2-3 A Simplified Correlation Chart

	Type of Vibration		Frequency (cm^{-1})	Wavelength (μ)	Intensity
C–H	Alkanes	(stretch)	3000–2850	3.33–3.51	s
	$-CH_3$	(bend)	1450 and 1375	6.90 and 7.27	m
	$-CH_2-$	(bend)	1465	6.83	m
	Alkenes	(stretch)	3100–3000	3.23–3.33	m
		(out-of-plane bend)	1000–650	10.0–15.3	s
	Aromatics	(stretch)	3150–3050	3.17–3.28	s
		(out-of-plane bend)	900–690	11.1–14.5	s
	Alkyne	(stretch)	ca. 3300	ca. 3.03	s
	Aldehyde		2900–2800	3.45–3.57	w
			2800–2700	3.57–3.70	w
C–C	Alkane	not interpretatively useful			
C=C	Alkene		1680–1600	5.95–6.25	m–w
	Aromatic		1600 and 1475	6.25 and 6.78	m–w
C≡C	Alkyne		2250–2100	4.44–4.76	m–w
C=O	Aldehyde		1740–1720	5.75–5.81	s
	Ketone		1725–1705	5.80–5.87	s
	Carboxylic Acid		1725–1700	5.80–5.88	s
	Ester		1750–1730	5.71–5.78	s
	Amide		1670–1640	6.00–6.10	s
	Anhydride		1810 and 1760	5.52 and 5.68	s
	Acid Chloride		1800	5.56	s
C–O	Alcohols, Ethers, Esters, Carboxylic Acids, Anhydrides		1300–1000	7.69–10.0	s
O–H	Alcohols, Phenols				
	Free		3650–3600	2.74–2.78	m
	H–Bonded		3500–3200	2.86–3.13	m
	Carboxylic Acids		3400–2400	2.94–4.17	m
N–H	Primary and Secondary Amines and Amides	(stretch)	3500–3100	2.86–3.23	m
		(bend)	1640–1550	6.10–6.45	m–s
C–N	Amines		1350–1000	7.4–10.0	m–s
C=N	Imines and Oximes		1690–1640	5.92–6.10	w–s
C≡N	Nitriles		2260–2240	4.42–4.46	m
X=C=Y	Allenes, Ketenes, Isocyanates, Isothiocyanates		2270–1950	4.40–5.13	m–s
N=O	Nitro ($R-NO_2$)		1550 and 1350	6.45 and 7.40	s
S–H	Mercaptans		2550	3.92	w
S=O	Sulfoxides		1050	9.52	s
	Sulfones, Sulfonyl Chlorides, Sulfates, Sulfonamides		1375–1300 and 1200–1140	7.27–7.69 and 8.33–8.77	s s
C–X	Fluoride		1400–1000	7.14–10.0	s
	Chloride		800–600	12.5–16.7	s
	Bromide, Iodide		<667	>15.0	s

2.8 HOW TO APPROACH THE ANALYSIS OF A SPECTRUM (OR, WHAT YOU CAN TELL AT A GLANCE)

In trying to analyze the spectrum of an unknown, you should concentrate your first efforts toward determining the presence (or absence) of a few major functional groups. The C=O, O–H, N–H, C–O, C=C, C≡C, C≡N, and NO_2 peaks are the most conspicuous and give immediate structural information if they are present. Do not try to make a detailed analysis of the CH absorptions near 3000 cm^{-1} (3.33 μ); almost all compounds have these absorptions. Do not worry about subtleties of the exact type of environment in which the functional group is found. Below is a major check list of the important gross features.

1. Is a carbonyl group present?
   The C=O group gives rise to a strong absorption in the region 1820–1660 cm^{-1} (5.5–6.1 μ). The peak is often the strongest in the spectrum and of medium width. You can't miss it.

2. If C=O is present, check the following types (if absent, go to 3).

ACIDS	is OH also present? – *broad* absorption near 3400–2400 cm^{-1} (usually overlaps C–H)
AMIDES	is NH also present? – medium absorption near 3500 cm^{-1} (2.85 μ) sometimes a double peak, with equivalent halves
ESTERS	is C–O also present? – strong intensity absorptions near 1300–1000 cm^{-1} (7.7–10 μ)
ANHYDRIDES	have *two* C=O absorptions near 1810 and 1760 cm^{-1} (5.5 and 5.7 μ)
ALDEHYDES	is aldehyde CH present? –two weak absorptions near 2850 and 2750 cm^{-1} (3.50 and 3.65 μ) on the right-hand side of CH absorptions
KETONES	The above 5 choices have been eliminated

3. If C=O is absent

ALCOHOLS PHENOLS	Check for OH – *broad* absorption near 3600–3300 cm^{-1} (2.8–3.0 μ) – confirm this by finding C–O near 1300–1000 cm^{-1} (7.7–10 μ)
AMINES	Check for NH – medium absorptions(s) near 3500 cm^{-1} (2.85 μ)
ETHERS	Check for C–O (and absence of OH) near 1300–1000 cm^{-1} (7.7–10 μ)

4. Double Bonds and/or Aromatic Rings
   – C=C is a *weak* absorption near 1650 cm^{-1} (6.1 μ)
   – medium to strong absorptions in the region 1650–1450 cm^{-1} (6–7 μ) often imply an aromatic ring
   – confirm the above by consulting the CH region; aromatic and vinyl CH occurs to the left of 3000 cm^{-1} (3.33 μ) (aliphatic CH occurs to the right of this value)

5. Triple Bonds
   – C≡N is a medium, sharp absorption near 2250 cm^{-1} (4.5 μ)
   – C≡C is a weak but sharp absorption near 2150 cm^{-1} (4.65 μ)
   Check also for acetylenic CH near 3300 cm^{-1} (3.0 μ)

6. Nitro Groups
   – *two* strong absorptions at 1600–1500 cm^{-1} (6.25–6.67 μ) and 1390–1300 cm^{-1} (7.2–7.7 μ)

7. Hydrocarbons
   - none of the above are found
   - major absorptions are in CH region near 3000 cm^{-1} (3.33 μ)
   - very simple spectrum, only other absorptions near 1450 cm^{-1} (6.90 μ) and 1375 cm^{-1} (7.27 μ)

The beginning student should resist the idea of trying to assign or interpret *every* peak in the spectrum. You simply will not be able to do this. Concentrate first on learning these *major* peaks and recognizing their presence or absence. This is done best by carefully studying the illustrative spectra in the sections that follow.

NOTE: In describing the shifts of absorption peaks or their relative positions, we have used the terms "to the left" and "to the right." This was done to save space when using *both* microns and reciprocal centimeters. The meaning is clear since all spectra are conventionally presented left to right from 4000 cm^{-1} to 600 cm^{-1} or from 2.5 μ to 16 μ. "To the right" avoids saying each time "to lower frequency (cm^{-1}) or to longer wavelength (μ)" which is confusing since cm^{-1} and μ have an inverse relationship; as one goes up, the other goes down.

A SURVEY OF THE IMPORTANT FUNCTIONAL GROUPS WITH EXAMPLES

The behavior of important functional groups toward infrared radiation will be described in the following sections. These sections will be organized as follows:

1. The *basic* information about the functional group or type of vibration will be abstracted and placed in a box where it may be consulted easily.
2. Examples of spectra will follow the basic section. The *major* absorptions of diagnostic value will be indicated on each spectrum.
3. Following the spectral examples, a discussion section will give *details* about the various functional groups, and other information which may be of use in identifying organic compounds. This section might be omitted on your first reading of this material.

2.9 HYDROCARBONS: ALKANES, ALKENES, AND ALKYNES

A. ALKANES

Spectrum is usually simple with few peaks.

C–H stretch occurs around 3000 cm^{-1} (3.33 μ)

a) In alkanes (except strained ring compounds) absorption always occurs to the right of 3000 cm^{-1} (3.33 μ)

b) If a compound has vinylic, aromatic, acetylenic, or cyclopropyl hydrogens, the CH absorption is to the left of 3000 cm^{-1} (3.33 μ)

CH_2 methylene groups have a characteristic absorption of approximately 1450 cm^{-1} (6.90 μ)

CH_3 methyl groups have a characteristic absorption at approximately 1375 cm^{-1} (7.27 μ)

C–C stretch not interpretatively useful –many peaks

Examples: Decane (Fig. 2–7) and cyclohexane (Fig. 2–8)

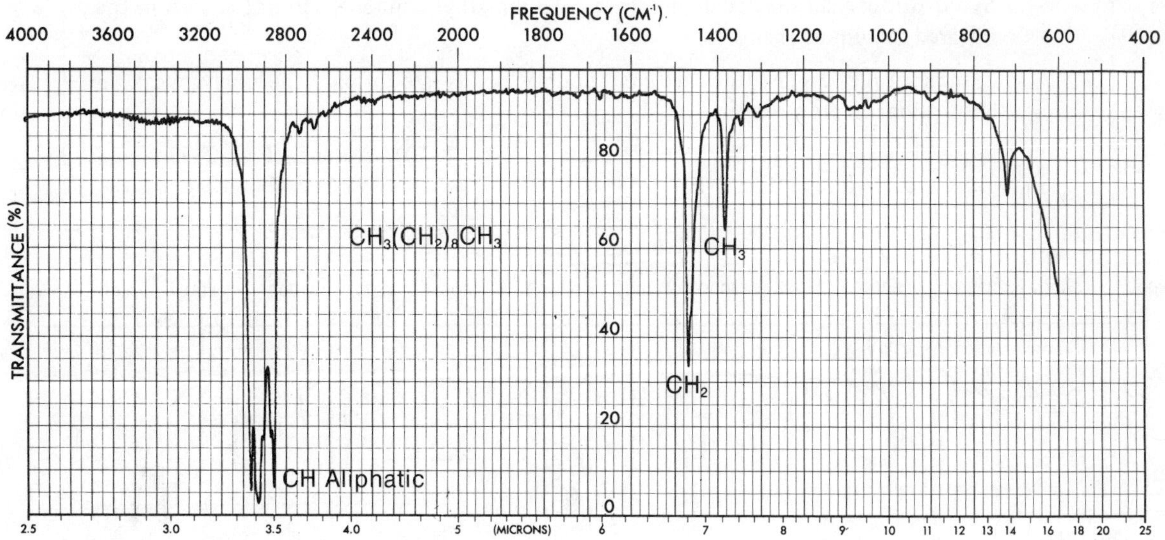

FIGURE 2–7 The Infrared Spectrum of Decane (Neat Liquid, Salt Plates)

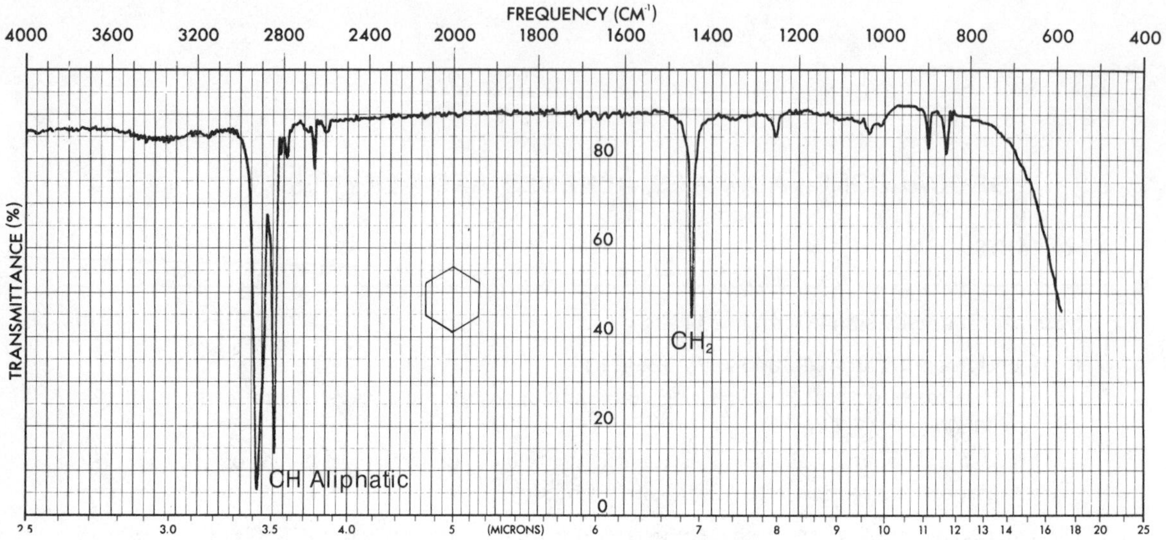

FIGURE 2-8 The Infrared Spectrum of Cyclohexane (Neat Liquid, Salt Plates)

B. ALKENES

=C–H stretch occurs to the left of 3000 cm^{-1} (3.33 μ)

=C–H out-of-plane (oop) bending 1000–650 cm^{-1} (10–15 μ)

C=C stretch 1660–1600 cm^{-1} (6.02–6.25 μ); often weak conjugation moves C=C stretch to the right and increases the intensity.

Symmetrically substituted bonds (*e.g.*, 2,3-dimethyl-2-butene) do not absorb in the infrared (no dipole change).

Symmetrically disubstituted (*trans*) double bonds are often vanishingly weak in absorption; *cis* are stronger.

Examples: Cyclohexene (Fig. 2-9), vinyl acetate (Fig. 2–10), styrene (Fig. 2–11), and *trans*-cyclododecene (Fig. 2-12)

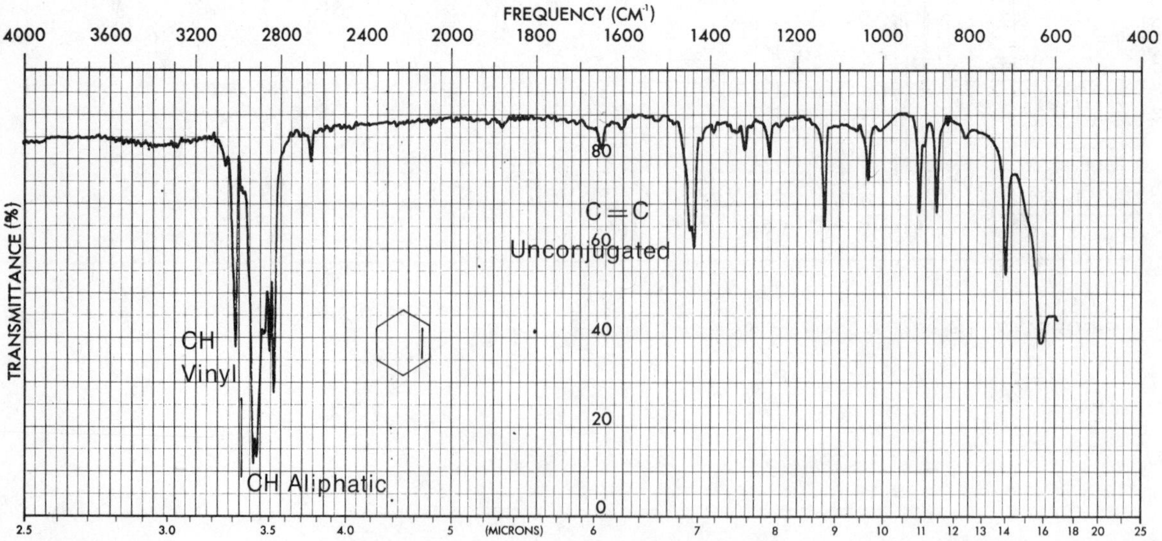

FIGURE 2-9 The Infrared Spectrum of Cyclohexene (Neat Liquid, Salt Plates)

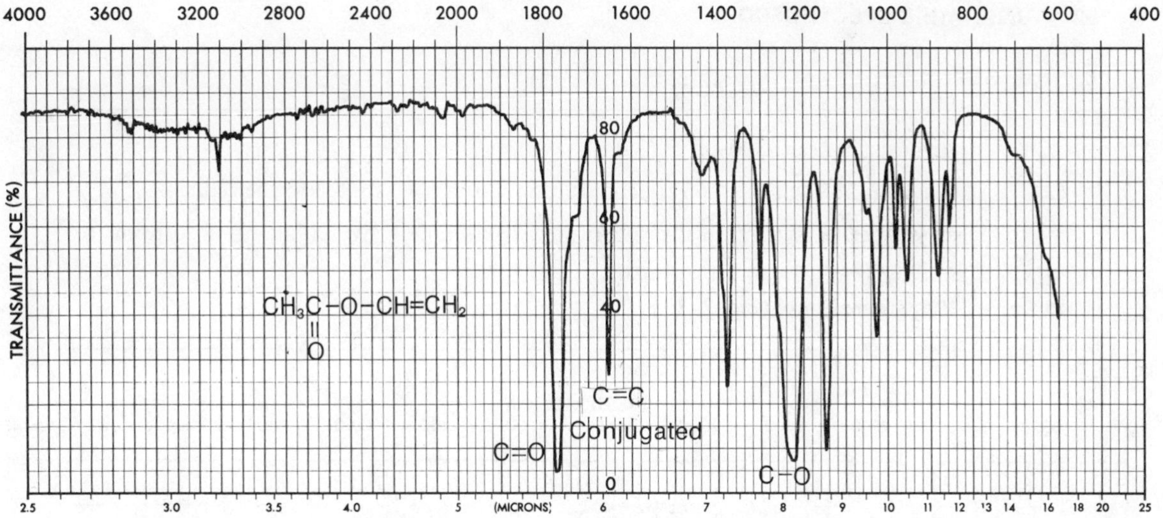

FIGURE 2–10 The Infrared Spectrum of Vinyl Acetate (Neat Liquid, Salt Plates)

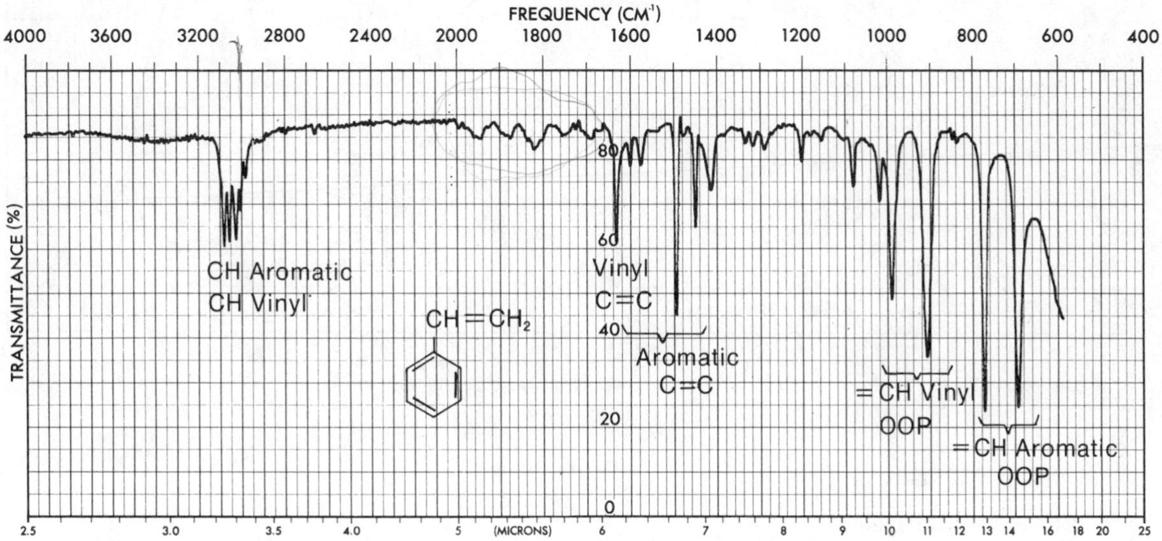

FIGURE 2–11 The Infrared Spectrum of Styrene (Neat Liquid, Salt Plates)

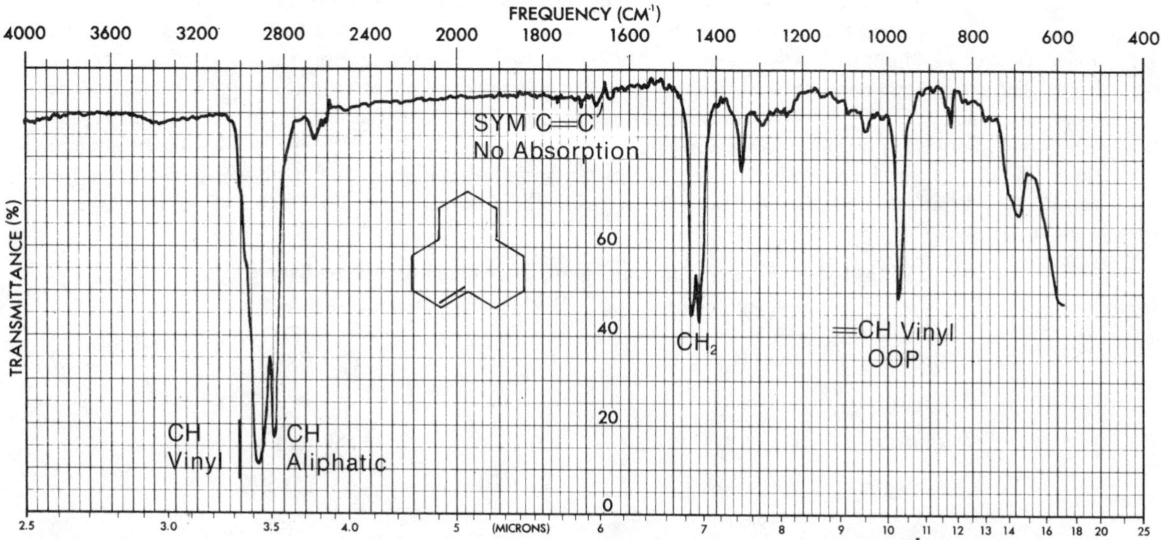

FIGURE 2–12 The Infrared Spectrum of *trans*-Cyclododecene (Neat Liquid, Salt Plates)

C. ALKYNES

≡C–H	stretch usually near 3300 cm^{-1} (3.0 μ)
C≡C	stretch near 2150 cm^{-1} (4.65 μ); conjugation moves C≡C stretch to the right.
	Disubstituted or symmetrically substituted triple bonds give either no absorption or weak absorption

Example: Propargyl alcohol (Fig. 2–13)

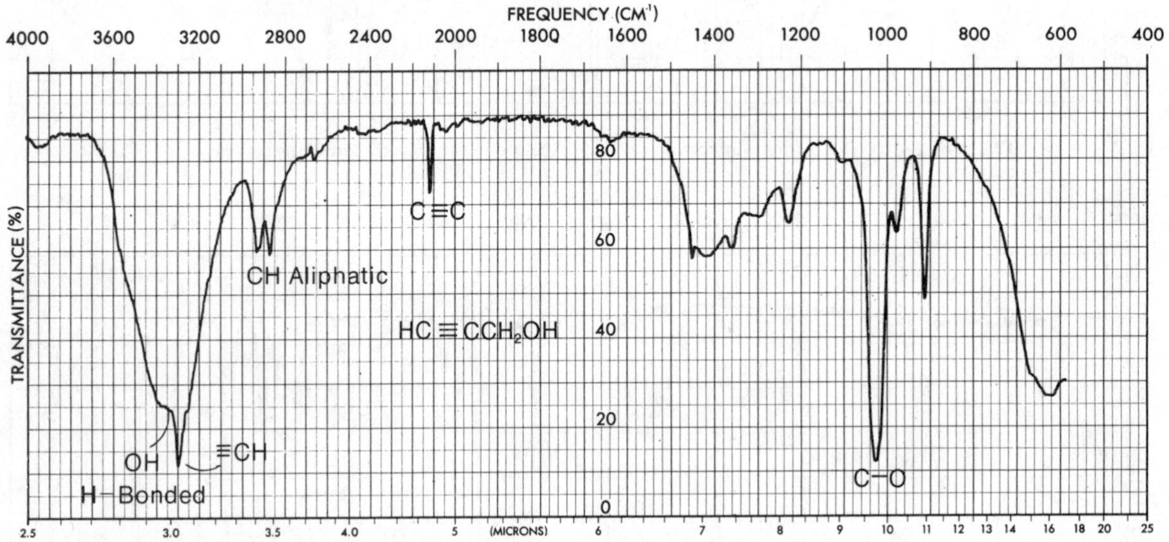

FIGURE 2–13 The Infrared Spectrum of Propargyl Alcohol (Neat Liquid, Salt Plates)

DISCUSSION SECTION

C–H Stretch Region

The C–H stretching and bending regions are two of the most difficult regions to interpret in infrared spectra. The C–H stretching region, which ranges from 3300 to 2750 cm^{-1} (3.0 to 3.6 μ), is generally the more useful of the two regions. As discussed in Section 2.4, the frequency of the absorption of C–H bonds is a function mostly of the type of hybridization that is attributed to the bond. The *sp*-1*s* C–H bond present in acetylenic compounds is stronger than the sp^2-1*s* bond present in vinyl compounds. This results in a larger vibrational force constant and a greater frequency of vibration. Likewise, the sp^2-1*s* C–H absorption in vinyl compounds will occur at a higher frequency than the sp^3-1*s* C–H absorption in saturated aliphatic compounds. Some physical constants for various C–H bonds involving *sp*, sp^2, and sp^3 hybridized carbon are given in Table 2–5.

TABLE 2–5 Physical Constants for *sp*, sp^2, and sp^3 Hybridized Carbon and the Resulting C–H Absorption Values

Bond	≡C–H	=C–H	–C–H
Type	*sp* - 1*s*	sp^2-1*s*	sp^3-1*s*
Length	1.08 Å	1.10 Å	1.12 Å
Strength	121 Kcal	106 Kcal	101 Kcal
IR Freq.	3300 cm^{-1}	~ 3100 cm^{-1}	~ 2900 cm^{-1}

As shown in Table 2–5, the frequency at which the C–H absorption occurs indicates the type of carbon to which the hydrogen is attached. The entire C–H stretching region is shown in Figure 2–14. Except for the aldehyde hydrogen, an absorption frequency of less than 3000 cm^{-1} usually implies a saturated compound (only sp^3-1*s* hydrogens). An absorption frequency greater than 3000 cm^{-1}, but not above about 3150 cm^{-1}, usually implies aromatic or vinyl hydrogens. However, cyclopropyl C–H bonds, which have extra *s* character because of the need to put more *p* character into the ring C–C bonds to reduce angle distortion, also give rise to absorption in the region of 3100 cm^{-1}. Cyclopropyl hydrogens can be distinguished easily from aromatic hydrogens or vinyl hydrogens by cross-reference to the C=C and C–H out-of-plane regions. The aldehyde C–H stretch appears to the right of saturated C–H absorptions and normally consists of two weak absorptions at about 2850 and 2750 cm^{-1}. The 2850 cm^{-1} band usually appears as a shoulder on the saturated C–H absorption bands. The band at 2750 cm^{-1} is far to the right of the usual range for aliphatic sp^3-1*s* C–H stretch and, when not too weak to distinguish, is a very diagnostic band.

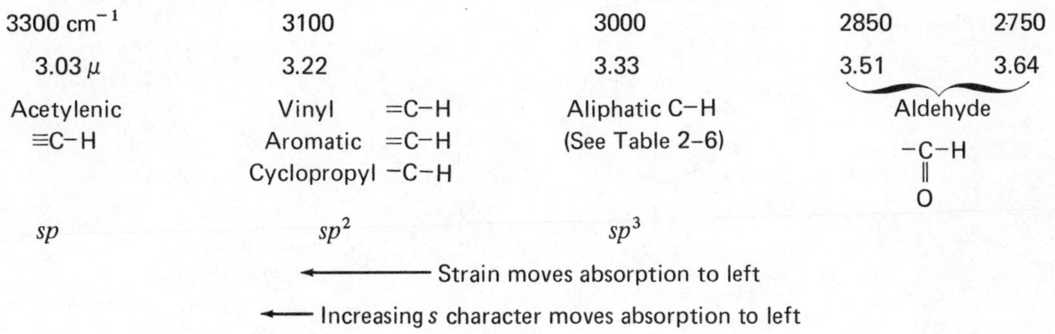

FIGURE 2–14 The C–H Stretch Region

The sp^3 hybridized C–H stretching vibrations for methyl, methylene, and methine are given in Table 2–6. The tertiary C–H (methine hydrogen) gives only one weak C–H stretch absorption, usually near 2890 cm^{-1}. Methylene hydrogens ($-CH_2-$), however, give rise to two C–H stretching bands, representing the symmetric and asymmetric stretching modes of the group. In effect, the 2890 cm^{-1} methine absorption is split into two bands at 2926 cm^{-1} (asym) and 2853 cm^{-1} (sym). The asymmetric mode generates a larger dipole moment and is of greater intensity than the symmetric mode. The splitting of the 2890 cm^{-1} methine absorption in the case of a methyl group is larger. Peaks appear at about 2962 and 2872 cm^{-1}. The asymmetric and symmetric stretching modes for methylene and methyl are shown in Section 2.3.

Since a number of bands may appear in the C–H stretch region, it is probably a good idea to decide only whether the absorptions are acetylenic (3300 cm^{-1}), vinylic or aromatic (>3000 cm^{-1}), aliphatic (<3000 cm^{-1}), or aldehydic (2850 and 2750 cm^{-1}). Further interpretation of C–H *stretching* vibrations may not be worth the effort. The C–H *bending* vibrations are much more useful in deciding whether methyl and/or methylene groups are present in a molecule.

TABLE 2-6 The Stretching Vibrations for Various sp^3 Hybridized C–H Bonds

Group		Stretching Vibration, cm^{-1} (μ)	
		Asymmetric	*Symmetric*
Methyl	CH_3-	2962 (3.38)	2872 (3.48)
Methylene	$-CH_2-$	2926 (3.42)	2853 (3.51)
Methine	$-\overset{\vert}{\underset{\vert \atop H}{C}}-$	2890 (3.46) very weak	

C–H Bending Vibrations for Methyl and Methylene

The presence of methyl and methylene groups may be determined, when not obscured by other absorptions, by analyzing the region from 1465 to 1370 cm^{-1} (6.8 to 7.3 μ). As shown in Figure 2–15, the band due to CH_2 scissoring usually occurs at 1465 cm^{-1} (6.83 μ). One of the bending modes for CH_3 usually absorbs strongly near 1375 cm^{-1} (7.27 μ). These bands can often be used to detect methylene and methyl groups, respectively. Further, the 1375 cm^{-1} (7.27 μ) methyl band is usually split into *two* peaks of nearly equal intensity (symmetric and asymmetric modes) if a geminal dimethyl group is present. Often this doublet will be observed in compounds with isopropyl groups. A *tert*-butyl group results in an even wider splitting of the 1375 cm^{-1} (7.27 μ) band into two peaks. The 1370 cm^{-1} (7.3 μ) band is more intense than the 1390 cm^{-1} (7.20 μ) one. The expected pattern for the isopropyl and *t*-butyl groups are shown in Figure 2–16. It should be noted that some variation from these idealized patterns may occur. Nuclear magnetic resonance spectroscopy (Chapter 3) may be used to confirm the presence of these groups. In cyclic hydrocarbons, which do not have attached methyl groups, the 1375 cm^{-1} (7.27 μ) band will be missing, as can be seen in the spectrum of cyclohexane (Fig. 2–8). Finally, a rocking band (Section 2.3) appears near 720 cm^{-1} (13.9 μ) for straight chain alkanes (see Fig. 2–7).

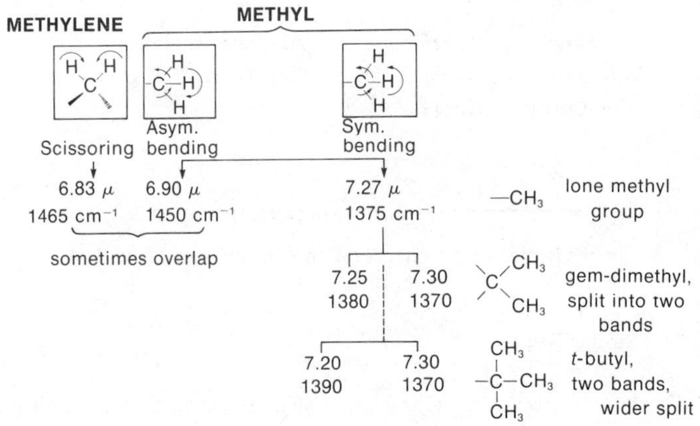

FIGURE 2–15 The C–H Bending Vibrations for Methyl and Methylene Groups

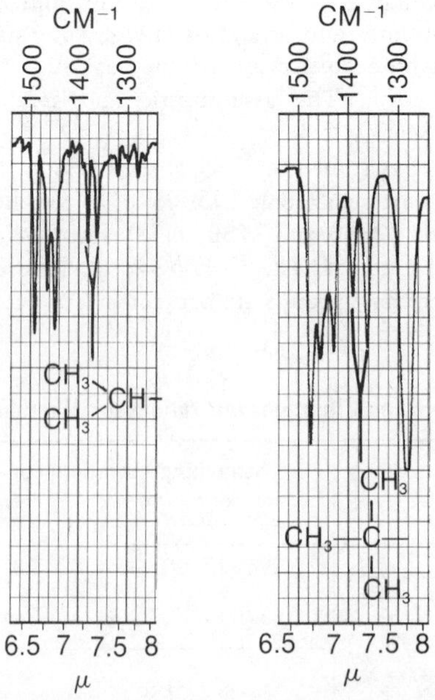

FIGURE 2–16 C–H Bending Patterns for the Isopropyl and *t*-Butyl Groups

C=C Stretching Vibrations

Conjugation Effects. The C=C stretching absorption usually appears between 1666 and 1640 cm^{-1} (6.0 to 6.1 μ) for unconjugated alkenes. The intensity of the absorption is generally much lower than that of a typical carbonyl group and gives rise to a medium to weak band. If the double bond is symmetrically substituted, such as in a tetrasubstituted alkene, this absorption may be very weak or even non-existent. Recall from Section 2.1 that if the attached groups are symmetrically arranged no change in dipole moment occurs during stretching and, hence, no infrared absorption is observed. *cis*-Alkenes, which have less symmetry than *trans*-alkenes, generally absorb more strongly than the latter. Double bonds in rings, because they are often symmetric, or nearly so, absorb more weakly than those not contained in rings. Terminal double bonds generally have stronger absorption.

Conjugation of a C=C double bond with either a carbonyl group or another double bond provides the multiple bond with more single bond character (through resonance as shown below), a lower force constant K, and thus a lower frequency of vibration. For example, the vinyl double bond in styrene (Figure 2–11) gives an absorption at 1630 cm^{-1} (6.13 μ). With several double bonds, the number of C=C absorptions often corresponds to the number of conjugated double bonds. An example of this is found in 1,3-pentadiene, where absorptions are observed at 1600 cm^{-1} (6.25 μ) and 1650 cm^{-1} (6.06 μ). In the exception to the rule, butadiene gives only one band near 1600 cm^{-1}. If the double bond is conjugated with a carbonyl group, its absorption intensity is usually enhanced somewhat by the strong dipole of the carbonyl group.

$$\left[\quad C=C-C=C \quad \longleftrightarrow \quad C^{+}-C=C-C^{-} \quad \right]$$

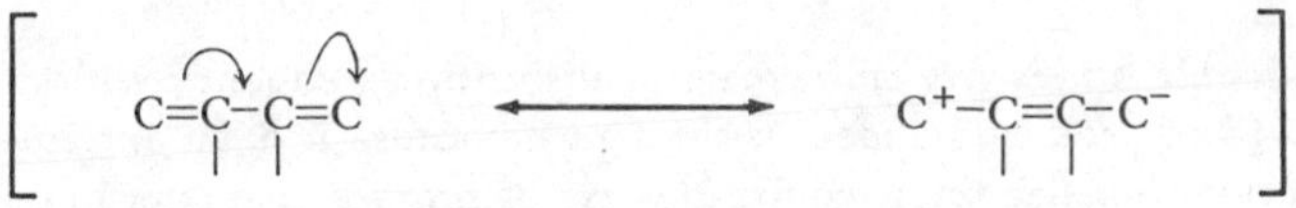

Ring Size Effects. The absorption frequency of *internal (endo)* double bonds in cyclic compounds is very sensitive to ring size. As shown in Figure 2–17, the absorption frequency decreases as the internal angle decreases until it reaches a minimum at 90° in cyclobutene. The frequency increases again for cyclopropene when the angle drops to 60°. This intially unexpected increase in frequency occurs because the C=C vibration in cyclopropene is strongly coupled to the attached C–C single bond vibration. When the attached C–C bonds are perpendicular to the C=C axis, as in cyclobutene, their vibrational mode is orthogonal to that of the C=C bond (*i.e.*, on a different axis) and does not couple. When the angle is greater than 90° (120° in the example shown below), the C–C single bond stretching vibration can be resolved into two components, one of which is coincident with the direction of the C=C stretch. In the diagram, components **a** and **b** of the C–C stretching vector are shown. Since component **a** is in line with the C=C stretching vector, the C–C and C=C bonds are coupled, leading to a higher frequency of absorption. A similar pattern exists for cyclopropene, which has an angle less than 90°.

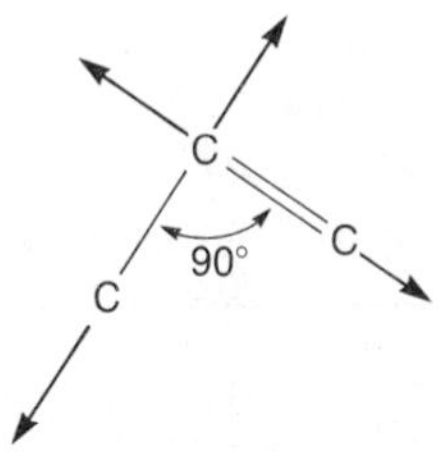

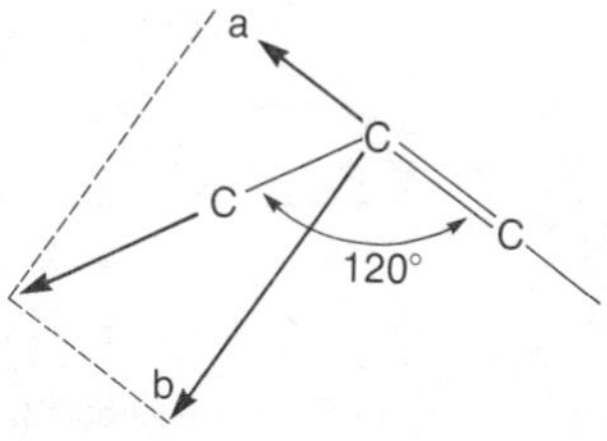

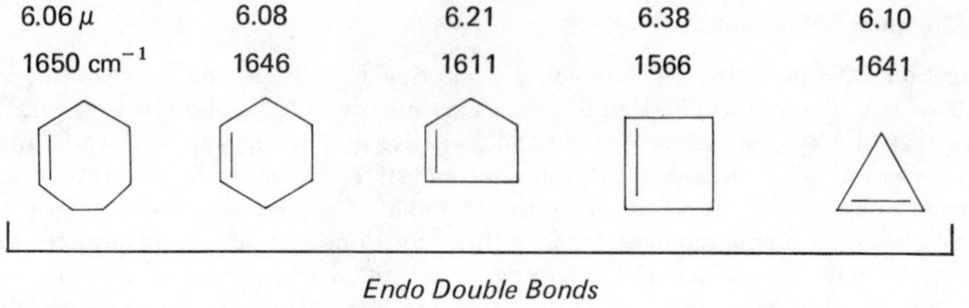

a) Strain moves the peak to the right
Anomaly: Cyclopropene

b) If an endo double bond is at a ring fusion, the absorption moves to the right an amount equivalent to the change that would occur if one carbon were removed from the ring.

e.g.: 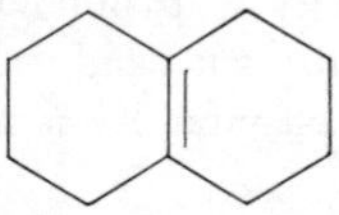 $\sim 1611\ cm^{-1}$

FIGURE 2–17 C=C Stretching Vibrations in Endocyclic Systems

External (exo) double bonds give an increase in absorption frequency with decreasing ring size, as shown in Figure 2–18. Allene is included in the figure because it is an extreme example of an exo double bond absorption. Smaller rings require the use of more *p* character in making the C–C bonds meet the requisite small angles (recall the trend: $sp = 180°$, $sp^2 = 120°$, $sp^3 = 109°$, $sp^{>3} = <109°$). This removes *p* character from the sigma bond of the double bond, but gives it more *s* character, thus strengthening and stiffening the double bond. The force constant *K* is then increased and the absorption frequency increases.

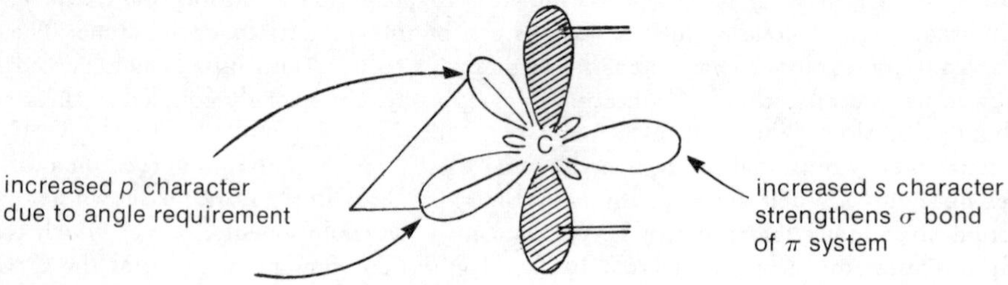

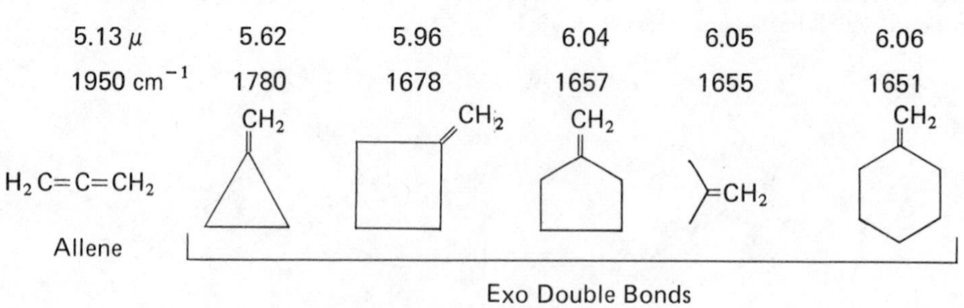

(a) Strain moves the peak to the left

(b) Ring fusion moves the absorption to the left

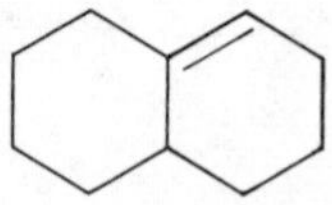

FIGURE 2–18 C=C Stretching Vibrations in Exocyclic Systems

C–H Bending Vibrations for Alkenes

The C–H bonds in alkenes can vibrate by bending both in-plane and out-of-plane when they absorb infrared radiation. The scissoring in-plane vibration for terminal alkenes occurs at about 1415 cm^{-1} (7.07 μ). This band appears at this value as a medium to weak absorption for both monosubstituted and 1,1-disubstituted alkenes.

The most valuable information for alkenes is obtained from analysis of the C–H out-of-plane region of the spectrum, which extends from 1000 to 650 cm^{-1} (10.0 to 15.4 μ). These bands are generally the strongest peaks in the spectrum. The number of absorptions and their positions in the spectrum can be used to indicate the substitution pattern on the double bond.

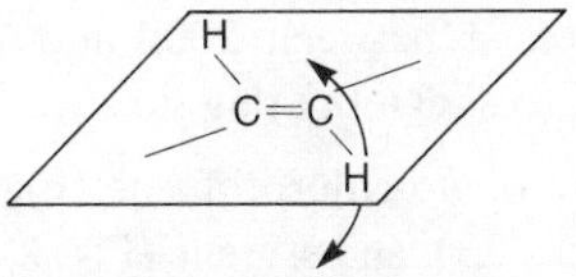

C—H OUT-OF-PLANE BENDING

Monosubstituted Double Bonds. This substitution pattern gives rise to two strong bands, one near 990 cm^{-1} (10.1 μ) and the other near 910 cm^{-1} (11.0 μ). The 910 cm^{-1} band is shifted to the right, to as low as 810 cm^{-1} (12.3 μ), when electron-withdrawing groups are attached to the double bond. For example, in the spectrum of vinyl acetate (Figure 2–10), the 910 cm^{-1} band is shifted to 880 cm^{-1}. The use of the out-of-plane vibrations to confirm the monosubstituted structure is considered very reliable. The absence of these bands almost certainly indicates that this structural feature is not present within the molecule.

cis- and trans-1,2-Disubstituted Double Bonds. A *cis* arrangement about a double bond will give one strong band near 700 cm^{-1} (14.3 μ), while a *trans* double bond will absorb near 970 cm^{-1} (10.3 μ). This kind of information can be valuable in assigning stereochemistry about the double bond (see Figure 2–12).

1,1-Disubstituted Double Bonds. One strong band near 890 cm^{-1} (11.2 μ) is obtained for a *gem*-disubstituted double bond.

Trisubstituted Double Bonds. One medium intensity band near 815 cm^{-1} (12.3 μ) is obtained. Tetrasubstituted alkenes will not give any absorption in this region because of the absence of a hydrogen atom on the double bond. In addition, the C=C stretching vibration will be very weak (or absent) at about 1670 cm^{-1} (6.0 μ) in these highly substituted systems.

The C–H out-of-plane bending vibrations for substituted alkenes are shown in Figure 2–19, together with the frequency (wavelength) ranges.

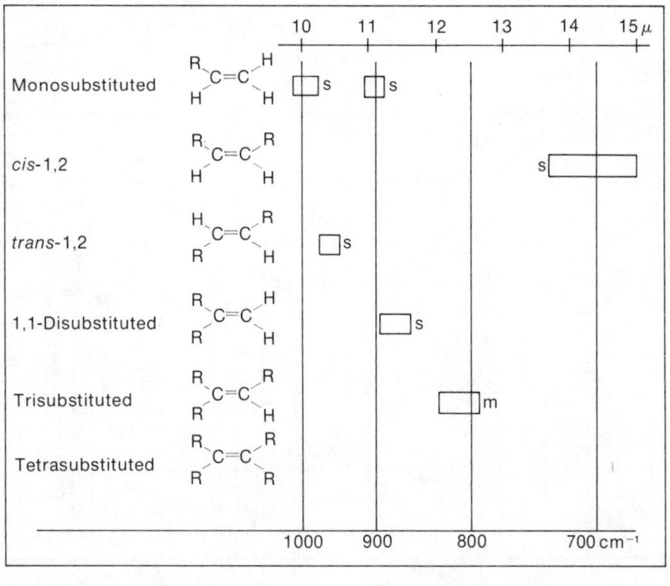

FIGURE 2–19 The C–H Out-of-plane Bending Vibrations for Substituted Alkenes

2.10 AROMATIC RINGS

=C–H stretch is always to the left of 3000 cm^{-1} (3.33 μ)

=C–H out-of-plane (oop) bending, 900–690 cm^{-1} (11.1–14.5 μ).

These bands can be used with great utility to assign the ring substitution pattern (see discussion).

C=C ring stretch absorptions often occur in pairs at 1600 cm^{-1} (6.25 μ) and 1475 cm^{-1} (6.78 μ).

Overtone/combination bands appear between 2000 and 1667 cm^{-1} (5 to 6 μ). These *weak* absorptions can be used to assign the ring substitution pattern (see discussion).

Examples: Toluene (Fig. 2–20), *ortho*-chlorotoluene (Fig. 2–21), *meta*-toluidine (Fig. 2–22), *para*-chlorotoluene (Fig. 2-23), and styrene (Fig. 2–11).

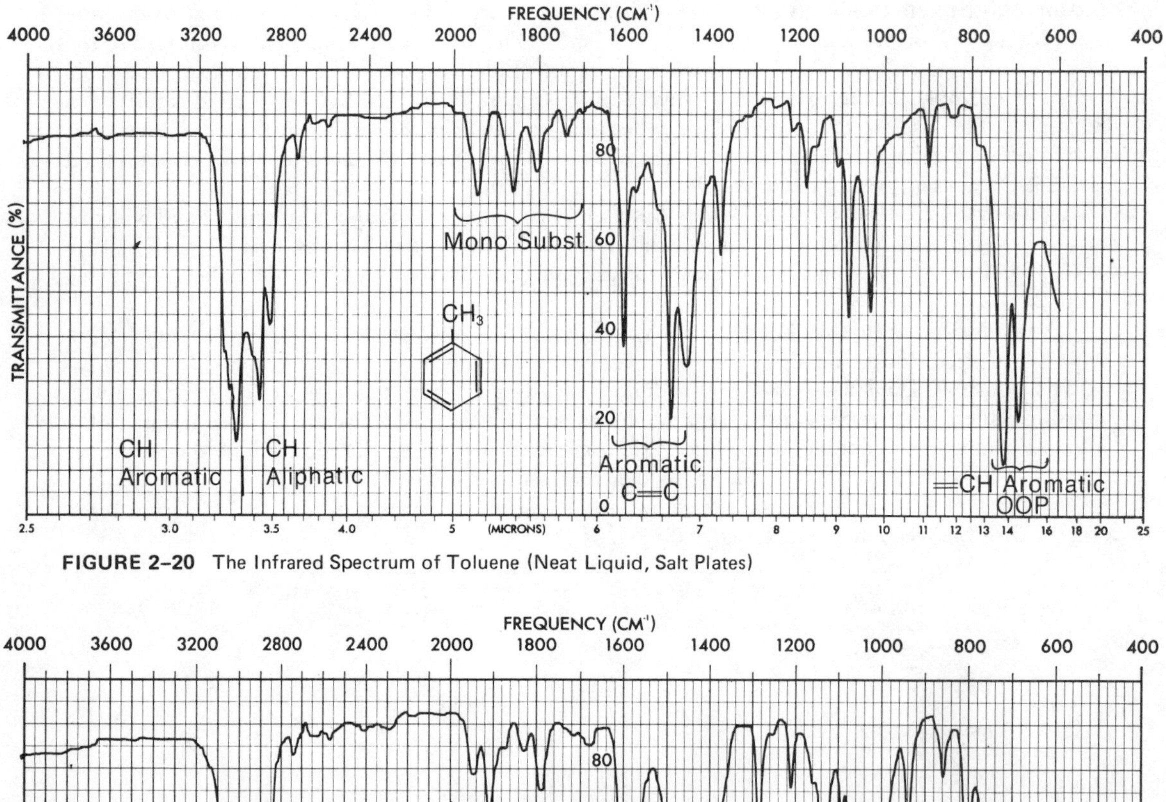

FIGURE 2–20 The Infrared Spectrum of Toluene (Neat Liquid, Salt Plates)

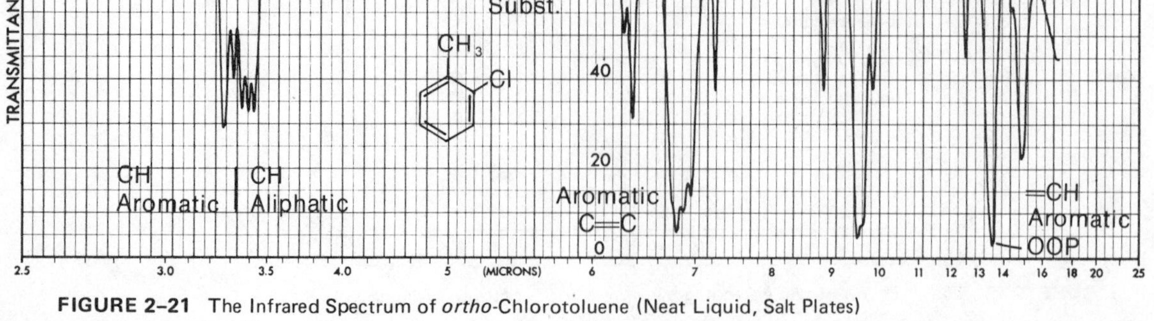

FIGURE 2–21 The Infrared Spectrum of *ortho*-Chlorotoluene (Neat Liquid, Salt Plates)

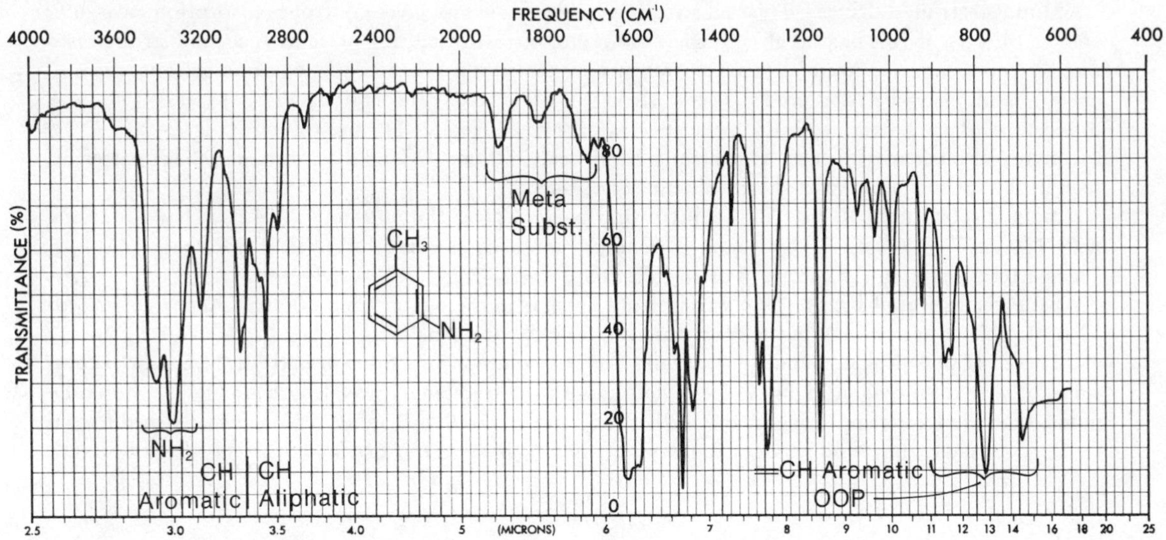

FIGURE 2–22 The Infrared Spectrum of *meta*-Toluidine (Neat Liquid, Salt Plates)

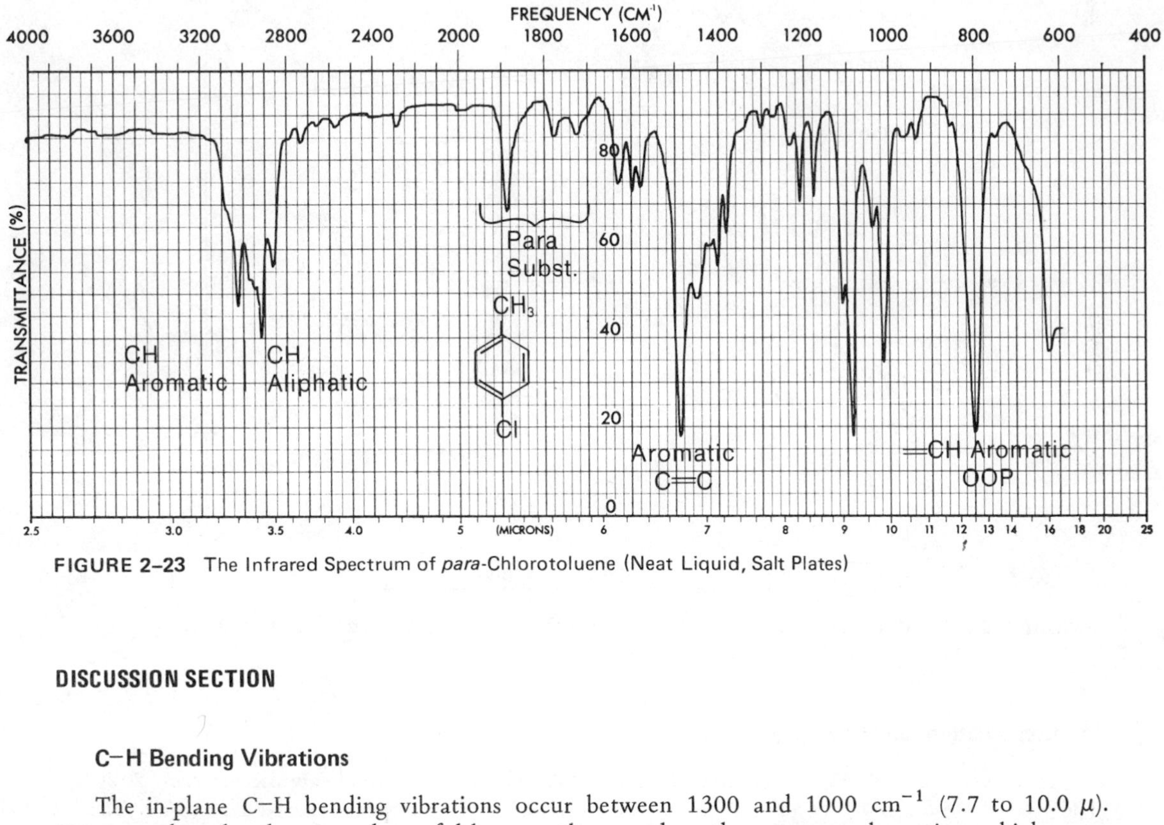

FIGURE 2–23 The Infrared Spectrum of *para*-Chlorotoluene (Neat Liquid, Salt Plates)

DISCUSSION SECTION

C–H Bending Vibrations

The in-plane C–H bending vibrations occur between 1300 and 1000 cm^{-1} (7.7 to 10.0 μ). However, these bands are rarely useful because they overlap other, stronger absorptions which occur in this region.

The out-of-plane C–H bending vibrations, which appear between 900 and 690 cm^{-1} (11.1 to 14.5 μ), are far more useful than the in-plane bands. These extremely intense absorptions, resulting from strong coupling with adjacent hydrogen atoms, can be used to assign the position of substituents on the aromatic ring. The assignment of structure based upon these out-of-plane bending vibrations is most reliable for alkyl substituted aromatic compounds or for molecules without certain polar groups. Aromatic nitro compounds and the derivatives of aromatic carboxylic acids (esters, amides, etc.) often lead to unsatisfactory interpretation.

Monosubstituted Rings. This substitution pattern always gives a strong absorption near 690 cm^{-1} (14.5 μ). If this band is absent, there is no monosubstituted ring present. A second strong band usually appears near 750 cm^{-1} (13.3 μ). When the spectrum is taken in a halocarbon solvent, the 690 cm^{-1} band may be obscured by the strong C–X stretch absorptions. The typical two-peak monosubstitution pattern appears in the spectra of styrene (Fig. 2–11) and toluene (Fig. 2–20).

ortho-Disubstituted Rings. One strong band near 750 cm^{-1} (13.3 μ) is obtained. This pattern is seen in the spectrum of *ortho*-chlorotoluene (Fig. 2–21).

meta-Disubstituted Rings. This substitution pattern gives the 690 cm^{-1} (14.5 μ) band plus one near 780 cm^{-1} (12.8 μ). A third band of medium intensity is often found near 880 cm^{-1} (11.4 μ). This pattern is seen in the spectrum of *m*-toluidine (Fig. 2–22).

para-Disubstituted Rings. One strong band appears in the region from 800 to 850 cm^{-1} (12.5 to 11.8 μ). This pattern is seen in the spectrum of *para*-chlorotoluene (Fig. 2–23).

The C–H out-of-plane bending vibrations for the common substitution patterns given above, plus some others, are shown in Figure 2–24, together with the frequency (wavelength) ranges. It should be noted that the bands appearing in the 720 to 667 cm^{-1} (13.9 to 15.0 μ) region actually result from C=C out-of-plane ring bending vibrations, rather than from C–H out-of-plane bending (shaded boxes).

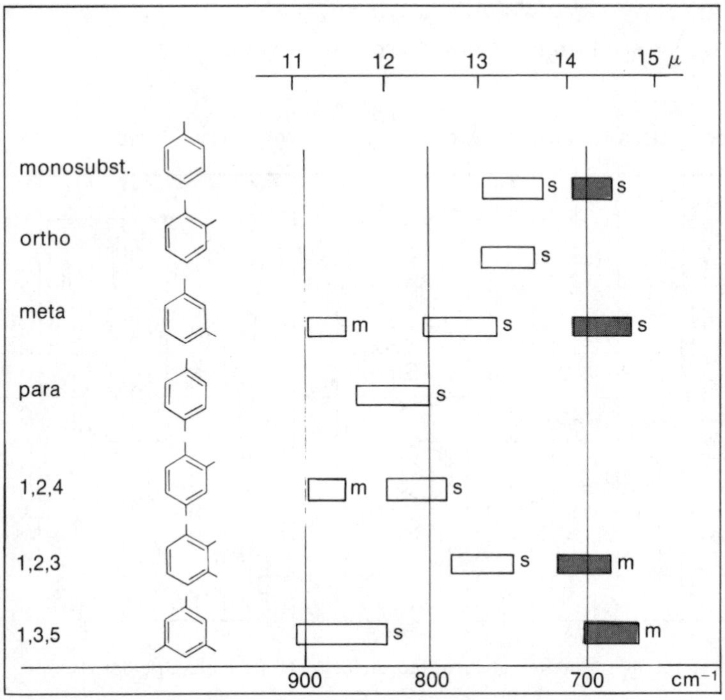

FIGURE 2–24 The C–H Out-of-plane Bending Vibrations for Substituted Benzenoid Compounds

Combinations and Overtone Bands

There are many *weak* combination and overtone absorptions which appear between 2000 and 1667 cm^{-1} (5 to 6 μ). The relative shapes and number of these peaks can be used to tell whether an aromatic ring is mono-, di-, tri-, tetra-, penta-, or hexa-substituted. Positional isomers can also be distinguished. Since the absorptions are weak, these bands are best observed by using neat liquids or concentrated solutions. If the compound has a high frequency carbonyl group, this absorption will overlap the weak overtone bands so that no useful information can be obtained from the analysis of this region.

The various patterns that are obtained in this region are shown in Figure 2–25. The monosubstitution pattern shown in the spectrum of toluene (Fig. 2–20) is particularly useful and helps to confirm the out-of-plane data given in the preceding section. Likewise, the *ortho*, *meta*, and *para* disubstituted patterns should be consistent with the out-of-plane bending vibrations discussed above. The spectra of *ortho*-chlorotoluene (Fig. 2–21), *meta*-toluidine (Fig. 2–22), and *para*-chlorotoluene (Fig. 2–23) each show bands in *both* the 2000 to 1667 cm^{-1} (5 to 6 μ) *and* 900 to 690 cm^{-1} (11.1 to 14.5 μ) regions, consistent with their structures. It should be noted, however, that the out-of-plane vibrations are generally more useful for diagnostic purposes.

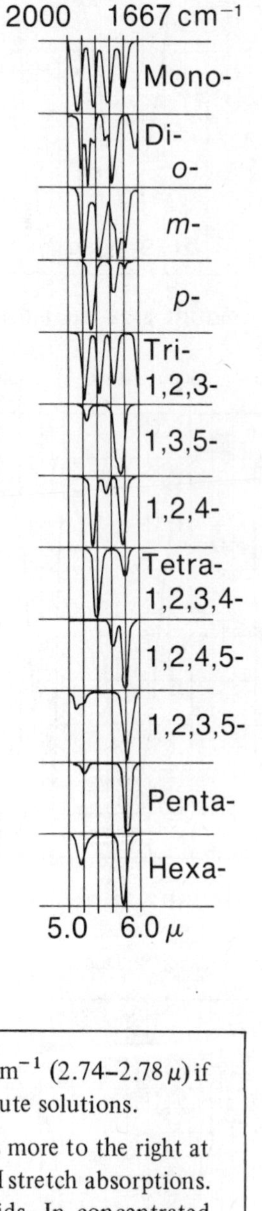

FIGURE 2–25 The 2000 to 1667 cm^{-1} (5 to 6 μ) Region for Substituted Benzenoid Compounds (From Dyer, John R., "Applications of Absorption Spectroscopy of Organic Compounds," Prentice-Hall, Inc., Englewood Cliffs, NJ, 1965)

2.11 ALCOHOLS AND PHENOLS

O–H The "free" O–H stretch is a *sharp* peak appearing at 3650–3600 cm^{-1} (2.74–2.78 μ) if no hydrogen bonding takes place. This band is observed only in dilute solutions.

The hydrogen bonded O–H stretch is a *broad* peak which occurs more to the right at 3500–3200 cm^{-1} (2.86–3.13 μ), sometimes overlapping with C–H stretch absorptions. This band is usually the only one present in neat (pure) liquids. In concentrated solutions, both the "free" and hydrogen bonded bands are usually present.

C–O This stretching vibration usually occurs in the range 1250–1000 cm^{-1} (8–10 μ). This band can be used to assign a primary, secondary, or tertiary structure to an alcohol (see discussion).

Examples: The hydrogen bonded O–H stretch is present in the pure liquid samples of 1-hexanol (Fig. 2–26) and 2-butanol (Fig. 2–27). A solution of the phenol, 2-naphthol (Fig. 2–28), shows both the "free" and hydrogen bonded O–H stretch.

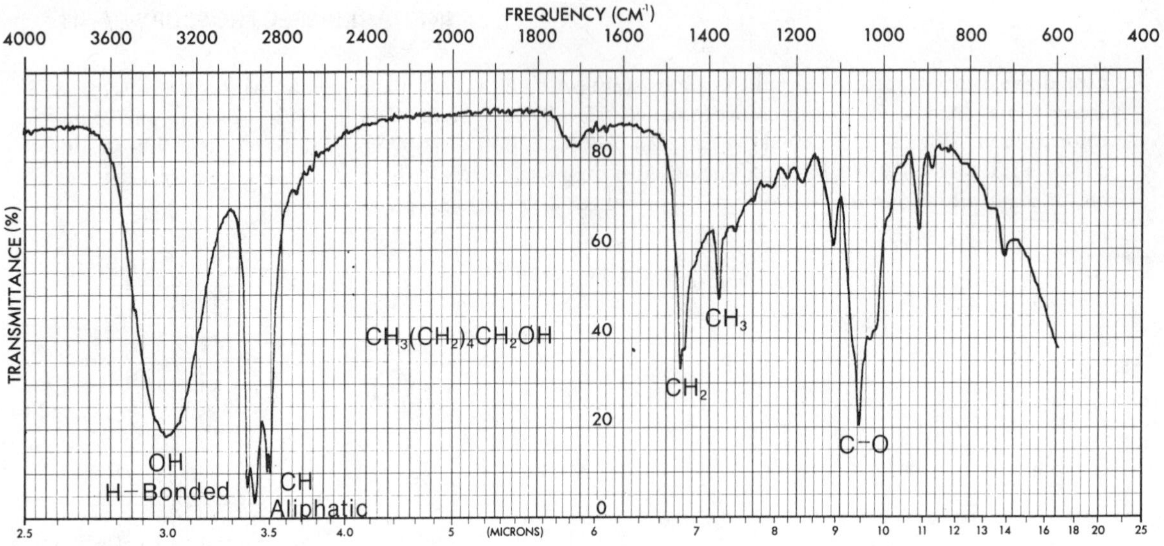

FIGURE 2–26 The Infrared Spectrum of 1–Hexanol (Neat Liquid, Salt Plates)

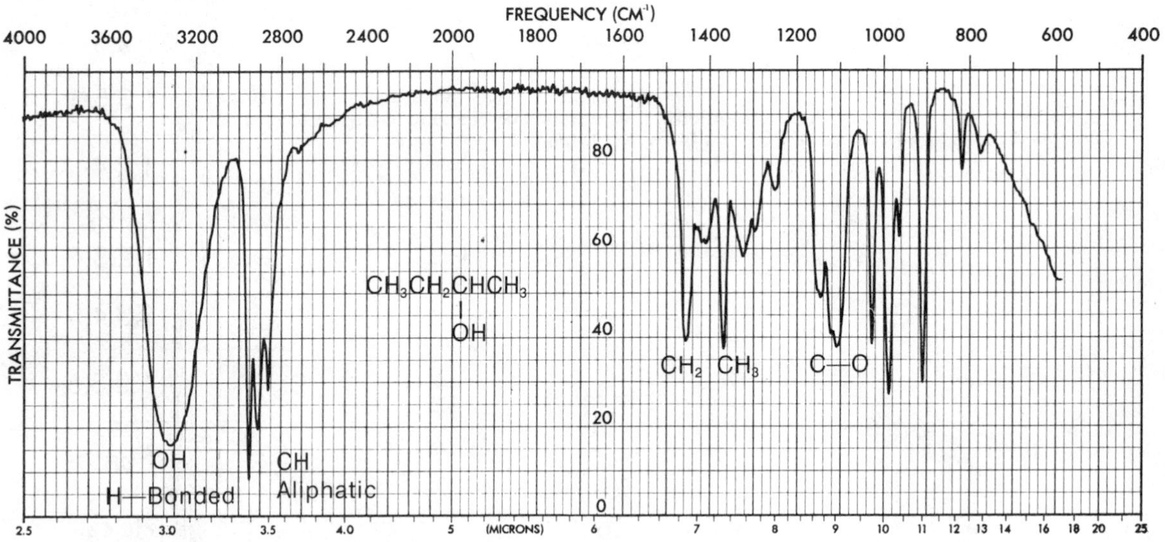

FIGURE 2–27 The Infrared Spectrum of 2-Butanol (Neat Liquid, Salt Plates)

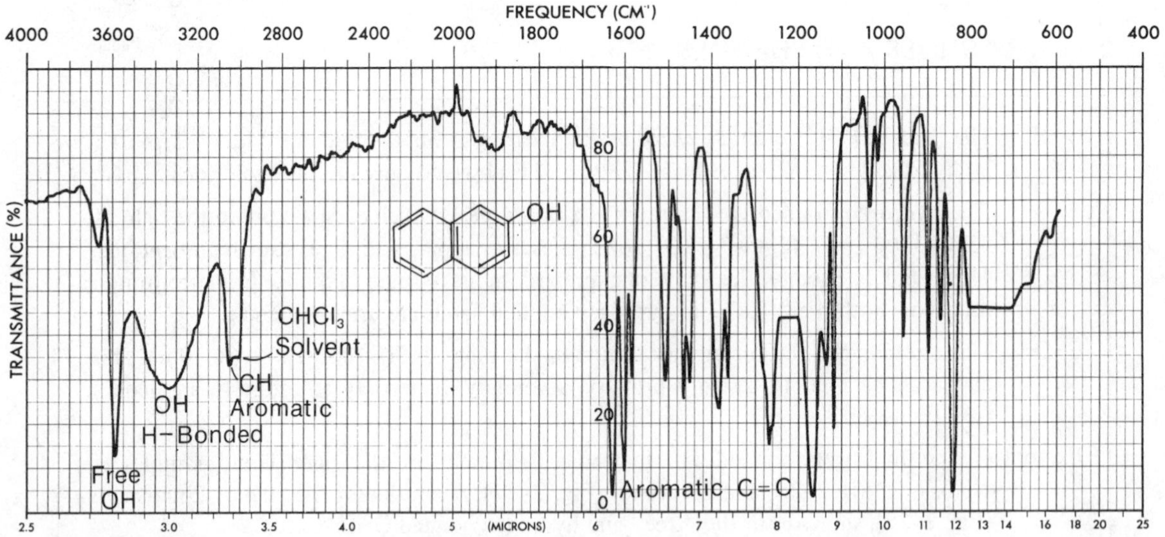

FIGURE 2–28 The Infrared Spectrum of 2-Naphthol Showing Both Free and Hydrogen-Bonded OH ($CHCl_3$ Solution)

DISCUSSION SECTION

O–H Stretching Vibrations

When alcohols or phenols are determined as pure (neat) liquid films, as is common practice, one obtains a broad O–H stretching vibration from about 3500 to 3200 cm^{-1} (2.86 to 3.13 μ) owing to intermolecular hydrogen bonding. This band is observed in the spectra of 1-hexanol (Fig. 2–26) and 2-butanol (Fig. 2-27) or in Figure 2–29A. As the alcohol is diluted with carbon tetrachloride, a sharp "free" (non-hydrogen bonded) O–H stretching band appears at about 3600 cm^{-1} (2.78 μ) to the left of the broad band, as shown in the spectrum of 2-naphthol (Fig. 2–28) or Figure 2–29B. When the solution is further diluted, the broad intermolecular hydrogen bonded band is reduced considerably, leaving as the major band the "free" O–H stretching absorption (Fig. 2-29C). Intermolecular hydrogen bonding weakens the O–H bond, thereby shifting the band to lower frequency (lower energy).

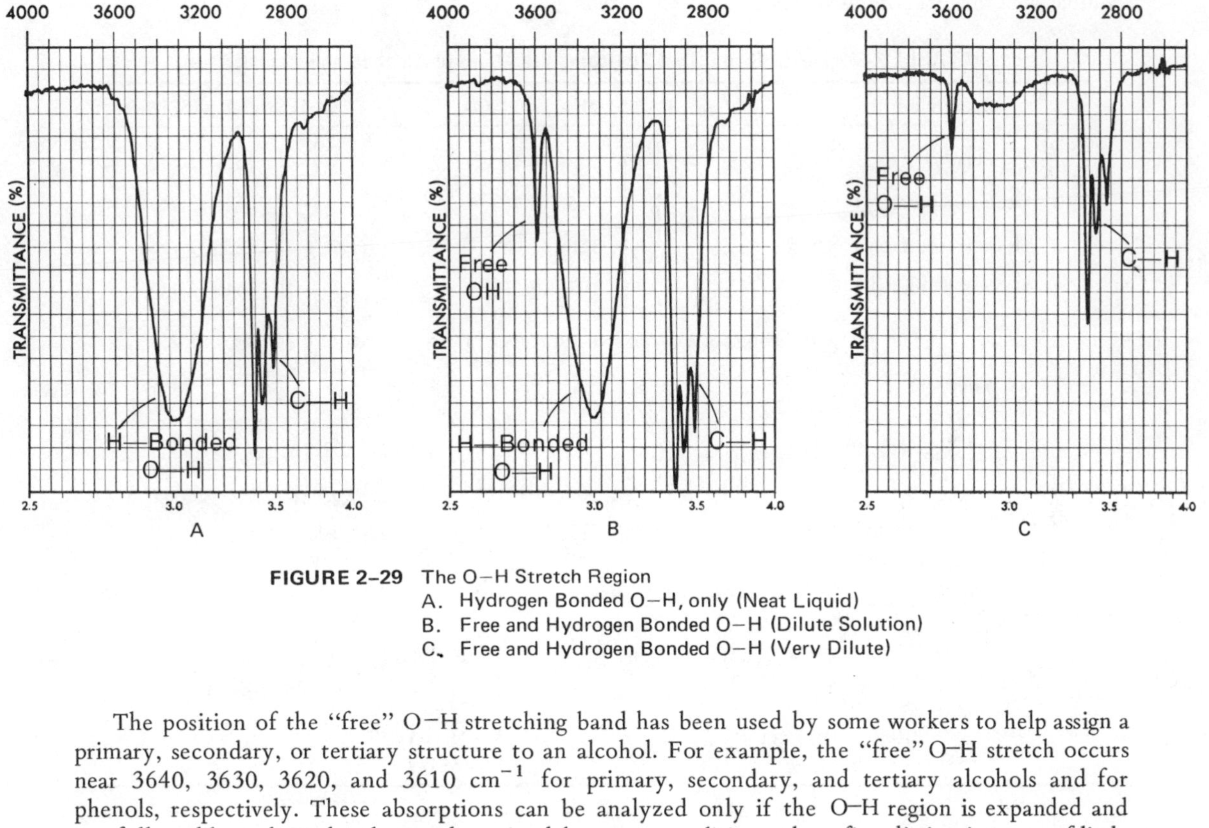

FIGURE 2–29 The O–H Stretch Region
A. Hydrogen Bonded O–H, only (Neat Liquid)
B. Free and Hydrogen Bonded O–H (Dilute Solution)
C. Free and Hydrogen Bonded O–H (Very Dilute)

The position of the "free" O–H stretching band has been used by some workers to help assign a primary, secondary, or tertiary structure to an alcohol. For example, the "free" O–H stretch occurs near 3640, 3630, 3620, and 3610 cm^{-1} for primary, secondary, and tertiary alcohols and for phenols, respectively. These absorptions can be analyzed only if the O–H region is expanded and carefully calibrated. Under the usual routine laboratory conditions, these fine distinctions are of little use.

Intramolecular hydrogen bonding, present in *ortho*-carbonyl-substituted phenols, usually shifts the broad O–H band to the right. For example, the O–H band is centered at about 3200 cm^{-1} (3.13 in the neat spectrum of methyl salicylate, while O–H bands from normal phenols are centered at about 3330 cm^{-1} (3.0 μ). This band does not change its position significantly even at high dilution because the internal bonding is not altered by a change in concentration.

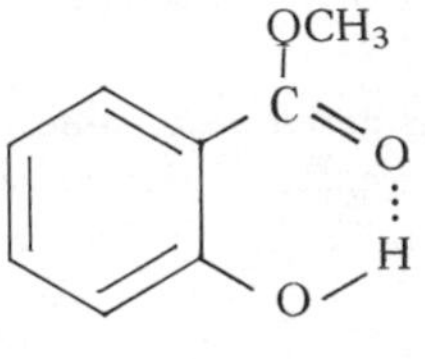

Methyl Salicylate

Although phenols often have broader O–H bands than alcohols, it is difficult to assign a structure based upon this absorption. One should use the aromatic C=C region and the C–O stretching vibration (see below) for assigning a phenolic structure. Finally, the O–H stretching vibrations in carboxylic acids also occur in this region. They may be distinguished easily from alcohols and phenols by the presence of a *very broad* band extending from 3400 to 2400 cm^{-1} (2.94 to 4.17 μ) *and* the presence of a carbonyl absorption (see Section 2.13D).

C–O Stretching Vibrations

The strong C–O single bond stretching vibrations are observed in the range from 1250 to 1000 cm^{-1} (8.0 to 10.0 μ). Since the C–O absorptions are coupled with the adjacent C–C stretching vibrations, the position of the band may be used to assign a primary, secondary, or tertiary structure to an alcohol or to determine whether a phenolic compound is present. The expected absorption bands for the C–O stretching vibrations in alcohols and phenols are given in Table 2–7. For comparison, the OH stretch values are also tabulated.

TABLE 2–7 The C–O and O–H Stretching Vibrations in Alcohols and Phenols

Compound		C–O Stretch, cm^{-1} (μ)		O–H Stretch, cm^{-1} (μ)
Phenols	DECREASE ↓	1220 (8.20)	INCREASE ↓	3610 (2.77)
3° Alcohols (saturated)		1150 (8.70)		3620 (2.76)
2° Alcohols (saturated)		1100 (9.09)		3630 (2.755)
1° Alcohols (saturated)		1050 (9.52)		3640 (2.75)

Unsaturation on adjacent carbons or a cyclic structure lowers the frequency of C–O absorption

2° examples:

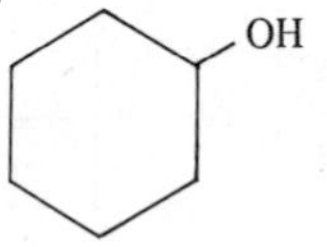

1100 → 1070 cm^{-1}

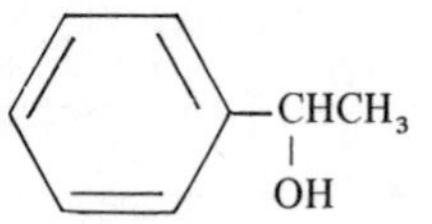

1100 → 1070 cm^{-1}

$CH_2 = CHCHCH_3$ (with OH on the CH)

OH

1100 → 1060 cm^{-1}

1° examples:

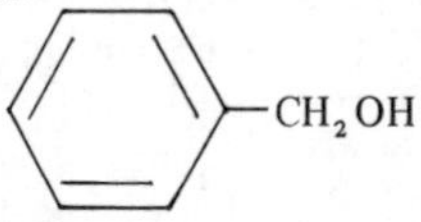

1050 → 1017 cm^{-1}

$CH{\equiv}CCH_2OH$

1050 → 1030 cm^{-1}
(Fig. 2–13)

The spectrum of 1-hexanol, a primary alcohol, has its C–O absorption at 1055 cm^{-1} (Fig. 2–26), while 2-butanol, a secondary alcohol, has its C–O absorption at 1105 cm^{-1} (Fig. 2–27). Thus, both alcohols have their C–O bands near the expected value given in Table 2–7. Phenols give a C–O absorption at about 1220 cm^{-1} because of conjugation of the oxygen with the ring, which shifts the band to higher energy (more double bond character). In addition to this band, an O–H in-plane bending absorption is usually found near 1360 cm^{-1} (7.35 μ) for neat samples. This latter band is also found in alcohols determined as neat liquids. It usually overlaps the C–H bending vibration for the methyl group at 1375 cm^{-1} (7.27 μ).

The numbers given in Table 2–7 should be considered *base* values. These C–O absorptions are shifted to lower frequencies when unsaturation is present on adjacent carbon atoms or when the OH is attached to a ring. Shifts of 30 to 40 cm^{-1} from the base values are common, as seen in some selected examples in Table 2–7.

2.12 ETHERS

C–O The most prominent band is that due to C–O stretch, 1300–1000 cm^{-1} (7.7–10.0 μ). Absence of C=O and O–H is required to be sure that C–O stretch is not due to an alcohol or ester. Phenyl and vinyl ethers give two strong bands at each end of this range; aliphatic ethers give one strong band to the right.

Examples: Butyl ether (Fig. 2–30) and anisole (Fig. 2–31)

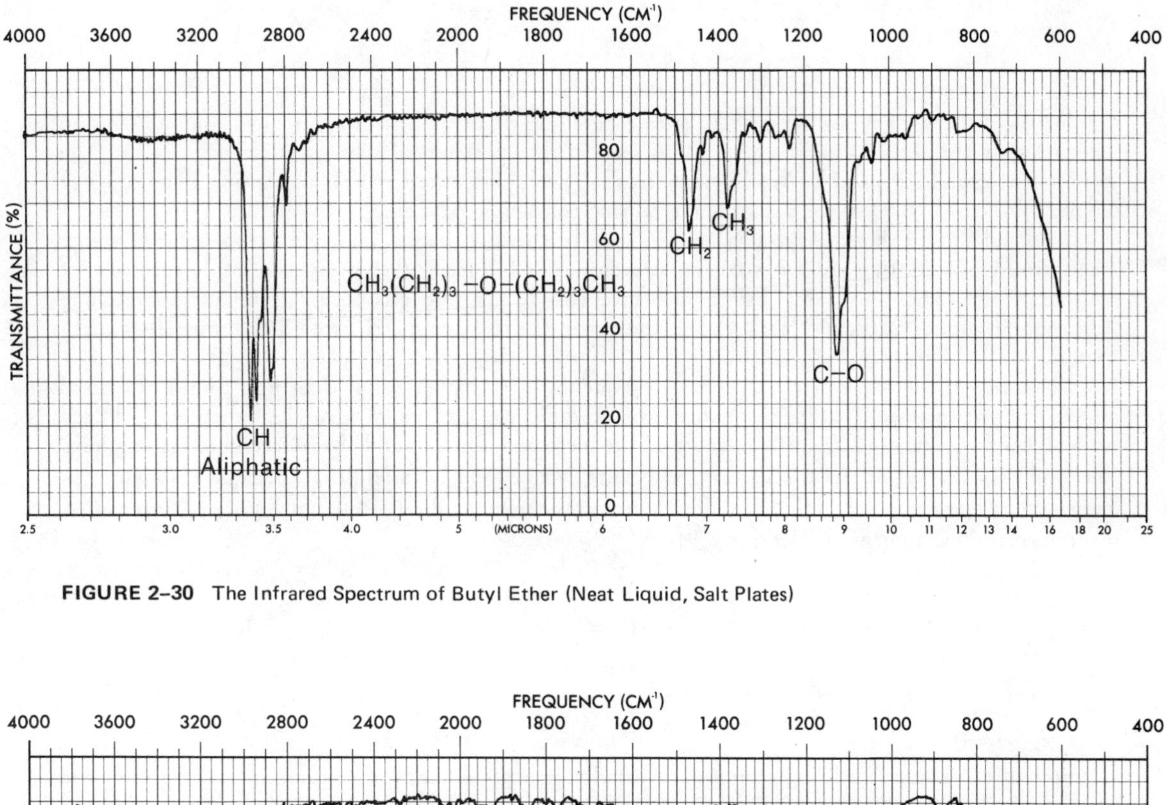

FIGURE 2–30 The Infrared Spectrum of Butyl Ether (Neat Liquid, Salt Plates)

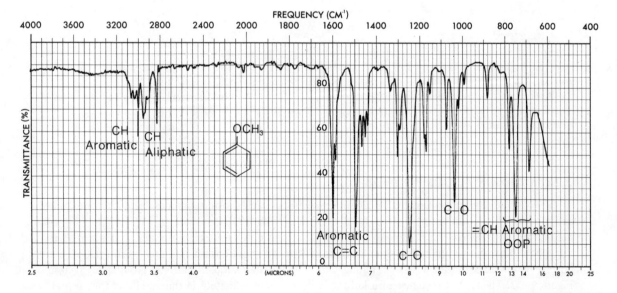

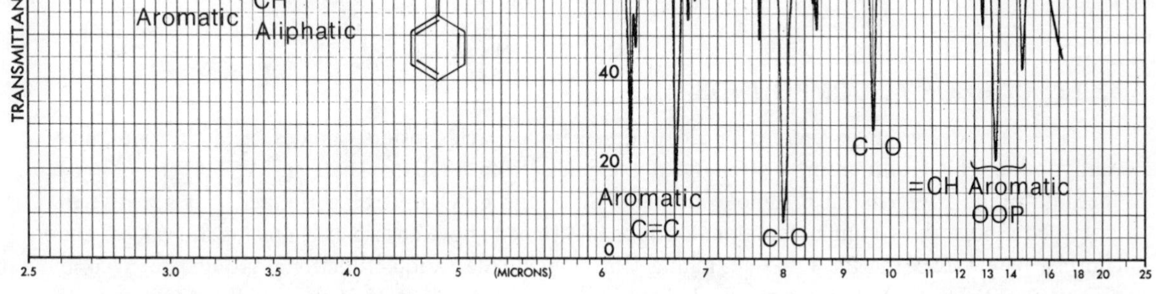

FIGURE 2–31 The Infrared Spectrum of Anisole (Neat Liquid, Salt Plates)

DISCUSSION SECTION

Ethers and related compounds such as epoxides, acetals, and ketals give rise to C–O–C stretching absorptions in the range from 1300 to 1000 cm^{-1} (7.7 to 10.0 μ). Since alcohols and esters also give strong C–O absorptions in this region, these possibilities must be eliminated by observing the absence of bands in the O–H stretch region (Section 2–11) and in the C=O stretch region (Section 2–13), respectively. Ethers are generally encountered more often than epoxides, acetals, and ketals.

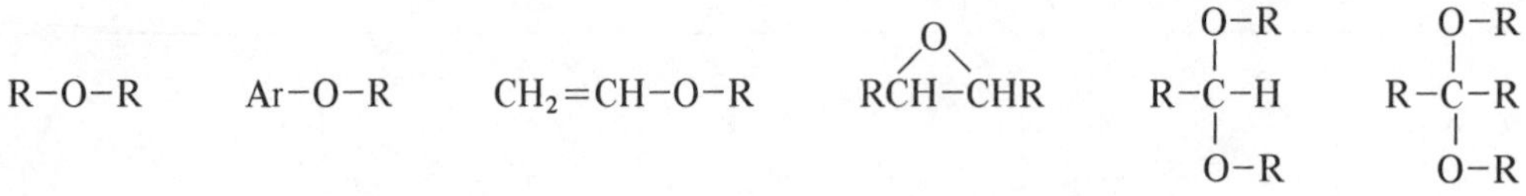

Dialkyl Ethers *Aryl Ethers* *Vinyl Ethers* *Epoxides* *Acetals* *Ketals*

Dialkyl Ethers. The asymmetric C–O–C stretching vibration leads to a single strong absorption appearing at about 1120 cm^{-1} (8.93 μ), as seen in the spectrum of dibutyl ether (Fig. 2–30). The symmetric stretching band is usually very weak. The asymmetric C–O–C absorption also occurs at about 1120 cm^{-1} (8.93 μ) for a six-membered ring containing oxygen.

Aryl and Vinyl Ethers. Aryl alkyl ethers give rise to *two* bands, an *asymmetric* C–O–C stretch near 1250 cm^{-1} (8.0 μ) and a *symmetric* stretch near 1040 cm^{-1} (9.6 μ),as seen in the spectrum of anisole (Fig. 2–31). Vinyl alkyl ethers also give two bands, one due to an asymmetric stretching vibration at about 1220 cm^{-1} (8.2 μ) and the other due to a symmetric stretch at about 1040 cm^{-1} (9.6 μ).

The shift in the asymmetric stretching frequency to a higher value than was found in dialkyl ethers can be explained by using resonance. For example, the C–O band in vinyl alkyl ethers is shifted to a higher frequency (1220 cm^{-1}) because of the increased double bond character, which strengthens the bond. In dialkyl ethers the absorption occurs at 1120 cm^{-1}. In addition, because resonance increases the polar character of the C=C double bond, the band at about 1640 cm^{-1} (6.1μ) is intensified over normal C=C absorption (Section 2.9B).

Resonance
1220 cm^{-1}

No resonance
1120 cm^{-1}

Epoxides. These small ring compounds usually give *three* bands, a medium intensity symmetric stretching vibration at about 1250 cm^{-1} (8.0 μ), a medium intensity asymmetric band between 950 and 815 cm^{-1} (10.5 to 12.3 μ), and a strong band between 850 and 750 cm^{-1} (11.8 to 13.3 μ).

Acetals and Ketals. Molecules which contain ketal or acetal linkages will often give four or five strong bands, respectively, in the region from 1200 to 1020 cm^{-1} (8.33 to 9.80 μ). These bands are often unresolved.

2.13 CARBONYL COMPOUNDS

The carbonyl group is present in aldehydes, ketones, acids, esters, amides, acid chlorides, and anhydrides. This group absorbs strongly in the range from 1850 to 1650 cm^{-1} (5.41 to 6.06 μ) because of its large change in dipole moment. Since the C=O stretching frequency is sensitive to attached atoms, the common functional groups mentioned above absorb at characteristic values. The normal base values for the C=O stretching vibrations of the various functional groups are given in Figure 2–32. The C=O frequency of a ketone, which is approximately in the middle of the range, is usually considered the reference point for comparisons of the values given in Figure 2–32.

5.52	5.56	5.68	5.76	5.80	5.83	5.85	5.92 μ
1810	1800	1760	1735	1725	1715	1710	1690 cm^{-1}
	ACID CHLORIDE		ESTER		KETONE		AMIDE
ANHYDRIDE (Band 1)		ANHYDRIDE (Band 2)		ALDEHYDE		CARBOXYLIC ACID	

FIGURE 2–32 Normal Base Values for the C=O Stretching Vibrations for Carbonyl Groups

The range of values given in Figure 2–32 may be explained by using electron-withdrawing effects (inductive effects), resonance effects, and hydrogen bonding. The first two effects operate in opposite ways to influence the C=O stretching frequency. First, an electronegative element may tend to draw in the electrons between the carbon and oxygen atoms through its electron-withdrawing effect, so that the C=O bond becomes somewhat stronger. A higher frequency (higher energy) absorption then results. Second, the unpaired electrons on a heteroatom can conjugate with the carbonyl group, resulting in increased single bond character and a lowering of the C=O absorption frequency. Oxygen is more electronegative than nitrogen, and the first effect dominates in an ester to raise the C=O frequency above that of a ketone. In an amide, the second effect is dominant since nitrogen is not as electronegative as oxygen and more willingly "back-donates" its unshared electrons, resulting in an absorption frequency lower than that of a ketone.

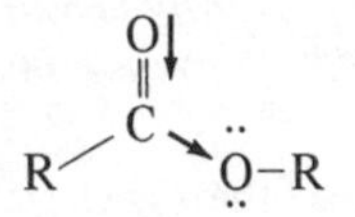

Electron-Withdrawing Effect Raises C=O Frequency

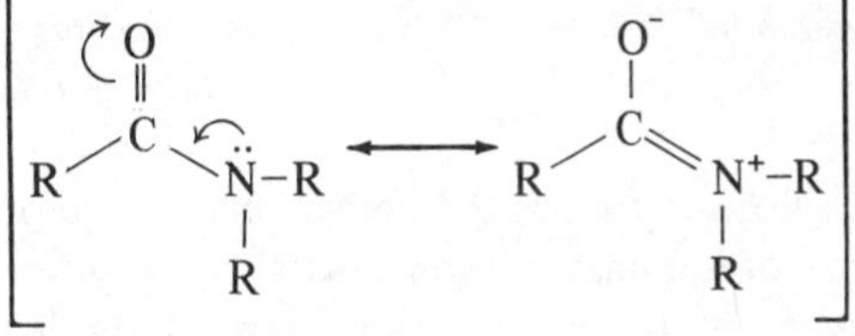

Resonance Effect Lowers C=O Frequency

In acid chlorides, the highly electronegative halogen atom strengthens the C=O bond through an enhanced inductive effect and shifts the frequency to even higher values than found in esters. Anhydrides are likewise shifted to higher frequencies than found in esters because of a concentration of electronegative oxygen atoms. In addition, anhydrides give two absorption bands that are due to symmetric and asymmetric stretching vibrations (Section 2.3).

A carboxylic acid exists in a monomeric form in very dilute solution, and it absorbs at about 1760 cm^{-1} because of the electron-withdrawing effect discussed above. However, in concentrated solution, as a neat liquid, or in the solid state (KBr pellet), acids tend to dimerize *via* hydrogen bonding. This weakens the C=O bond and lowers the stretching force constant K, resulting in a lowering of the carbonyl frequency of saturated acids to about 1710 cm^{-1}.

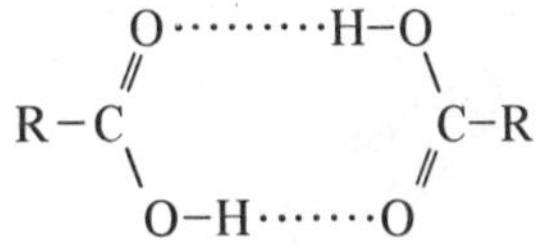

Ketones absorb at a lower frequency than aldehydes because of the additional alkyl group present on a ketone. This second alkyl group is electron-donating (compared to H) and supplies electrons to the C=O bond. This electron-releasing effect weakens the C=O bond in the ketone and lowers the force constant and the absorption frequency.

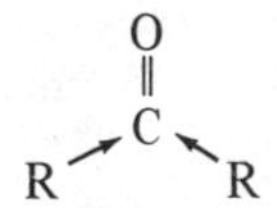

versus

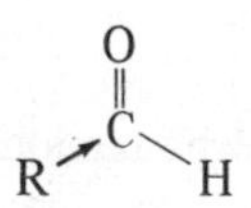

A. FACTORS WHICH INFLUENCE THE C=O STRETCHING VIBRATION

1. Conjugation Effects. The introduction of a C=C bond adjacent to a carbonyl group results in delocalization of the π electrons in the carbonyl and double bonds. This conjugation increases the single bond character of the C=O bond and, hence, lowers its force constant, resulting in a *lowering of the frequency* of carbonyl absorption.

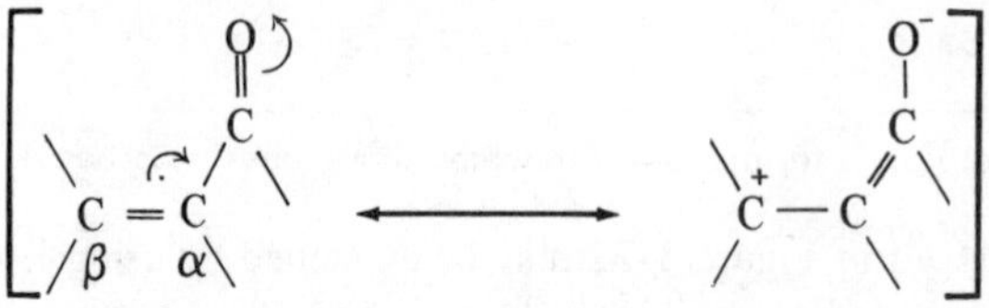

Generally, the introduction of an α,β-double bond in a carbonyl compound results in a shift of about 30 cm^{-1} to *lower* frequency from the *base* value given in Figure 2-32. A similar lowering occurs when an adjacent aryl group is introduced. Further addition of unsaturation (γ,δ) results in a further shift to lower frequency, but only by about 15 cm^{-1} more. Some examples are:

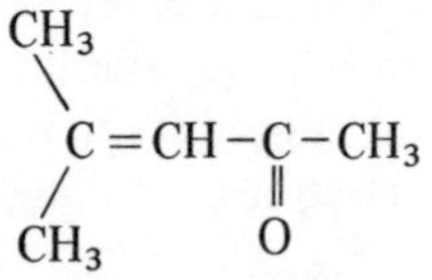

α,β-Unsaturated Ketone
1715 → 1690 cm^{-1}

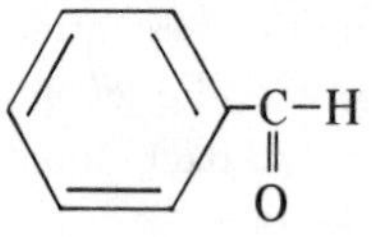

Aryl Substituted Aldehyde
1725 → 1700 cm^{-1}

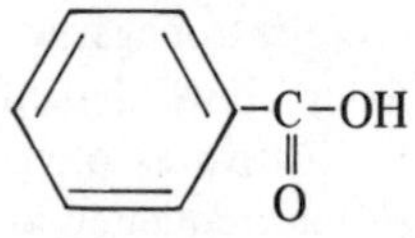

Aryl Substituted Acid
1710 → 1680 cm^{-1}

Amides do not follow the above rule. Conjugation does not reduce the C=O frequency. The introduction of α,β-unsaturation causes an *increase* in frequency from the base value given in Figure 2–32. Apparently, the introduction of sp^2 hybridized carbon atoms removes electron density from the carbonyl group and strengthens the bond instead of interacting by resonance as in other carbonyl examples. Since the parent amide group is *already* highly stabilized (see page 47), the introduction of the C=C unsaturation does not overcome this resonance.

2. Ring Size Effects. Six-membered rings with carbonyl groups are unstrained and absorb at about the same values given in Figure 2–32. Decreasing the ring size *increases the frequency* of the C=O absorption for the same reasons discussed in Section 2.9 (C=C stretching vibrations, exocyclic double bonds). *All* of the functional groups listed in Figure 2–32, which can form rings, give increased frequencies of absorption with increased angle strain. For ketones and esters, there is often a 30 cm^{-1} increase in frequency for each carbon removed from the unstrained six-membered ring values. Some examples are:

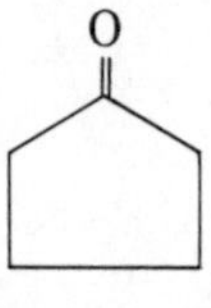

Cyclic Ketone
1715 → 1745 cm^{-1}

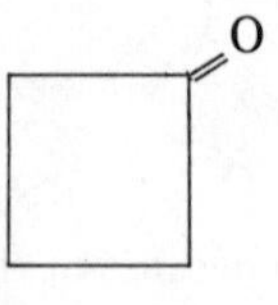

Cyclic Ketone
1715 → 1780 cm^{-1}

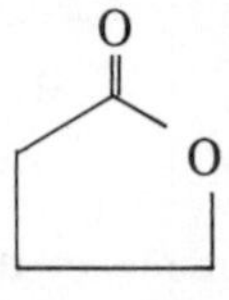

Cyclic Ester
(Lactone)
1735 → 1770 cm^{-1}

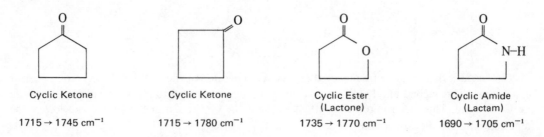

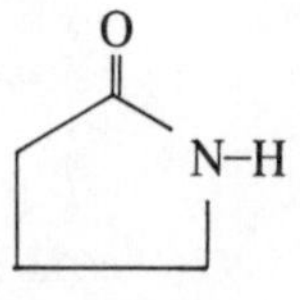

Cyclic Amide
(Lactam)
1690 → 1705 cm^{-1}

In ketones, larger rings have frequencies which range from nearly the same value as cyclohexanone (1715 cm^{-1}) to values that are slightly less than 1715 cm^{-1}. For example, cycloheptanone absorbs at about 1705 cm^{-1}.

3. Alpha-Substitution Effects. When the carbon next to the carbonyl is substituted with a chlorine (or other halogen) atom, the carbonyl band will shift to a *higher frequency.* The electron-withdrawing effect removes electrons from the carbon of the C=O bond. This is compensated for by a tightening of the π bond (shortening) which increases the force constant and leads to an increase in the absorption frequency. This effect holds for all carbonyl compounds.

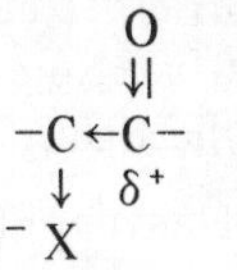

In ketones, two bands result from the substitution of an adjacent chlorine atom: one is due to the conformation in which the chlorine is rotated next to the carbonyl, and one from the conformation in which the chlorine is away from the group. When the chlorine is next to the carbonyl, non-bonded electrons on the oxygen atom are repelled. This results in a stronger bond and a higher absorption frequency. Information of this kind can be used to establish a structure in rigid ring systems, such as in the following examples:

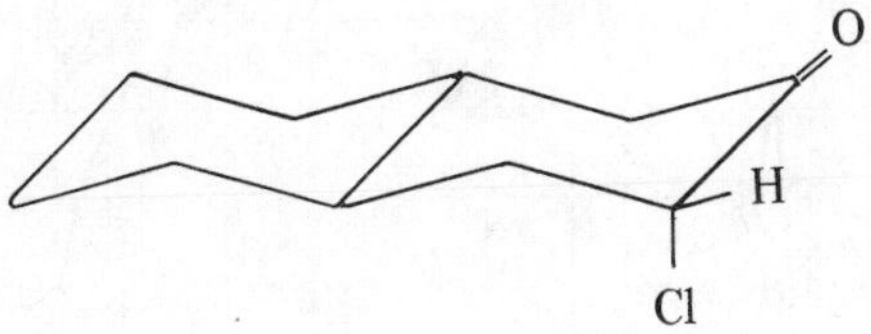

Axial Chlorine

~1725 cm^{-1}

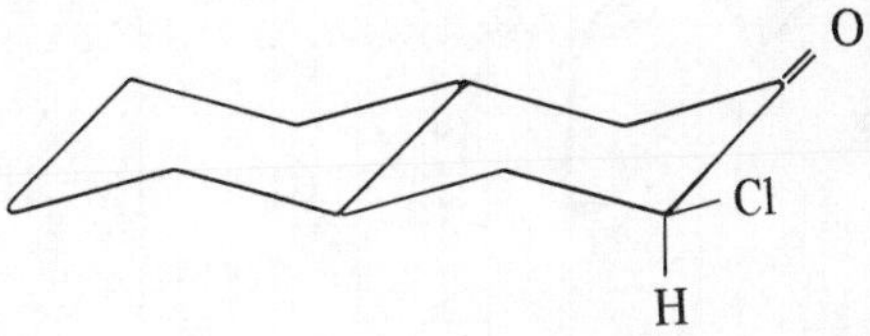

Equatorial Chlorine

~1750 cm^{-1}

4. Hydrogen Bonding Effects. Hydrogen bonding to a carbonyl group lengthens the C=O bond and lowers the stretching force constant *K*. This results in a *lowering of the absorption frequency*. An example of this effect is seen in the decrease in the C=O frequency of the carboxylic acid dimer (p. 47) and the lowering of the ester C=O frequency in methyl salicylate caused by intramolecular hydrogen bonding:

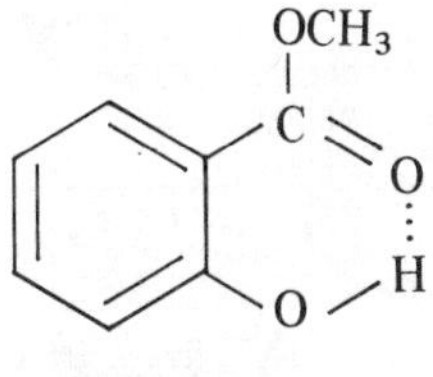

1680 cm^{-1}
Methyl Salicylate

An extreme example of intramolecular hydrogen bonding is seen in enolic β-diketones, which have broad and intense absorptions at values of 1640 cm^{-1} (6.1 μ) or lower. The keto tautomer absorbs at higher frequencies, near where normal ketones absorb.

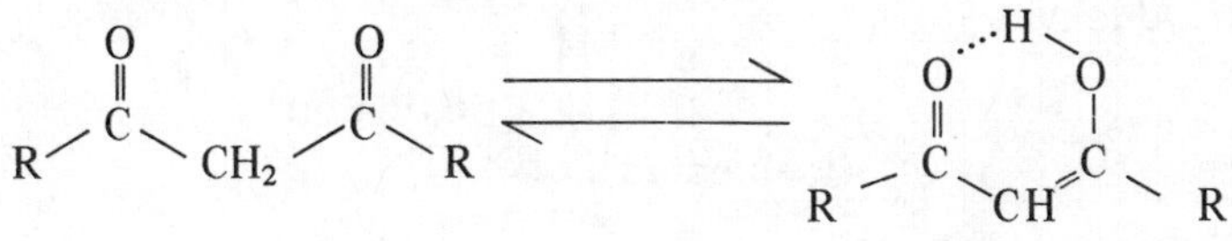

Keto Tautomer of a β-Diketone ~1720 cm^{-1} (doublet)

Enol Tautomer of a β-Diketone ~1640 cm^{-1}

B. ALDEHYDES

C=O	stretch at approximately 1725 cm^{-1} (5.80 μ) is normal. Conjugation moves the absorption to the right.
C–H	stretch, aldehyde hydrogen (–CHO), consists of *weak* bands at about 2750 cm^{-1} (3.65 μ) and 2850 cm^{-1} (3.50 μ). Note that the C–H stretch in alkyl chains does not usually extend this far to the right.

Examples: Nonanal (Fig. 2–33), crotonaldehyde (Fig. 2–34), and benzaldehyde (Fig. 2–35).

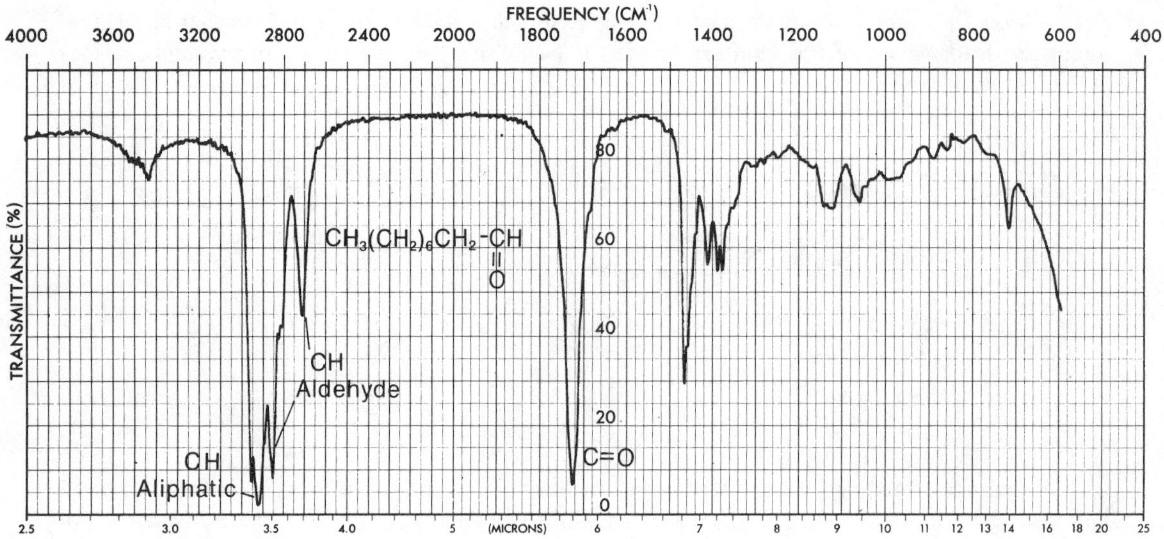

FIGURE 2–33 The Infrared Spectrum of Nonanal (Neat Liquid, Salt Plates)

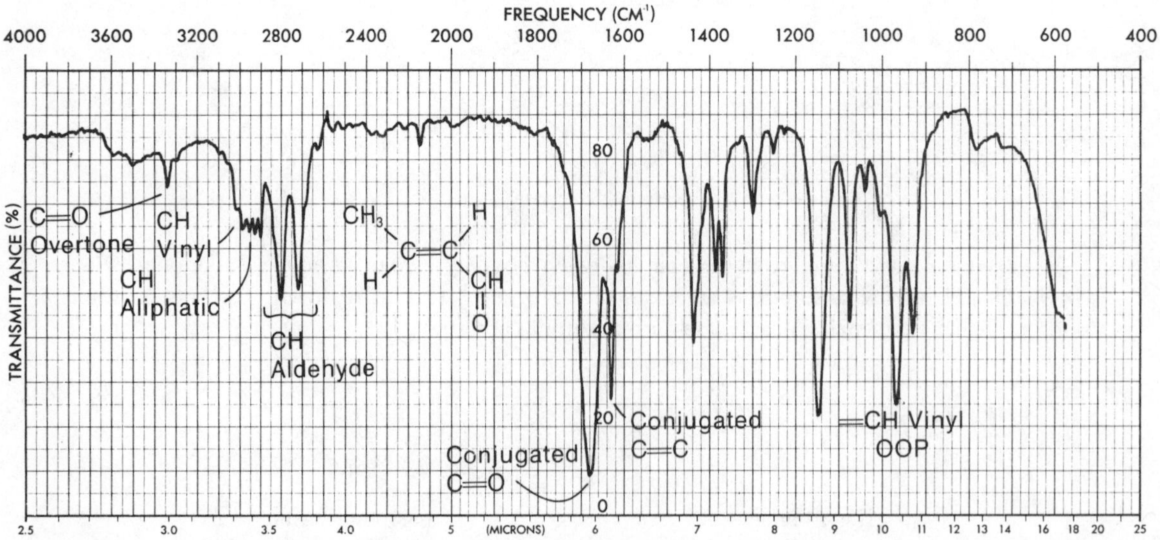

FIGURE 2–34 The Infrared Spectrum of Crotonaldehyde (Neat Liquid, Salt Plates)

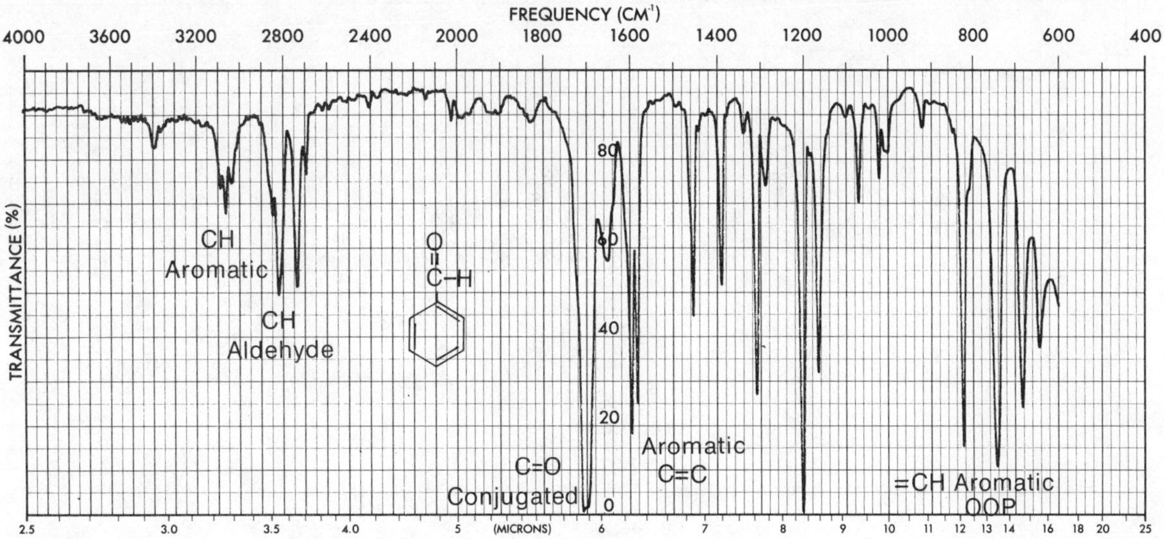

FIGURE 2–35 The Infrared Spectrum of Benzaldehyde (Neat Liquid, Salt Plates)

DISCUSSION SECTION

Conjugation of the carbonyl group with an aryl or an α,β-double bond shifts the normal C=O stretching band to the right as predicted in Section 2.13A-1. This effect is seen in crotonaldehyde (Fig. 2–34), which has α,β-unsaturation, and in benzaldehyde (Fig. 2–35), in which an aryl group is attached directly to the carbonyl group. The normal aldehyde stretching frequency at 1725 cm^{-1} (5.80 μ) is shown in the spectrum of nonanal (Fig. 2–33). Since the positions of these absorptions are not that different from those of ketones, one may not easily distinguish between aldehydes and ketones on this basis.

The C–H stretching vibrations found in aldehydes at about 2750 cm^{-1} (3.65 μ) and 2850 cm^{-1} (3.50 μ) are extremely important for distinguishing between ketones and aldehydes. The band at 2750 cm^{-1} (3.65 μ) is probably the more useful of the pair because it appears in a region where other C–H absorptions (CH_3, CH_2, etc.) are absent. If this band is present, together with the proper C=O absorption value, an aldehyde functional group is almost certainly indicated.

The medium intensity absorption in nonanal (Fig. 2–33) at 1460 cm^{-1} (6.85 μ) is due to the scissoring (bending) vibration of the CH_2 group next to the carbonyl group. Methylene groups often absorb more strongly when they are attached directly to a carbonyl group.

C. KETONES

C=O	stretch at approximately 1715 cm^{-1} (5.83 μ) is normal. Conjugation moves the absorption to the right. Ring strain moves the absorption to the left in cyclic ketones.

Examples: Methyl isopropyl ketone (Fig. 2–3), mesityl oxide (Fig. 2–36), acetophenone (Fig. 2–37), and cyclopentanone (Fig. 2–38).

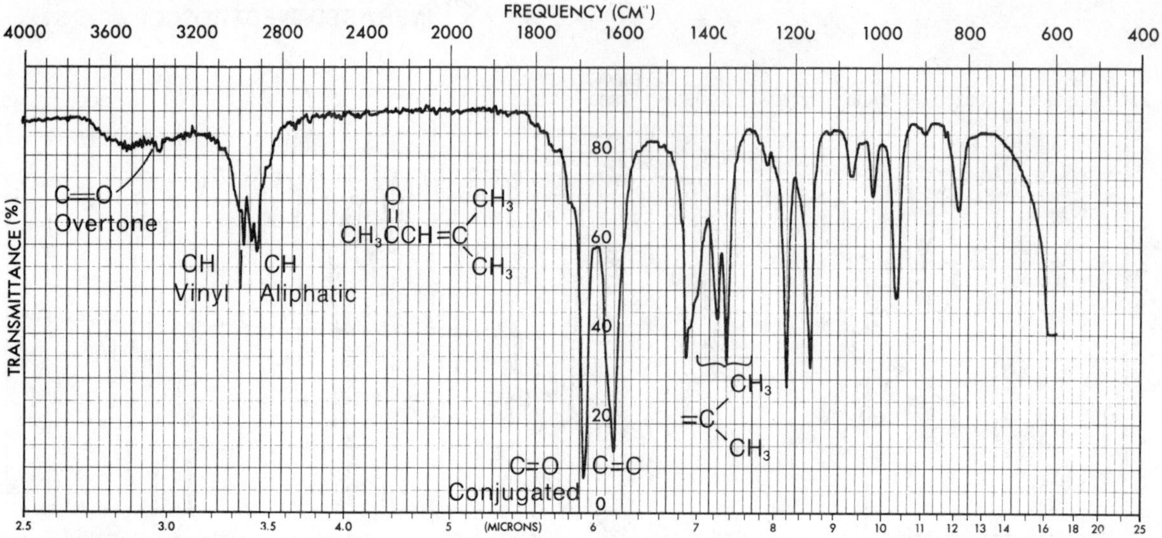

FIGURE 2–36 The Infrared Spectrum of Mesityl Oxide (Neat Liquid, Salt Plates)

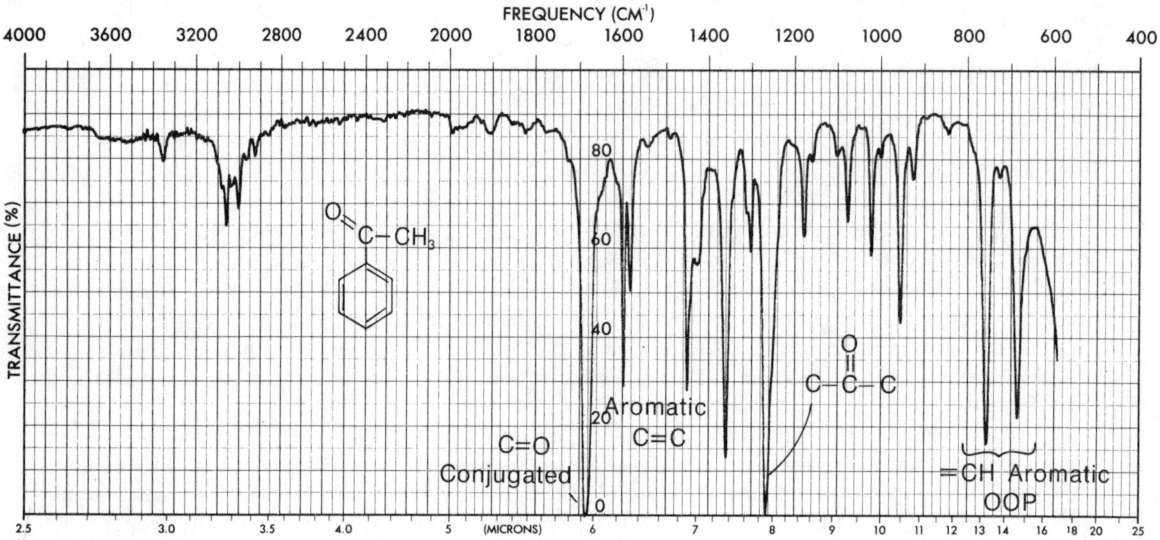

FIGURE 2–37 The Infrared Spectrum of Acetophenone (Neat Liquid, Salt Plates)

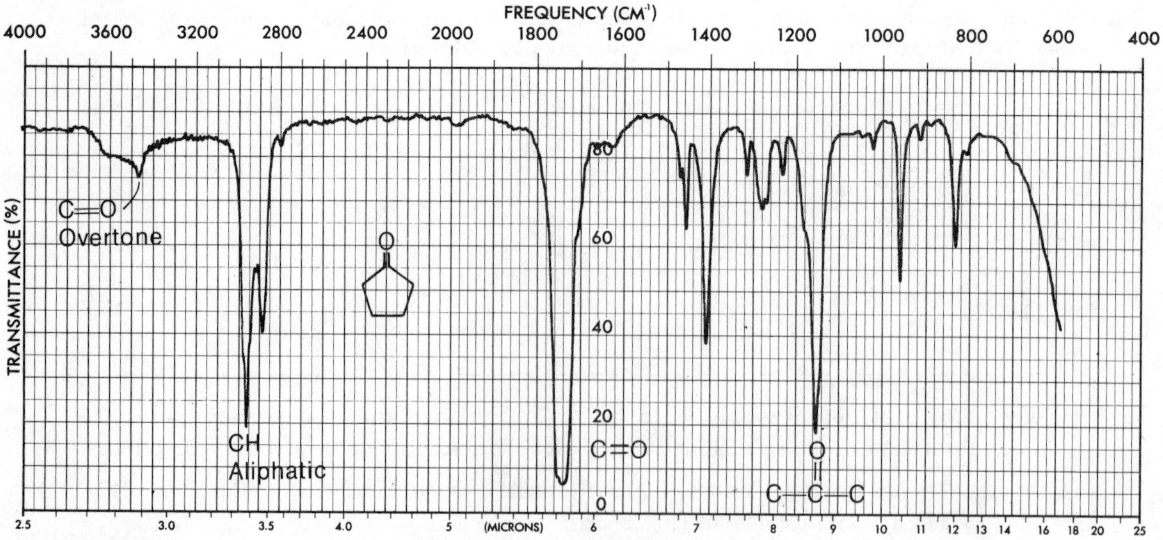

FIGURE 2–38 The Infrared Spectrum of Cyclopentanone (Neat Liquid, Salt Plates)

DISCUSSION SECTION

Conjugation of the carbonyl group with an aryl or an α,β-double bond shifts the normal C=O stretching band to the right, as predicted in Section 2.13A-1. This shift is seen in mesityl oxide (Fig. 2–36), which has α,β-unsaturation, and in acetophenone (Fig. 2–37), in which an aryl group is attached to the carbonyl group. The normal ketone stretching frequency at 1715 cm^{-1} (5.83 μ) is shown in the spectrum of methyl isopropyl ketone (Fig. 2–3). Overtones from the C=O band appear at twice the frequency of the C=O absorption in the range from 3500 to 3350 cm^{-1} (2.86 to 3.00 μ). These small bands should not be confused with O–H absorptions, which also appear in this range. The O–H stretching absorptions are *much more* intense. Some typical C=O stretching vibrations are given in Figure 2–39, which demonstrates the influence of conjugation. A value for enolic β-diketones is also shown in this figure. The substantial shift is due to the hydrogen bonding, as discussed in Section 2.13A-4.

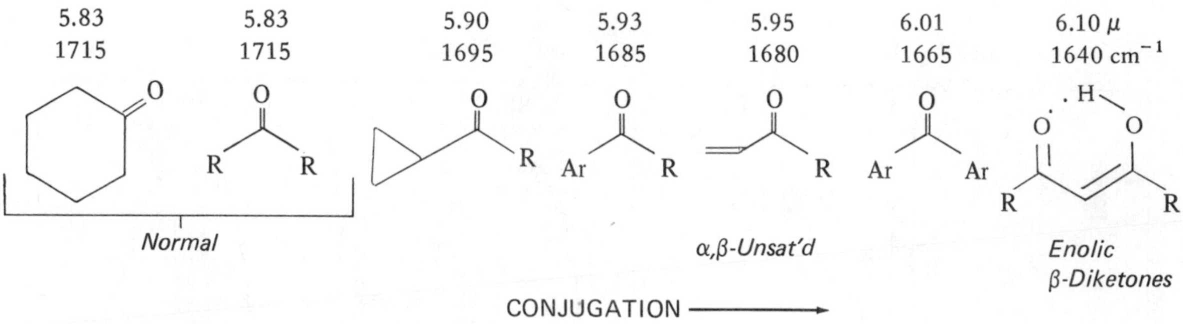

FIGURE 2–39 The C=O Stretching Vibrations for Conjugated Ketones

Some values for the C=O absorptions for cyclic ketones are given in Figure 2–40. Note that ring strain shifts the absorption values to the left, as was predicted in Section 2.13A-2. The spectrum of cyclopentanone (Fig. 2–38) shows this behavior. The substitution of a chlorine atom alpha to the carbonyl group also shifts the absorption to the left (Section 2.13A-3).

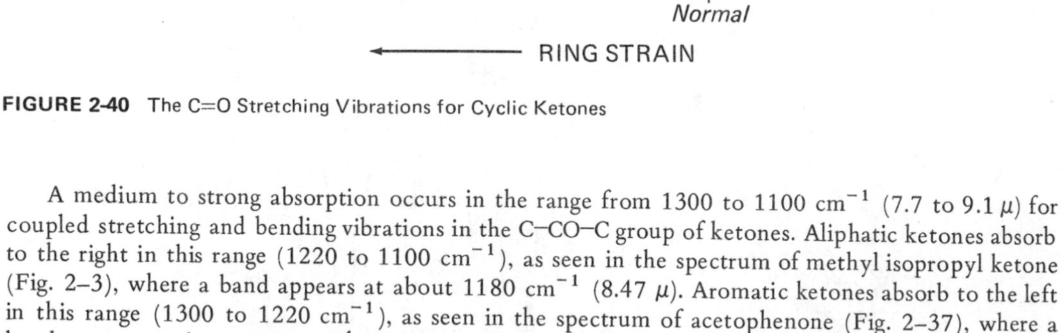

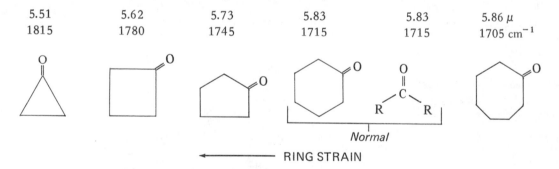

FIGURE 2-40 The C=O Stretching Vibrations for Cyclic Ketones

A medium to strong absorption occurs in the range from 1300 to 1100 cm^{-1} (7.7 to 9.1 μ) for coupled stretching and bending vibrations in the C–CO–C group of ketones. Aliphatic ketones absorb to the right in this range (1220 to 1100 cm^{-1}), as seen in the spectrum of methyl isopropyl ketone (Fig. 2–3), where a band appears at about 1180 cm^{-1} (8.47 μ). Aromatic ketones absorb to the left in this range (1300 to 1220 cm^{-1}), as seen in the spectrum of acetophenone (Fig. 2–37), where a band appears at about 1260 cm^{-1} (7.94 μ).

A medium intensity band appears for a methyl group adjacent to a carbonyl at about 1370 cm^{-1} (7.30 μ), for the symmetric bending vibration. These methyl groups absorb with a greater intensity than methyl groups found in hydrocarbons (see Figure 2–7).

D. CARBOXYLIC ACIDS

O–H	stretch, usually *very broad* (strongly H-bonded), 3400 to 2400 cm^{-1} (2.94 to 4.17 μ), often interferes with C–H absorptions.
C=O	stretch, broad, 1730 to 1700 cm^{-1} (5.8 to 5.9 μ). Conjugation moves the absorption to the right.
C–O	stretch, in range of 1320 to 1210 cm^{-1} (7.6 to 8.3 μ), medium intensity.

Examples: Isobutyric acid (Fig. 2–41) and benzoic acid (Fig. 2–42).

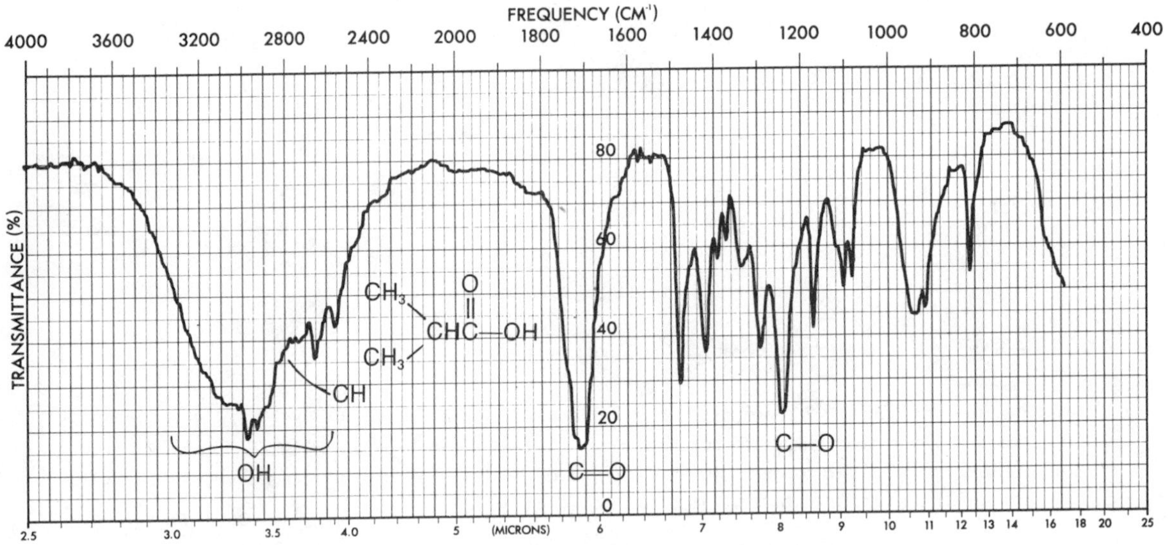

FIGURE 2–41 The Infrared Spectrum of Isobutyric Acid (Neat Liquid, Salt Plates)

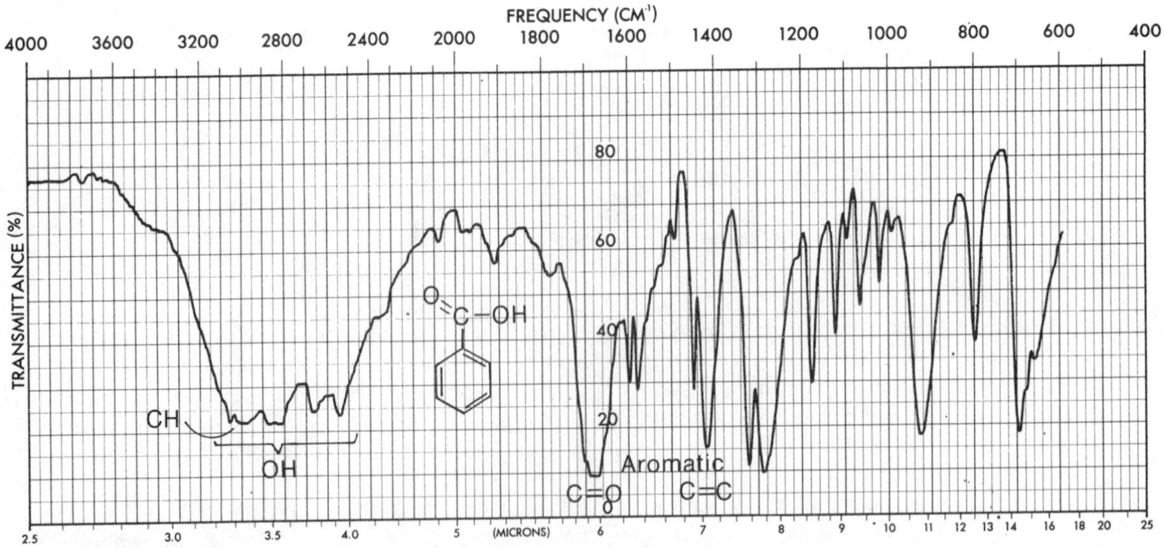

FIGURE 2–42 The Infrared Spectrum of Benzoic Acid (Solid phase, KBr)

DISCUSSION SECTION

The most characteristic feature in the spectrum of a carboxylic acid is the extremely *broad* O–H absorption occurring in the region from 3400 to 2400 cm^{-1} (2.94 to 4.17 μ). This band is attributed to the *strong* hydrogen bonding present in the dimer, discussed in the introduction to Section 2.13. The absorption will often obscure the C–H stretching vibrations that occur in the same region. If this broad hydrogen bonded band is present *together* with the proper C=O absorption value, a carboxylic acid is almost certainly indicated. The spectra of an aliphatic and an aromatic carboxylic acid are shown in Figures 2–41 and 2–42, respectively.

The carbonyl stretching absorption, which occurs at about 1730 to 1700 cm^{-1} (5.78 to 5.88 μ) for the dimer, is usually broader and more intense than that present in an aldehyde or ketone. For most acids, when the acid is diluted with a solvent, the C=O absorption appears between 1760 and 1730 cm^{-1} (5.68 to 5.78 μ) for the monomer. However, experimentally one does not often see the monomer, since it is usually easier to run the spectrum as a neat liquid. Under these conditions, as well as in a potassium bromide pellet, the dimer exists. It should be noted that some acids exist as dimers even at high dilution. Conjugation usually shifts the absorption band to the right (lower frequency) as predicted in Section 2.13A-1, and as shown in the spectrum of benzoic acid (Fig. 2–42). Salts of carboxylic acids are discussed in Section 2.17.

The C–O stretching vibration for acids (dimer) appears near 1260 cm^{-1} (7.94 μ) as a medium intensity band. A broad band, attributed to the hydrogen bonded O–H out-of-plane bending vibration, appears at about 930 cm^{-1} (10.75 μ). This latter band is usually of low to medium intensity.

E. ESTERS $\left(\mathrm{R{-}\overset{\overset{\displaystyle O}{\|}}{C}{-}OR'}\right)$

C=O	stretch occurs at about 1735 cm^{-1} (5.76 μ) in normal esters. a. Conjugation in the R part moves the absorption to the right. b. Conjugation with the O in the R′ part moves the absorption to the left. c. Ring strain (lactones) moves the absorption to the left.
C–O	stretch, two or more bands, one, stronger and broader than the other, is in the range from 1300 to 1000 cm^{-1} (7.7 to 10.0 μ).

Examples: Vinyl acetate (Fig. 2–10), ethyl butyrate (Fig. 2–43), methyl methacrylate (Fig. 2–44), methyl benzoate (Fig. 2–45), methyl salicylate (Fig. 2–46), and γ-butyrolactone (Fig. 2–47).

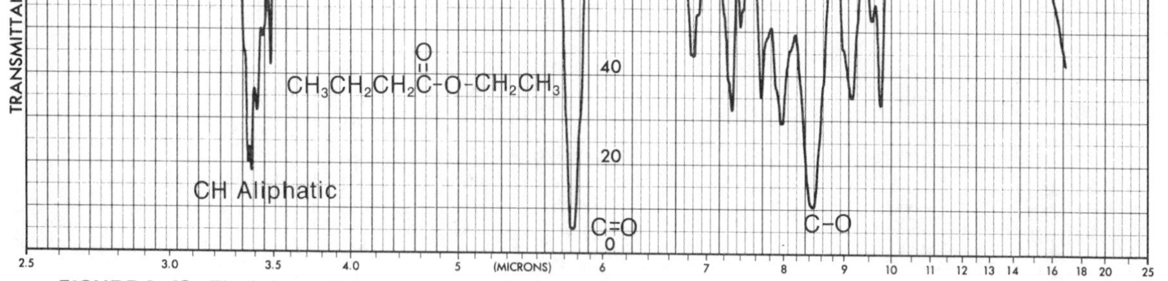

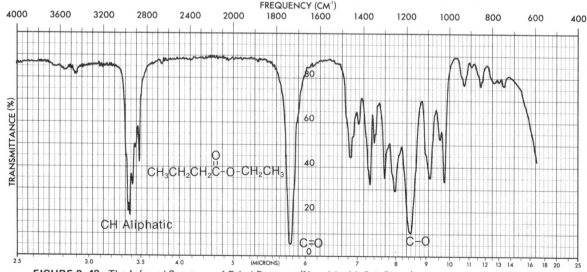

FIGURE 2–43 The Infrared Spectrum of Ethyl Butyrate (Neat Liquid, Salt Plates)

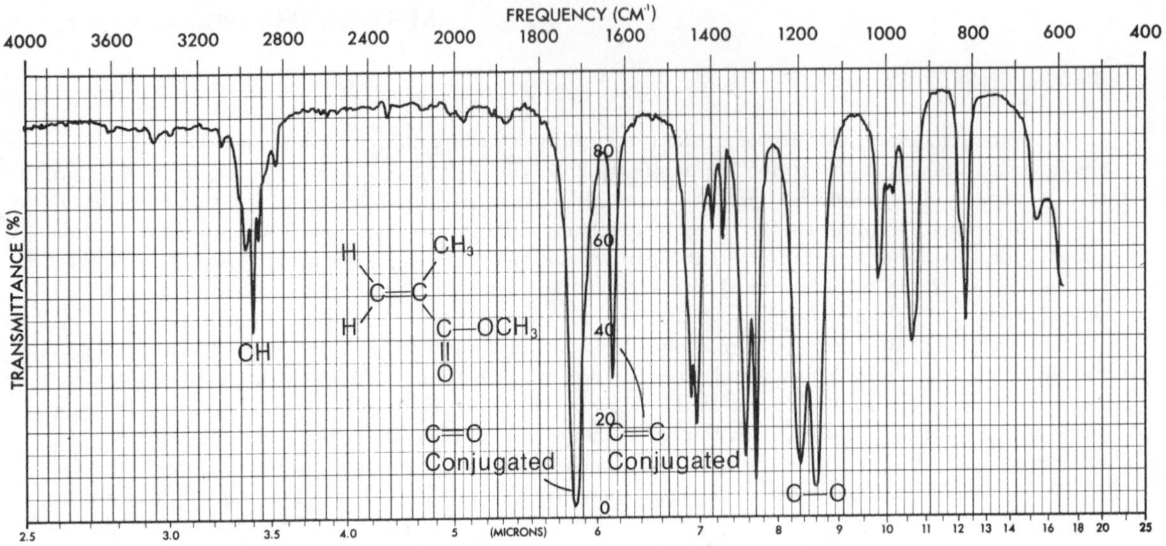

FIGURE 2–44 The Infrared Spectrum of Methyl Methacrylate (Neat Liquid, Salt Plates)

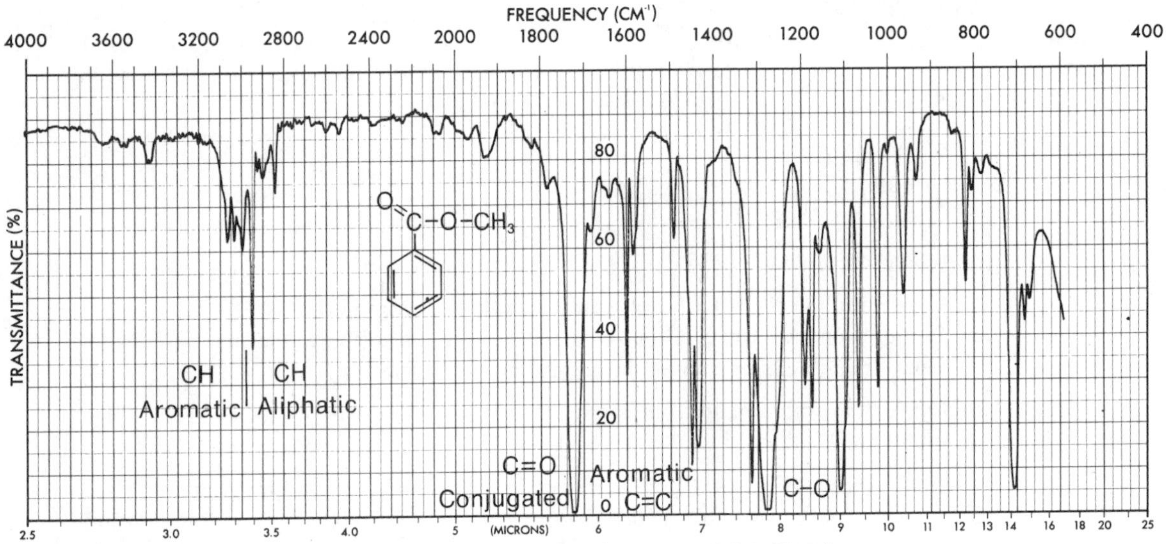

FIGURE 2–45 The Infrared Spectrum of Methyl Benzoate (Neat Liquid, Salt Plates)

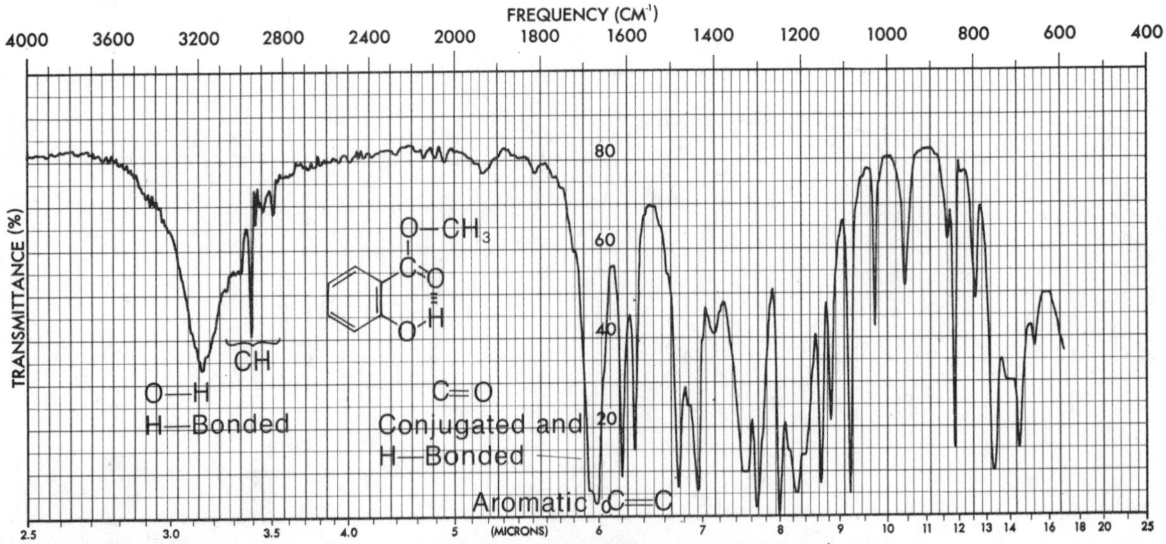

FIGURE 2–46 The Infrared Spectrum of Methyl Salicylate (Neat Liquid, Salt Plates)

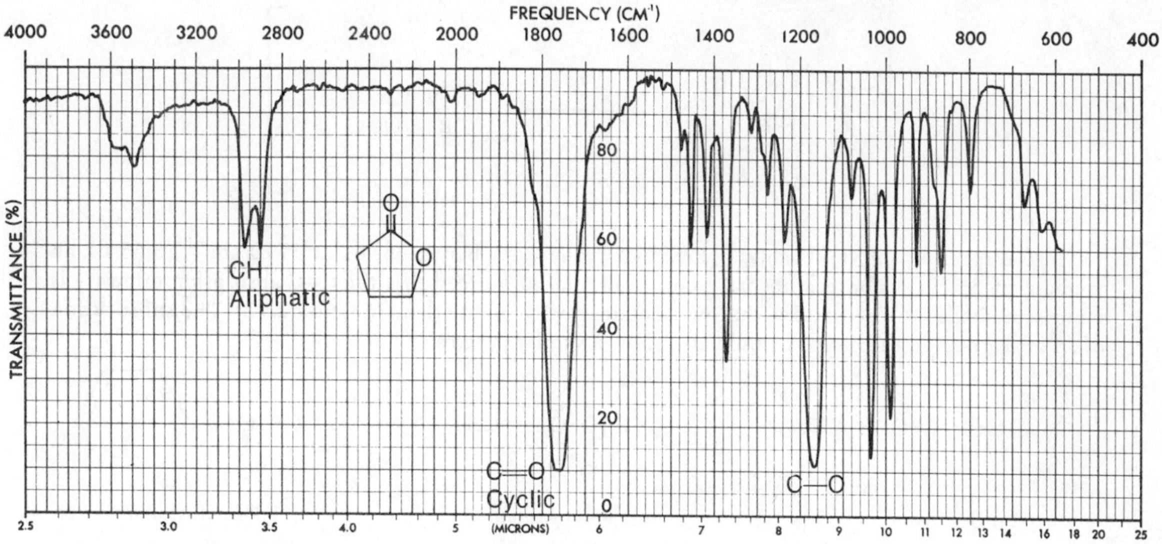

FIGURE 2–47 The Infrared Spectrum of γ-Butyrolactone (Neat Liquid, Salt Plates)

DISCUSSION SECTION

The two most characteristic features in the spectrum of an ester are the strong C=O and C–O stretching absorptions. Although some conjugated esters may appear in the same general area as ketones, one can usually eliminate ketones from consideration by observing the *strong* and *broad* C–O stretching vibrations which appear in a region (1300 to 1000 cm^{-1}) where ketonic absorptions appear as *weaker* and *narrower* bands. For example, one should compare the spectra of ketones (Figs. 2–36, 2–37, and 2–38) with those of esters (Figs. 2–43 through 2–47) in the 1300 to 1000 cm^{-1} (7.7 to 10.0 μ) region.

There are basically three important structural features in esters that can alter the position of the C=O absorption:

1. **α,β-Unsaturation or Aryl Substitution.** The C=O stretching vibrations are shifted to the right (lower frequencies) with α,β-unsaturation or aryl substitution, as predicted in Section 2.13A-1. The spectra of both methyl methacrylate (Fig. 2–44) and methyl benzoate (Fig. 2–45) show the C=O absorption shift from the position in a normal ester, ethyl butyrate (Fig. 2–43). Further shifts to the right occur when intramolecular (internal) hydrogen bonding is present, as predicted in Section 2.13A-4, and as shown in the spectrum of methyl salicylate (Fig. 2–46). It should also be noticed that the C=C absorption band at 1630 cm^{-1} (6.13 μ) in methyl methacrylate has been intensified over what is obtained with a non-conjugated double bond (Section 2.9B).

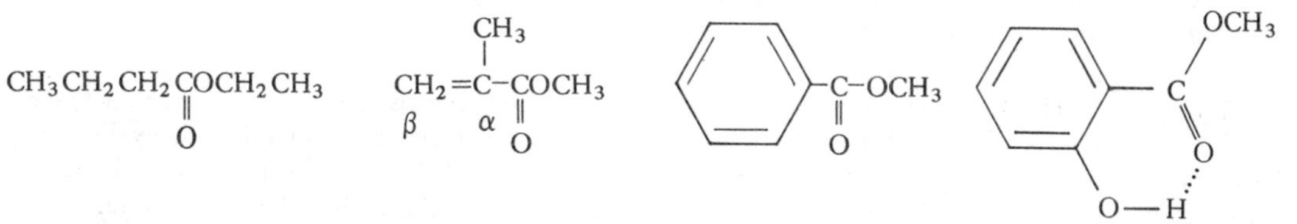

$$CH_3CH_2CH_2\underset{\underset{O}{\|}}{C}OCH_2CH_3$$

Ethyl Butyrate
1735 cm^{-1}

$$\underset{\beta}{CH_2}=\underset{\alpha}{\overset{\overset{CH_3}{|}}{C}}-\underset{\underset{O}{\|}}{C}OCH_3$$

Methyl Methacrylate
1720 cm^{-1}

Methyl Benzoate
1720 cm^{-1}

Methyl Salicylate
1680 cm^{-1}

2. **Conjugation with the Ester Single-Bonded Oxygen.** The C=O vibrations are shifted to the left (higher frequencies) by conjugation involving the single-bonded oxygen. Apparently, the conjugation interferes with possible resonance with the carbonyl group, leading to an *increase* in the absorption frequency for the C=O band.

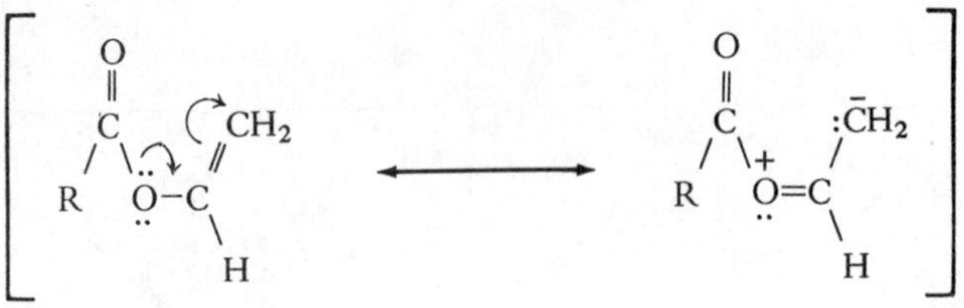

In the spectrum of vinyl acetate (Fig. 2–10), the C=O band appears at 1770 cm^{-1} (5.65 μ), an increase of 35 cm^{-1} above a normal ester. It should be noticed that the C=C absorption intensity is increased, similar to the pattern that was obtained with vinyl ethers (Section 2.12). The substitution of an aryl group on the oxygen would also show a similar pattern.

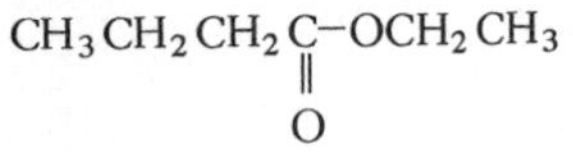

Ethyl Butyrate
1735 cm^{-1}

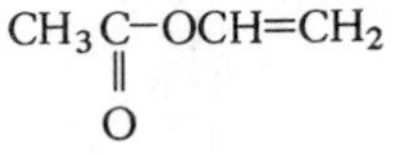

Vinyl Acetate
1770 cm^{-1}

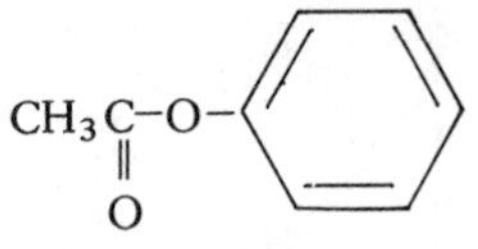

Phenyl Acetate
1770 cm^{-1}

The general effect of α,β-unsaturation or aryl substitution *and* conjugation with oxygen on the C=O vibrations is shown in Figure 2–48.

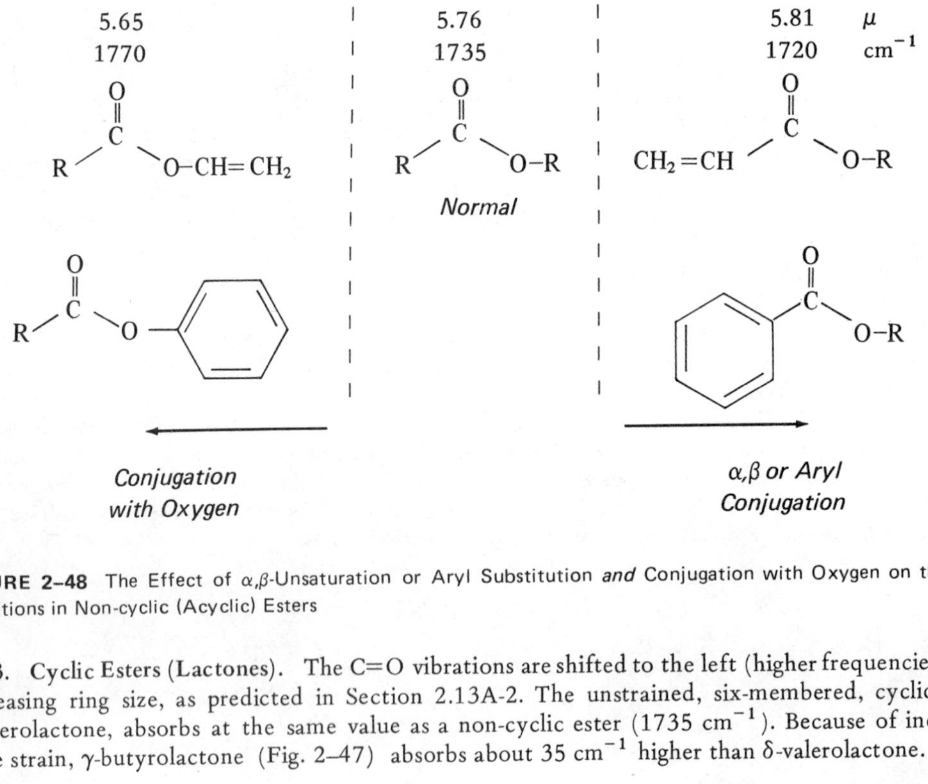

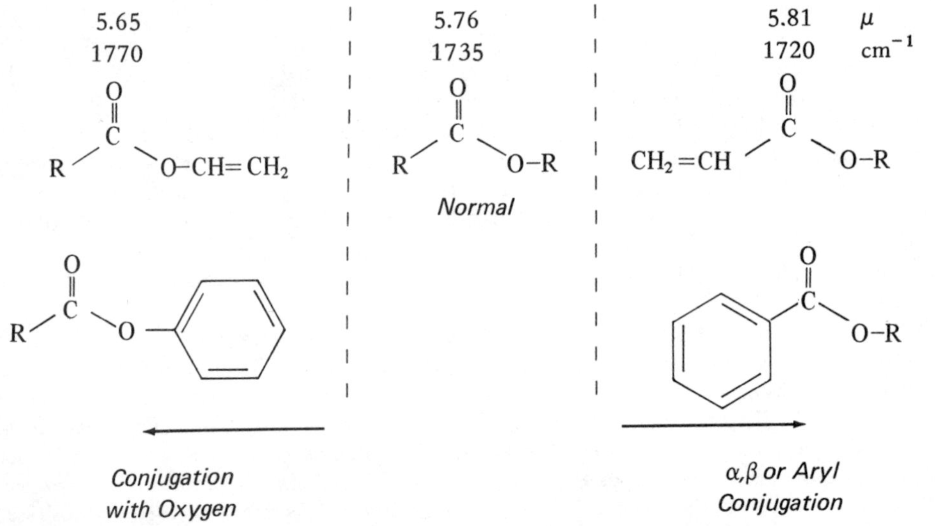

FIGURE 2–48 The Effect of α,β-Unsaturation or Aryl Substitution *and* Conjugation with Oxygen on the C=O Vibrations in Non-cyclic (Acyclic) Esters

3. Cyclic Esters (Lactones). The C=O vibrations are shifted to the left (higher frequencies) with decreasing ring size, as predicted in Section 2.13A-2. The unstrained, six-membered, cyclic ester, δ-valerolactone, absorbs at the same value as a non-cyclic ester (1735 cm^{-1}). Because of increased angle strain, γ-butyrolactone (Fig. 2–47) absorbs about 35 cm^{-1} higher than δ-valerolactone.

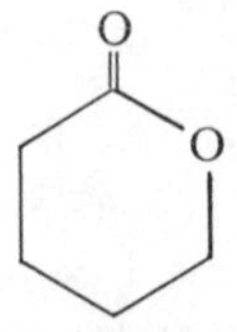

δ-Valerolactone
1735 cm^{-1}

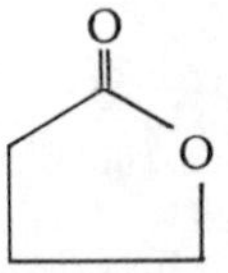

γ-Butyrolactone
1770 cm^{-1}

Some typical lactones are given in Table 2–8, together with their C=O stretching absorption values. The influence of the three factors discussed above can be seen by inspecting these values.

TABLE 2–8 The Effects of Ring Size, α,β-Unsaturation, and Conjugation with Oxygen on the C=O Vibrations in Lactones

Ring Size Effects, cm^{-1} (μ)		α,β-Conjugation, cm^{-1} (μ)		Conjugation with Oxygen, cm^{-1} (μ)	
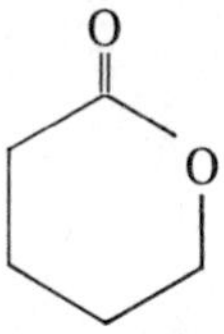	1735 (5.76)	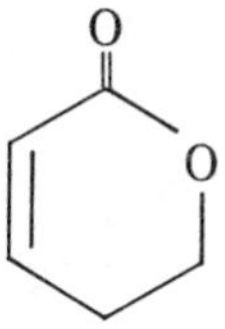	1720 (5.81)	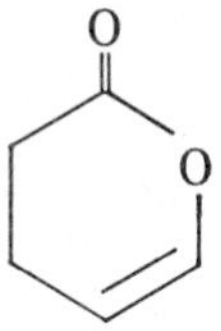	1760 (5.68)
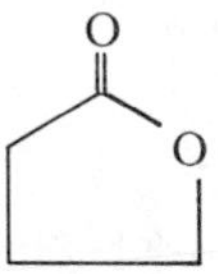	1770 (5.65)	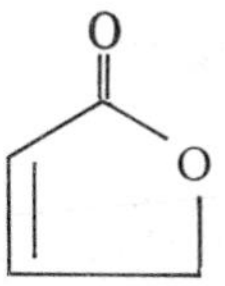	1750 (5.71)	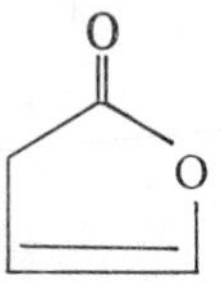	1800 (5.56)
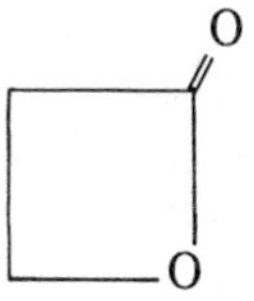	1820 (5.49)				

Two (or more) bands appear for the C–O stretching vibrations in esters in the range from 1300 to 1000 cm^{-1} (7.7 to 10.0 μ). Generally, the C–O stretch on the "acid" part of the ester will appear as one of the strongest and broadest bands in the spectrum. This absorption appears between 1300 and 1150 cm^{-1} (7.7 to 8.7 μ) for most common esters; esters of aromatic acids absorb to the left and esters of saturated acids absorb to the right in this range. The C–O stretch for the "alcohol" part of the ester may appear as a weaker band in the range from 1150 to 1000 cm^{-1} (8.7 to 10.0 μ). In analyzing the 1300 to 1000 cm^{-1} (7.7 to 10.0 μ) region, one should not worry about fine details. It is usually sufficient to find at least *one* very strong and broad absorption to help identify the compound as an ester.

F. AMIDES

C=O	stretch is at approximately 1670 to 1640 cm^{-1} (6.0 to 6.1 μ).
N–H	stretch (primary or secondary amides), 3500 and 3100 cm^{-1} (2.86 to 3.23 μ). Primary amides have two bands ($-NH_2$) in this region.
N–H	bending around 1640 to 1550 cm^{-1} (6.10 to 6.45 μ).

Example: Benzamide (Fig. 2–49).

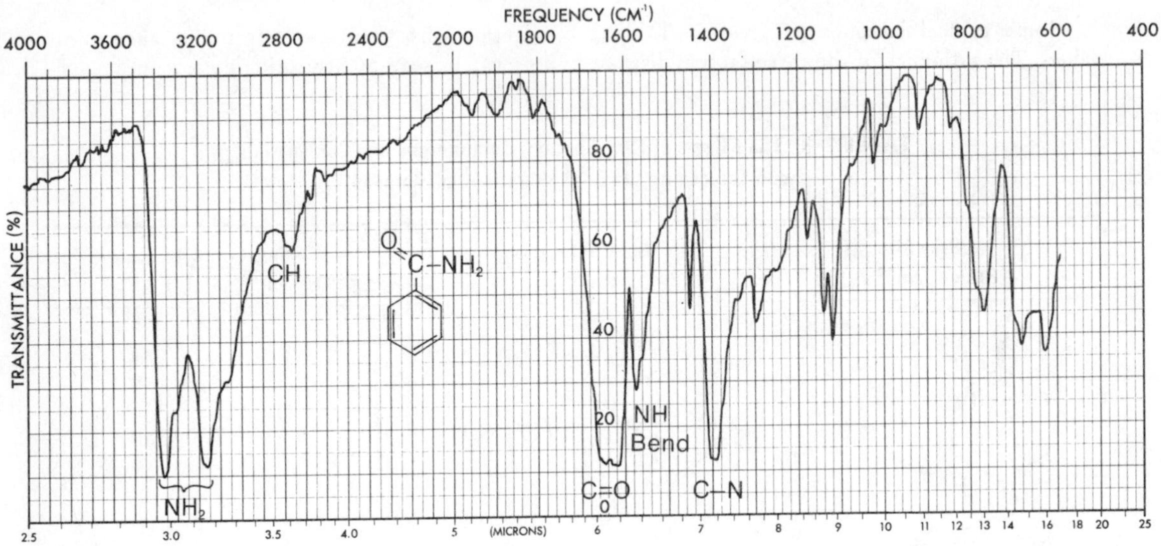

FIGURE 2–49 The Infrared Spectrum of Benzamide (Solid Phase, KBr)

DISCUSSION SECTION

Primary and secondary amides have broad C=O absorptions in the range from 1670 to 1640 cm^{-1} (6.0 to 6.1 μ) when determined in the solid phase (potassium bromide pellet). In very dilute solution, the band appears to the left at about 1690 cm^{-1} (5.92 μ). This effect is similar to that observed for carboxylic acids, in which hydrogen bonding reduces the frequency in the solid state or in concentrated solution. Tertiary amides, which cannot form hydrogen bonds, have C=O frequencies that are not influenced by the physical state and absorb in about the same range as that given above. Cyclic amides (lactams) give the expected increase in frequency for decreasing ring size, as shown for lactones (Table 2–8).

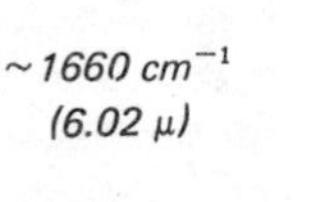

~1660 cm^{-1}
(6.02 μ)

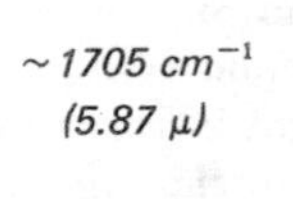

~1705 cm^{-1}
(5.87 μ)

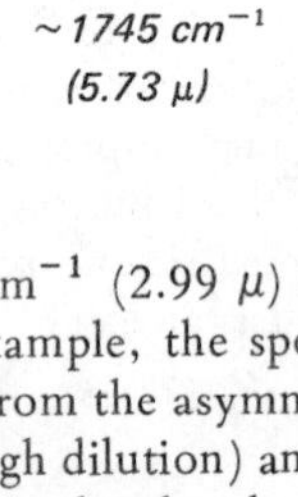

~1745 cm^{-1}
(5.73 μ)

A pair of fairly intense N–H stretching bands appear at about 3350 cm^{-1} (2.99 μ) and 3150 cm^{-1} (3.17 μ) for a primary amide in the solid state (KBr pellet). An example, the spectrum of benzamide, is shown in Figure 2–49. The 3350 and 3150 cm^{-1} bands result from the asymmetric and symmetric vibrations, respectively (Section 2.3). Secondary amides (under high dilution) and lactams give one band in the 3500 to 3100 cm^{-1} (2.86 to 3.23 μ) region. On the other hand, secondary amides in the solid phase give a complex absorption pattern.

Primary and secondary amides give medium intensity bands to the right (1640 to 1550 cm^{-1}) of the C=O absorption, caused by the N–H bending vibration. Unfortunately, in the solid phase this N–H bending vibration may often overlap the C=O absorption. Finally, a C–N stretching vibration occurs at about 1400 cm^{-1} (7.14 μ) for primary amides.

G. ACID CHLORIDES

C=O	stretch occurs at about 1800 cm^{-1} (5.56 μ) in normal acid chlorides. Conjugation moves the absorption to the right.

Examples: Acetyl chloride (Fig. 2–50) and benzoyl chloride (Fig. 2–51).

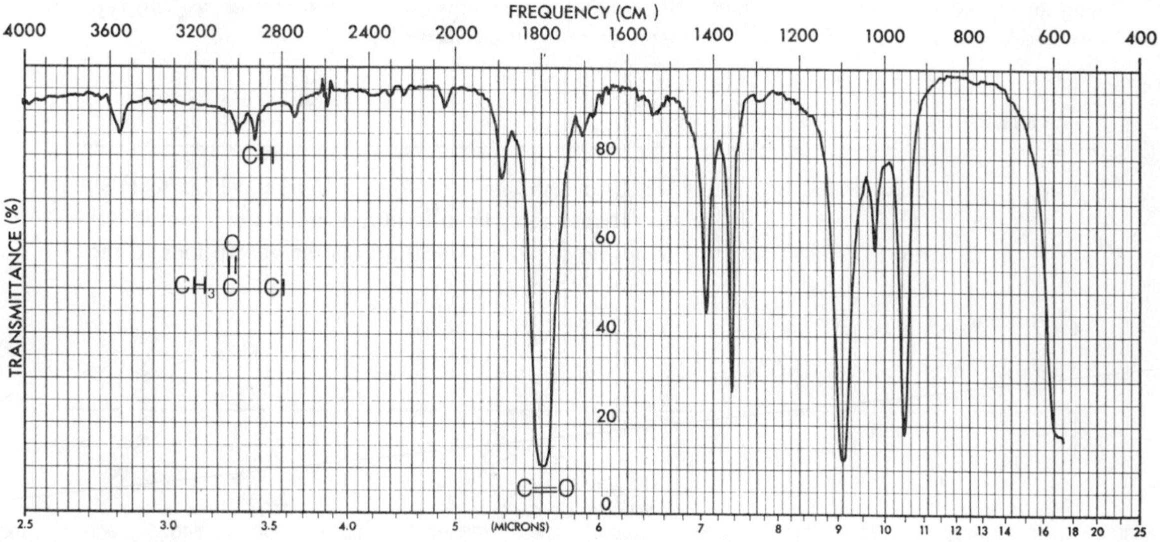

FIGURE 2–50 The Infrared Spectrum of Acetyl Chloride (Neat Liquid, Salt Plates)

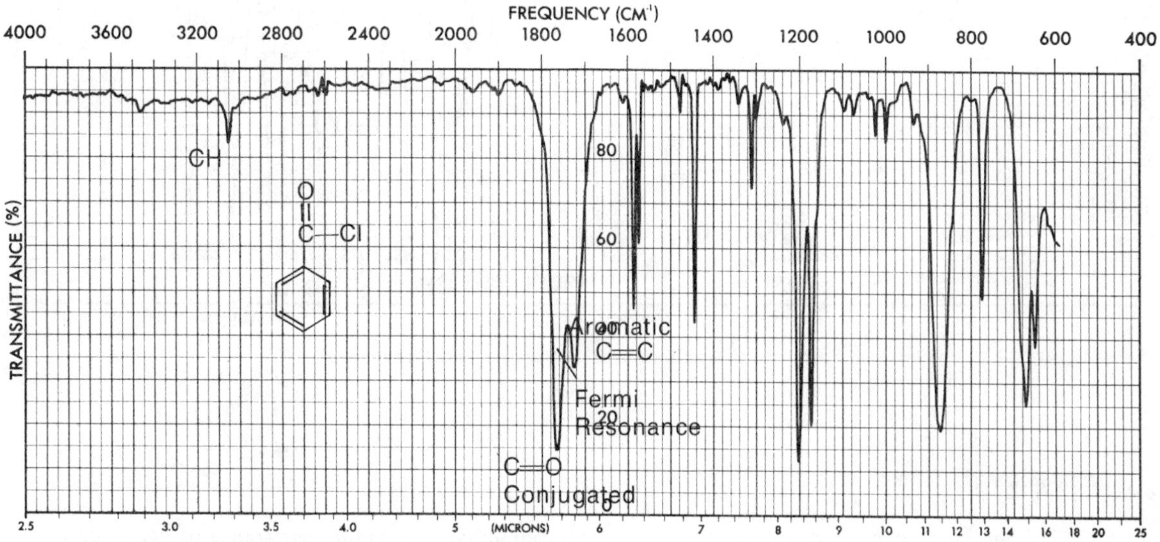

FIGURE 2–51 The Infrared Spectrum of Benzoyl Chloride (Neat Liquid, Salt Plates)

DISCUSSION SECTION

The strong carbonyl absorption appears at a characteristically high frequency of about 1800 cm^{-1} (5.56 μ) for saturated acid chlorides. The spectrum of acetyl chloride is shown in Figure 2–50. Conjugated acid chlorides absorb to the right (lower frequency), as predicted in Section 2.13A-1. An example of an aryl substituted acid chloride, benzoyl chloride, is shown in Figure 2–51. In this spectrum, the main absorption occurs at 1770 cm^{-1} (5.65 μ), but a shoulder appears at 1740 cm^{-1} (5.78 μ) on the main band. The shoulder probably is a result of an interaction between the C=O band and an overtone of a band appearing at a lower frequency. When a fundamental vibration couples with an overtone or combination band, the coupled vibration is called *Fermi resonance.* This type of interaction can lead to splitting in other carbonyl compounds, as well.

H. ANHYDRIDES

C=O	stretch always has *two* bands, 1830 to 1800 cm^{-1} (5.46 to 5.56 μ) and 1775 to 1740 cm^{-1} (5.63 to 5.75 μ), with variable relative intensity. Unsaturation moves the absorption to the right. Ring strain (cyclic anhydrides) moves the absorptions to the left.
C–O	stretch is at 1300 to 900 cm^{-1} (7.7 to 11.1 μ).

Example: Acetic anhydride (Fig. 2–52).

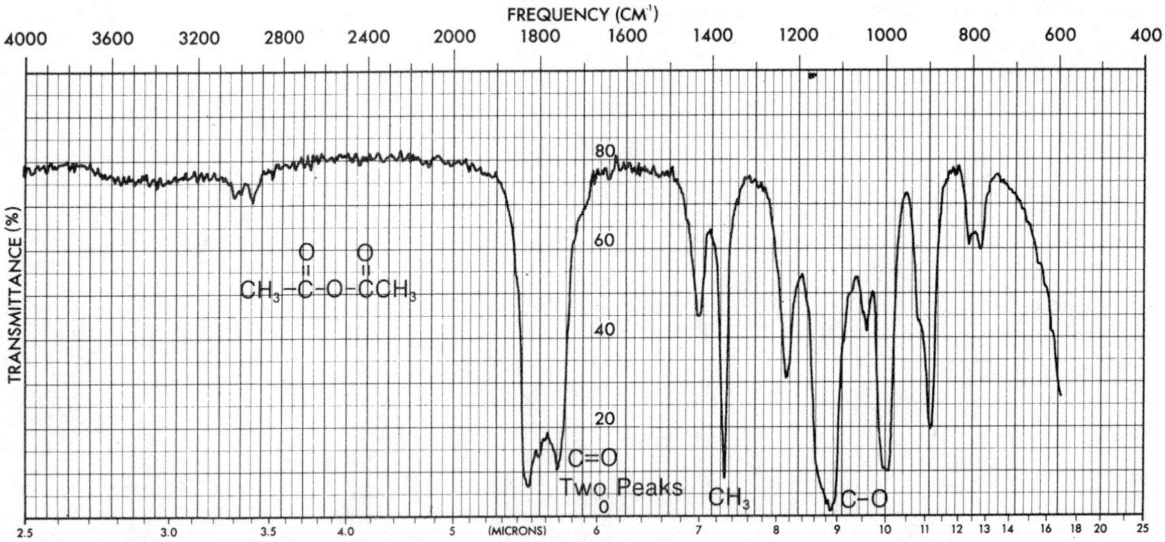

FIGURE 2–52 The Infrared Spectrum of Acetic Anhydride (Neat Liquid, Salt Plates)

DISCUSSION SECTION

The characteristic pattern for non-cyclic and saturated anhydrides is the appearance of *two strong* bands, not necessarily of equal intensities, in the region from 1830 to 1800 cm^{-1} (5.46 to 5.56 μ) and 1775 to 1740 cm^{-1} (5.63 to 5.75 μ). The two bands result from asymmetric and symmetric stretch (Section 2.3). Conjugation shifts the absorption to the right, while cyclization shifts the absorption to the left. The *strong* and *broad* C–O stretching vibration occurs in the region from 1300 to 900 cm^{-1} (7.7 to 11.1 μ).

2.14 AMINES

N–H stretch occurs in the range from 3500 to 3300 cm^{-1} (2.86 to 3.03 μ). Primary amines have *two* bands. Secondary amines have *one* band; a vanishingly weak one for aliphatic compounds and a stronger band for aromatic secondary amines. Tertiary amines have no N–H stretch.

N–H bend in primary amines results in a broad band appearing in the range from 1640 to 1560 cm^{-1} (6.1 to 6.4 μ). Secondary amines absorb near 1500 cm^{-1} (6.67 μ).

N–H out-of-plane bending absorption sometimes can be observed near 800 cm^{-1} (12.5 μ).

C–N stretch occurs in the range from 1350 to 1000 cm^{-1} (7.4 to 10.0 μ).

Examples: Butylamine (Fig. 2–53), dibutylamine (Fig. 2–54), tributylamine (Fig. 2-55), N-methylaniline (Fig. 2–56), and *meta*-toluidine (Fig. 2–22).

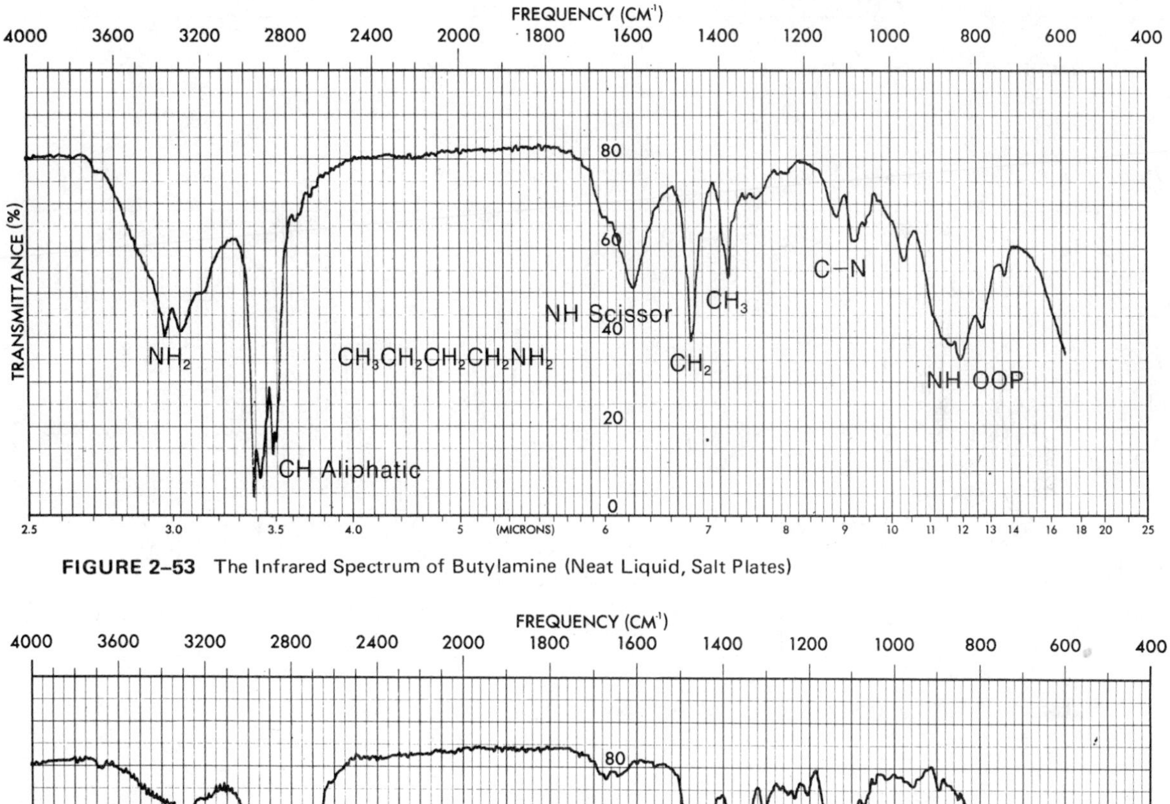

FIGURE 2–53 The Infrared Spectrum of Butylamine (Neat Liquid, Salt Plates)

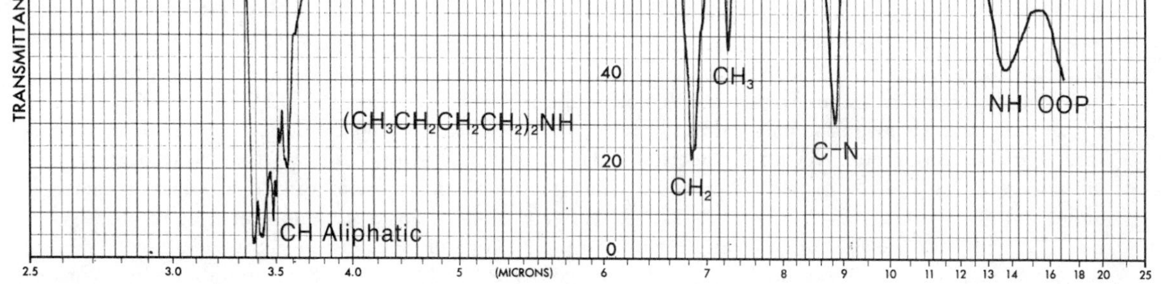

FIGURE 2–54 The Infrared Spectrum of Dibutylamine (Neat Liquid, Salt Plates)

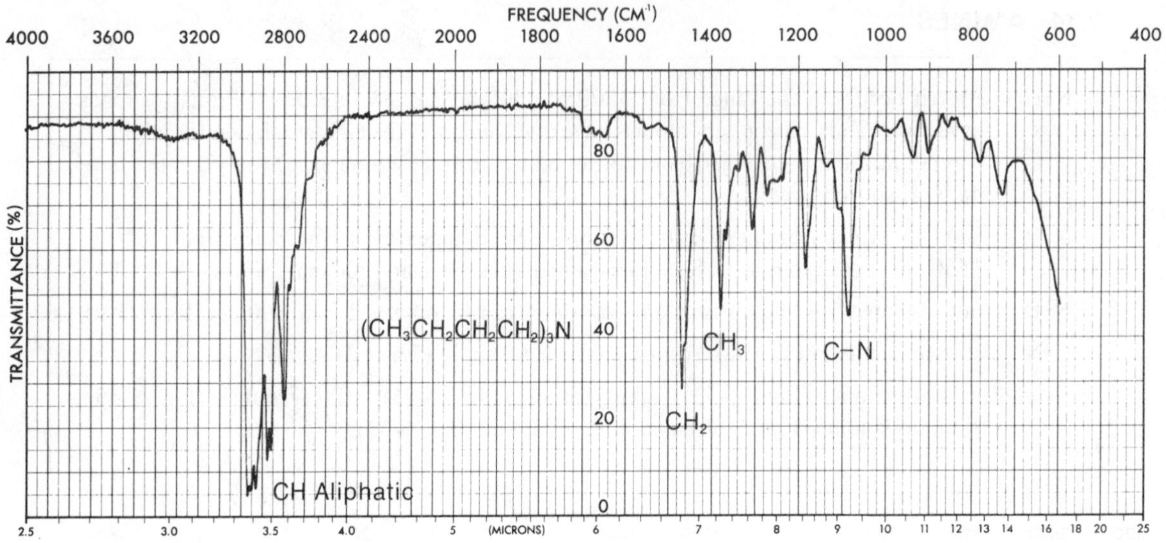

FIGURE 2–55 The Infrared Spectrum of Tributylamine (Neat Liquid, Salt Plates)

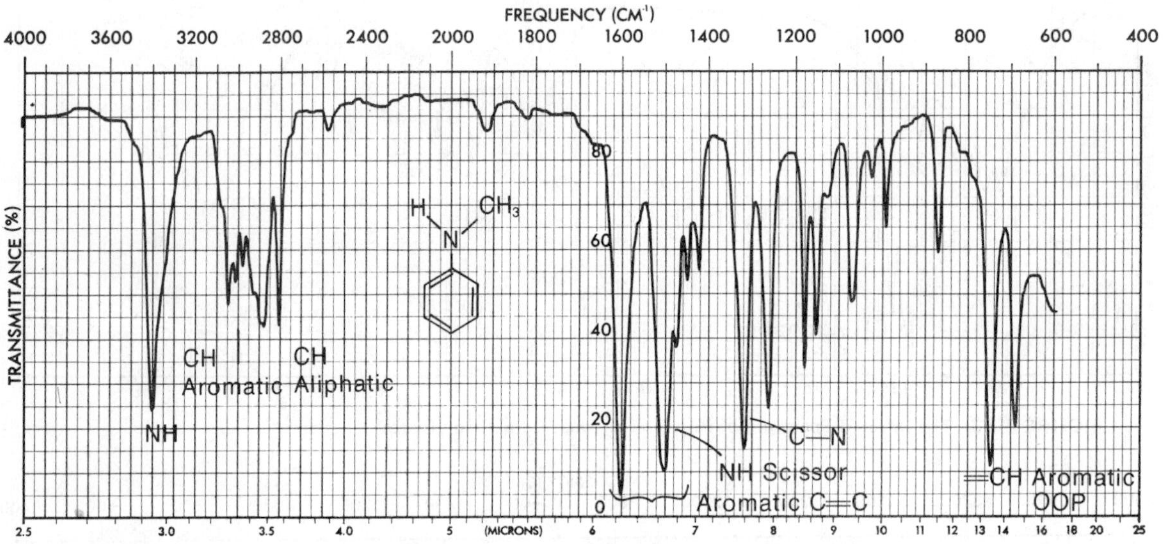

FIGURE 2–56. The Infrared Spectrum of N-Methylaniline (Neat Liquid, Salt Plates)

DISCUSSION SECTION

The N–H stretching vibrations occur in the range from 3500 to 3300 cm^{-1} (2.86 to 3.03 μ). In neat liquid samples, the N–H bands are often weaker and sharper than an O–H band (see Figure 2–5). Amines may sometimes be differentiated from alcohols on that basis. Primary amines, determined as neat liquids (hydrogen bonded), give *two* bands at about 3400 and 3300 cm^{-1} (2.94 and 3.03 μ). The higher frequency band in the pair is due to the asymmetric vibration, while the lower frequency band results from a symmetric vibration (Section 2.3). In dilute solution, the two free N–H stretching vibrations are shifted to higher frequencies. The spectra of an aliphatic and an aromatic primary amine are given in Figures 2–53 and 2–22, respectively. A small absorption appearing at about 3200 cm^{-1} (3.13 μ) may be attributed to an overtone of the N–H bending vibration (about 1600 cm^{-1}).

Secondary amines determined as neat liquids give *one* band in the N–H stretching region. With aliphatic secondary amines, the band may often be vanishingly weak. On the other hand, an aromatic secondary amine gives a somewhat stronger band. The spectra of an aliphatic and an aromatic secondary amine are given in Figures 2–54 and 2-56, respectively. Tertiary amines do not absorb in this region, as shown in Figure 2–55.

The N–H bending mode (scissoring) appears as a medium to strong intensity band (broad) in the range from 1640 to 1560 cm^{-1} (6.1 to 6.4 μ) for primary amines. In aromatic secondary amines, the band shifts to the right and appears near 1500 cm^{-1} (6.67 μ). However, in aliphatic secondary amines the N–H bending vibration is very weak and usually is not observed. The N–H vibrations in aromatic compounds often overlap the aromatic C=C ring absorptions, which also appear in this region. An out-of-plane N–H bending vibration appears as a broad band near 800 cm^{-1} (12.5 μ) for primary and secondary amines. These bands appear in spectra of compounds determined as neat liquids and are seen most easily in aliphatic amines (Fig. 2–53 and 2–54).

The C–N stretching absorption occurs in the region from 1350 to 1000 cm^{-1} (7.4 to 10.0 μ) as a medium to strong band for all amines. Aliphatic amines absorb from 1250 to 1000 cm^{-1} (8.0 to 10.0 μ), while aromatic amines absorb from 1350 to 1250 cm^{-1} (7.4 to 8.0 μ). The C–N absorption occurs at a higher frequency in aromatic amines because resonance increases the double bond character between the ring and the attached nitrogen atom.

2.15 NITRILES, ISOCYANATES AND IMINES

–C≡N	stretch is a sharp absorption near 2250 cm^{-1} (4.44 μ).
	Conjugation with double bonds or aromatic rings moves the absorption to the right.
–N=C=O	stretch in an isocyanate gives a broad and intense absorption near 2270 cm^{-1} (4.4 μ).
–N=C=S	stretch in an isothiocyanate gives a broad and intense absorption near 2125 cm^{-1} (4.7 μ).
–C(–)=N–	stretch in an imine, oxime, etc. gives a variable intensity absorption in the range from 1690 to 1640 cm^{-1} (5.9 to 6.1 μ).

Examples: Butyronitrile (Fig. 2–57) and benzonitrile (Fig. 2–58).

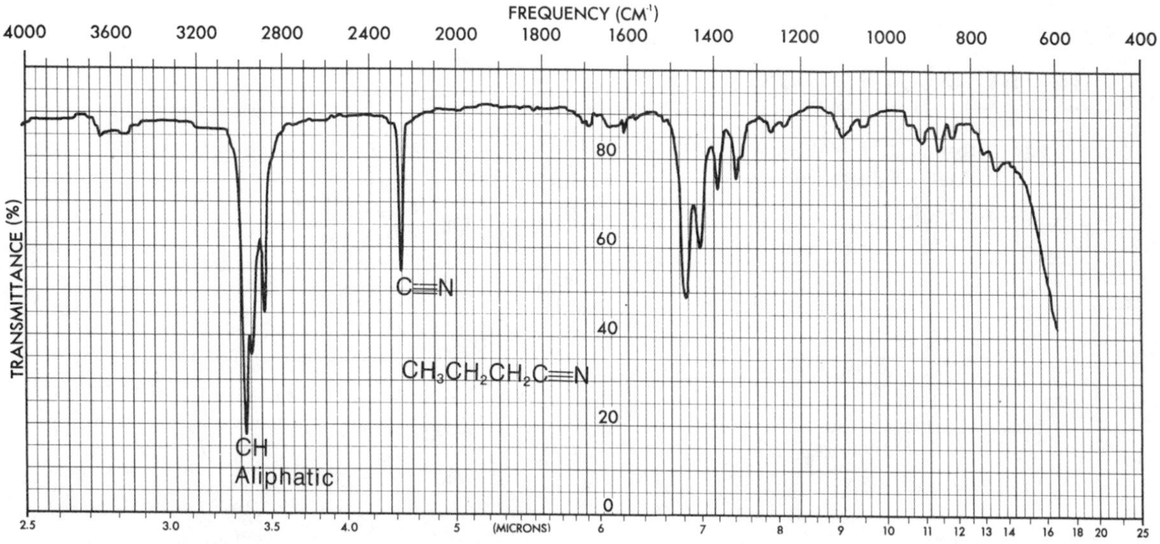

FIGURE 2–57 The Infrared Spectrum of Butyronitrile (Neat Liquid, Salt Plates)

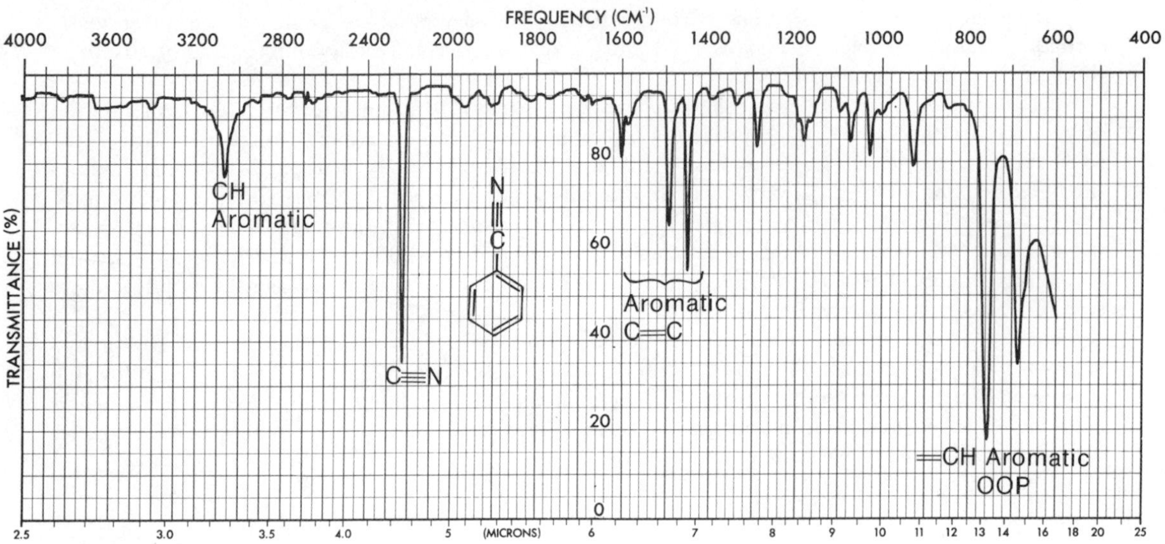

FIGURE 2–58 The Infrared Spectrum of Benzonitrile (Neat Liquid, Salt Plates)

DISCUSSION SECTION

The C≡N group in a nitrile gives a medium intensity band in the triple bond region of the spectrum (2500 to 2000 cm^{-1}). The C≡C bond, which also absorbs in this region, usually gives a much weaker band. An isocyanate (R–N=C=O), on the other hand, gives a broad intense band in this region. Aliphatic nitriles absorb at about 2250 cm^{-1} (4.44 μ), while their aromatic counterparts absorb at lower frequencies near 2230 cm^{-1} (4.48 μ). The spectra of an aliphatic and an aromatic nitrile are shown in Figures 2–57 and 2–58, respectively. Aromatic nitriles absorb at lower frequencies with increased intensity because of conjugation of the triple bond with the ring.

The C=N bond absorbs in about the same range as a C=C bond. Although the C=N band varies in intensity from compound to compound, it usually is more intense than that obtained from the C=C bond. An oxime (R–CH=N–OH) gives a C=N absorption in the range from 1690 to 1640 cm^{-1} (5.92 to 6.10μ) and a broad O–H absorption between 3650 and 2600 cm^{-1} (2.74 to 3.85 μ). An imine (R–CH=N–R) gives a C=N absorption in the range from 1690 to 1650 cm^{-1} (5.92 to 6.06 μ).

2.16 NITRO COMPOUNDS

N=O stretch gives two strong bands, 1600–1500 cm^{-1} (6.25–6.67 μ) and 1390–1300 cm^{-1} (7.19–7.69 μ).

Examples: 1-Nitropropane (Fig. 2–59) and nitrobenzene (Fig. 2–60)

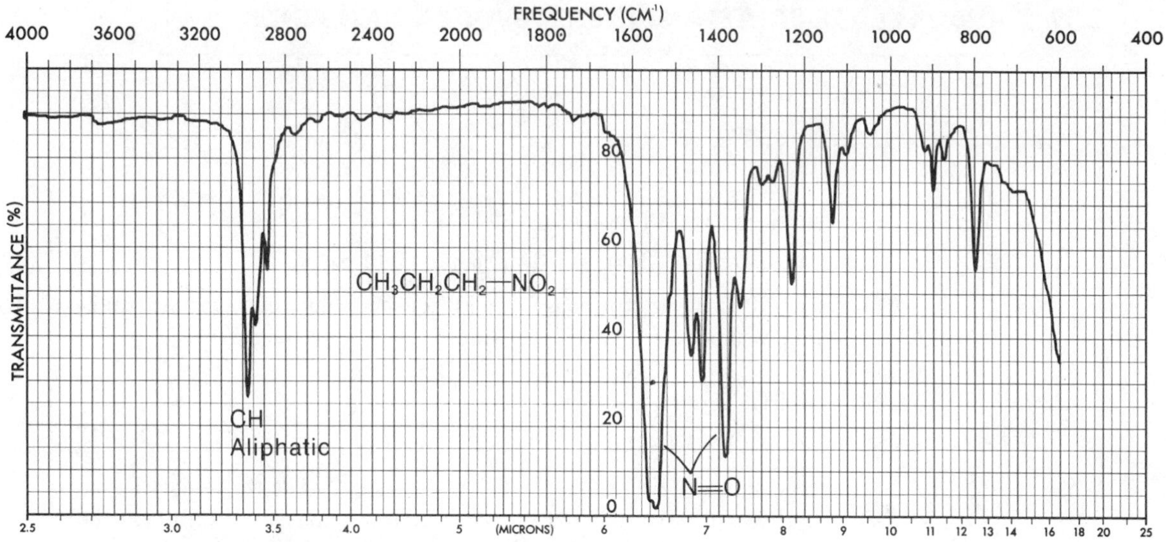

FIGURE 2–59 The Infrared Spectrum of 1-Nitropropane (Neat Liquid, Salt Plates)

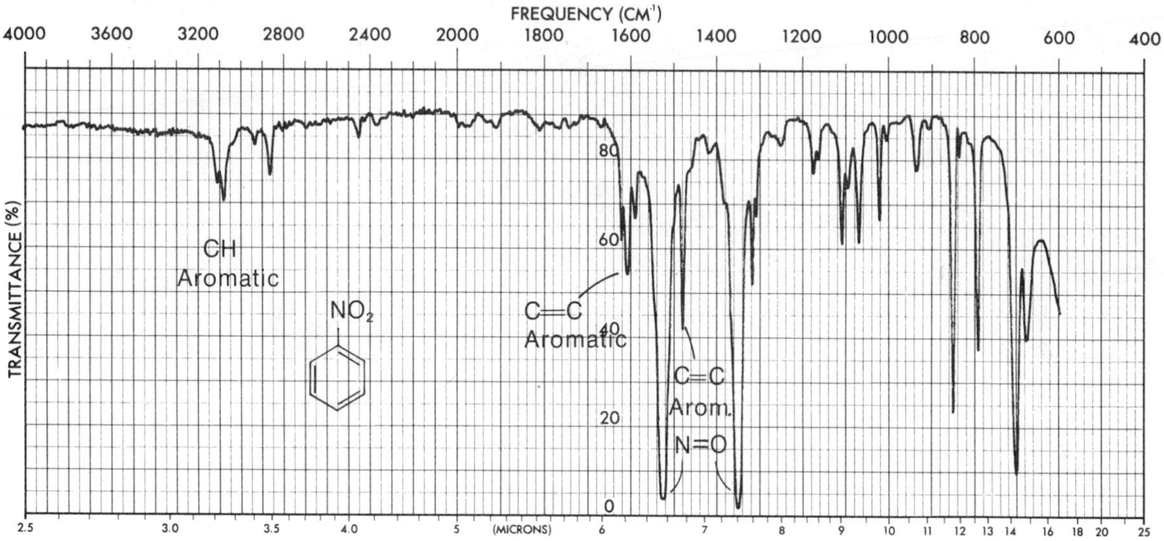

FIGURE 2–60 The Infrared Spectrum of Nitrobenzene (Neat Liquid, Salt Plates)

DISCUSSION SECTION

The nitro group (NO_2) gives two strong bands in the infrared spectrum. The asymmetric stretching vibration (Section 2.3) occurs in the range from 1600 to 1500 cm^{-1} (6.25 to 6.67 μ), while the symmetric stretching band appears between 1390 and 1300 cm^{-1} (7.19 to 7.69 μ). An aliphatic nitro compound, for example 1-nitropropane (Fig. 2–59), absorbs at about 1550 cm^{-1} (6.45 μ) and 1380 cm^{-1} (7.25 μ). Conjugation of a nitro group with a double bond or a ring shifts the bands to lower frequencies. For example, nitrobenzene (Fig. 2–60) absorbs strongly at 1530 cm^{-1} (6.53 μ) and 1350 cm^{-1} (7.41 μ). The nitroso group (R–N=O) gives only one strong band that appears in the range 1600 to 1500 cm^{-1} (6.25 to 6.67 μ).

2.17 CARBOXYLATE SALTS, AMINE SALTS AND AMINO ACIDS

CARBOXYLATE SALTS $R-C(=O)-O^-$

– Asymmetric stretch (strong), near 1600 cm^{-1} (6.25 μ).

Symmetric stretch (strong), near 1400 cm^{-1} (7.14 μ).

Frequency of C=O absorption is lowered from the value found for the parent carboxylic acid because of resonance (more single bond character).

AMINE SALTS NH_4^+, RNH_3^+, $R_2NH_2^+$, R_3NH^+

N–H Stretch (broad), 3300–2600 cm^{-1} (3.03–3.85 μ).

The ammonium ion absorbs to the left in this range, while the tertiary amine salt absorbs to the right. Primary and secondary amine salts absorb in the middle of the range, 3000–2700 cm^{-1} (3.33–3.70 μ).

N–H Bend (strong), 1610–1500 cm^{-1} (6.21–6.67 μ).

Primary (two bands); asymmetric at 1610 cm^{-}1, symmetric at 1500 cm^{-1}.

Secondary absorbs in the range from 1610 to 1550 cm^{-1} (6.21 to 6.45 μ).

Tertiary absorbs only weakly.

AMINO ACIDS These compounds exist as zwitterions, and show spectra that are combinations of carboxylate and primary amine salts (above). Amino acids show N–H stretch, NH_3^+ (asymmetric/symmetric) bend, and COO^- (asymmetric/symmetric) stretch.

2.18 SULFUR COMPOUNDS

MERCAPTANS R–SH

S–H Stretch, one weak band near 2550 cm^{-1} (3.92 μ).

Virtually confirms the presence of this group, since few other absorptions appear here.

Example: Benzenethiol (Fig. 2–61).

SULFIDES R–S–R

Little useful information is obtained from the infrared spectrum.

SULFOXIDES R–S(=O)–R

S=O	Stretch, one strong band near 1050 cm^{-1} (9.52 μ).

SULFONES R–S(=O)(=O)–R

S=O	Asymmetric stretch (strong), 1300 cm^{-1} (7.69 μ).
	Symmetric stretch (strong), 1150 cm^{-1} (8.70 μ).

SULFONYL CHLORIDES R–S(=O)(=O)–Cl

S=O	Asymmetric stretch (strong), 1375 cm^{-1} (7.27 μ).
	Symmetric stretch (strong), 1200 cm^{-1} (8.33 μ).

Example: Benzenesulfonyl chloride (Fig. 2–62).

SULFONATES R–S(=O)(=O)–OR

S=O	Asymmetric stretch (strong), 1350 cm^{-1} (7.41 μ).
	Symmetric stretch (strong), 1175 cm^{-1} (8.51 μ).
S–O	Stretch, several strong bands, 1000–750 cm^{-1} (10.0–13.3 μ).

Example: Methyl *p*-toluenesulfonate (Fig. 2–63).

SULFONAMIDES (Solid State) R–S(=O)(=O)–NH_2, R–S(=O)(=O)–NHR

S=O	Asymmetric stretch (strong), 1325 cm^{-1} (7.55 μ).
	Symmetric stretch (strong), 1140 cm^{-1} (8.77 μ).
N–H	Stretch, primary 3350 and 3250 cm^{-1} (2.99 and 3.08 μ).
	Stretch, secondary 3250 cm^{-1} (3.08 μ).

Examples: Benzenesulfonamide (Fig. 2–64) and sulfanilamide (Fig. 2–65).

SULFONIC ACIDS R–S(=O)(=O)–OH

S=O	Asymmetric stretch (strong), 1200 cm^{-1} (8.33 μ).
	Symmetric stretch (strong), 1050 cm^{-1} (9.52 μ).
S–O	Stretch (strong), 650 cm^{-1} (15.4 μ).

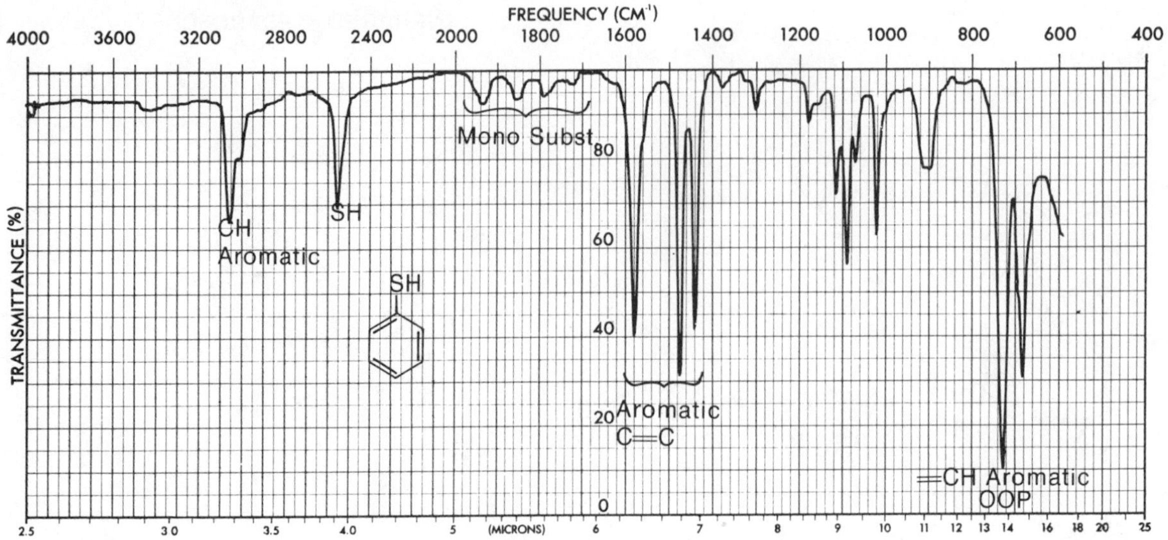

FIGURE 2–61 The Infrared Spectrum of Benzenethiol (Neat Liquid, Salt Plates)

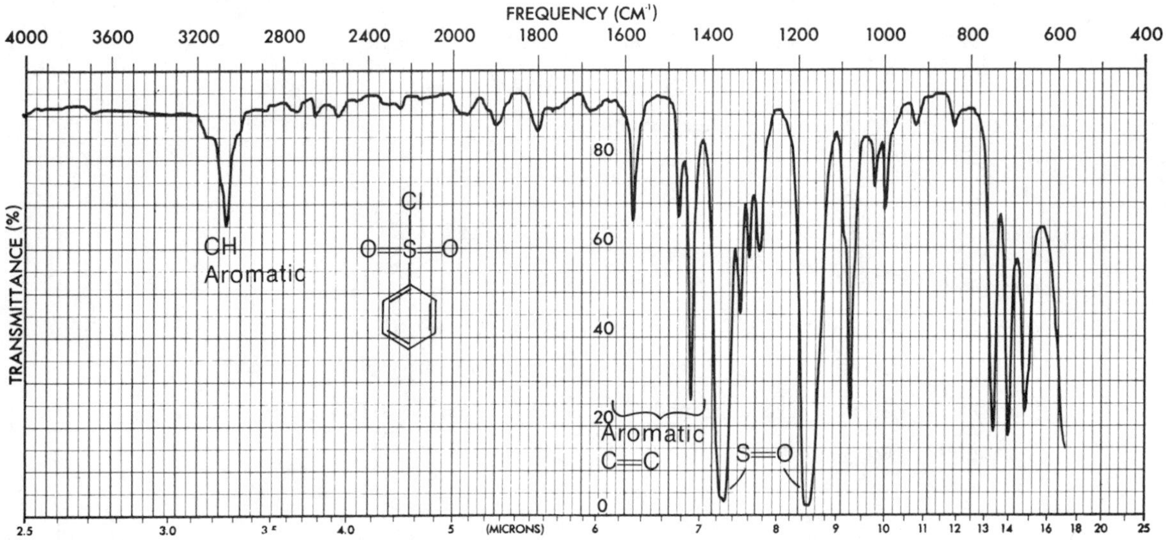

FIGURE 2–62 The Infrared Spectrum of Benzenesulfonyl Chloride (Neat Liquid, Salt Plates)

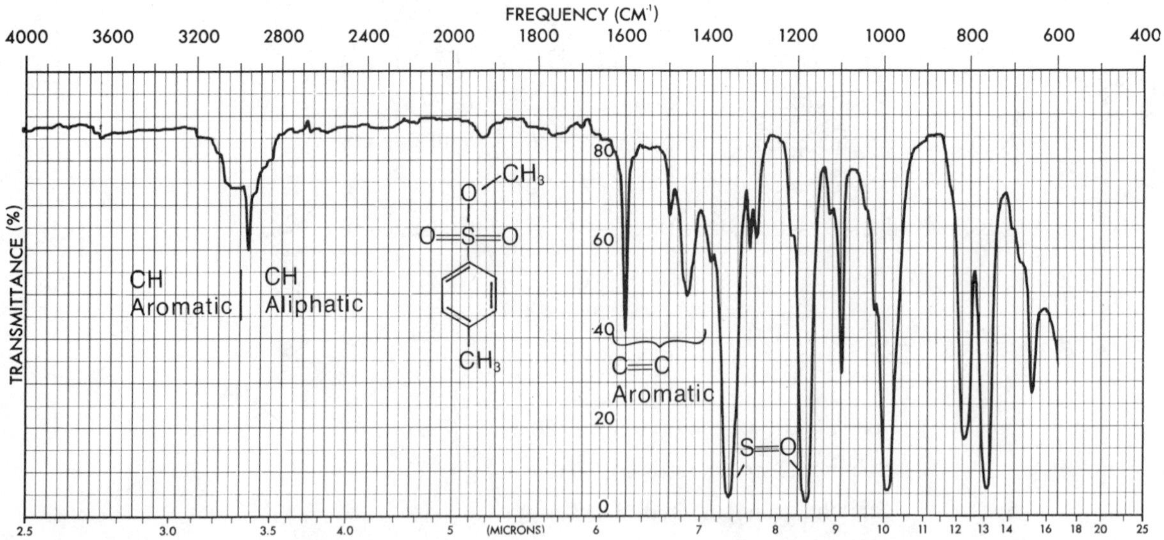

FIGURE 2–63 The Infrared Spectrum of Methyl *p*-Toluenesulfonate (Neat Liquid, Salt Plates)

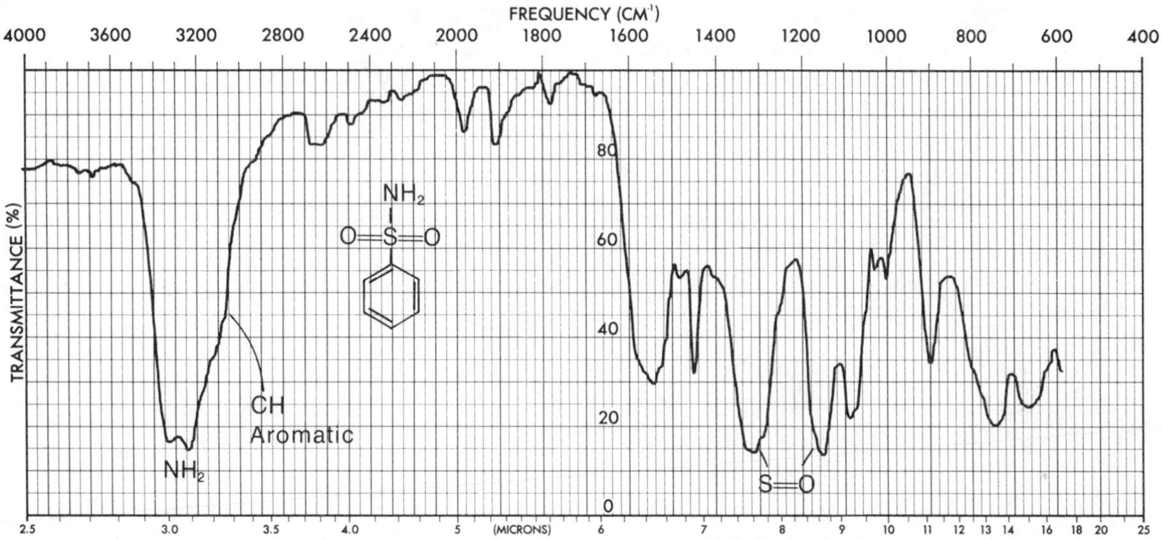

FIGURE 2-64 The Infrared Spectrum of Benzenesulfonamide (Solid Phase, KBr)

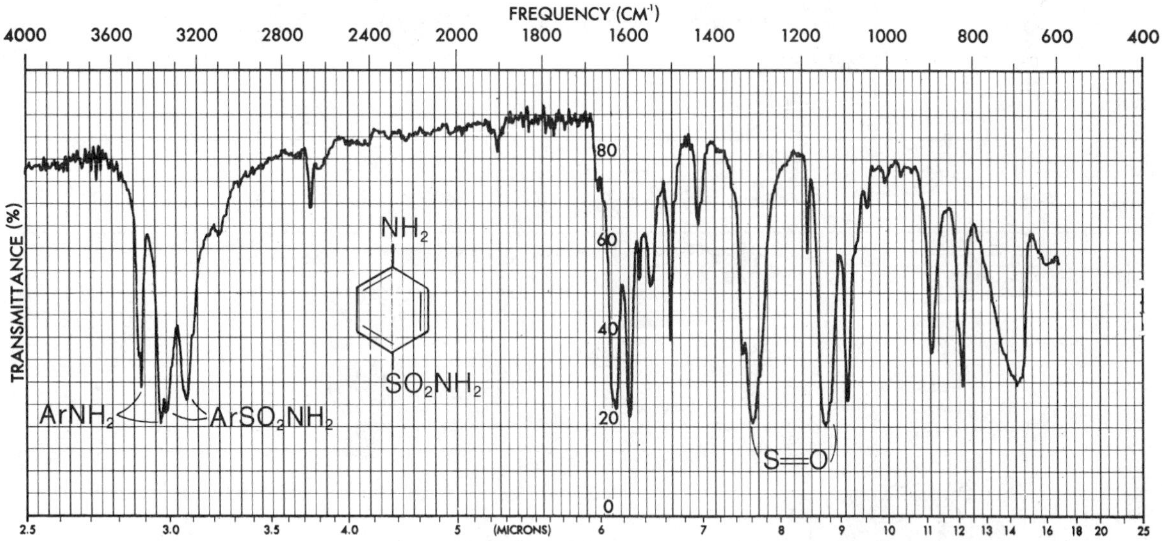

FIGURE 2–65 The Infrared Spectrum of Sulfanilamide (Solid Phase, KBr)

2.19 ALKYL AND ARYL HALIDES

It is difficult to determine either the presence or the absence of a halide in a compound by infrared spectroscopy. Elemental analysis (Section 1.1) or mass spectral methods (Section 6.4 and 6.5M) give more reliable information.

FLUORIDES

C–F — Stretch (strong), 1400–1000 cm^{-1} (7.14–10.0 μ).

Monofluoroalkanes absorb to the right in the range, while polyfluoroalkanes give multiple strong bands over the range from 1350 to 1100 cm^{-1} (7.41 to 9.09 μ). Aryl fluorides absorb between 1250 and 1100 cm^{-1} (8.0 and 9.09 μ).

CHLORIDES

C–Cl — Stretch (strong) in aliphatic chlorides, 800–600 cm^{-1} (12.5–16.7 μ).

Multiple substitution on a single carbon atom results in an intense absorption to the left in the range. Aryl chlorides absorb between 1100 and 1050 cm^{-1} (9.09 and 9.52 μ).

CH_2–Cl — Bend (wagging), 1300–1200 cm^{-1} (7.69–8.33 μ).

Examples: Carbon tetrachloride (Fig. 2–66), *ortho*-chlorotoluene (Fig. 2–21), *para*-chlorotoluene (Fig. 2–23), and chloroform (Fig. 2–67).

BROMIDES

C–Br — Stretch occurs to the right of 667 cm^{-1} (15.0 μ), out of the range of routine spectroscopy.

CH_2–Br — Bend (wagging), 1250–1150 cm^{-1} (8.00–8.70 μ).

IODIDES

C–I — Stretch occurs to the right of 667 cm^{-1} (15.0 μ), out of the range of routine spectroscopy.

CH_2–I — Bend (wagging), 1200–1150 cm^{-1} (8.33–8.70 μ).

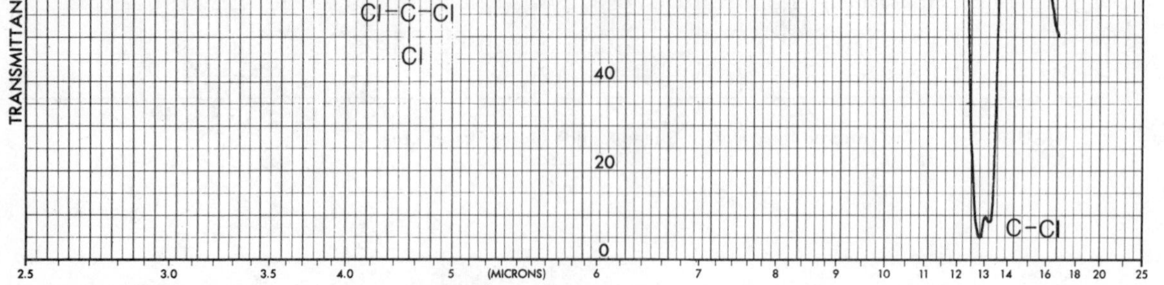

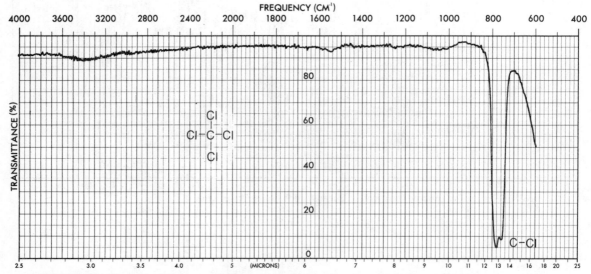

FIGURE 2–66 The Infrared Spectrum of Carbon Tetrachoride (Neat Liquid, Salt Plates)

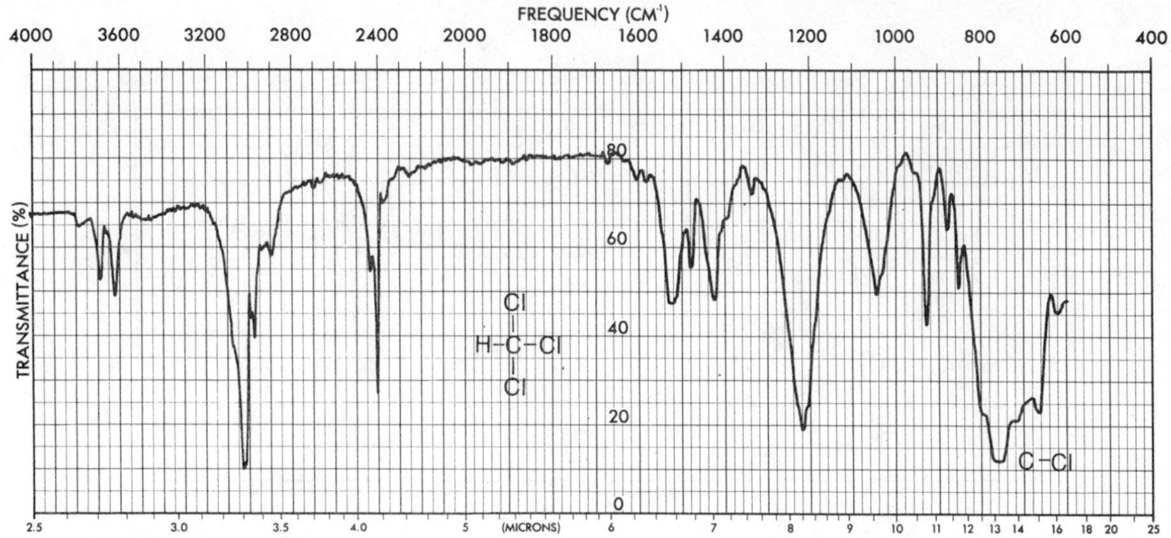

FIGURE 2–67 The Infrared Spectrum of Chloroform (Neat Liquid, Salt Plates)

REFERENCES

N. L. Alpert, W. E. Kaiser, and H. A. Szymanski, "Theory and Practice of Infrared Spectroscopy," 2nd ed., Plenum Press, New York, 1970.

L. J. Bellamy, "The Infra-red Spectra of Complex Molecules," 2nd ed., John Wiley, New York, 1958.

R. T. Conley, "Infrared Spectroscopy," 2nd ed., Allyn and Bacon, Boston, 1972.

A. D. Cross and R. A. Jones, "Introduction to Practical Infrared Spectroscopy," 3rd ed., Plenum Press, New York, 1969.

J. R. Dyer, "Applications of Absorption Spectroscopy of Organic Compounds," Prentice-Hall, Englewood Cliffs, N.J., 1965.

R. Mecke and F. Langenbucher, "Infrared Spectra of Selected Chemical Compounds", 8 Volumes, Heyden and Son, London, 1965.

K. Nakanishi, "Infrared Absorption Spectroscopy–Practical," Holden-Day, San Francisco, 1962.

D. J. Pasto and C. R. Johnson, "Organic Structure Determination," Prentice-Hall, Englewood Cliffs, N.J., 1969.

C. J. Pouchert, "Aldrich Library of Infrared Spectra," 2nd ed., Aldrich Chemical Co., Milwaukee, Wis., 1975.

Sadtler Standard Spectra, Sadtler Research Laboratories, Inc., Philadelphia.

R. M. Silverstein, G. C. Bassler, and T. C. Morrill, "Spectrometric Identification of Organic Compounds," 3rd ed., John Wiley, New York, 1974.

H. A. Szymanski, "Interpreted Infrared Spectra," Vols. 1–3, Plenum Press, New York, 1964–1967.

PROBLEMS

When a molecular formula is given, it is suggested that an index of hydrogen deficiency be calculated (Section 1.4). The index often gives useful information about the functional group or groups that may be present in the molecule.

1. In each part, a molecular formula is given. Deduce the structure or structures which would be consistent with the infrared spectrum.

a) $C_5H_{12}O$

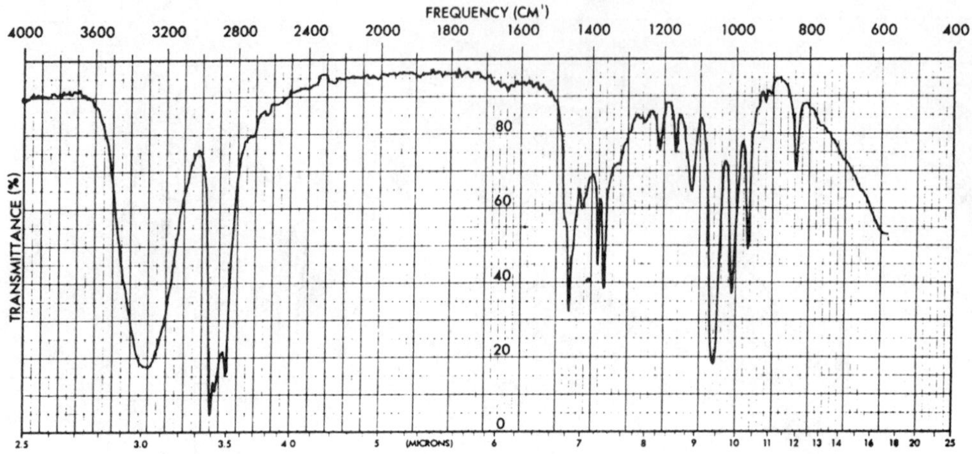

b) C_3H_3Br

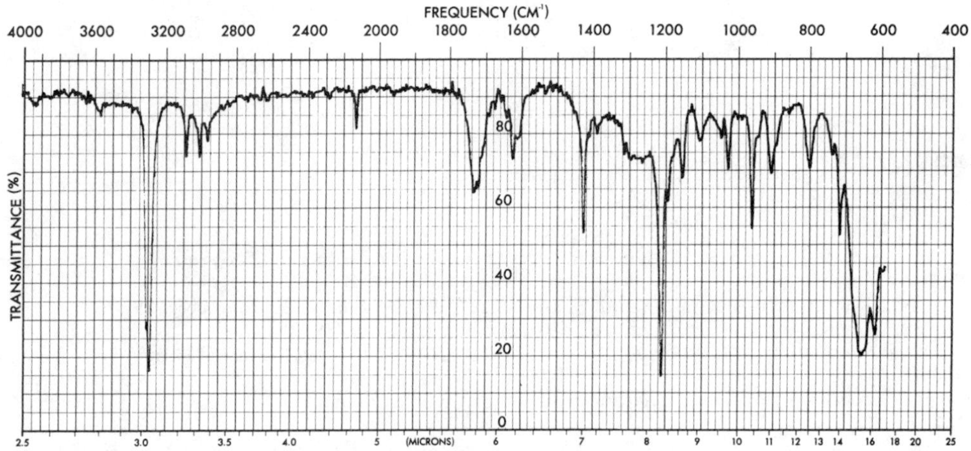

c) C_5H_8O (Assign the proper stereochemistry)

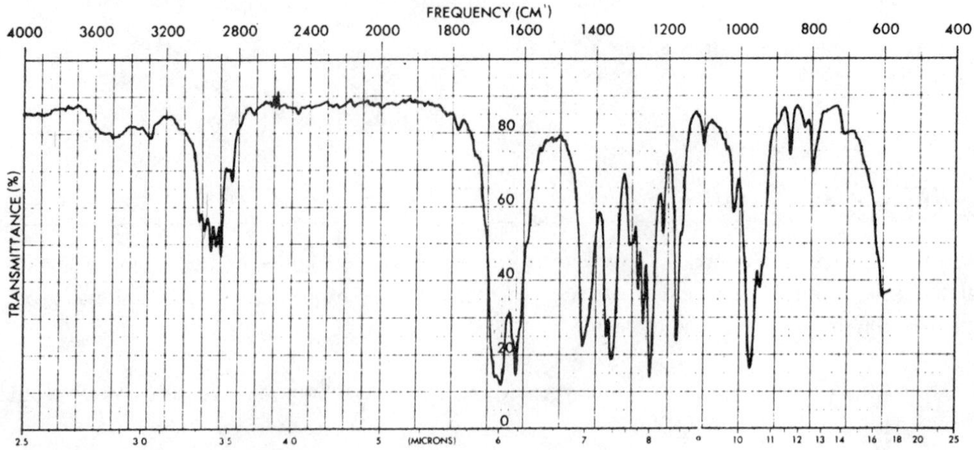

d) $C_{10}H_{14}$

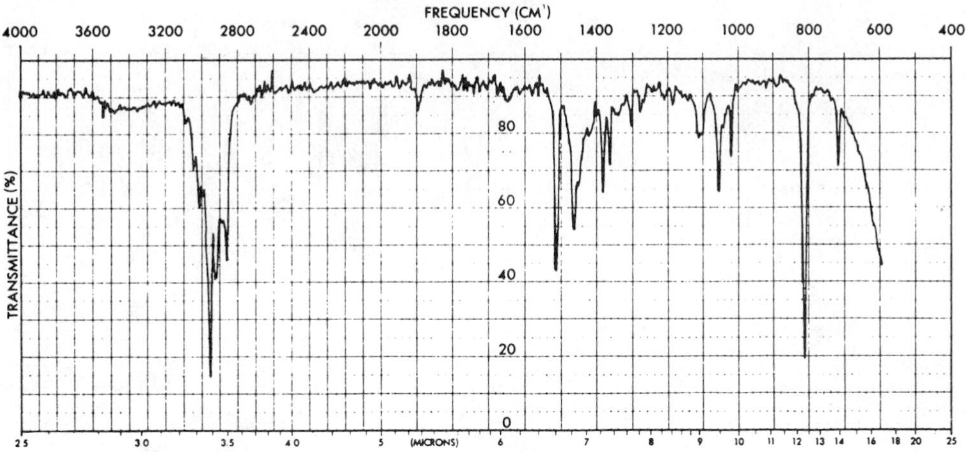

e) C_7H_9N

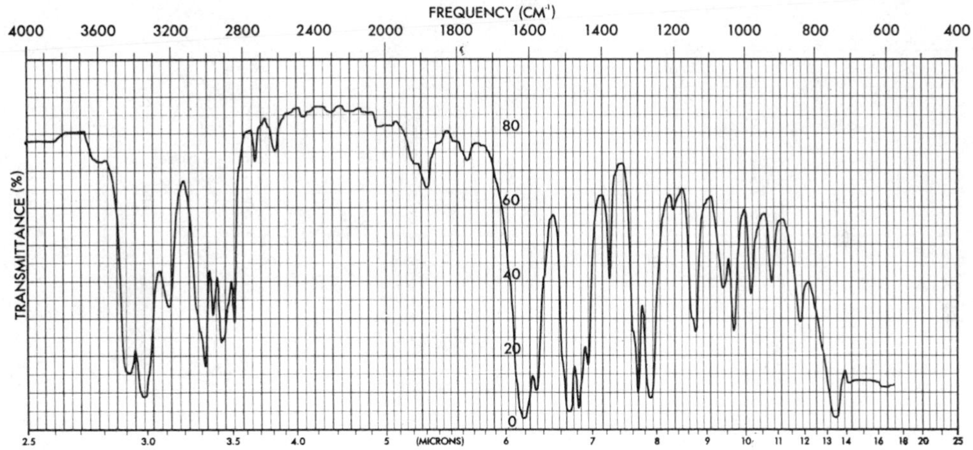

f) C_7H_8O

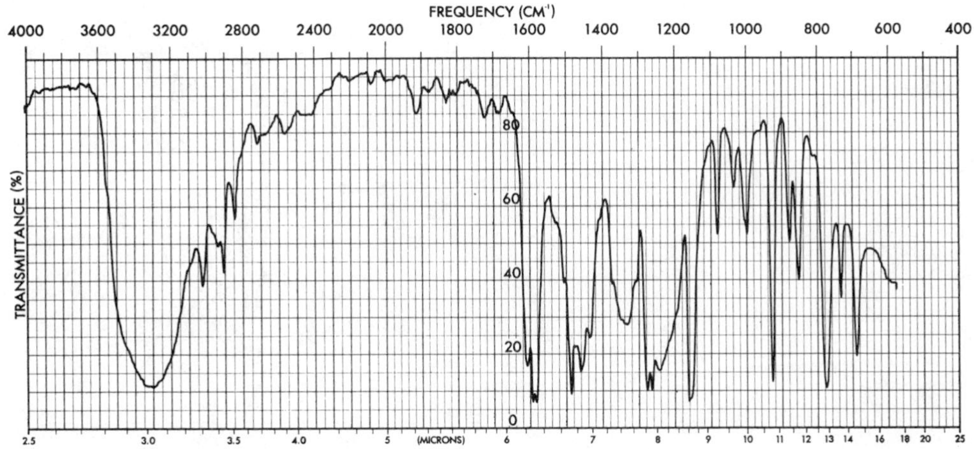

g) $C_9H_{10}O$

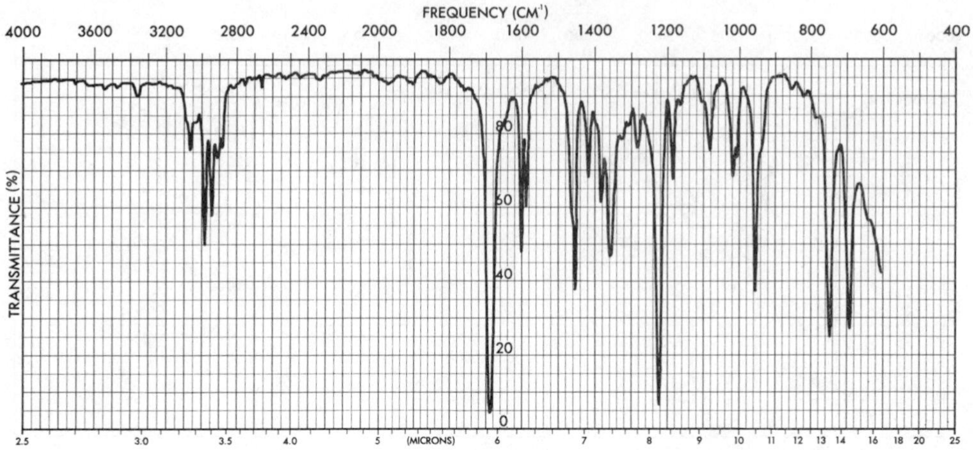

h) $C_9H_{10}O$

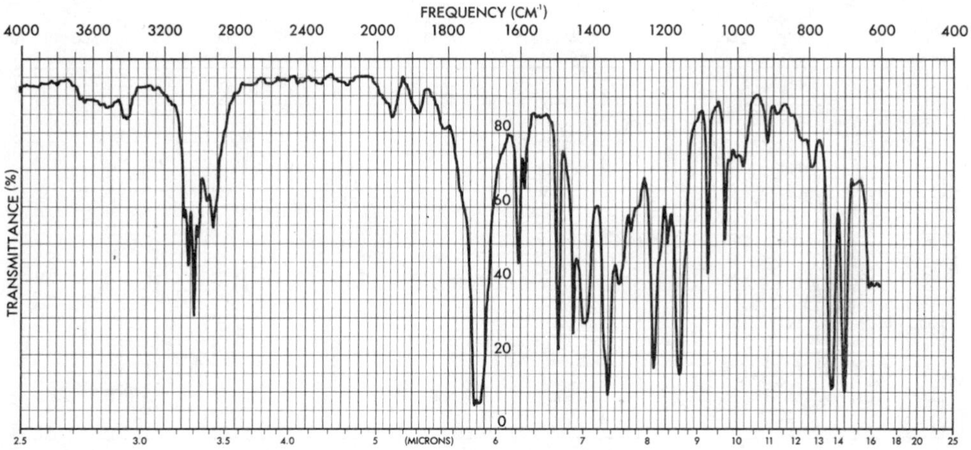

i) $C_4H_4O_3$

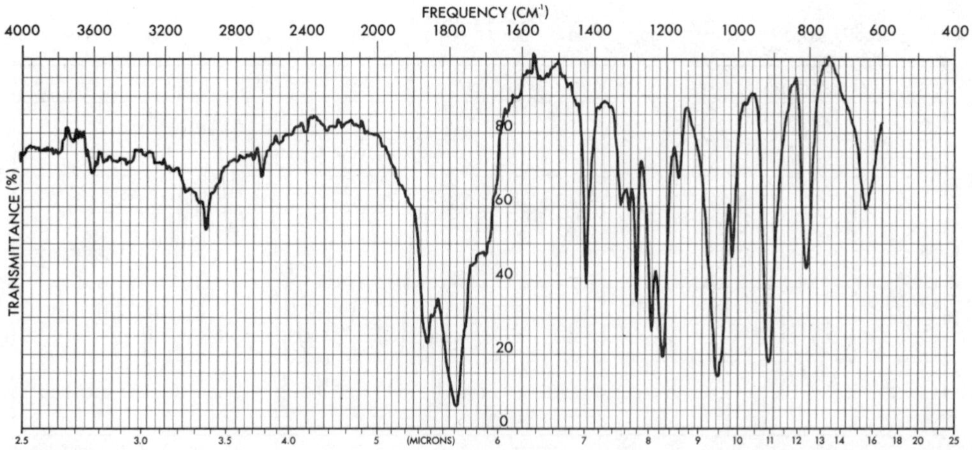

j) $C_8H_{11}N$

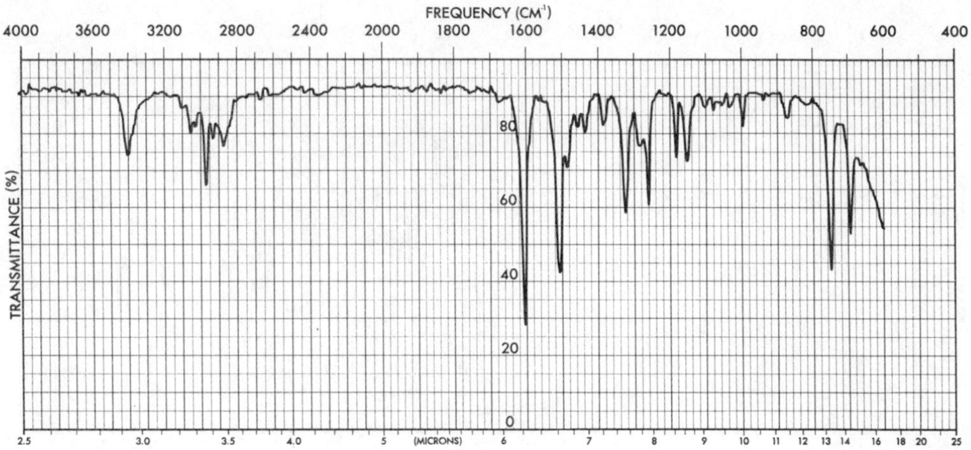

k) $C_3H_5O_2Cl$

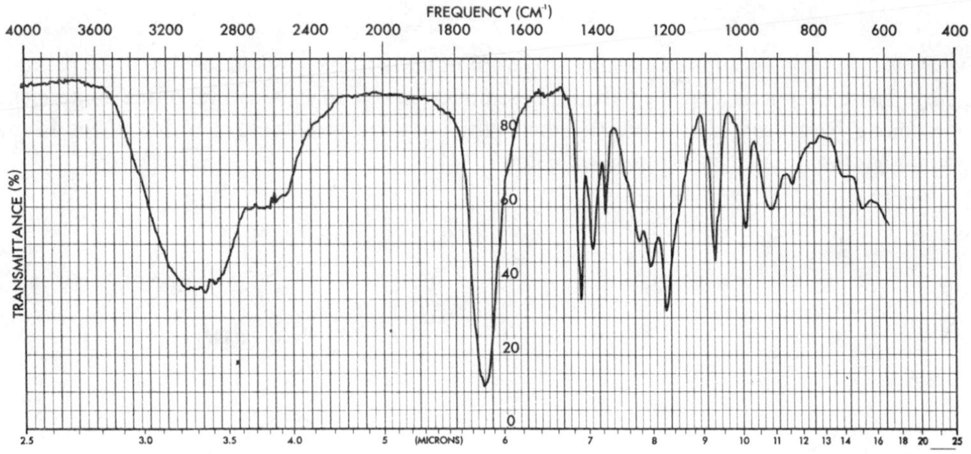

l) $C_5H_{10}N_2$

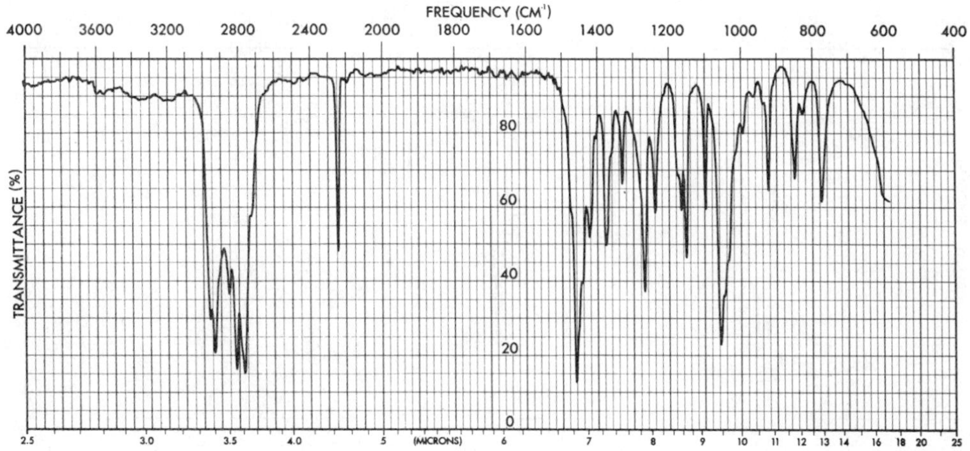

m) $C_{11}H_{12}O_2$

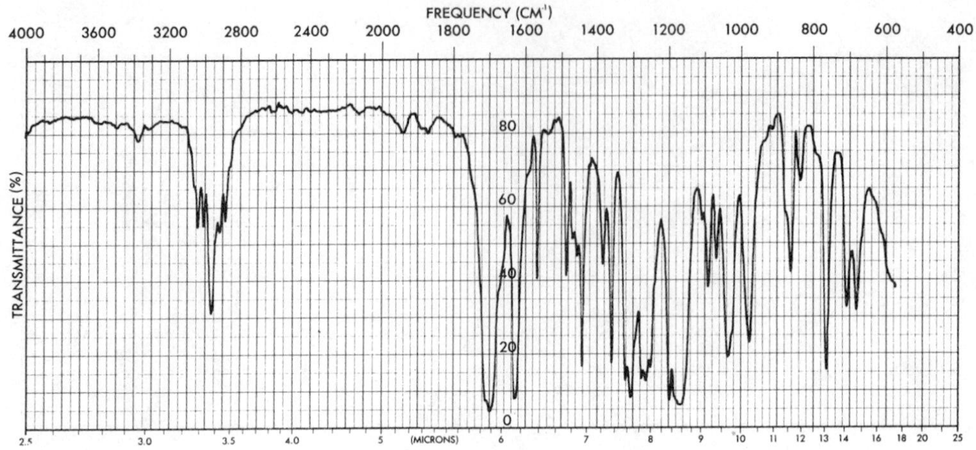

2. Ants emit tiny amounts of chemicals to warn other ants of the same species of the presence of an enemy. These chemicals are called alarm pheromones. In one species, several of the components of the pheromone have been identified, and their structures are given below. Which of the following compounds has the infrared spectrum given below?

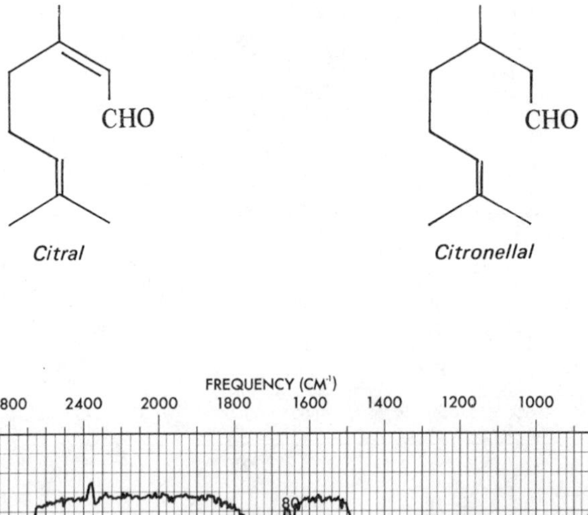

Citral *Citronellal*

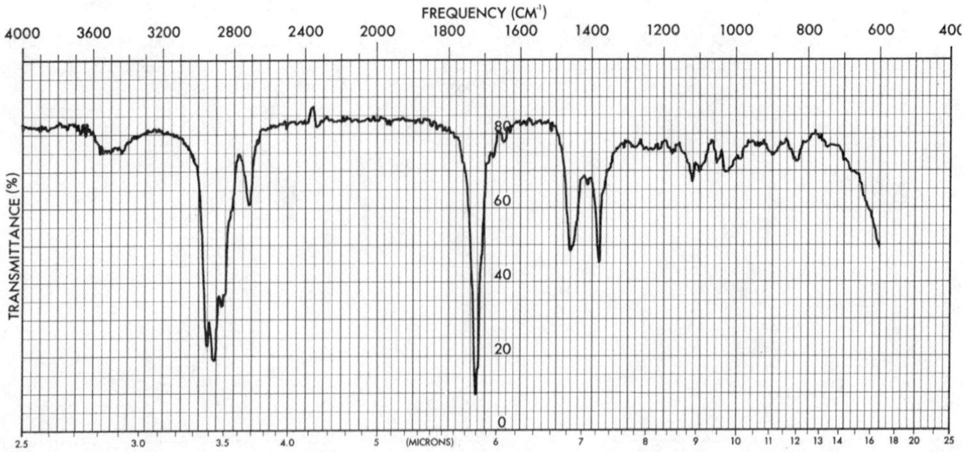

3. A compound with the formula $C_7H_7NO_2$ reacts with both dilute acid and base. This substance is one of the components in folic acid and plays a vital role in the growth of bacteria. From the following infrared spectrum, deduce its structure.

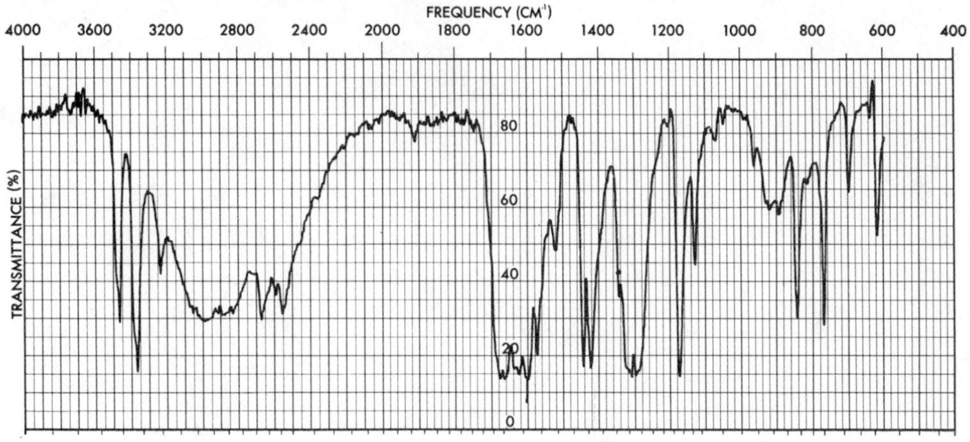

4. The main constituent of cinnamon oil has the formula C_9H_8O. From the following infrared spectrum, deduce the structure of this component.

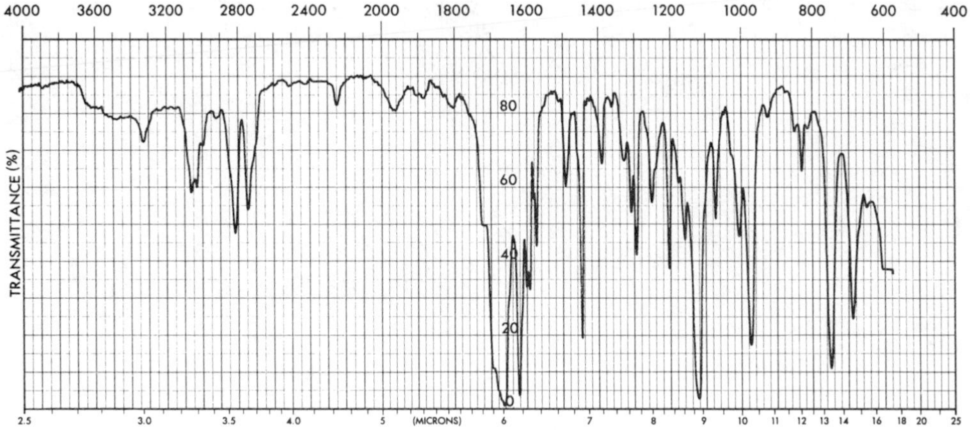

5. The infrared spectra of some polymeric films are shown below. Assign a structure to each of them, selected from the following choices: polyamide (Nylon), polymethyl methacrylate, polyethylene, polystyrene, and polypropylene. You may need to look up the structures of these materials.

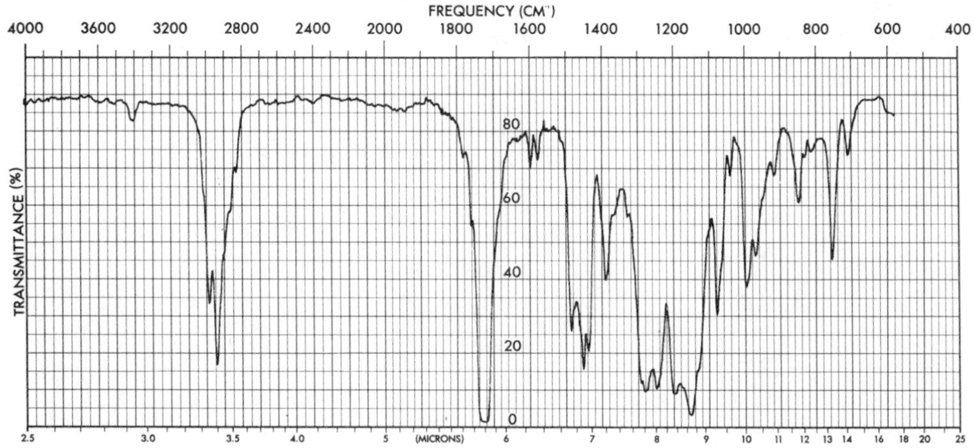

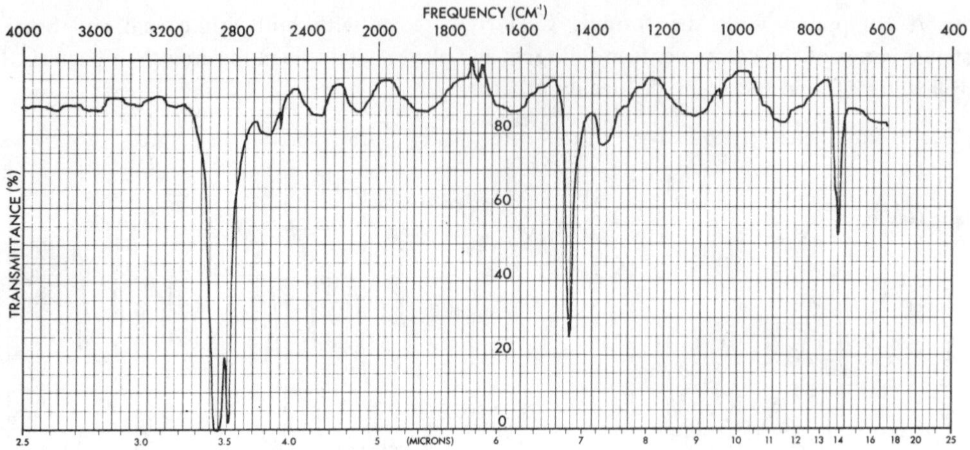
FREQUENCY (CM⁻¹)
4000 3600 3200 2800 2400 2000 1800 1600 1400 1200 1000 800 600 400
TRANSMITTANCE (%)
80 60 40 20 0
2.5 3.0 3.5 4.0 5 (MICRONS) 6 7 8 9 10 11 12 13 14 16 18 20 25

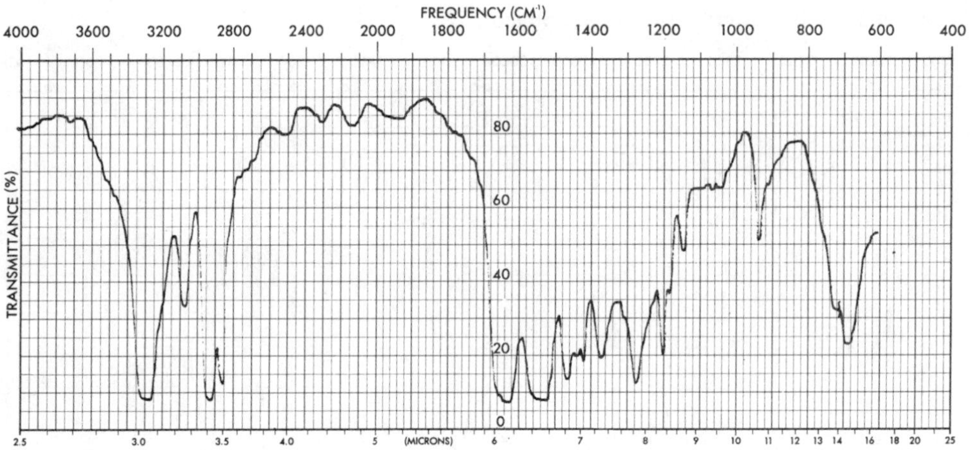
FREQUENCY (CM⁻¹)
4000 3600 3200 2800 2400 2000 1800 1600 1400 1200 1000 800 600 400
TRANSMITTANCE (%)
80 60 40 20 0
2.5 3.0 3.5 4.0 5 (MICRONS) 6 7 8 9 10 11 12 13 14 16 18 20 25

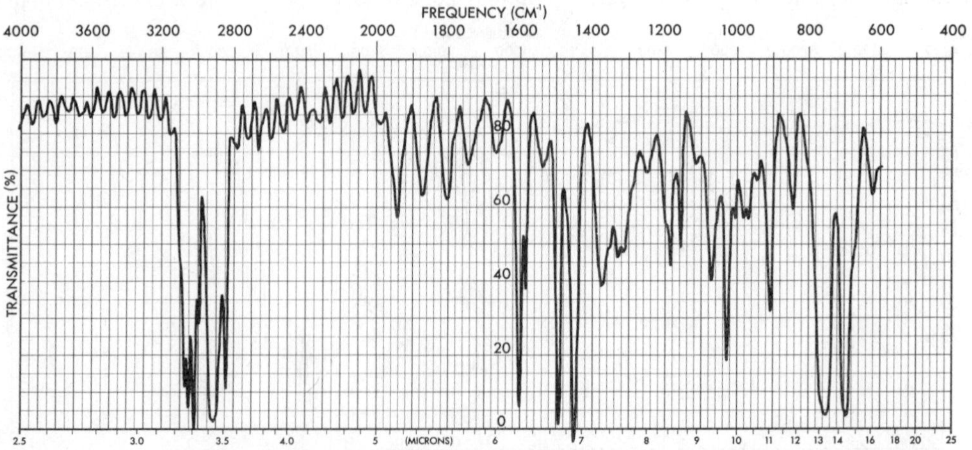
FREQUENCY (CM⁻¹)
4000 3600 3200 2800 2400 2000 1800 1600 1400 1200 1000 800 600 400
TRANSMITTANCE (%)
80 60 40 20 0
2.5 3.0 3.5 4.0 5 (MICRONS) 6 7 8 9 10 11 12 13 14 16 18 20 25

CHAPTER 3

NUCLEAR MAGNETIC RESONANCE SPECTROSCOPY PART ONE: BASIC CONCEPTS

After infrared spectroscopy, nuclear magnetic resonance is the most important spectroscopic method available to the organic chemist. While infrared spectroscopy reveals information as to the types of functional groups present in a molecule, nuclear magnetic resonance (nmr) gives information about the number of each type of hydrogen. It also gives information regarding the nature of the immediate environment of each of these types of hydrogen atoms. The combination of ir and nmr data is often sufficient to determine completely the structure of an unknown molecule.

3.1 NUCLEAR SPIN STATES

Many atomic nuclei behave as if they were spinning. In fact, any atomic nucleus which possesses either *odd mass* or *odd atomic number*, or both, has a quantized spin angular momentum and a magnetic moment. Among the more common nuclei which possess "spin," one may include $^{1}_{1}H$, $^{2}_{1}H$, $^{13}_{6}C$, $^{14}_{7}N$, $^{17}_{8}O$, and $^{19}_{9}F$. Notice that the nuclei of the ordinary (most abundant) isotopes of carbon and oxygen, $^{12}_{6}C$ and $^{16}_{8}O$, are *not* included among those with the spin property. However, the nucleus of the ordinary hydrogen atom, the proton, does have spin. For each of the nuclei with spin, the number of allowed spin states which it may adopt is quantized and is determined by its nuclear spin quantum number, I. This number is a physical constant for each nucleus. For a nucleus of spin quantum number I, there are $2I + 1$ allowed spin states which range with integral differences from $+I$ to $-I$. The individual spin states fit into the sequence:

$$-I, (-I + 1), \ldots\ldots, (I - 1), I$$

For instance, a proton (hydrogen nucleus) has the spin quantum number $I = 1/2$, and has two allowed spin states $[2(1/2) + 1 = 2]$ for its nucleus: $-1/2$ and $+1/2$. For chlorine, $I = 3/2$, there are four allowed spin states $[2(3/2) + 1 = 4]$ for its nucleus: $-3/2$, $-1/2$, $+1/2$, and $+3/2$. The spin quantum numbers of several nuclei are given in Table 3–1.

TABLE 3–1 The Spin Quantum Numbers of Some Common Nuclei

ELEMENT	$^{1}_{1}H$	$^{2}_{1}H$	$^{12}_{6}C$	$^{13}_{6}C$	$^{14}_{7}N$	$^{16}_{8}O$	$^{17}_{8}O$	$^{19}_{9}F$	$^{31}_{15}P$	$^{35}_{17}Cl$
NUCLEAR SPIN QUANTUM NO.	1/2	1	0	1/2	1	0	5/2	1/2	1/2	3/2
NUMBER OF SPIN STATES	2	3	0	2	3	0	6	2	2	4

In the absence of an applied magnetic field, all the spin states of a given nucleus are of equivalent energy (degenerate) and, given a collection of atoms, all of the spin states should be almost equally populated, with equal numbers of atoms having each of the allowed spins.

3.2 NUCLEAR MAGNETIC MOMENTS

Spin states are not of equivalent energy in an applied magnetic field because the nucleus is a charged particle, and any moving charge generates a magnetic field of its own. Thus, the nucleus has a magnetic moment (μ) generated by its charge and spin. A hydrogen nucleus may have a clockwise (+1/2) or a counterclockwise (-1/2) spin, and the nuclear magnetic moments (μ) in the two cases are in opposite directions. In an applied magnetic field, all protons will have their magnetic moments either aligned with the field or opposed to it. These two situations are illustrated in Figure 3–1.

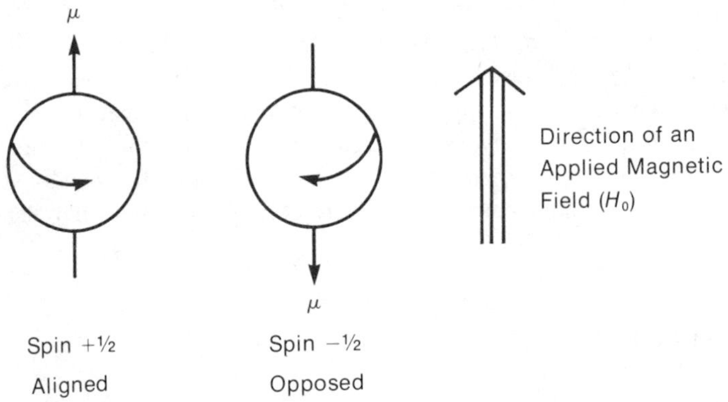

FIGURE 3–1 The Two Allowed Spin States for a Proton

Hydrogen nuclei can adopt only one or the other of these orientations with respect to the applied field. The spin state +1/2 is of lower energy because it is aligned with the field, while the spin state -1/2 is of higher energy since it is opposed to the applied field. This should be intuitively obvious to anyone who thinks a little about the two situations involving magnets depicted in Figure 3–2.

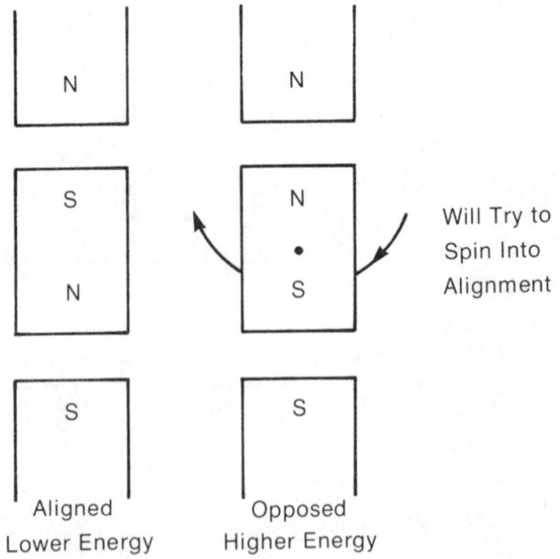

FIGURE 3–2 Aligned and Opposed Arrangements of Bar Magnets

Hence, as an external magnetic field is applied, the degenerate spin states *split* into two states of unequal energy, as shown in Figure 3–3.

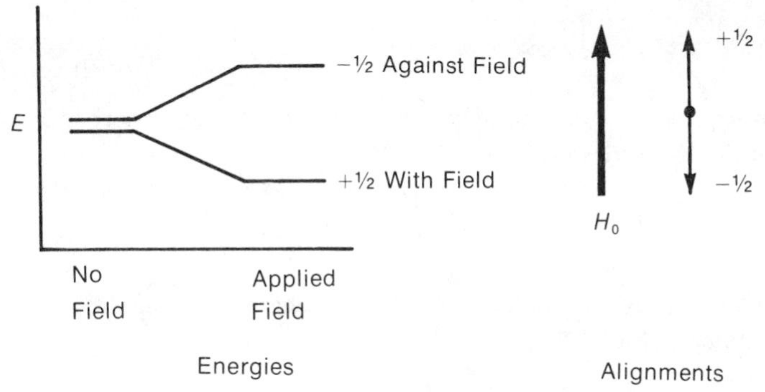

FIGURE 3–3 The Spin States of a Proton Both in the Absence and in the Presence of an Applied Magnetic Field

In the case of a chlorine nucleus there will be four energy levels, as shown in Figure 3–4. The +3/2 and −3/2 spin states are aligned with the applied field and opposed to the applied field, respectively. The +1/2 and −1/2 spin states have intermediate orientations, as indicated by the vector diagram to the right of the figure.

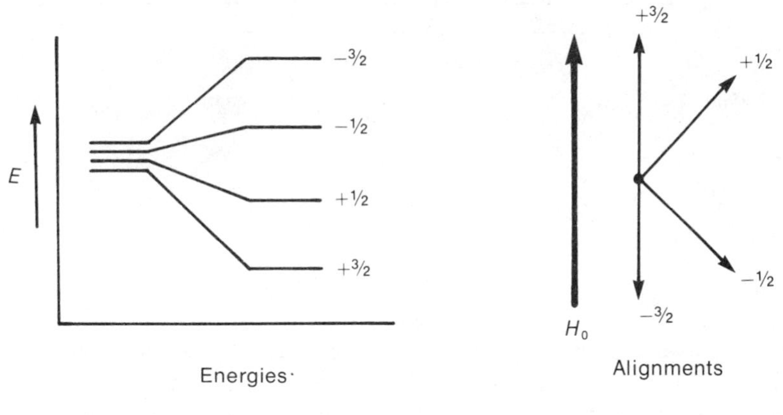

FIGURE 3–4 The Spin States of a Chlorine Atom Both in the Presence and in the Absence of an Applied Magnetic Field

3.3 ABSORPTION OF ENERGY

The nuclear magnetic resonance phenomenon occurs when nuclei aligned with an applied field are induced to absorb energy and change their spin orientation with respect to the applied field. This process is illustrated in Figure 3–5 for a hydrogen nucleus.

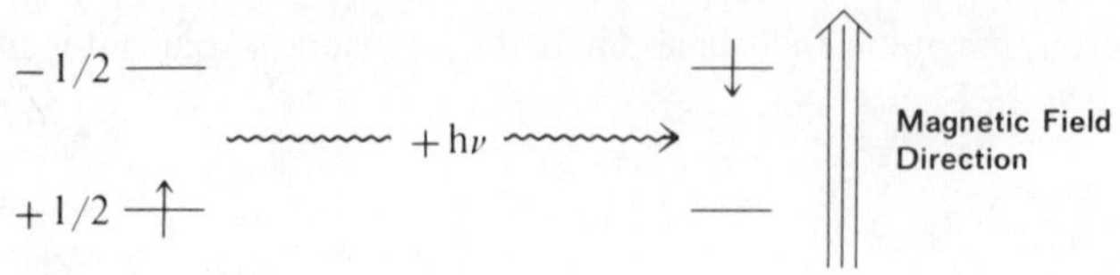

FIGURE 3–5 The NMR Absorption Process for a Proton

The energy absorption is a quantized process, and the energy absorbed must equal the energy difference between the two states involved.

$$E_{\text{absorbed}} = (E_{-1/2\text{ state}} - E_{+1/2\text{ state}}) = h\nu$$

In practice this energy difference is a function of the strength of the applied magnetic field, H_0. This dependence is illustrated in Figure 3–6.

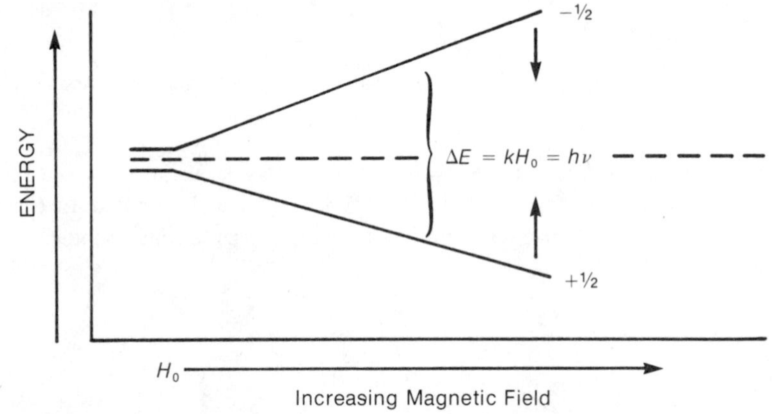

FIGURE 3–6 The Spin State Energy Separation as a Function of the Strength of the Applied Magnetic Field

The stronger the applied magnetic field, the greater the energy difference between the possible spin states:

$$\Delta E = f(H_0)$$

The magnitude of the energy level separation also depends on the particular nucleus involved. Each nucleus (hydrogen, chlorine, etc.) has a different ratio of its magnetic moment to its angular momentum, since each has different charge and mass. This ratio, called the *magnetogyric ratio*, γ, is a constant for each nucleus and determines the energy dependence on the magnetic field:

$$\Delta E = f(\gamma H_0) = h\nu$$

Since the angular momentum of the nucleus is quantized in units of $h/2\pi$, the final equation takes the form:

$$\Delta E = \gamma \left(\frac{h}{2\pi}\right) H_0 = h\nu$$

and solving for the frequency of the absorbed energy:

$$\nu = \left(\frac{\gamma}{2\pi}\right) H_0$$

If the correct value of γ for the proton is substituted, one finds that an unshielded proton should absorb radiation of frequency 42.6 MHz in a field of strength 10,000 Gauss, or radiation of frequency 60.0 MHz in a field of strength 14,100 Gauss. Table 3–2 shows the field strengths and frequencies at which several nuclei have resonance (*i.e.*, absorb energy and make transitions).

TABLE 3–2 Frequencies and Field Strengths at Which Selected Nuclei Have Their Nuclear Resonance

Isotope	Field Strength (H_0)	Frequency (ν)
^1H	10,000 Gauss	42.6 MHz
	14,100	60.0
	23,500	100.0
	51,480	220.0
^2H	10,000	6.5
^{13}C	10,000	10.7
^{19}F	10,000	40.0
^{35}Cl	10,000	4.2

Although many nuclei are capable of exhibiting magnetic resonance, the organic chemist is mainly interested in proton resonances. Accordingly, in this chapter emphasis will be placed on hydrogen. However, interest in other nuclei is currently increasing, particularly in the case of carbon-13. Certain nuclei other than hydrogen, for example deuterium, carbon-13, nitrogen, and the halogens, will be discussed in Chapter 4.

For a proton, if the applied magnetic field has a strength of approximately 14,100 Gauss, the difference in energy between its two spin states will be about 5.72×10^{-6} kcal/mole. Radiation with a frequency of about 60 MHz (60,000,000 Hz), which lies in the radiofrequency region of the electromagnetic spectrum, corresponds to this energy difference. Other nuclei have energy differences between their spin states which are both larger and smaller than those for hydrogen. The most common nuclear magnetic resonance spectrometers apply a variable magnetic field with a range of strengths near 14,100 Gauss and supply a constant radiofrequency radiation of 60 MHz. They are effectively able to induce transitions only among *proton* spin states in a molecule, and are not useful for other nuclei. Usually, separate instruments are required to observe transitions in other nuclei, although some of the newer, more expensive instruments are equipped to observe several different nuclei.

3.4 THE MECHANISM OF ABSORPTION (RESONANCE)

To understand the nature of a nuclear spin transition, the analogy of a child's spinning top is useful. Energy is absorbed by protons because of the fact that they begin to *precess* in an applied magnetic field. This phenomenon is similar to that which everyone has seen in a spinning top. Owing to the influence of the earth's gravitational field, the top begins to "wobble," or precess, about its axis. This is illustrated in Figure 3–7A.

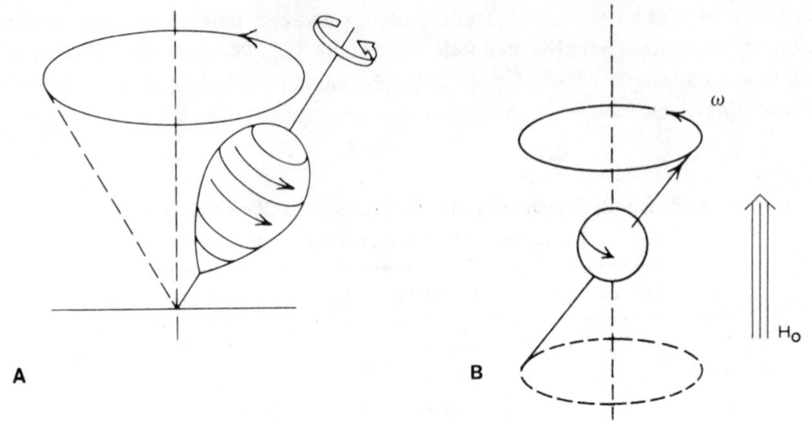

FIGURE 3–7 (A) A Top Precessing in the Earth's Gravitational Field; (B) The Precession of a Spinning Nucleus Owing to the Influence of an Applied Magnetic Field

A spinning nucleus will behave similarly under the influence of an applied magnetic field. This is illustrated in Figure 3–7B. When the magnetic field is switched on, the nucleus will begin to precess about its own axis of spin with an angular frequency ω. The frequency at which a proton will precess is directly proportional to the strength of the applied magnetic field. The stronger the applied field, the faster the rate (angular frequency ω) of precession. For a proton, if the applied field is 14,100 Gauss, the frequency of precession will be approximately 60 MHz. Since the nucleus has a charge, the precession generates an oscillating electric field of the same frequency. If radiofrequency waves of this same frequency are supplied to the precessing proton, the energy can be absorbed. That is, when the frequency of the oscillating electric field component of the incoming radiation just matches the frequency of the electric field generated by the precessing nucleus, the two fields can couple, and energy can be transferred from the incoming radiation to the nucleus, thus causing a spin change. This condition is called *resonance*, and the nucleus is said to have resonance with the incoming electromagnetic wave. The resonance process is schematically illustrated in Figure 3–8.

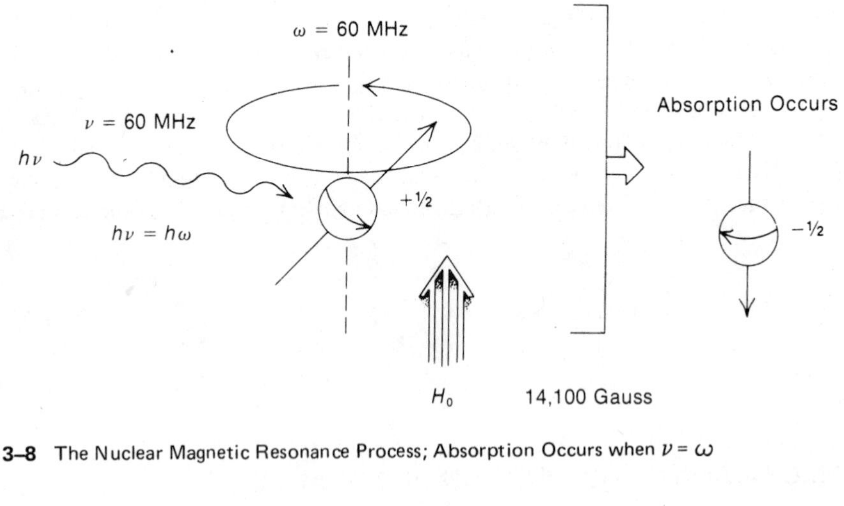

FIGURE 3–8 The Nuclear Magnetic Resonance Process; Absorption Occurs when $\nu = \omega$

3.5 THE CHEMICAL SHIFT AND SHIELDING

The great utility of nuclear magnetic resonance arises because not every proton in a molecule has resonance at identically the same frequency. This is due to the fact that the various protons in a molecule are surrounded by electrons and exist in slightly different electronic environments from one another. The protons are *shielded* by the electrons which surround them. In a magnetic field, the circulating valence electrons of the protons generate counter magnetic fields which *oppose* the applied magnetic field. Thus, each proton in a molecule is shielded from the applied magnetic field to an extent that depends on the electron density that surrounds it. The greater the electron density around a nucleus, the greater is the induced counter field which opposes the applied field. The net field which the nucleus experiences is *diminished* by the counter field which shields it. Since the nucleus experiences a lower applied magnetic field, it will precess at a lower frequency. This means that it will also absorb radiofrequency radiation at this lower frequency. Each proton in a molecule will be in a slightly different chemical environment and will, in consequence, have a slightly different amount of electronic shielding which will result in a slightly different resonance frequency. The actual instrumental method achieves resonance by a variation on this approach (see Section 3.6).

These differences in resonance frequency are very small. For instance, the difference between the resonance frequencies of the protons in chloromethane and of those in fluoromethane is only 72 Hz when the applied field is 14,100 Gauss. Since the radiation used to induce proton spin transitions at that magnetic field strength is of a frequency near 60 MHz, this difference between chloromethane and fluoromethane represents a change of frequency of only slightly more than *one part in a million*! It is very difficult to measure exact frequencies to that precision; hence, no attempt is made to measure the exact resonance frequency of any proton. Instead, a reference compound is placed in the solution of the substance to be measured, and the resonance frequency of each proton in the sample is measured *relative to* the resonance frequency of the protons of the reference substance. In other words, the frequency *difference* is measured directly. The standard reference substance which is universally used is tetramethylsilane, $(CH_3)_4Si$, also called TMS. This compound was chosen because the protons of its methyl groups are more shielded than those of most other known compounds. Thus, when another compound is measured, the resonances of its protons are reported in terms of how far (in Hz) they are shifted from those of TMS.

The number of Hz shift from TMS for a given proton will depend on the strength of the applied magnetic field. The resonance of a proton in an applied field of 14,100 Gauss is at approximately 60 MHz, whereas, in an applied field of 23,500 Gauss, the resonance appears at approximately 100 MHz. The ratio of the resonance frequencies is the same as the ratio of the two field strengths:

$$\frac{100 \text{ MHz}}{60 \text{ MHz}} = \frac{23{,}500 \text{ Gauss}}{14{,}100 \text{ Gauss}} = \frac{5}{3}$$

Hence, in the 100 MHz range (H_0 = 23,500 Gauss), the shift (in Hz) from TMS will be 5/3 larger for a given proton than it will be for the same proton in the 60 MHz range (H_0 = 14,100 Gauss). This could be confusing for workers trying to compare data if they have spectrometers that differ in the strength of the applied magnetic field. The confusion is easily overcome if one defines a new parameter that is independent of field strength. This can be done by dividing the shift in Hz of a given proton by the frequency in MHz of the spectrometer with which the shift value was obtained. In this manner a field-independent measure, called the *chemical shift* (δ), is obtained.

$$\delta = \frac{\text{(shift in Hz)}}{\text{(spectrometer frequency in MHz)}}$$

The chemical shift in δ units expresses the amount by which a proton resonance is shifted from TMS in parts per million (ppm) of the spectrometer's basic operating frequency. Values of δ for a given proton will always be the same irrespective of whether the measurement was made at 60 MHz (H_0 =

14,100 Gauss) or at 100 MHz (H_0 = 23,500 Gauss). For instance, at 60 MHz the shift of the protons in CH_3Br is 162 Hz from TMS, while at 100 MHz the shift is 270 Hz. However, both of these correspond to the same value of δ ($\delta = 2.70$):

$$\delta = \frac{162\ \text{Hz}}{60\ \text{MHz}} = \frac{270\ \text{Hz}}{100\ \text{MHz}} = 2.70\ \text{ppm}$$

By agreement, most workers report chemical shifts in delta (δ) units, or parts per million (ppm) of the main spectrometer frequency. On this scale the resonance of the protons in TMS comes at exactly 0.00 (by definition). Some workers prefer to use a related scale of chemical shift, called the tau (τ) scale. On this scale, the resonance position of TMS is defined to be 10.00. To convert δ values to τ values, one merely subtracts them from 10:

Direction of Scan ⇒

$\tau = (10 - \delta)$

12	11	10	9	8	7	6	5	4	3	2	1	0	−1	−2	δ scale
−2	−1	0	1	2	3	4	5	6	7	8	9	10	11	12	τ scale
												TMS			

The nmr spectrometer actually scans from high δ values to low ones (as will be discussed in Section 3.6), and the two scales are arranged above in the sequence in which they would be found on a typical nmr spectrum chart.

3.6 THE NUCLEAR MAGNETIC RESONANCE SPECTROMETER

The basic elements of an nmr spectrometer are schematically illustrated in Figure 3–9. The sample is dissolved in a solvent having no protons (usually CCl_4), and a small amount of TMS is added to serve as an internal reference. The sample cell is a small cylindrical glass tube which is suspended in the gap between the faces of the pole pieces of the magnet. The sample is spun around its axis to ensure that all parts of the solution experience a relatively uniform magnetic field.

Also in the magnet gap is a coil attached to a 60 MHz radiofrequency (RF) generator. This coil supplies the electromagnetic energy used to change the spin orientations of the protons. Perpendicular to the RF oscillator coil is a detector coil. When no absorption of energy is taking place, the detector coil picks up none of the energy given off by the RF oscillator coil. When the sample absorbs energy, however, the reorientation of the nuclear spins induces a radiofrequency signal in the plane of the detector coil, and the instrument responds by recording this as a resonance signal or peak.

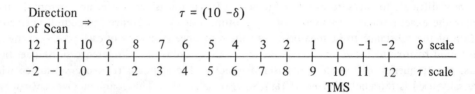

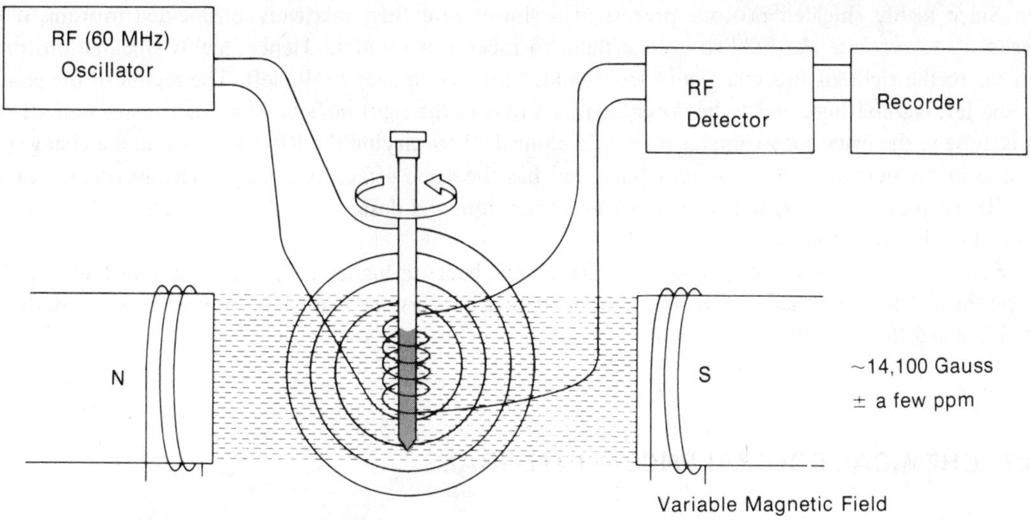

FIGURE 3–9 The Basic Elements of the Nuclear Magnetic Resonance Spectrometer

Rather than changing the frequency of the RF oscillator to bring various protons in a molecule into resonance, the typical nmr spectrometer uses a constant frequency RF signal and *varies the magnetic field strength* to bring each proton in turn into resonance. The magnet is actually a two-part device. There is a main magnet whose strength is about 14,100 Gauss, and it is capped by electromagnet pole pieces. By varying the current through the pole pieces, the main field strength can be increased by several parts per million. Changing the field in this way brings the various protons in a sample into resonance.

The field strength is increased in a linear fashion while a pen travels across a recording chart. A typical spectrum will be recorded as in Figure 3–10. As the pen travels from left to right, the magnetic field is increasing. As each chemically distinct type of proton comes into resonance, it is recorded as a peak on the chart. The peak at δ = 0 is due to the internal reference compound (TMS).

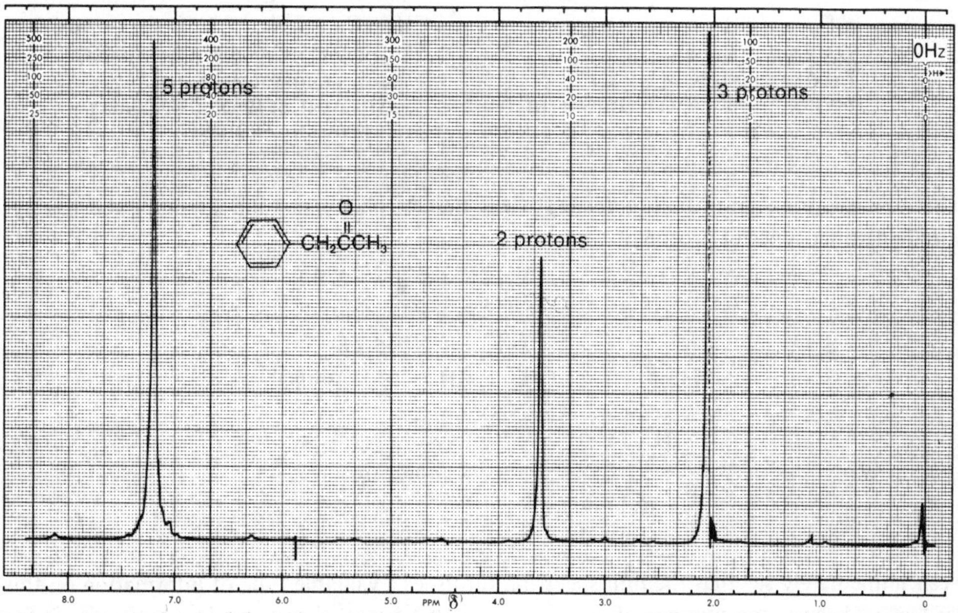

FIGURE 3–10 The Nuclear Magnetic Resonance Spectrum of Phenylacetone (the Absorption Peak at the Far Right is Caused by the Added Reference Substance, TMS)

Since highly shielded protons precess at a slower rate than relatively unshielded protons, it is necessary to *increase the field* to induce them to precess at 60 MHz. Hence, highly shielded protons appear to the right of this chart, and less shielded protons appear to the left. The region of the chart to the left is sometimes said to be *downfield*, and that to the right *upfield*. Varying the magnetic field as is done in the usual spectrometer is exactly equivalent to varying the RF frequency, and a change of 1 ppm in the magnetic field strength (increase) has the same effect as a 1 ppm change (decrease) in the RF frequency. Hence, it is only a matter of instrumental design that the field strength is changed instead of the RF frequency.

Some workers favor the τ scale over the δ scale because higher values of τ *parallel* higher field strengths and higher shielding. On the δ scale there is a *reverse* relationship between field strength or shielding and the δ value.

3.7 CHEMICAL EQUIVALENCE – INTEGRALS

All of the protons in a molecule which are in chemically identical environments will often exhibit the same chemical shift. Thus, all the protons in tetramethylsilane (TMS), or all the protons in benzene, cyclopentane, or acetone have their own respective resonance values all at the same δ value. Each compound gives rise to a single absorption peak in its nmr spectrum. The protons are said to be *chemically equivalent*. On the other hand, molecules which have sets of protons which are chemically distinct from one another may give rise to a different absorption peak from each set.

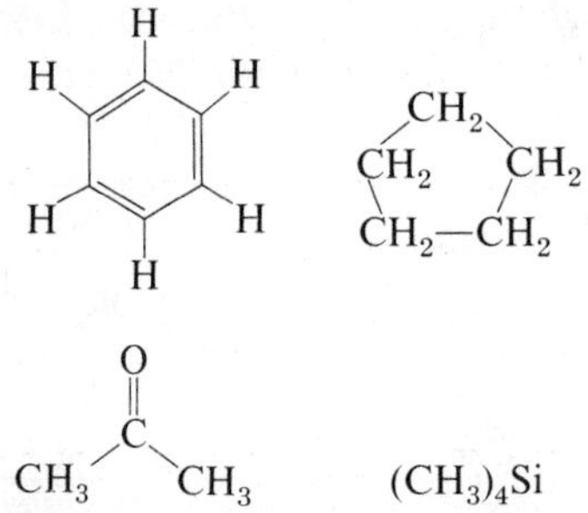

MOLECULES GIVING RISE TO ONE NMR ABSORPTION PEAK–ALL PROTONS CHEMICALLY EQUIVALENT

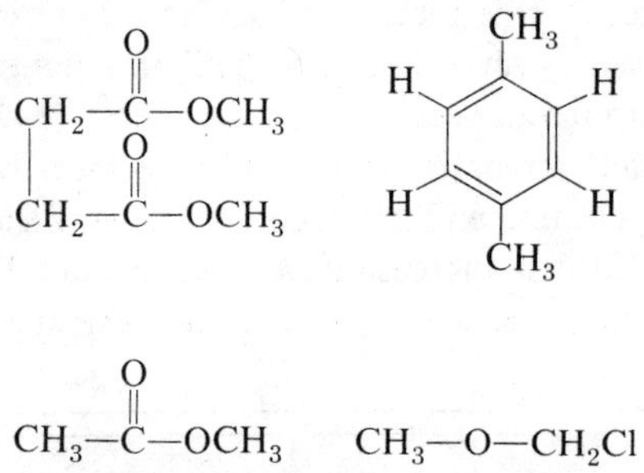

MOLECULES GIVING RISE TO TWO NMR ABSORPTION PEAKS–TWO DIFFERENT SETS OF CHEMICALLY EQUIVALENT PROTONS

The nmr spectrum given in Figure 3–10 was that of phenylacetone, a compound having *three* chemically distinct types of protons:

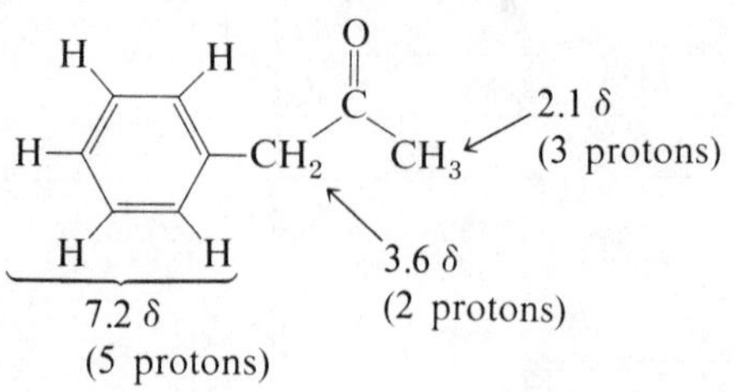

One can immediately see that the nmr spectrum furnishes a valuable type of information on this basis alone; that is, it gives the number of chemically distinct types of protons in a molecule. In fact, the nmr spectrum cannot only distinguish how many different types of protons a molecule has, but it can also reveal *how many* of each different type are contained within the molecule.

In the nmr spectrum, the area under each peak is proportional to the number of hydrogens generating that peak. Hence, in phenylacetone, the area ratio of the three peaks is 5:2:3, the same as the ratio of the numbers of the three types of hydrogen. The nmr spectrometer has the capability to electronically "integrate" the area under each peak. It does this by tracing over each peak a vertically rising line which rises in height by an amount proportional to the area under the peak. Figure 3–11 is an nmr spectrum of benzyl acetate showing each of the peaks integrated in this way.

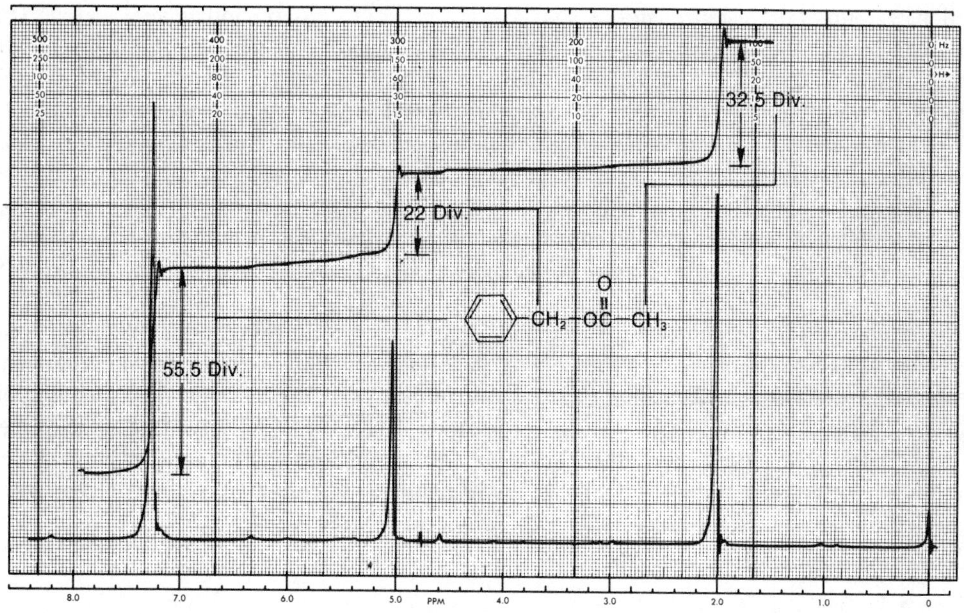

FIGURE 3–11 Determination of the Integral Ratios for Benzyl Acetate

It is important to note that the height of the integral line does not give the absolute number of hydrogens. It gives the *relative* number of each type of hydrogen. For a given integral to be of any use, there must be a second integral to which it may be referred. Benzyl acetate gives a good example of this. The first integral rises for 55.5 divisions on the chart paper, the second 22.0 divisions, and the third 32.5 divisions. These numbers are relative and give the *ratios* of the various types of protons. One can find these ratios by dividing each of the larger numbers by the smallest number:

$$\frac{55.5\text{ div}}{22.0\text{ div}} = 2.52 \qquad \frac{22.0\text{ div}}{22.0\text{ div}} = 1.00 \qquad \frac{32.5\text{ div}}{22.0\text{ div}} = 1.48$$

Thus, the number ratio of the protons of each type is 2.52 : 1.00 : 1.48. If we assume that the peak at 5.1 δ is really due to two hydrogens, and if we assume the integrals are slightly in error (this can be as much as 10 per cent), then we arrive at the true ratio by multiplying each figure by two and rounding off to 5 : 2 : 3. Clearly the peak at 7.3 δ, which integrates for 5 protons, arises from the resonance of the aromatic ring protons, while that at 2.0 δ, which integrates for 3 protons, is due to the methyl protons. The two-proton resonance at 5.1 δ arises from the benzyl protons. Notice then that the integrals give the simplest ratio, but not necessarily the true ratio, of numbers of protons of each type.

3.8 CHEMICAL ENVIRONMENT AND CHEMICAL SHIFT

If the resonance frequencies of all protons in a molecule were the same, nmr would be of little use to the organic chemist. However, not only do different types of protons have different chemical shifts, but they also have a value of chemical shift which is characteristic of the type of proton they represent. Every type of proton has only a limited range of δ values over which it gives resonance. Hence, the numerical value (in δ or τ units) of the chemical shift for a proton gives a clue as to the *type of proton* originating the signal, just as an infrared frequency gives a clue to the type of bond or functional group. For instance, notice that the aromatic protons of both phenylacetone (Figure 3–10) and benzyl acetate (Figure 3–11) have resonance near 7.3 δ and that both of the methyl groups attached directly to a carbonyl have resonance at about 2.1 δ. Aromatic protons characteristically have resonance near 7 to 8 δ, while acetyl groups (methyl groups of this type) have their resonance near 2 δ. These values of chemical shift are diagnostic. Notice also how the resonance of the benzyl ($-CH_2-$) protons comes at a higher value of chemical shift (5.1 δ) in benzyl acetate than in phenylacetone (3.6 δ). Being attached to the electronegative element oxygen, these protons are more deshielded (see Section 3.9) than those in phenylacetone. A trained chemist would have recognized readily the probable presence of the oxygen by the value of chemical shift shown by these protons.

It is important for the student to learn the ranges of chemical shifts over which the most common types of protons have resonance. In Figure 3–12 a correlation chart is presented which contains the most essential and frequently encountered types of protons. For the beginner it is often difficult to memorize a large body of numbers relating to chemical shifts and proton types. One actually need do this only crudely. It is more important to "get a feel" for the regions and the types of protons than to know a string of actual numbers. To do this, study Figure 3–12 carefully. A more detailed listing of chemical shifts is given in Appendix Two.

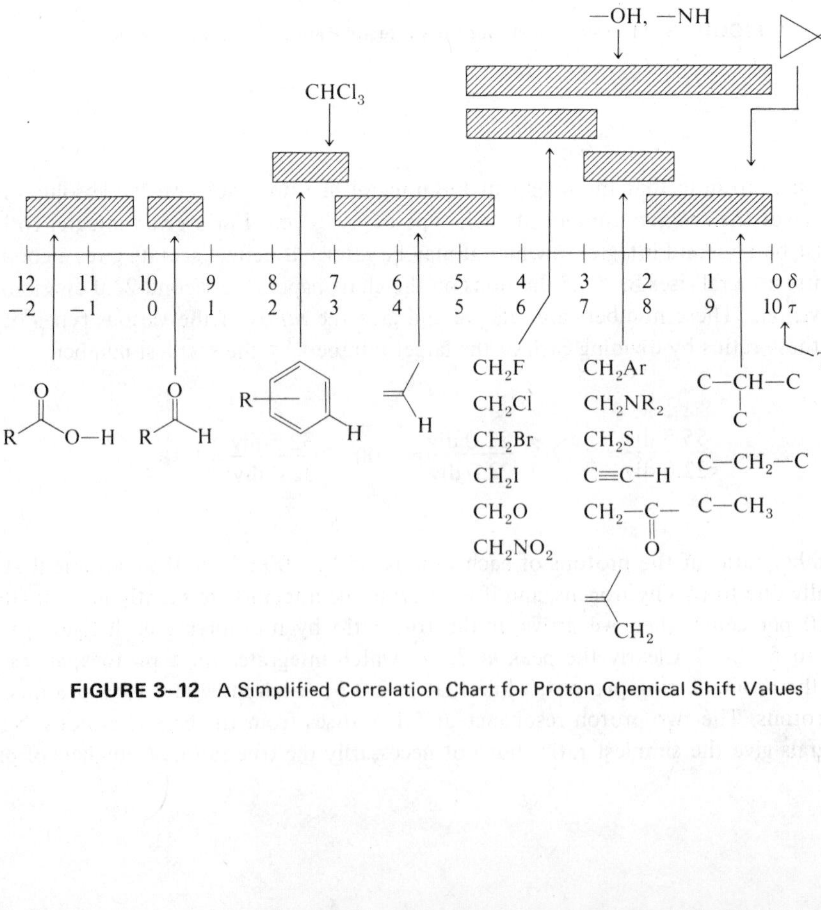

FIGURE 3–12 A Simplified Correlation Chart for Proton Chemical Shift Values

3.9 LOCAL DIAMAGNETIC SHIELDING

A. ELECTRONEGATIVITY EFFECTS

The trend of chemical shifts which is easiest to explain is that involving electronegative elements substituted on the same carbon to which the protons of interest are also attached. The chemical shift simply increases as the electronegativity of the attached element increases. This is illustrated in Table 3–3 for several compounds of the type CH_3X.

TABLE 3–3 The Dependence of the Chemical Shift of CH_3X on the Element X

COMPOUND CH_3X		CH_3F	CH_3OH	CH_3Cl	CH_3Br	CH_3I	CH_4	$(CH_3)_4Si$
ELEMENT X		F	O	Cl	Br	I	H	Si
ELECTRONEGATIVITY OF X		4.0	3.5	3.1	2.8	2.5	2.1	1.8
CHEMICAL	δ	4.26	3.40	3.05	2.68	2.16	0.23	0
SHIFT	τ	5.74	6.60	6.95	7.32	7.84	9.77	10

TABLE 3–4 Substitution Effects

	$CHCl_3$	CH_2Cl_2	CH_3Cl	$-\underline{CH_2}Br$	$-\underline{CH_2}-CH_2Br$	$-\underline{CH_2}-CH_2CH_2Br$
δ	7.27	5.30	3.05	3.30	1.69	1.25
τ	2.73	4.70	6.95	6.70	8.31	8.75

Multiple substituents have a stronger effect than a single substituent. The influence of the substituent drops off rapidly with distance, an electronegative element having little effect on protons which are more than three carbons distant. These effects are illustrated in Table 3–4 for the protons which are underlined.

The origin of the electronegativity effect was briefly discussed in Section 3.5. Electronegative substituents attached to a carbon atom, because of their electron-withdrawing effects, reduce the valence electron density around the protons attached to that carbon. These electrons, it will be recalled, *shield* the proton from the applied magnetic field. This effect, called *local diamagnetic shielding*, is illustrated in Figure 3–13. Electronegative substituents on carbon reduce the local diamagnetic shielding in the vicinity of the attached protons because they reduce the electron density around those protons. Substituents which have this type of effect are said to *deshield* the proton. The greater the electronegativity of the substituent, the more it deshields protons, and, hence, the greater is the chemical shift of those protons.

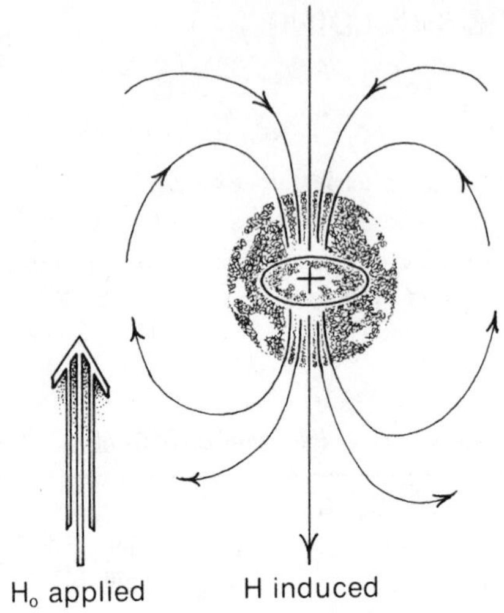

FIGURE 3–13 Local Diamagnetic Shielding of a Proton Owing to its Valence Electrons

B. HYBRIDIZATION EFFECTS

The second important set of trends is that due to differences in the hybridization of the atom to which hydrogen is attached.

"sp^3" Hydrogens

Referring to Figure 3–12, notice that all hydrogens attached to purely sp^3 carbon atoms ($C-CH_3$, $C-CH_2-C$, $C-\underset{\underset{C}{|}}{C}H-C$, cycloalkanes) have resonance in the limited range from 0 to 2 δ. At the extreme right of this range is TMS (0 δ) and hydrogens attached to carbons in highly strained rings (0–1δ), as for example with cyclopropyl hydrogens. Most methyl groups occur near 1 δ if they are attached to other sp^3 carbons. Methylene group hydrogens (attached to sp^3 carbons) appear at larger chemical shifts (near 1.2–1.4 δ) than do methyl group hydrogens. Tertiary methine hydrogens occur at higher chemical shift than secondary hydrogens, which, in turn, have a greater chemical shift than do primary or methyl hydrogens. The following diagram illustrates these relationships.

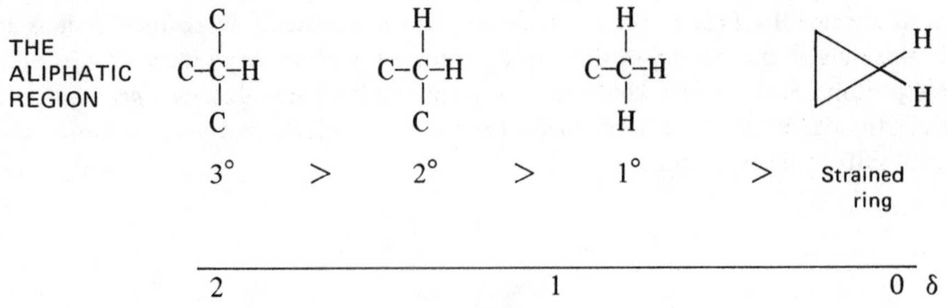

Of course, hydrogens on an sp^3 carbon which is attached to a heteroatom ($-O-CH_2-$, etc.) or an unsaturated carbon ($\diagdown C{=}\underset{|}{C}-CH_2-$) do not fall in this region, but have greater chemical shifts.

"sp^2" Hydrogens

Simple vinyl hydrogens ($=\backslash_{H}$) have resonance in the range from 4.5 to 7 δ.

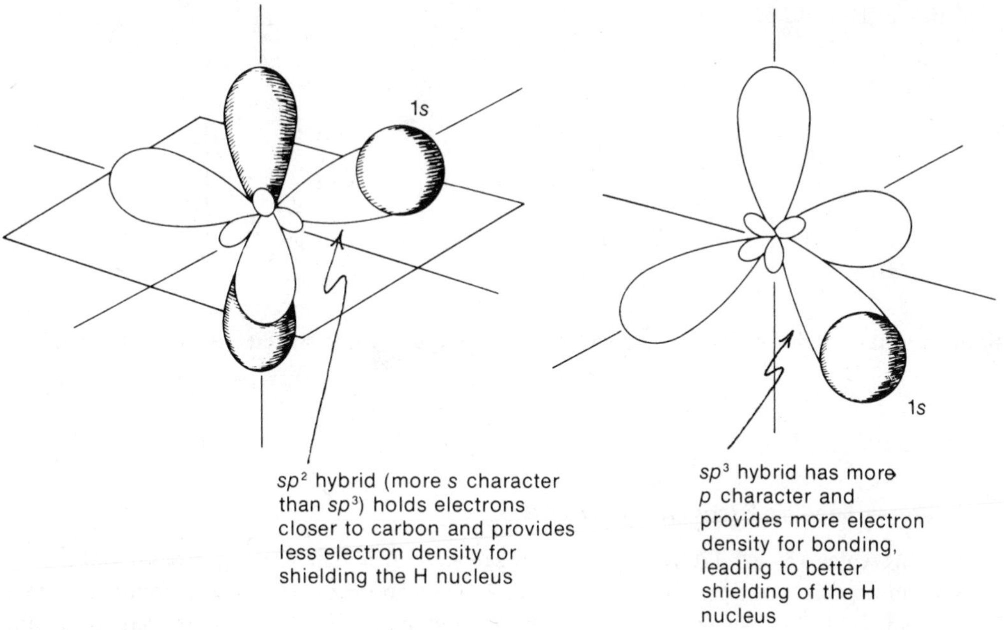
1s
1s
sp² hybrid (more s character than sp³) holds electrons closer to carbon and provides less electron density for shielding the H nucleus
sp³ hybrid has more p character and provides more electron density for bonding, leading to better shielding of the H nucleus

In an sp^2-1*s* C–H bond the carbon atom, having more *s* character, behaves as if it were more electronegative than an sp^3 carbon (*s* orbitals hold electrons closer to the nucleus than do the carbon *p* orbitals). The sp^2 carbon atom holds its electrons more tightly, resulting in less shielding for the H nucleus than in an sp^3-1*s* bond. Thus, vinyl hydrogens have a greater chemical shift than aliphatic hydrogens on sp^3 carbons. Aromatic hydrogens appear in a range even further downfield (7–8 δ). Their position is, in fact, anomalous and will be discussed below. Aldehyde protons (also attached to sp^2 carbon) are even further downfield (9–10 δ), since the inductive effect of the electronegative oxygen atom further decreases the electron density on the attached proton. Aldehydes, just as with aromatic protons and those on alkenes, also exhibit an anomalously large chemical shift, which is due to anisotropy. This latter effect is discussed in Section 3.10.

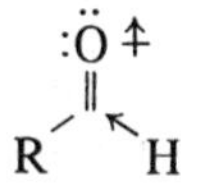

An Aldehyde

"*sp*" Hydrogens

Acetylenic hydrogens (C–H, *sp*-1*s*) appear anomalously at 2 to 3 δ owing to anisotropy, which is discussed in Section 3.10. On the basis of hybridization alone, as discussed above, one would expect the acetylenic proton to have a chemical shift *greater* than that of the vinyl proton. An *sp* carbon should behave as if it were more electronegative than an sp^2 carbon. This is the opposite of what is actually observed.

C. ACIDIC AND EXCHANGEABLE PROTONS; HYDROGEN BONDING

Acidic Hydrogens

The least shielded protons are those attached to carboxylic acids. These protons have their resonance at 10 to 12 δ.

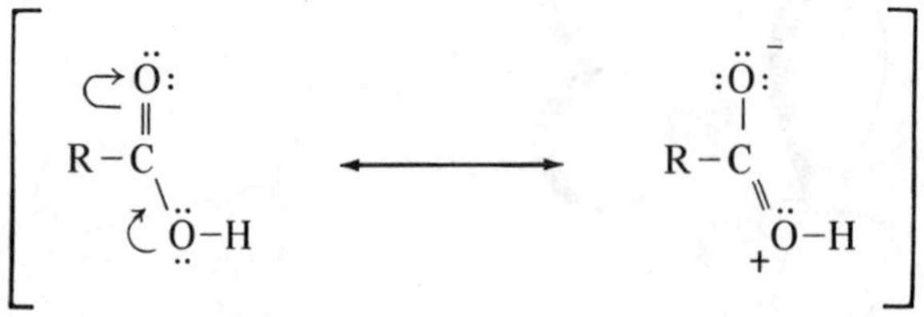

Both resonance and the electronegativity effect of oxygen withdraw electrons from the acid proton.

Hydrogen Bonding; Exchangeable Hydrogens

Protons which can exhibit hydrogen bonding (*e.g.*, hydroxyl or amino protons) exhibit extremely variable absorption positions over a wide range. These protons are usually found attached to a heteroatom. Table 3–5 lists the ranges over which a number of these types of protons are found.

TABLE 3–5 Typical Ranges for Protons with Variable Chemical Shift

Acids	RCOOH	10.5–12.0 δ
Phenols	ArOH	4.0–7.0
Alcohols	ROH	0.5–5.0
Amines	RNH_2	0.5–5.0
Amides	$RCONH_2$	5.0–8.0
Enols	CH=CH–OH	≥15

The more hydrogen bonding that takes place, the more deshielded a proton becomes. The amount of hydrogen bonding is often a function of concentration and temperature. The more concentrated the solution, the more molecules can come into contact with each other and hydrogen bond. At high dilution (no H–bonding), hydroxyl protons absorb near 0.5–1.0 δ; in concentrated solution their absorption is closer to 4–5 δ. Protons on other heteroatoms show similar tendencies.

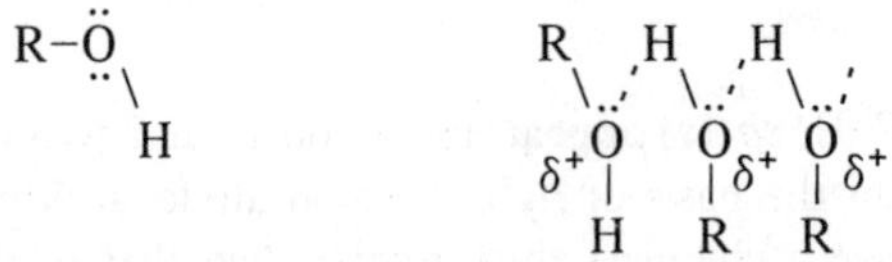

Free (Dilute Solution) Hydrogen Bonded (Concentrated Solution)

Hydrogens which can exchange either with the solvent medium or with one another also tend to be variable in their absorption positions. Possible situations are illustrated in the equations below.

$$\text{R–O–H}_a + \text{R}'\text{–O–H}_b \rightleftharpoons \text{R–O–H}_b + \text{R}'\text{–O–H}_a$$

$$\text{R–O–H} + \text{H:SOLV} \rightleftharpoons \text{R–}\overset{+}{\text{O}}(\text{H})\text{–H} + \text{:SOLV}^-$$

$$\text{R–O–H} + \text{:SOLV} \rightleftharpoons \text{H:SOLV}^+ + \text{R–O:}^-$$

All of these situations are discussed in more detail in Chapter 4.

3.10 MAGNETIC ANISOTROPY

Examination of Figure 3–12 clearly shows that there are a number of types of protons whose chemical shifts are not easily explained by simple considerations of the electronegativity of the attached groups. For instance, consider the protons of benzene or other aromatic systems. Aryl protons generally have a chemical shift which is as large as that for the proton of chloroform! Alkenes, alkynes, and aldehydes also have protons whose resonance values are not in line with the expected magnitude of any electron-withdrawing or hybridization effects. In each of these cases, the effect is due to the presence of an unsaturated system (π electrons) in the vicinity of the proton in question. Take benzene for an example. When placed in a magnetic field, the π electrons in the aromatic ring system are induced to circulate around the ring. This circulation is called a *ring current*. Moving electrons (the ring current) generate a magnetic field much like that generated in a loop of wire through which a current is induced to flow. The magnetic field which is generated covers a spatial volume large enough that it influences the shielding of the benzene hydrogens. This is illustrated in Figure 3–14.

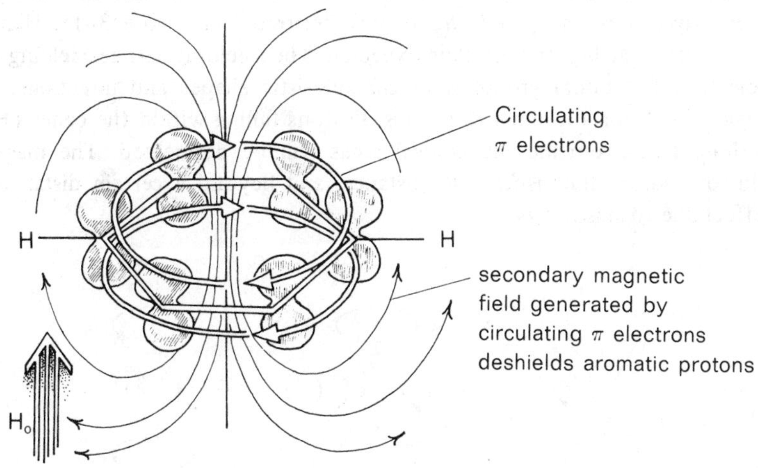

FIGURE 3–14 Diamagnetic Anisotropy in Benzene

The benzene hydrogens are said to be deshielded by the *diamagnetic anisotropy* of the ring. In electromagnetic terminology, an *isotropic* field is one of uniform density everywhere or of spherically symmetric distribution; an *anisotropic* field is “not isotropic” or non-uniform. An applied magnetic field is anisotropic in the vicinity of a benzene molecule because the labile electrons in the ring interact with the applied field. This creates a non-homogeneity in the immediate vicinity of the molecule. Thus, a proton attached to a benzene ring is influenced by *three* magnetic fields: the strong magnetic field applied by the electromagnets of the nmr spectrometer; and two weaker fields, one due to the usual shielding by the valence electrons around the proton, and the other due to the anisotropy generated by the ring system electrons. It is this anisotropic effect that gives the benzene protons a chemical shift which is greater than expected. These protons just happen to lie in a *deshielding* region of this anisotropic field. If a proton were placed in the center of the ring rather than on its periphery, it would be found to be shielded since the field lines there would have the opposite direction.

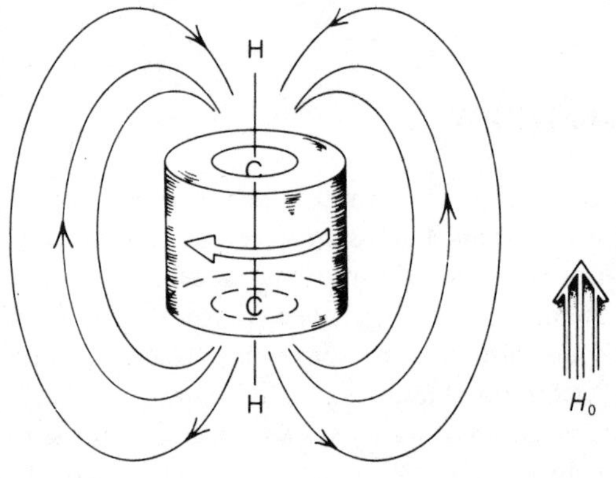

FIGURE 3–15 Diamagnetic Anisotropy in Acetylene

All groups in a molecule which have π electrons generate secondary anisotropic fields. In acetylene the magnetic field generated by induced circulation of the π electrons has a geometry such that the acetylenic hydrogens are *shielded*. This is illustrated in Figure 3–15. Hence, acetylenic hydrogens have resonance at higher field than expected. The shielding and deshielding regions due to the various π-electron functional groups have characteristic shapes and directions, and these are illustrated in Figure 3–16 for a number of groups. Protons falling within the cones (+ areas) will be shielded, and those falling outside the conical areas will be deshielded. The magnitude of the anisotropic field, of course, diminishes with distance, and beyond a certain distance there will be essentially no effect due to anisotropy.

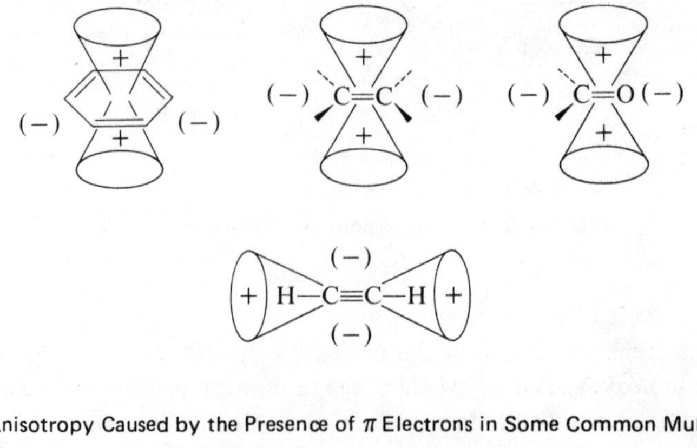

FIGURE 3–16 Anisotropy Caused by the Presence of π Electrons in Some Common Multiple Bond Systems

Figure 3–17 shows the effects of anisotropy in several actual molecules.

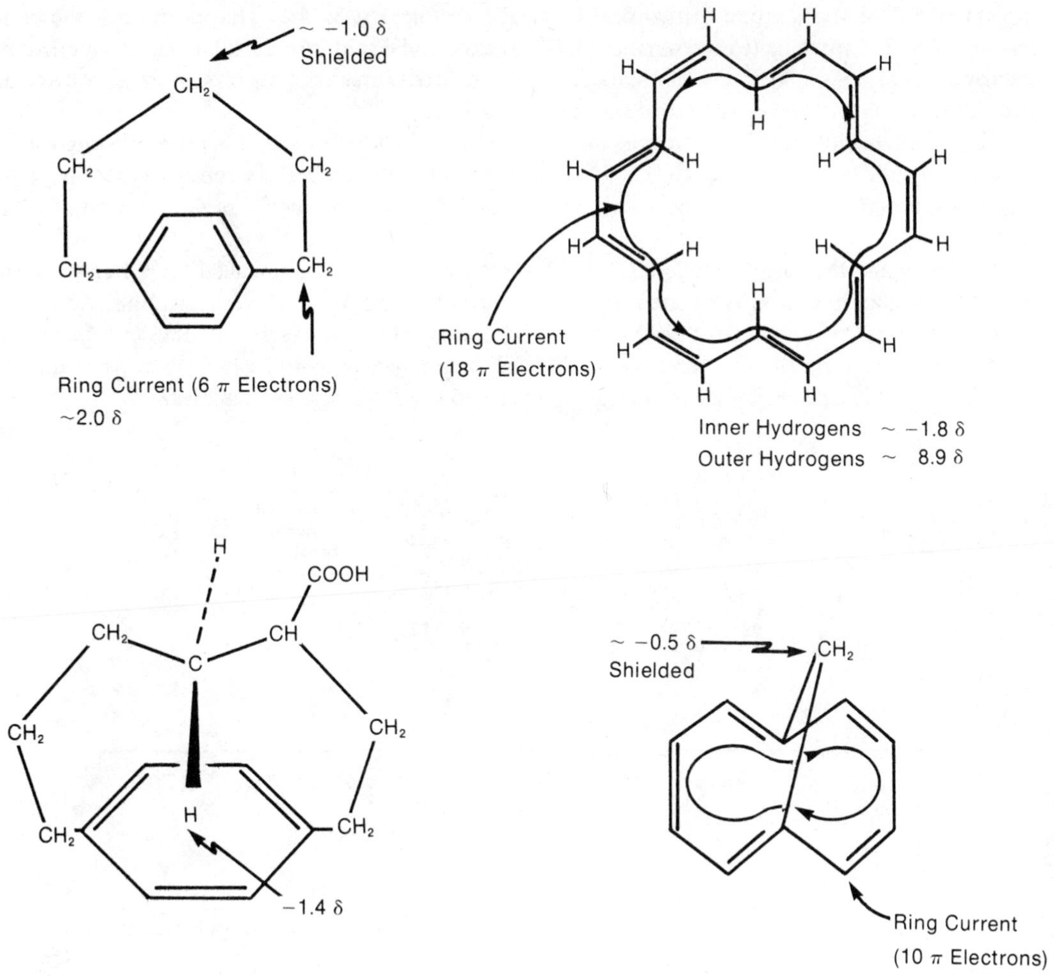

FIGURE 3-17 The Effects of Anisotropy in Some Actual Molecules

3.11 SPIN-SPIN SPLITTING (N+1) RULE

The manner in which the chemical shift and the integral (peak area) can give information about the number and type of hydrogens contained in a molecule has been discussed above. A third type of information that can be obtained from the nmr spectrum is that derived from the spin-spin splitting phenomenon. Even in simple molecules, one finds that each type of proton rarely gives a single resonance peak. For instance, in 1,1,2-trichloroethane there are two chemically distinct types of hydrogens:

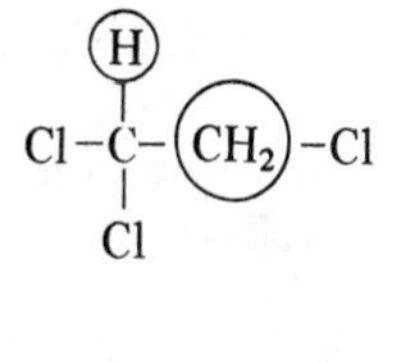

On the basis of the information given thus far, one would predict *two* resonance peaks in the nmr spectrum of 1,1,2-trichloroethane with an area ratio (integral ratio) of 2 : 1. In fact, the high resolution nmr spectrum of this compound has *five* peaks. There is a group of three peaks (called a triplet) at 5.77 δ and a group of two peaks (called a doublet) at 3.95 δ. The spectrum is shown in Figure 3–18. The methine (CH) resonance (5.77 δ) is said to be split into a triplet, and the methylene resonance (3.95 δ) is split into a doublet. The area under the three triplet peaks is *one*, relative to an area of *two* under the two doublet peaks.

This phenomenon is called *spin-spin splitting.* In an empirical fashion, spin-spin splitting can be explained by the "$n + 1$ rule." Each type of proton "senses" the number of equivalent protons (n) on the carbon atom(s) next to the one to which it is bonded, and its resonance peak is split into ($n + 1$) components.

Let's examine the case at hand, 1,1,2-trichloroethane, utilizing the so called "$n + 1$ rule." First, the lone methine hydrogen is situated next to a carbon bearing two methylene protons. According to the rule, it has two equivalent neighbors ($n = 2$) and is split into $n + 1 = 3$ peaks (a triplet). The methylene protons are situated next to a carbon bearing only one methine hydrogen. According to the rule, they have one neighbor ($n = 1$) and are split into $n + 1 = 2$ peaks (a doublet).

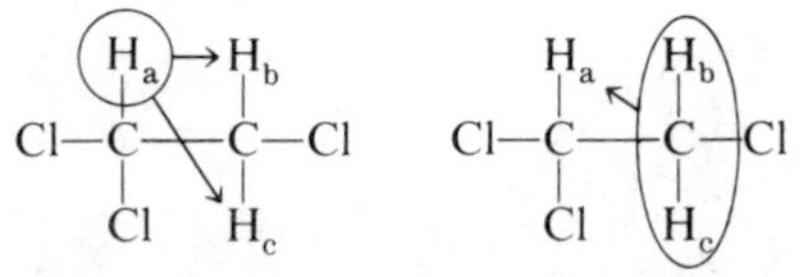

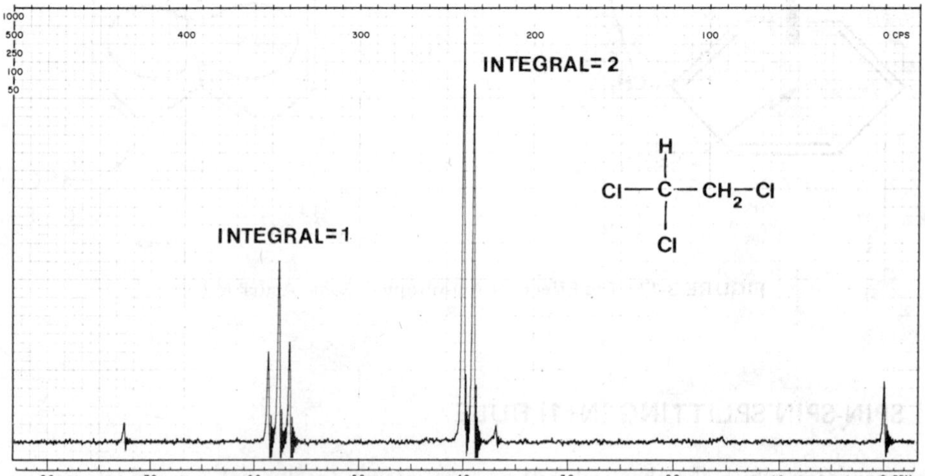

FIGURE 3–18 The NMR Spectrum of 1,1,2-Trichloroethane

Before proceeding to explain the origin of this effect, let us examine two more simple cases as predicted by the "$n + 1$ rule." The spectrum of ethyl iodide (CH_3CH_2I) is given in Figure 3–19. Notice that the methylene protons are split into a quartet (4 peaks) and that the methyl group is split into a triplet (3 peaks). This is explained as follows:

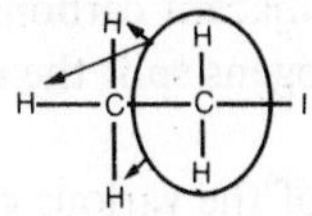

three equivalent neighbors give a quartet ($n + 1 = 4$) (area = 2)

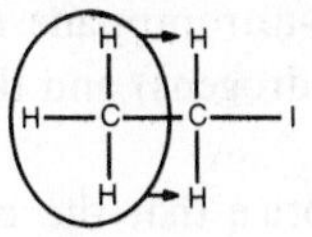

two equivalent neighbors give a triplet ($n + 1 = 3$) (area = 3)

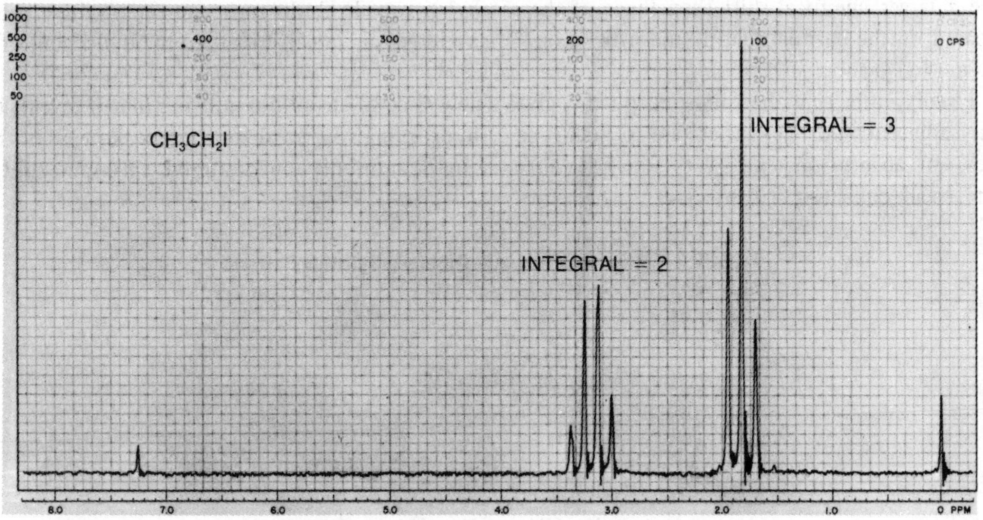

FIGURE 3–19 The NMR Spectrum of Ethyl Iodide

Finally, consider 2-nitropropane, which has the spectrum given in Figure 3–20.

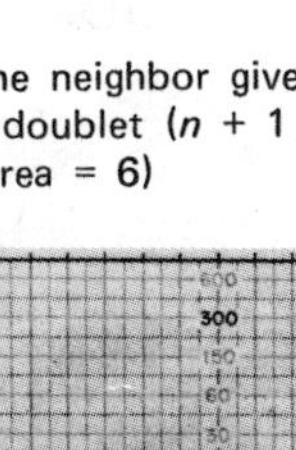

one neighbor gives a doublet ($n + 1 = 2$) (area = 6)

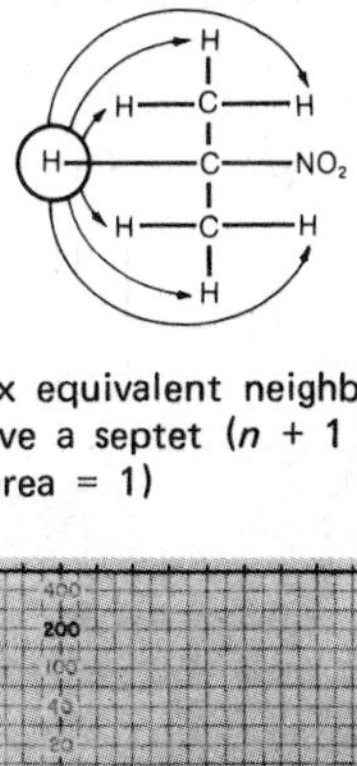

six equivalent neighbors give a septet ($n + 1 = 7$) (area = 1)

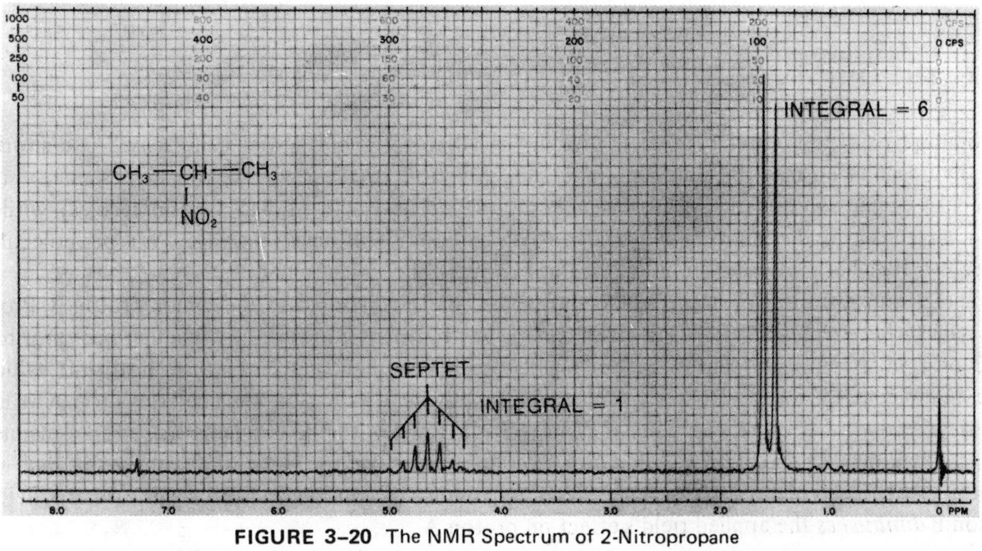

FIGURE 3–20 The NMR Spectrum of 2-Nitropropane

Notice that in the case of 2-nitropropane there are *two* adjacent carbons which bear hydrogens (two carbons, each with three hydrogens) and that all six hydrogens split the methine hydrogen *as a group* into a septet.

Before going on, also notice that the chemical shifts of the various groups of protons make sense according to the discussions in Sections 3.8 and 3.9. Thus, in 1,1,2-trichloroethane, the methine hydrogen (on a carbon bearing two Cl atoms) has a larger chemical shift than the methylene protons (on a carbon bearing only one Cl atom). In ethyl iodide the hydrogens on the carbon bearing iodine have a larger chemical shift than those of the methyl group. In 2-nitropropane the methine proton (located on the carbon bearing the nitro group) has a larger chemical shift than the hydrogens of the two methyl groups.

Finally, note that the spin-spin splitting gives a new type of structural information. It reveals how many adjacent hydrogens there are to each type of hydrogen giving an absorption peak or, as in these cases, an absorption multiplet. For reference, some commonly encountered spin-spin splitting patterns are collected in Table 3–6.

TABLE 3–6 Some Commonly Observed Splitting Patterns

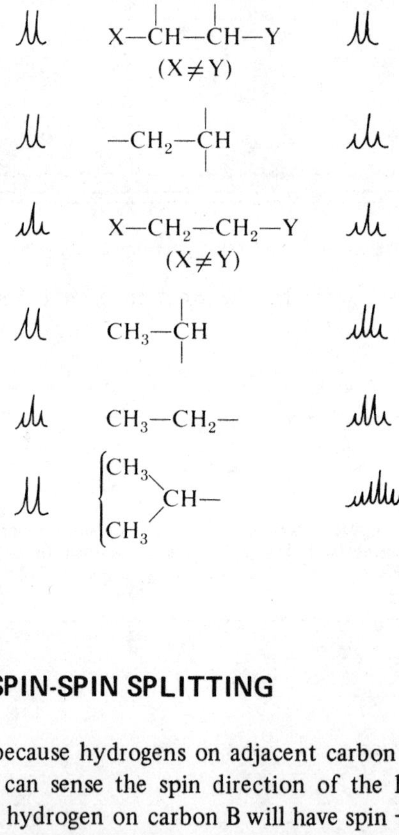

3.12 THE ORIGIN OF SPIN-SPIN SPLITTING

Spin-spin splitting arises because hydrogens on adjacent carbon atoms can "sense" one another. The hydrogen on carbon A can sense the spin direction of the hydrogen on carbon B. In some molecules of the solution the hydrogen on carbon B will have spin +1/2 (X type molecules), while in other molecules in the solution the hydrogen on carbon B will have spin −1/2 (Y type molecules). These two types of molecules are illustrated in Figure 3–21.

The chemical shift of proton A will be influenced by the direction of the spin in proton B. Proton A is said to be *coupled* to proton B. Its magnetic environment is affected by whether proton B has a +1/2 or a −1/2 spin state. Thus, proton A will absorb at a slightly different chemical shift value in type X molecules than it will in type Y molecules. In fact, in X type molecules, proton A will be slightly *deshielded* because the field of proton B is aligned with the applied field, and its magnetic moment will *add to* the applied field. In Y type molecules, proton A will be slightly shielded with respect to what its chemical shift would be in the absence of coupling. In this latter case, the field of proton B *diminishes* the applied field's effect on proton A.

Since in a given solution there will be approximately equal numbers of X and Y type molecules at any given time, *two* absorptions will be observed for proton A. Proton A's resonance is said to have been *split* by proton B, and the general phenomenon is called spin-spin splitting. Figure 3–22 summarizes the spin-spin splitting situation for proton A.

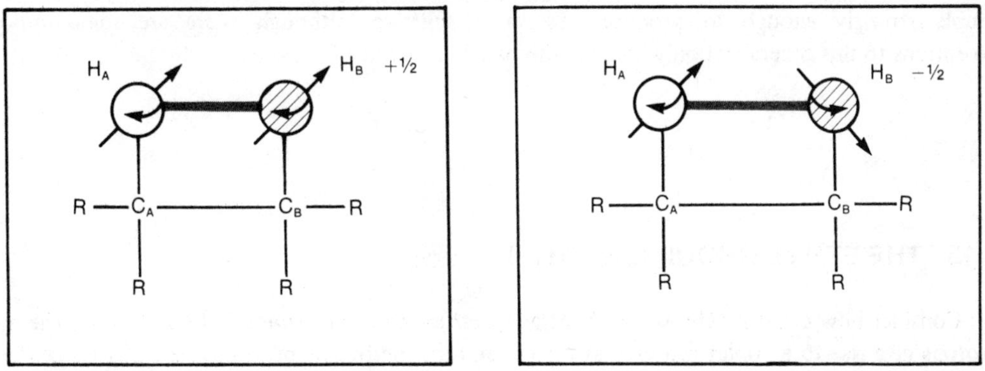

X Type Molecule Y Type Molecule

FIGURE 3–21 Two Different Molecules in a Solution With Differing Spin Relationships Between Protons H_A and H_B

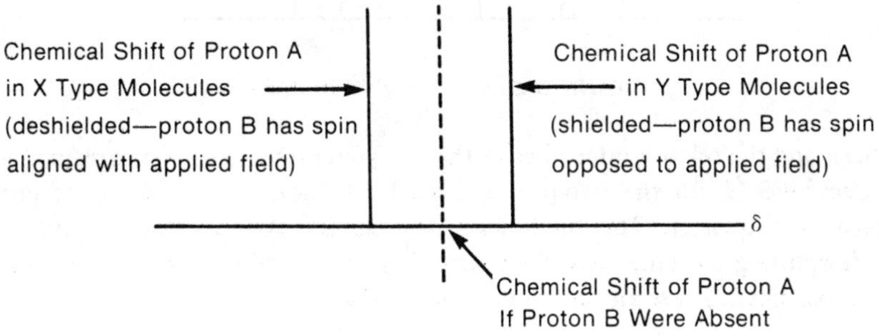

FIGURE 3–22 The Origin of Spin-Spin Splitting in Proton A's NMR Spectrum

Of course, proton B will be "split" by proton A also, since proton A can adopt two spin states as well. The final spectrum for this situation will consist of two doublets:

Two doublets will be observed for all such situations of this type, except for those in which both proton A and proton B are identical by symmetry. A situation of this type is represented in the case of the first molecule below.

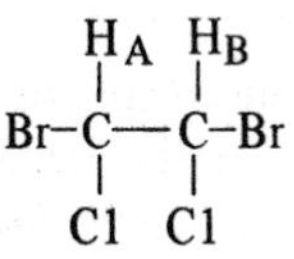

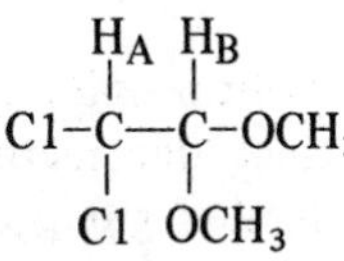

The first molecule would give only a single nmr peak, since both proton A and proton B have the same chemical shift value, and are in fact identical. The second molecule would probably exhibit the two-doublet spectrum, since protons A and B are not identical and would surely have different chemical shifts.

It should be noted that, except in unusual cases, coupling (spin-spin splitting) occurs only between hydrogens on *adjacent* carbons. Hydrogens on non-adjacent carbon atoms generally do not couple strongly enough to produce observable splitting, although there are some important exceptions to this generalization which are discussed in Chapter 4.

3.13 THE ETHYL GROUP (CH_3CH_2-)

Consider now ethyl iodide, which has the spectrum given in Figures 3–19 and 3–23. The methyl protons give rise to a triplet centered at 1.83 δ, and the methylene protons give a quartet centered at 3.20 δ.

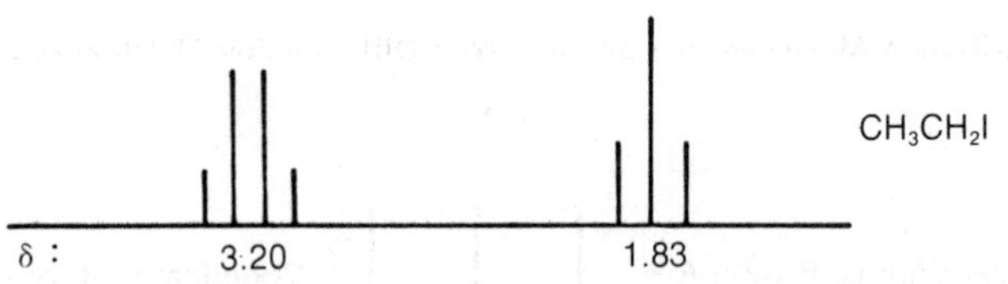

FIGURE 3–23 The Ethyl Splitting Pattern

This pattern and the relative intensities of the component peaks can be explained using the model outlined in Section 3.12 for the two-proton case. First, look at the methylene protons and their pattern, which is a quartet. The methylene protons are split by the methyl protons, and to understand the splitting pattern, the various possible spin arrangements of the protons for the methyl group must be examined. These are shown in Figure 3–24.

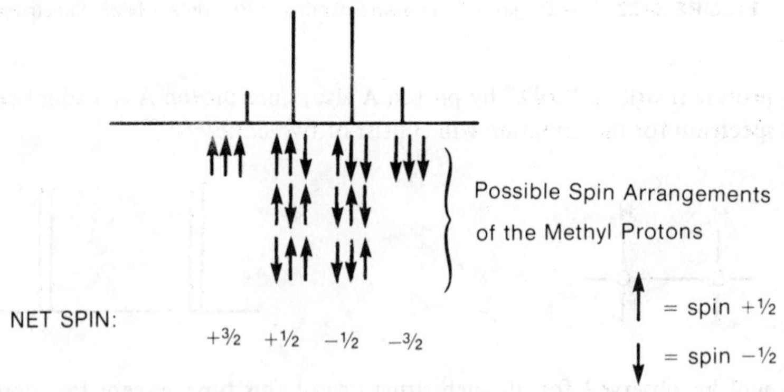

FIGURE 3–24 The Splitting Pattern of Methylene Protons Due to the Presence of an Adjacent Methyl Group

There are eight possible spin arrangements; however, some arrangements of spin are identical to one another since one methyl proton is indistinguishable from another, and since there is free rotation in a methyl group. Taking this into consideration, there are only *four* different types of arrangements. There are, however, three possible ways to obtain the arrangements with net spins of +1/2 and -1/2. Hence, these arrangements are three times more probable statistically than are the +3/2 and -3/2 spin arrangements. Thus one notes in the splitting pattern of the methylene protons that the center two peaks are more intense than the outer ones. In fact, the intensity ratios are 1 : 3 : 3 : 1. Each of these different spin arrangements of the methyl protons (except the sets of degenerate ones which are effectively identical) gives the methylene protons in that molecule a different chemical shift value. Each of the spins in the +3/2 arrangement tends to deshield the methylene proton with respect to its position in the absence of coupling. The +1/2 arrangement also deshields the methylene proton, but only slightly, since the two opposite spins cancel each other's effect. The -1/2 arrangement shields the methylene proton slightly, while the -3/2 arrangement shields the methylene proton more strongly. Keep in mind that there are actually four different "types" of molecules in the solution, each type having different methyl spin arrangements. Each type of spin arrangement causes the methylene protons to have a chemical shift in that molecule different from the chemical shift in a molecule with a different methyl spin arrangement, excepting, of course, those spin arrangements which are indistinguishable from one another (degenerate). Molecules having the +1/2 and -1/2 spin arrangement are three times more numerous in solution than those with the +3/2 and -3/2 spin arrangements.

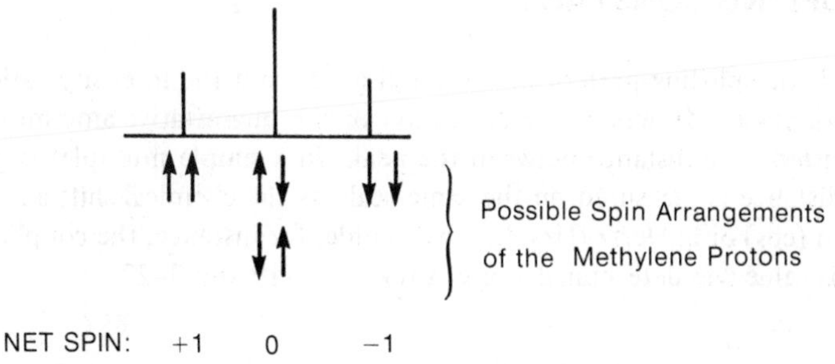

FIGURE 3–25 The Splitting Pattern of Methyl Protons Due to the Presence of an Adjacent Methylene Group

A similar analysis of the methyl splitting pattern is given in Figure 3–25, where the four possible spin arrangements of the methylene protons are shown. On examination of this figure, the origin of the triplet for the methyl group and the intensity ratios of 1 : 2 : 1 can easily be explained.

Now one can see the origin of the ethyl pattern and the explanation of its intensity ratios. The occurrence of spin-spin splitting is very important for the organic chemist, as it gives additional structural information about molecules. Namely, it reveals the *number of nearest proton neighbors* that each type of proton has. From the chemical shift, one can determine *what type of proton* is being split, and from the integral (area under the peaks) one can determine the *relative numbers* of each type of hydrogen. This is a large amount of structural information, and its use is invaluable to the chemist attempting to identify a particular compound.

3.14 PASCAL'S TRIANGLE

As can easily be verified, the intensity ratios of multiplets derived from the $n + 1$ rule follow the entries in the mathematical mnemonic device called Pascal's Triangle (Figure 3–26).

Singlet	1
Doublet	1 1
Triplet	1 2 1
Quartet	1 3 3 1
Quintet	1 4 6 4 1
Sextet	1 5 10 10 5 1
Septet	1 6 15 20 15 6 1

FIGURE 3–26 Pascal's Triangle

Each entry in the triangle is the sum of the two entries above it to the left and to the right. Notice that the intensities of the outer peaks of a multiplet such as a septet are so small compared to the inner peaks that they are often obscured in the baseline of the spectrum. In Figure 3–20 is an example of this occurrence.

3.15 THE COUPLING CONSTANT

In Section 3.13 the splitting pattern of the ethyl group and the intensity ratios of the multiplet components were discussed. However, the discussion of the quantitative amount by which the peaks were split was omitted. The distance between the peaks in a simple multiplet is called the *coupling constant*, *J*. This distance is measured on the same scale as the chemical shift and is expressed either in cycles per second (cps) or in Hertz (Hz). In ethyl iodide, for instance, the coupling constant *J* is 7.5 Hz. To see how this value was determined, consult Figures 3–19 and 3–27.

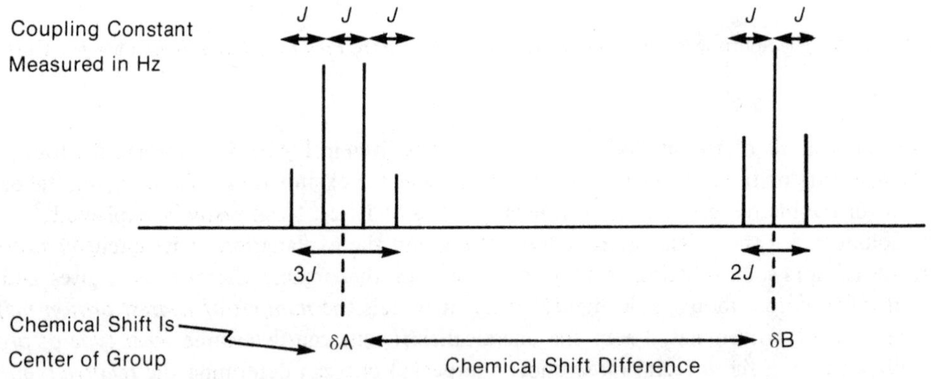

FIGURE 3–27 The Definition of the Coupling Constants in the Ethyl Splitting Pattern

The spectrum in Figure 3–19 was determined at 60 MHz and, thus, each ppm of chemical shift (δ unit) represents 60 Hz. Since there are 12 grid lines per ppm, each grid line represents 60 Hz/12 = 5 Hz. Notice the top of the spectrum. It is calibrated in "cps," which are the same as Hz, and since there are 20 chart divisions per 100 cps, one division equals 100 cps/20 = 5 cps = 5 Hz. Now examine the multiplets. The spacing between the component peaks is approximately 1.5 chart divisions, so

$$J = 1.5 \text{ div} \times \frac{5 \text{ Hz}}{1 \text{ div}} = 7.5 \text{ Hz}$$

That is, the coupling constant between the methyl and methylene protons is 7.5 Hz. When they interact, the magnitude (in ethyl iodide) will always be found to be of this same value, namely 7.5 Hz. The amount of coupling is *constant* and, hence, J can be called a *coupling constant.*

The constant nature of the coupling constant can be observed when the nmr spectrum of ethyl iodide is determined at both 60 MHz and 100 MHz. A comparison of the two spectra will indicate that the 100 MHz spectrum is greatly expanded over the 60 MHz spectrum. The chemical shift in Hz for the CH_3 and CH_2 protons will be much larger in the 100 MHz spectrum, although the chemical shifts in δ or τ units for these protons will remain identical to those in the 60 MHz spectrum. In spite of the expansion of the spectrum determined at the higher spectrometer frequency, a careful examination of the spectra will indicate that the coupling constant between the CH_3 and CH_2 protons is 7.5 Hz in both spectra! The spacing of the lines of the triplet, and of those of the quartet as well, does not expand when the spectrum of ethyl iodide is determined at 100 MHz. The extent of coupling between these two sets of protons remains constant irrespective of the spectrometer frequency at which the spectrum was determined. This is illustrated in Figure 3–28 below.

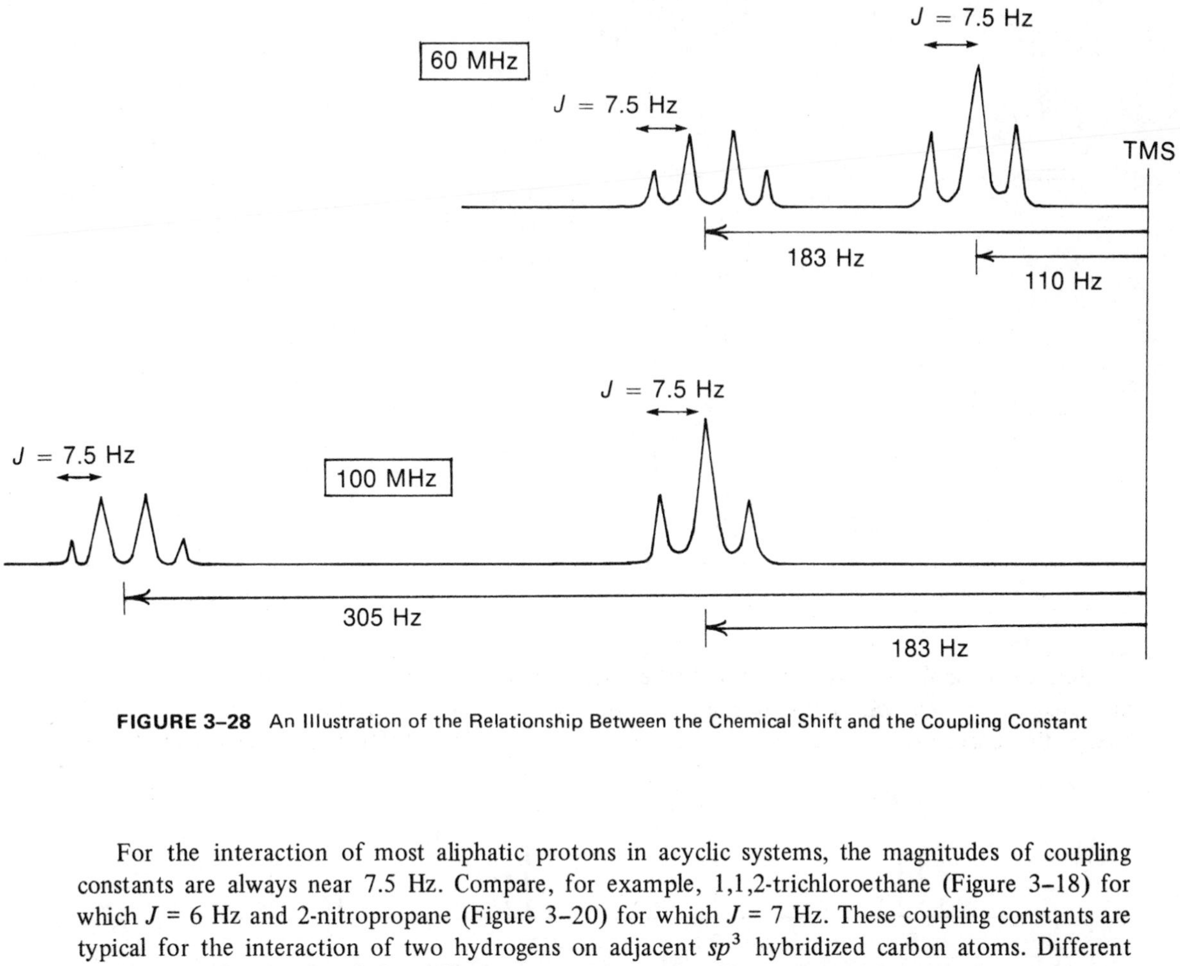

FIGURE 3–28 An Illustration of the Relationship Between the Chemical Shift and the Coupling Constant

For the interaction of most aliphatic protons in acyclic systems, the magnitudes of coupling constants are always near 7.5 Hz. Compare, for example, 1,1,2-trichloroethane (Figure 3–18) for which J = 6 Hz and 2-nitropropane (Figure 3–20) for which J = 7 Hz. These coupling constants are typical for the interaction of two hydrogens on adjacent sp^3 hybridized carbon atoms. Different magnitudes of J are found for different types of protons. For instance, the *cis* and *trans* protons substituted on a double bond commonly have values of approximately J_{cis} = 10 Hz and J_{trans} = 17 Hz. In ordinary compounds, coupling constants may range anywhere from 0 to 18 Hz. The magnitude of J will often provide structural clues. For instance, one can usually distinguish between a *cis* olefin and a *trans* olefin on the basis of the observed coupling constants for the vinyl protons. The approximate values of some representative coupling constants are given in Table 3–7. A more extensive list of coupling constants will be found in Appendix Three.

TABLE 3-7 Some Representative Coupling Constants and Their Approximate Values (Hz)

H H | | C–C	6 to 8	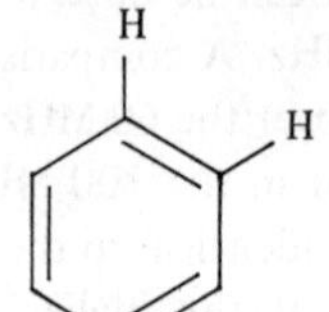	*ortho* 6 to 10	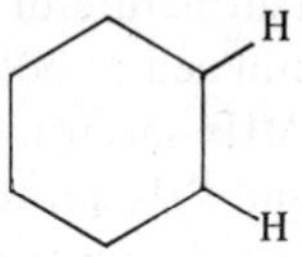	a,a 8 to 14 a,e 0 to 7 e,e 0 to 5
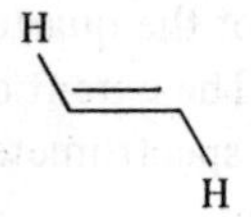	11 to 18	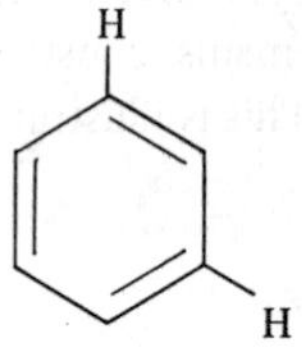	*meta* 1 to 4	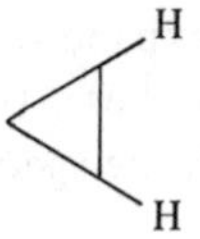	*cis* 6 to 12 *trans* 4 to 8
	6 to 15			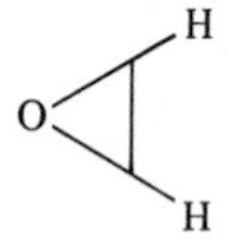	*cis* 2 to 5 *trans* 1 to 3
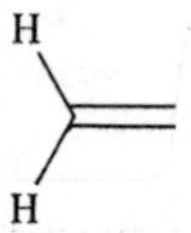	0 to 5	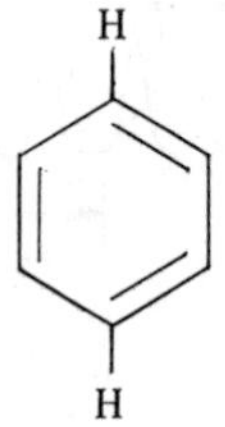	*para* 0 to 2	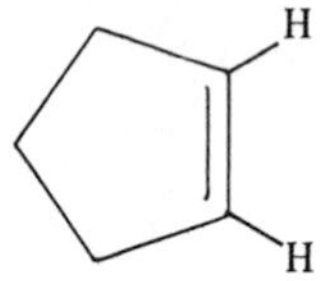	5 to 7
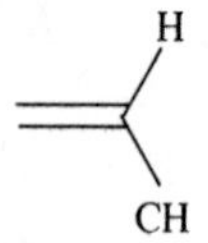	4 to 10	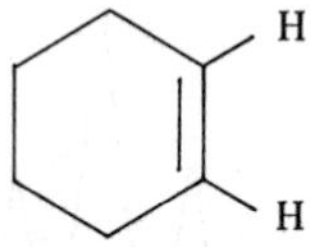	8 to 11		
H–C=C–CH	0 to 3				

Before closing this chapter, we should take note of a truism. Namely, the coupling constants of the groups of protons which split one another must be identical. This truism is extremely useful in interpreting a spectrum that may have several multiplets, each with a different coupling constant.

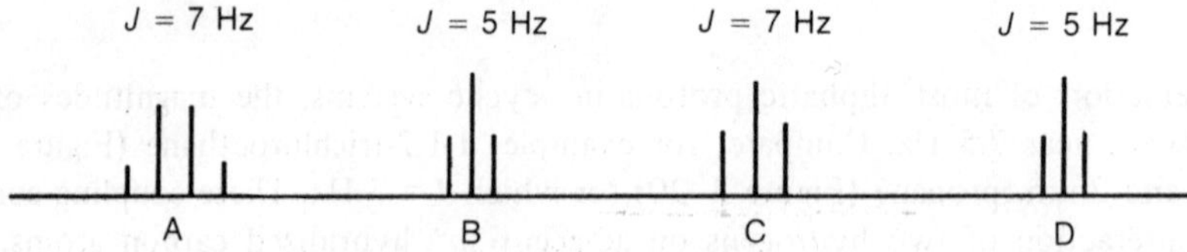

Take, for example, the spectrum shown above. There are three triplets and one quartet. Which triplet is associated with the quartet? It will, of course, be the one that has the same J values as are found in the quartet. The protons in each group interact *to the same extent*. In the example above, where the J values are given, it is obvious that quartet A ($J = 7$ Hz) is associated with triplet C ($J = 7$ Hz) and not with triplets B or D ($J = 5$ Hz). It is also obvious that triplets B and D are related to each other in the interaction scheme.

PROBLEMS

1. What are the allowed nuclear spin states for the following atoms?

   a) ^{14}N, b) ^{13}C, c) ^{17}O, d) ^{19}F.

2. Calculate the chemical shift in ppm (δ) for a proton that has resonance 128 Hz downfield from TMS on a spectrometer that operates at 60 MHz.

3. A proton has resonance 90 Hz downfield from TMS when the field strength is 14,100 Gauss and the oscillator frequency is 60 MHz.

   a) What will its shift in Hz be if the field strength is increased to 28,200 Gauss and the oscillator frequency to 120 MHz?
   b) What will be its chemical shift in ppm (δ)?

4. Acetonitrile (CH_3CN) has resonance at 1.97 δ while methyl chloride (CH_3Cl) has resonance at 3.05 δ, even though the dipole moment of acetonitrile is 3.92 D while that of methyl chloride is only 1.85 D. The larger dipole moment for the cyano group suggests that the electronegativity of this group is larger than that of the chlorine atom. Explain why the methyl hydrogens on acetonitrile are actually *more* shielded than those in methyl chloride, in contrast to the expected results based upon electronegativity. Hint: What kind of spatial pattern would you expect for the magnetic anisotropy of the cyano group (−C≡N)?

5. The position of the OH resonance of phenol varies with concentration in solution, as shown in the following table. On the other hand, the hydroxyl proton of *ortho*-hydroxyacetophenone appears at 12.05 δ and does not show any great shift upon dilution. Explain.

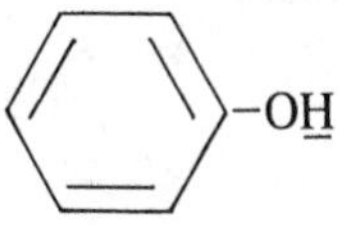

Phenol

Conc. w/v in CCl_4	δ (ppm)
100%	7.45
20%	6.75
10%	6.45
5%	5.95
2%	4.88
1%	4.37

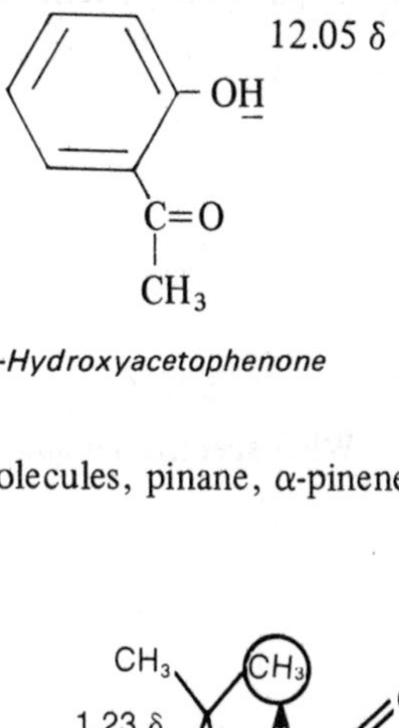

o-Hydroxyacetophenone

6. The chemical shifts of the methyl groups of three related molecules, pinane, α-pinene, and β-pinene, are given below.

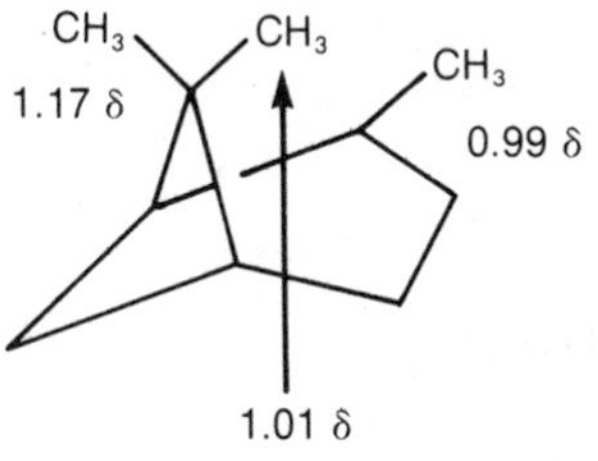

Pinane

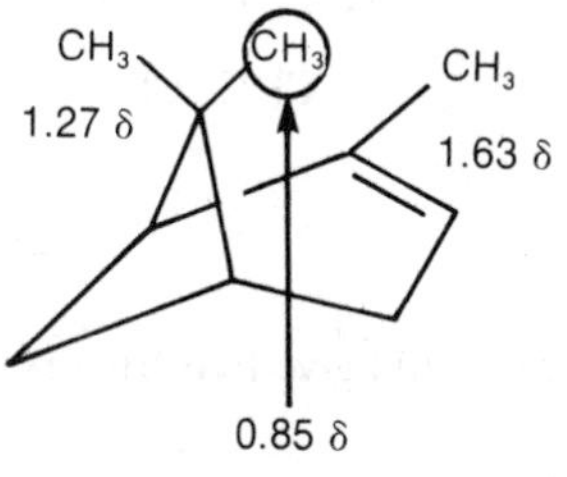

α-Pinene

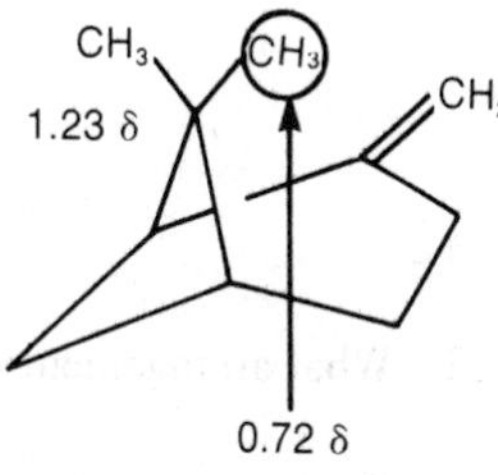

β-Pinene

Build models of these three compounds, and then give an explanation of why the two circled methyl groups have such small chemical shifts.

7. In benzaldehyde, two of the ring protons have resonance at 7.72 δ and the other three have resonance at 7.40 δ. Explain this.

8. Make a three-dimensional drawing illustrating the magnetic anisotropy in 15,16-dihydro-15,16-dimethylpyrene, and give an explanation of why the methyl groups have resonance at -4.2 δ!

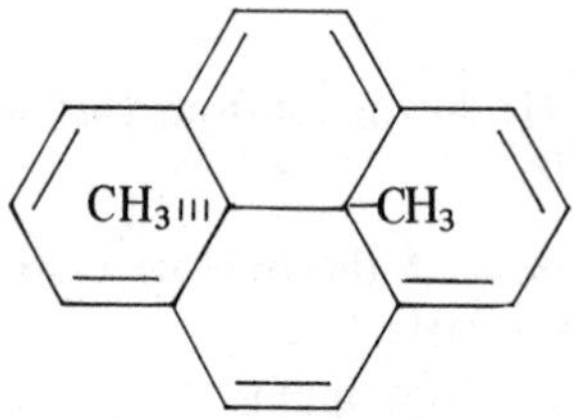

15,16-Dihydro-15,16-dimethylpyrene

9. Work out the spin arrangements and splitting patterns for the following spin system.

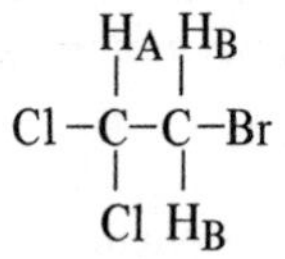

10. Explain the patterns and intensities of the isopropyl group in isopropyl iodide.

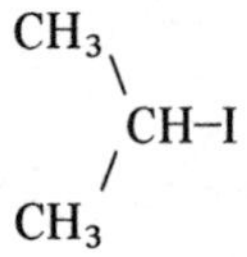

11. What spectrum would you expect for the following molecule?

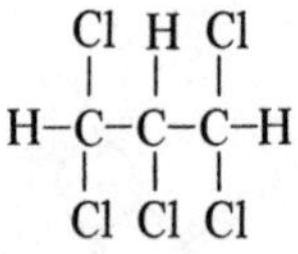

12. What arrangement of protons would give two triplets of equal area?

13. Predict the appearance of the nmr spectrum of *n*-propyl bromide.

14. The following compound, with the formula $C_4H_8O_2$, is an ester. Give its structure and assign the chemical shift values.

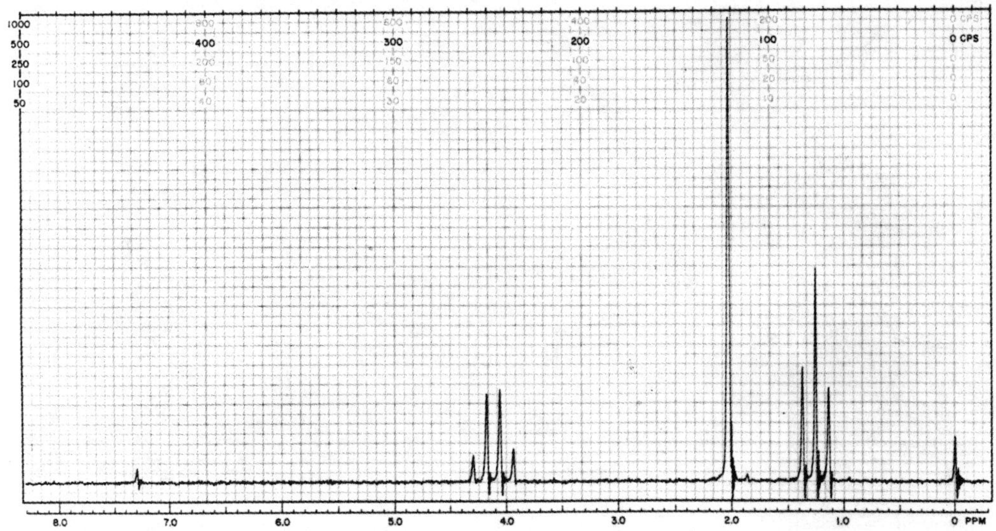

15. The following compound is a monosubstituted aromatic hydrocarbon with the formula C_9H_{12}. Give its structure and assign the chemical shift values.

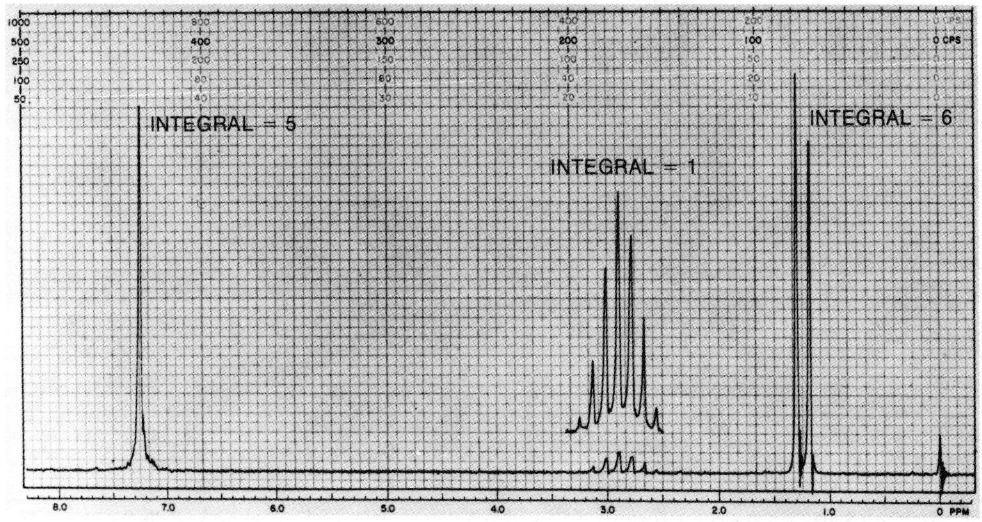

16. The functional group in this compound ($C_3H_7NO_2$) is a nitro group. What is its structure?

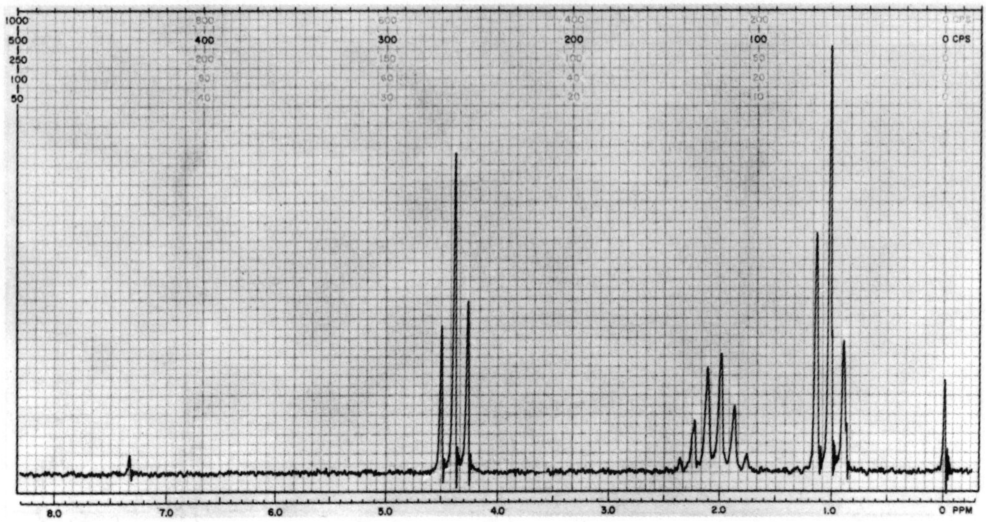

17. This compound is a carboxylic acid which contains a bromine atom: $C_4H_7O_2Br$. The peak at 10.97 δ was moved onto the chart (which runs only from 0 to 8 δ) for clarity. What is its structure?

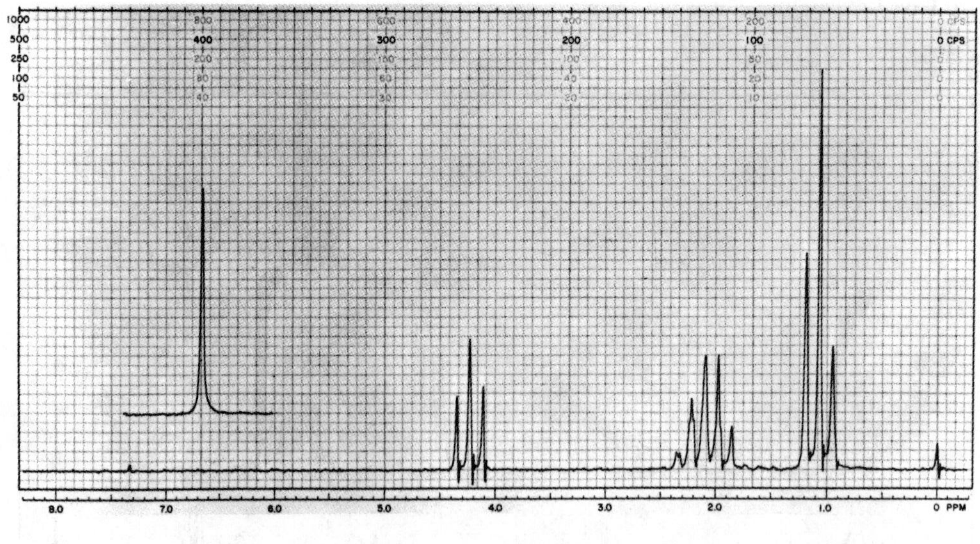

18. In each part, a molecular formula is given together with the nmr chemical shifts, peak shape, and number of hydrogens for each compound. The coupling constants are approximately 7 Hz. It is suggested that an index of hydrogen deficiency be calculated (Section 1.4); the index often gives useful information about the functional group or groups that are present in the molecule. Deduce the structure which is consistent with the nmr spectrum.

a) $C_4H_8O_2$: 1.2 δ (triplet, 3H), 2.3 δ (quartet, 2H), and 3.6 δ (singlet, 3H).
b) $C_9H_{10}O$: 1.2 δ (triplet, 3H), 3.0 δ (quartet, 2H), and 7.4 to 8.0 δ (multiplet, 5H).
c) $C_{10}H_{14}$: 1.3 δ (singlet, 9H), and 7.2 δ (multiplet, 5H).
d) $C_{10}H_{12}O_2$: 2.0 δ (singlet, 3H), 2.9 δ (triplet, 2H), 4.3 δ (triplet, 2H), and 7.3 δ (singlet, 5H).
e) C_8H_7N: 3.7 δ (singlet, 2H), and 7.2 δ (singlet, 5H).
f) $C_4H_6Cl_2O_2$: 1.4 δ (triplet, 3H), 4.3 δ (quartet, 2H), and 5.9 δ (singlet, 1H).
g) $C_7H_{14}O$: 0.9 δ (triplet, 6H), 1.6 δ (sextet, 4H), and 2.4 δ (triplet, 4H).
h) $C_5H_{10}O_2$: 1.2 δ (doublet, 6H), 2.0 δ (singlet, 3H), and 5.0 δ (septet, 1H).
i) $C_6H_{12}O_2$: 1.5 δ (singlet, 9H), and 2.0 δ (singlet, 3H).
j) $C_9H_{10}O_2$: 2.1 δ (singlet, 3H), 5.1 δ (singlet, 2H), and 7.3 δ (singlet, 5H).
k) $C_{10}H_{12}O_2$: 1.2 δ (triplet, 3H), 3.5 δ (singlet, 2H), 4.1 δ (quartet, 2H), and 7.1 δ (singlet, 5H).
l) $C_{10}H_{12}O_2$: 1.5 δ (doublet, 3H), 2.1 δ (singlet, 3H), 5.9 δ (quartet, 1H), and 7.3 δ (singlet, 5H).
m) $C_4H_8O_2$: 1.4 δ (doublet, 3H), 2.2 δ (singlet, 3H), 3.7 δ (broad singlet, 1H), and 4.3 δ (quartet, 1H). The broad singlet changes its position upon dilution.
n) $C_{10}H_{12}O$: 1.0 δ (triplet, 3H), 2.3 δ (quartet, 2H), 3.6 δ (singlet, 2H), and 7.2 δ (singlet, 5H).
o) $C_2H_4Br_2$: 2.5 δ (doublet, 3H), and 5.9 δ (quartet, 1H).
p) $C_3H_6Br_2$: 2.4 δ (quintet, 2H), and 3.5 δ (triplet, 4H).
q) C_8H_9Br: 2.0 δ (doublet, 3H), 5.0 δ (quartet, 1H), and 7.3 δ (multiplet, 5H).
r) $C_{14}H_{14}$: 2.9 δ (singlet, 4H) and 7.2 δ (singlet, 10H).
s) $C_{11}H_{17}N$: 1.0 δ (triplet, 6H), 2.5 δ (quartet, 4H), 3.6 δ (singlet, 2H), and 7.3 δ (singlet, 5H).
t) $C_3H_5ClO_2$: 1.7 δ (doublet, 3H), 4.5 δ (quartet, 1H), and 12.2 δ (singlet, 1H).
u) $C_3H_5ClO_2$: 2.8 δ (triplet, 2H), 3.8 δ (triplet, 2H), and 12.2 δ (singlet, 1H).
v) $C_{10}H_{14}$: 1.2 δ (doublet, 6H), 2.3 δ (singlet, 3H), 2.9 δ (septet, 1H), and 7.0 δ (singlet, 4H).
w) $C_7H_{12}O_4$: 1.3 δ (triplet, 6H), 3.4 δ (singlet, 2H), and 4.2 δ (quartet, 4H).
x) C_9H_8O: 3.4 δ (singlet, 4H), and 7.2 δ (singlet, 4H).
y) C_9H_{10}: 2.0 δ (quintet, 2H), 2.9 δ (triplet, 4H), and 7.2 δ (singlet, 4H).
z) $C_3H_4O_2$: 3.6 δ (triplet, 2H), and 4.3 δ (triplet, 2H).
aa) $C_5H_9BrO_2$: 1.2 δ (triplet, 3H), 2.9 δ (triplet, 2H), 3.5 δ (triplet, 2H), and 4.0 δ (quartet, 2H).
bb) $C_5H_9BrO_2$ (isomer of Problem aa): 1.3 δ (triplet, 3H), 1.8 δ (doublet, 3H), and 4.3 δ ("quintet", 3H). The latter peaks arise from two nearly overlapped quartets.
cc) $C_6H_{13}NO_2$ (contains an ester functional group): 1.3 δ (triplet, 3H), 2.4 δ (singlet, 6H), 3.2 δ (singlet, 2H), and 4.2 δ (quartet, 2H).

CHAPTER 4

NUCLEAR MAGNETIC RESONANCE SPECTROSCOPY PART TWO: MORE ADVANCED CONSIDERATIONS

In Chapter 3 only the most essential elements of nmr theory were covered. In this chapter, applications of the basic concepts to more complicated situations, including the study of other nuclei, are considered. Also considered are special techniques such as double resonance and the use of shift reagents.

4.1 THE MECHANISM OF COUPLING; COUPLING CONSTANTS

A physical picture of the way in which the spin of one proton influences that of another is not easy to develop. Although several theoretical models are available, the best theory we have is that of the Dirac vector model. This model has limitations but, nevertheless, it is substantially correct in the predictions which it develops. According to this picture, nuclear and electronic spins interact with one another, and these two particles have the lowest energy of interaction when the spin of the electron (small arrow) near the nucleus has its spin direction opposed to (or "paired" with) that of the nucleus (heavy arrow).

Spins of nucleus and electron paired or opposed (lower energy)

Spins of nucleus and electron parallel (higher energy)

In a given bond, then, it is assumed that the bonding electrons avoid each other such that when one is near nucleus A, the other is near nucleus B. According to the Pauli principle, pairs of electrons in the same orbital have opposed spins; therefore, the most stable condition in a bond where both nuclei have spin is as shown in Figure 4–1A.

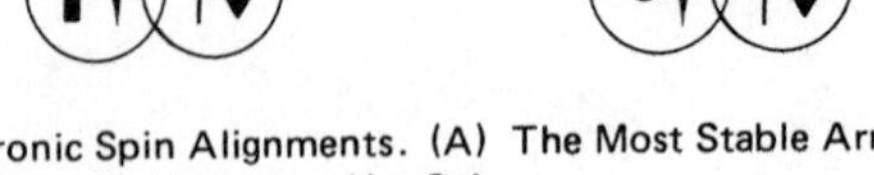

FIGURE 4–1 Nuclear and Electronic Spin Alignments. (A) The Most Stable Arrangement When Both Nuclei Have Spin. (B) A Typical C–H Bond; Only the H Nucleus Has Spin.

However, in the carbon-12 atom, the most abundant isotope of carbon, the spin quantum number is zero, and carbon exhibits no spin (see Table 3–1). Therefore, the usual spin coupling situation is as in Figure 4–1B. In a typical hydrocarbon, this spin coupling situation in one CH bond is coupled to that in adjacent CH bonds. Since the σ C–C bond is orthogonal (perpendicular) to the σ CH bonds, the electrons cannot interact through the sigma bond system. They apparently transfer the nuclear spin information via the small amount of parallel orbital *overlap* that exists between adjacent CH bond orbitals. This allows spin interaction between the electrons near the two carbon nuclei.

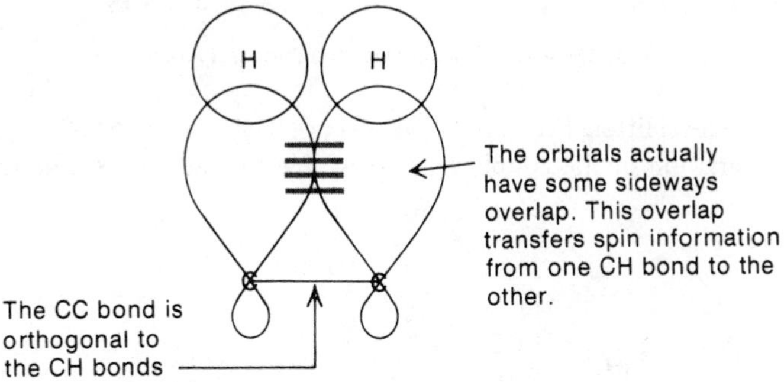

The two possible arrangements of nuclear and electronic spins for two coupled protons which are on adjacent carbon atoms are illustrated in Figure 4–2. Recall that the carbon nuclei have zero spin.

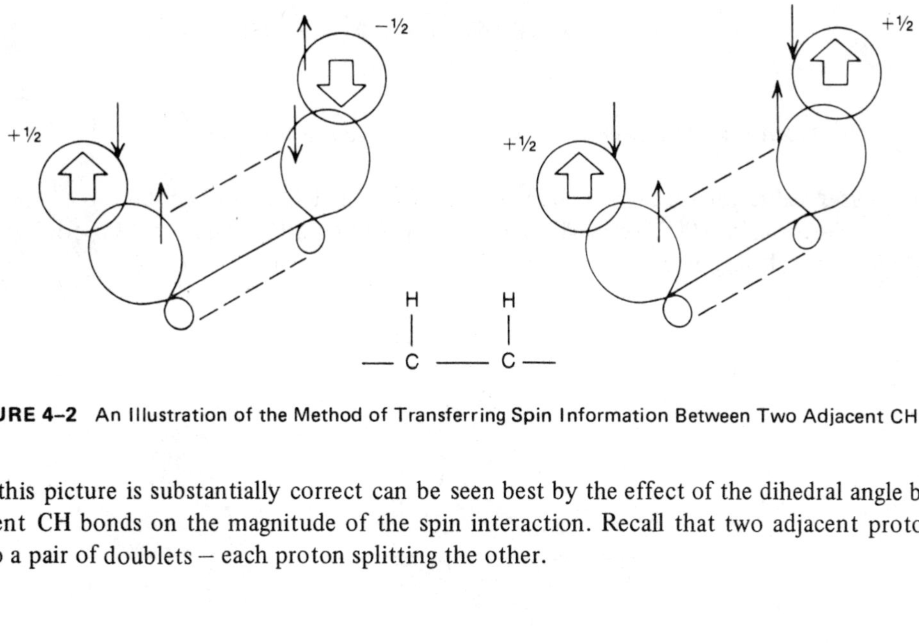

FIGURE 4–2 An Illustration of the Method of Transferring Spin Information Between Two Adjacent CH Bonds

That this picture is substantially correct can be seen best by the effect of the dihedral angle between adjacent CH bonds on the magnitude of the spin interaction. Recall that two adjacent protons give rise to a pair of doublets – each proton splitting the other.

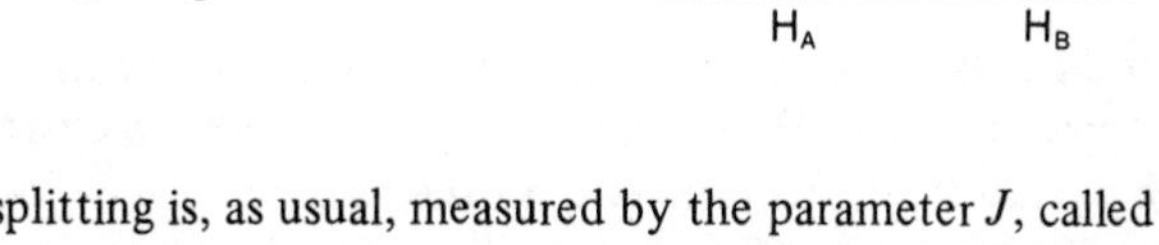

The magnitude of the splitting is, as usual, measured by the parameter J, called the *coupling constant*. The coupling constant measures the separation between the multiplet peaks in Hz (or cps). The magnitude of the coupling constant between two adjacent CH bonds can be shown to depend directly on the dihedral angle α between these two bonds. The definition of angle α is shown in Figure 4–3 in both a perspective drawing and a Newman diagram.

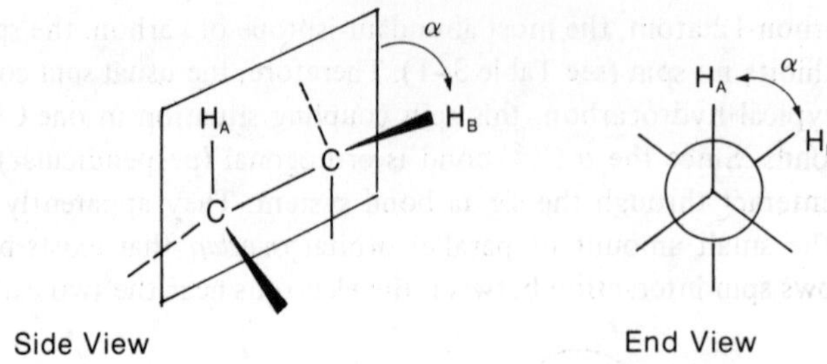

FIGURE 4–3 Definition of the Dihedral Angle α

The magnitude of the splitting between H_A and H_B is largest when $\alpha = 0°$ or $180°$, and is smallest when $\alpha = 90°$. The variation of the coupling constant J with angle α is shown in the graph in Figure 4–4.

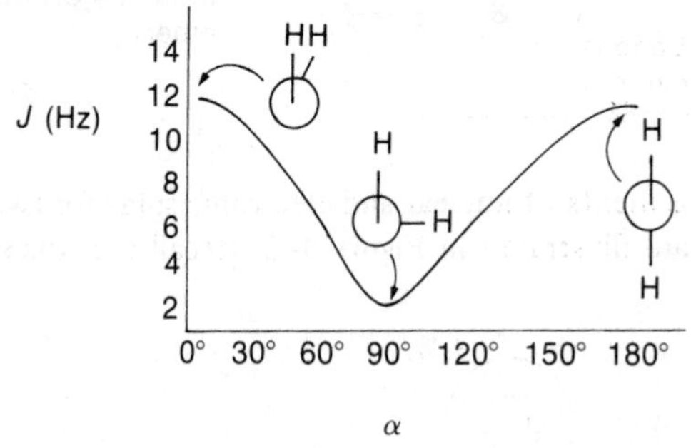

FIGURE 4–4 Variation of the Coupling Constant J with the Dihedral Angle α

This, of course, makes perfect sense according to the Dirac model, since when the two adjacent CH σ bonds are orthogonal ($\alpha = 90°$, perpendicular) there should be little or no orbital overlap and, as a result, little or no splitting of the absorption peaks.

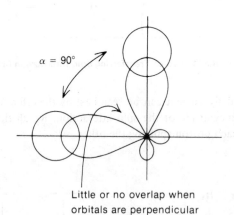

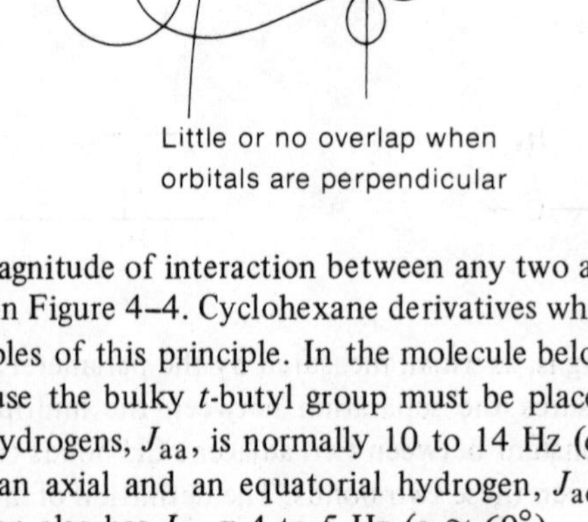

In any hydrocarbon the magnitude of interaction between any two adjacent CH bonds will always be close to those values given in Figure 4–4. Cyclohexane derivatives which are *conformationally rigid* are the best illustrative examples of this principle. In the molecule below, the ring is "locked" in the indicated conformation because the bulky *t*-butyl group must be placed equatorially. The coupling constant between two axial hydrogens, J_{aa}, is normally 10 to 14 Hz ($\alpha = 180°$), whereas the magnitude of interaction between an axial and an equatorial hydrogen, J_{ae}, is generally 4 to 5 Hz ($\alpha \cong 60°$). A diequatorial interaction also has J_{ee} = 4 to 5 Hz ($\alpha \cong 60°$).

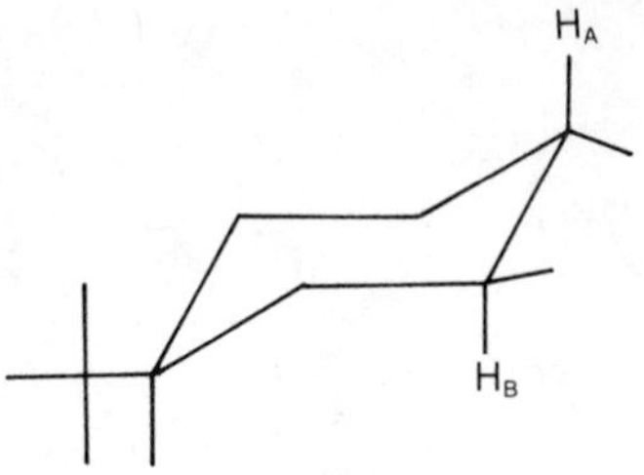

J = 10–14 Hz
α = 180°

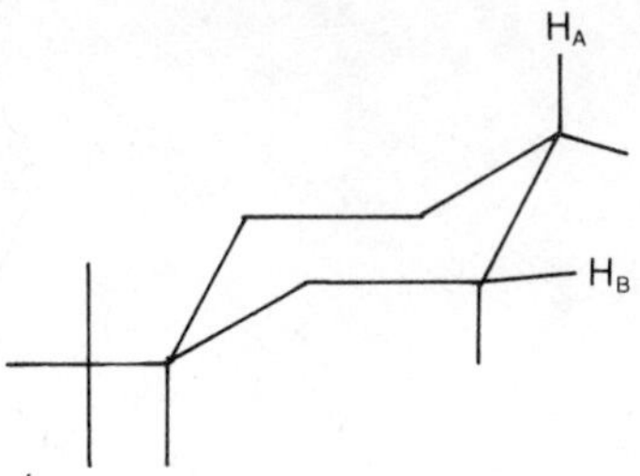

J = 4 - 5 Hz
α = 60°

Cyclopropane derivatives provide another conformationally rigid example. Notice that J_{cis} (α = 0°) is a larger coupling constant than J_{trans} ($\alpha \cong 120°$).

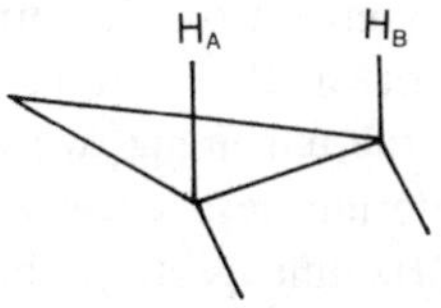

J = 10–12 Hz
α = 0°

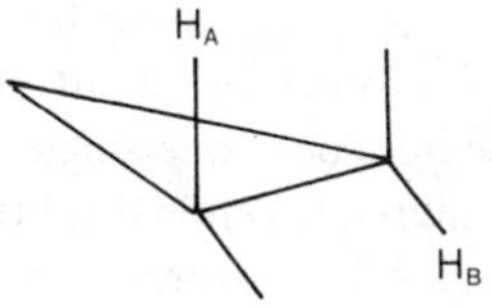

J = 7–8 Hz
$\alpha \cong$ 120°

4.2 MAGNETIC EQUIVALENCE

The coupling between protons on adjacent carbon atoms is called *vicinal coupling* (from the Latin, *vicinus*, meaning neighbor). The adjective "vicinal" refers to substituents which are 1,2-substituted. A related adjective is "geminal" (from the Latin, *geminus*, meaning twin). Geminal substituents are 1,1-substituted – that is, they are on the same carbon atom.

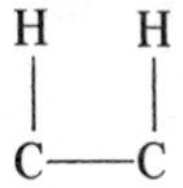

Vicinal Protons

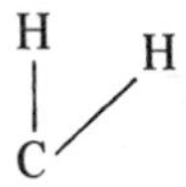

Geminal Protons

As discussed above, coupling is quite widespread between various types of vicinal protons. It is less obvious that there is also coupling between geminal protons. Normally one does not observe geminal coupling because it often does not lead to any spin-spin splitting. For instance, in methyl iodide the nmr absorption of the methyl protons is an unsplit singlet (2.16 δ). There is no spin-spin splitting observed within the group, but, nevertheless, there is coupling between the individual methyl protons. This coupling does not lead to spin-spin splitting because all three protons are equivalent by symmetry. All three protons are in an identical magnetic environment and have the same chemical shift. Such protons are said to be *magnetically equivalent*. Protons which are magnetically equivalent (have the same δ value) do not show spin-spin splitting even though they are coupled. As will be seen shortly, a second requirement for magnetic equivalence is that all the protons which are magnetically equivalent must be coupled to all other protons in the molecule by the same J values. The methyl group in methyl iodide easily meets this requirement since there are no other protons.

How can one say that coupling exists if no spin-spin splitting is observed? The answer comes from two sources. First is the fact that geminal coupling is often observed in cyclic compounds which are conformationally rigid; for instance, in cyclopropane, cyclobutane, and cyclopentane derivatives. In the cyclopropane compound shown below, the two geminal hydrogens are not magnetically equivalent.

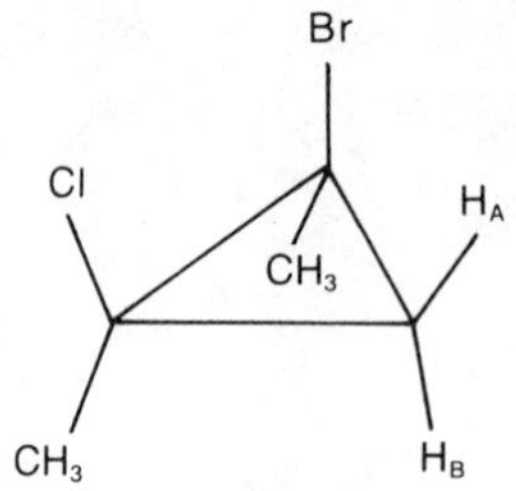

Proton H_A is on the same side of the ring as the two halogens; proton H_B is on the same side as the two methyl groups. Obviously, proton H_A will be perturbed more by the electric fields of the nearby halogens than will proton H_B. These two protons will have different chemical shifts, will couple with one another, and will show spin-spin splitting. Two doublets will be observed for H_A and H_B. For cyclopropane rings, $J_{geminal}$ is usually around 5 Hz.

The second source of information about geminal coupling comes from deuterium substitution experiments. In methyl iodide, only one singlet is observed for the methyl protons. However, if one proton is replaced by deuterium, geminal splitting is observed; the remaining two protons are split into a triplet (deuterium, $I = 1$, has three spin states). The deuterium resonance, which takes place near 9.22 MHz ($H_0 = 14{,}100$ Gauss), is not observed in the 60 MHz nmr spectrum, but is found to be a triplet (two adjacent hydrogens) when observed under the appropriate conditions. Here we have a clear case of geminal coupling between the deuterium and hydrogen substituents. This coupling is illustrated in Figure 4–5.

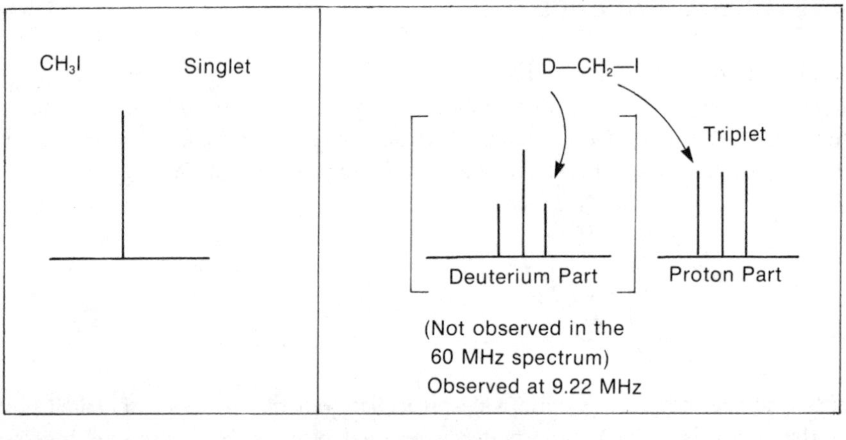

FIGURE 4–5 A Comparison of the Spectra of Methyl Iodide and its Monodeuterio Analog

The amount of coupling that takes place between hydrogen and deuterium, J_{HD}, can be measured easily. From theory and empirical data, it is known that the ratio of two coupling constants J_{HH} (proton–proton) and J_{HN} (proton–other nucleon) is given by the ratio of the magnetic moments of the two nuclei. From this relationship, one can calculate the magnitude of J_{HH} geminal from a measurement of J_{HD} geminal. Experiment has shown that $J_{HH} = 6.55 \times J_{HD}$, and thus we have the relationship:

$$\frac{J_{HH}}{J_{HD}} = \frac{\mu_H}{\mu_D} = 6.55$$

Most geminal coupling constants which are measured by this method are on the order of 12 to 15 Hz. Thus, the methyl protons are strongly coupled even though no spin-spin splitting is observed. One should, therefore, not confuse the term *spin-spin splitting* with *coupling.* Coupling is necessary, but not sufficient, for spin-spin splitting to occur. If coupled protons can be made magnetically equivalent either via a symmetry element of the molecule or through rotation or other rapid conformational changes, no spin-spin splitting will be observed.

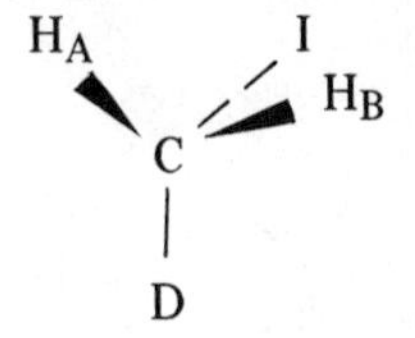

Now let us take a moment to analyze the spectrum of DCH_2I. The hydrogen nuclei are magnetically different from the deuterium nucleus; therefore, coupling between D and H results in spin-spin splitting. However, protons H_A and H_B are chemically and magnetically equivalent, and thus show no spin-spin splitting even though they are coupled to one another. The coupling constants J_{H_AD} and J_{H_BD} are equivalent, and the methylene protons behave as a single group. The two requirements for magnetic equivalence are handily met:

a) identical chemical shifts;
b) equal coupling constants to all other protons (in this case, deuterium).

Deuterium has three possible spin states (+1, 0, −1) since it has $I = 1$ (see Table 3–1). Thus, the proton part of the nmr spectrum is split into a triplet with all three component peaks of *equivalent intensity*, since the three spin states of deuterium have approximately equivalent statistical probability. The deuterium resonance (at 9.22 MHz) is split into the usual triplet, with intensity ratios 1 : 2 : 1, by the methylene protons. Note that the $n + 1$ rule does not apply for this compound, since deuterium has a different spin than hydrogen.

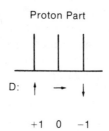

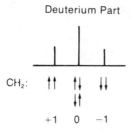

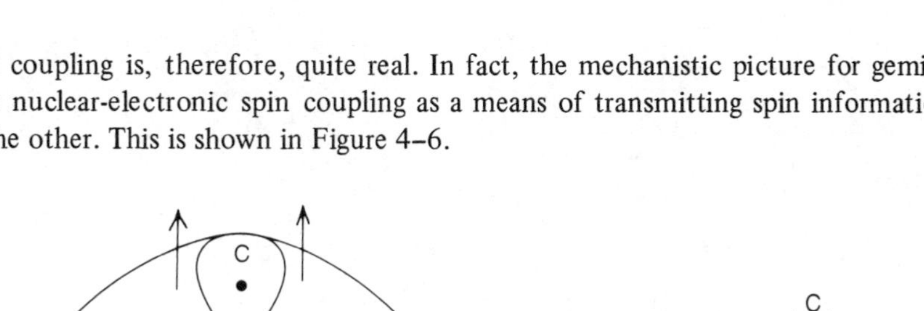

Geminal coupling is, therefore, quite real. In fact, the mechanistic picture for geminal coupling also invokes nuclear-electronic spin coupling as a means of transmitting spin information from one nucleus to the other. This is shown in Figure 4–6.

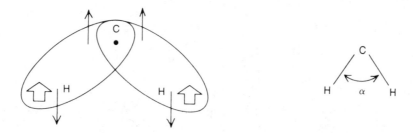

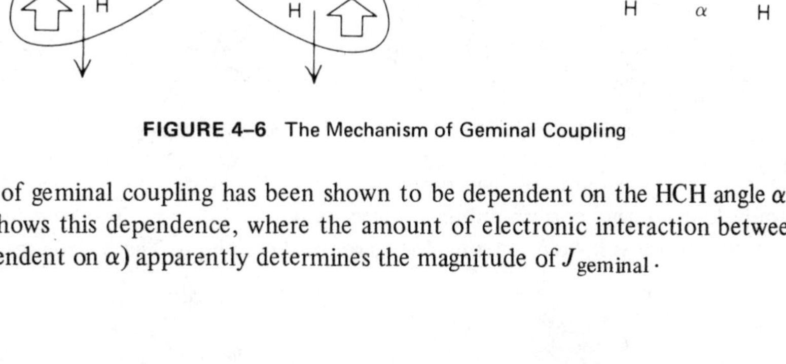

FIGURE 4–6 The Mechanism of Geminal Coupling

The amount of geminal coupling has been shown to be dependent on the HCH angle α. The graph in Figure 4–7 shows this dependence, where the amount of electronic interaction between the two CH orbitals (dependent on α) apparently determines the magnitude of $J_{geminal}$.

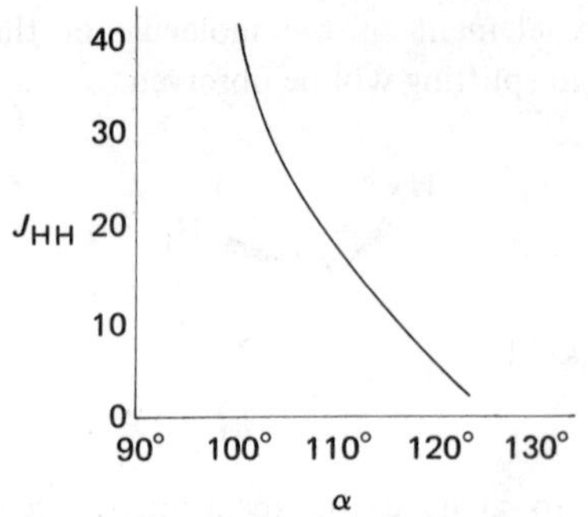

FIGURE 4–7 The Dependence of the Magnitude of the Geminal Coupling Constant on the HCH Bond Angle α

Some typical values for geminal coupling constants are given below.

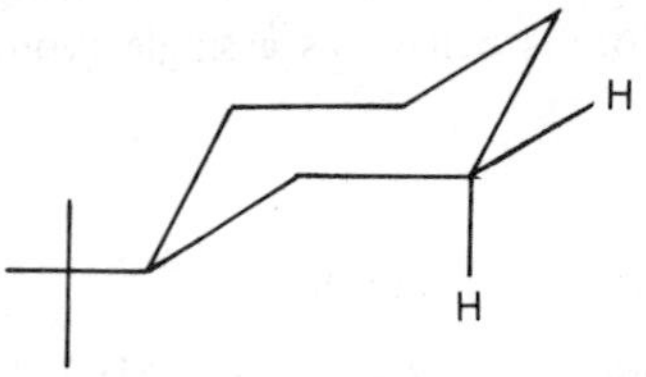

$\alpha \cong 109°$

$J_{HH} \cong$ 12–18 Hz

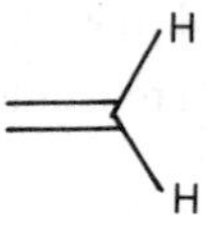

$\alpha \cong 120°$

$J_{HH} \cong$ 0–3 Hz

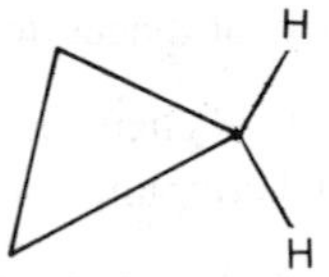

$\alpha \cong 118°$

$J_{HH} \cong$ 5 Hz

4.3 FAILURE OF THE N + 1 RULE DUE TO MAGNETIC NON-EQUIVALENCE WITHIN A GROUP

When the protons attached to a single carbon are magnetically non-equivalent (have different chemical shifts), the $n + 1$ rule no longer applies. That is, if one has a geminal coupling that leads to spin-spin splitting, the simple rule begins to fail. Let us examine two cases, one in which the rule applies (trimethylene oxide) and one in which it fails (styrene oxide). The spectrum of trimethylene oxide is shown in Figure 4–8.

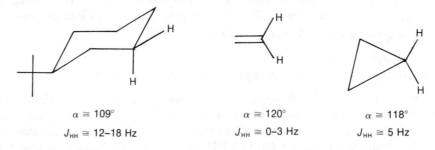

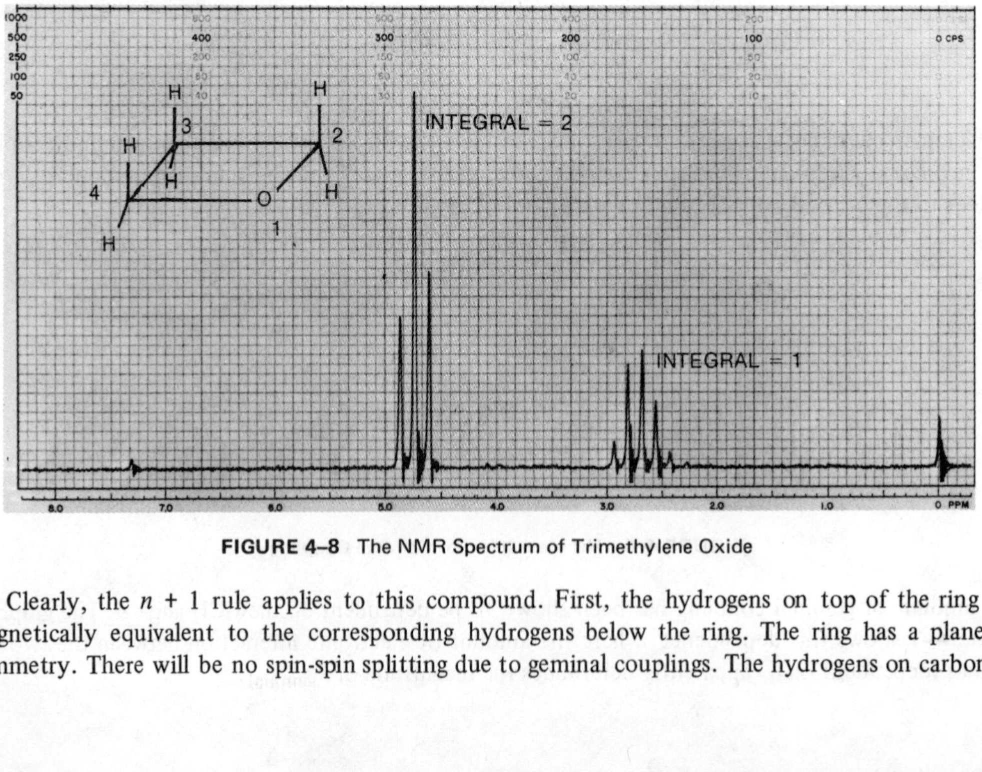

FIGURE 4–8 The NMR Spectrum of Trimethylene Oxide

Clearly, the $n + 1$ rule applies to this compound. First, the hydrogens on top of the ring are magnetically equivalent to the corresponding hydrogens below the ring. The ring has a plane of symmetry. There will be no spin-spin splitting due to geminal couplings. The hydrogens on carbons 2 and 4 are equivalent by symmetry and, hence, they have the same chemical shift and will couple identically to the protons on carbon 3. The four hydrogens on carbons 2 and 4 behave as a group and are split by the two hydrogens on carbon 3 into a triplet ($n + 1 = 3$). The protons on carbon 3 have four equivalent neighbors and are split into a quintet ($n + 1 = 5$). In this compound, the $n + 1$ rule successfully predicts the spectrum.

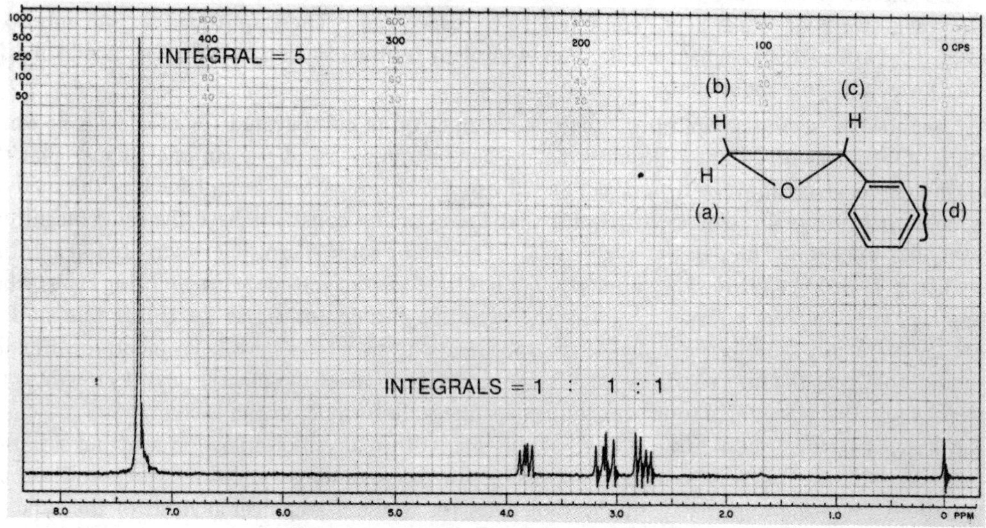

FIGURE 4–9 The NMR Spectrum of Styrene Oxide

The spectrum of styrene oxide is given in Figure 4–9 and shows how magnetic non-equivalence complicates the spectrum. Protons **a**, **b**, and **c** all have different chemical shift values and are magnetically non-equivalent. Hydrogen **a** (H_a) is on the same side of the ring as the phenyl group; hydrogen **b** (H_b) is on the opposite side of the ring. These hydrogens have slightly different chemical shift values, $H_a = 2.77\ \delta$ and $H_b = 3.12\ \delta$, and they will show spin-spin splitting with respect to one another. The third proton, **c** (H_c), has yet another value of chemical shift ($H_c = 3.83\ \delta$), and it will be coupled differently to H_a (which is *trans*) than it will be to H_b (which is *cis*). Because H_a and H_b are non-equivalent, and because H_c is coupled differently to H_a than to H_b ($J_{ac} \neq J_{bc}$), the $n + 1$ rule fails, and the spectrum of styrene oxide is more complicated. To explain the spectrum, one must examine each hydrogen individually and take into account its coupling with every other hydrogen. Each coupling is independent of the other.

An analysis of the splitting pattern in styrene oxide is given in Figure 4–10. Examine hydrogen **c** first. It is split first (J_{ac}) by the two possible spins of H_a into a doublet; second, each of the doublet

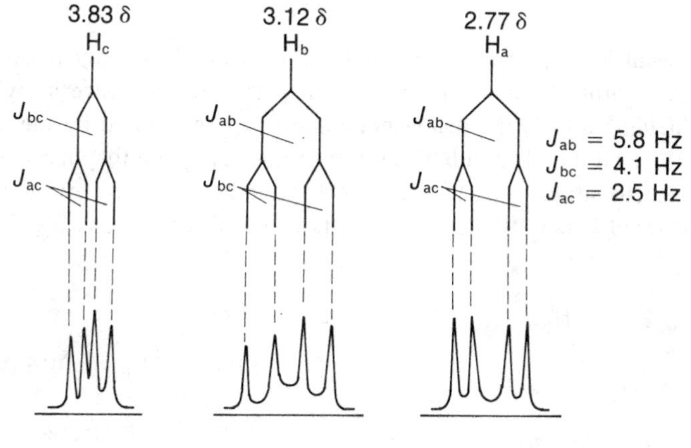

FIGURE 4–10 An Analysis of the Splitting Pattern in Styrene Oxide

peaks is split (J_{bc}) into another doublet by H_b. It is also possible to look at splitting from H_b first, and then H_a second. This is the way the analysis of the splitting has been done in the splitting diagram in Figure 4-10.

Note that J_{bc} is larger than J_{ac}. This is typical for small ring compounds where there is more interaction between protons which are *cis* to each other than for protons which are *trans* to each other (see Section 4.1). Thus, we see that H_c gives rise to *four* peaks centered at 3.83 δ. Similarly, H_a and H_b each give four peaks at 2.77 δ and 3.12 δ, respectively. These splittings are shown also in Figure 4–10. Notice that the magnetically non-equivalent protons H_a and H_b give rise to splitting J_{ab} (geminal) which is quite significant.

As can now be seen, the splitting situation becomes quite complicated for molecules in which there are non-equivalent groups of hydrogens. In fact, one may ask, how can one be sure that the graphical analysis given above is the correct one? First, the entire pattern is explained by this analysis and, second, it is internally consistent. Notice that the coupling constants have the same magnitude wherever they are used. Thus, in the analysis J_{bc} is given the same magnitude when used in splitting H_c as when it is used in splitting H_b. Similarly, J_{ac} has the same magnitude when splitting H_c as when splitting H_a. The coupling constant J_{ab} has the same magnitude for H_a as for H_b. If this kind of consistency were not seen in the analysis, the splitting analysis would have been incorrect. Finally, one should note that the nmr peak at 7.28 δ is due to the protons of the phenyl ring. It integrates for 5 protons, while the other three four-peak multiplets integrate for one proton each.

One note of caution should be entered at this point. In many molecules the splitting situation becomes so complicated that it is virtually impossible for the beginning student to derive it. There are also situations involving apparently simple molecules for which a graphical analysis of the type we have just completed will not suffice. A few of these cases are discussed in Section 4.11.

Two situations in which the $n + 1$ rule fails have now been discussed: (1) when the coupling involves nuclei other than hydrogen that do not have spin 1/2 (*e.g.*, deuterium), and (2) when there is magnetic non-equivalence in a set of protons attached to the same carbon. These latter protons may behave identically with respect to chemical reactions (chemical equivalence) but may, nevertheless, have different magnetic environments and chemical shifts (magnetic non-equivalence). Thus, there is a difference between *chemical equivalence* and *magnetic equivalence* which is important to recognize. The $n + 1$ rule is, in fact, a rule only in certain special situations: each proton in a group must be equivalent (chemically and magnetically); each proton in the group must be coupled to the same extent to every proton in the group with which it is interacting; and each proton in the group must have the same chemical shift. Another rather unexpected situation arises in which the $n + 1$ rule fails; it occurs when the chemical shift difference between two sets of protons is small compared to the coupling constant between them. This situation will be discussed in Section 4.11.

4.4 ALKENES

The nmr chemical shifts of protons attached to double bonds are much larger than those of protons attached to sp^3 carbon atoms. This is due, in part, to the change in hybridization. More important is the deshielding due to the diamagnetic anisotropy generated by the π electrons of the double bond. The protons on double bonds also differ in that rarely are they magnetically equivalent, and often they give rise to splitting patterns which cannot be explained by the $n + 1$ rule. In the typical alkene, as illustrated below, the chemical shifts of the three protons H_A, H_B, and H_C rarely coincide.

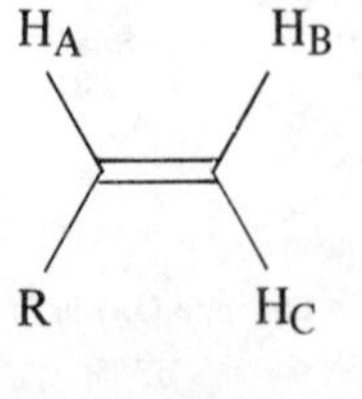

$$\delta_{H_A} \neq \delta_{H_B} \neq \delta_{H_C}$$

$$J_{AB} \neq J_{AC} \neq J_{BC}$$

Furthermore, the extent of coupling between the protons is usually quite different, and three distinct types of spin interaction are observed:

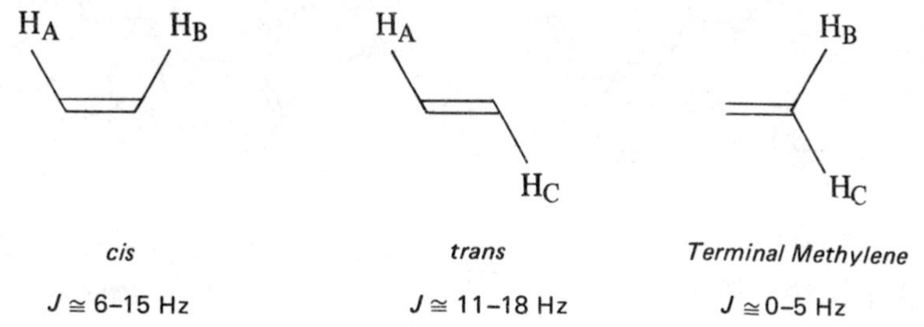

Protons substituted *trans* on a double bond couple most strongly, with a typical value for J of about 16 Hz. The *cis* coupling constant is commonly half this value, being about 8 Hz. Coupling between terminal methylene protons (geminal coupling) is much smaller yet. As an example of the nmr spectrum of a simple alkene, the spectrum of *trans*-cinnamic acid is shown in Figure 4–11.

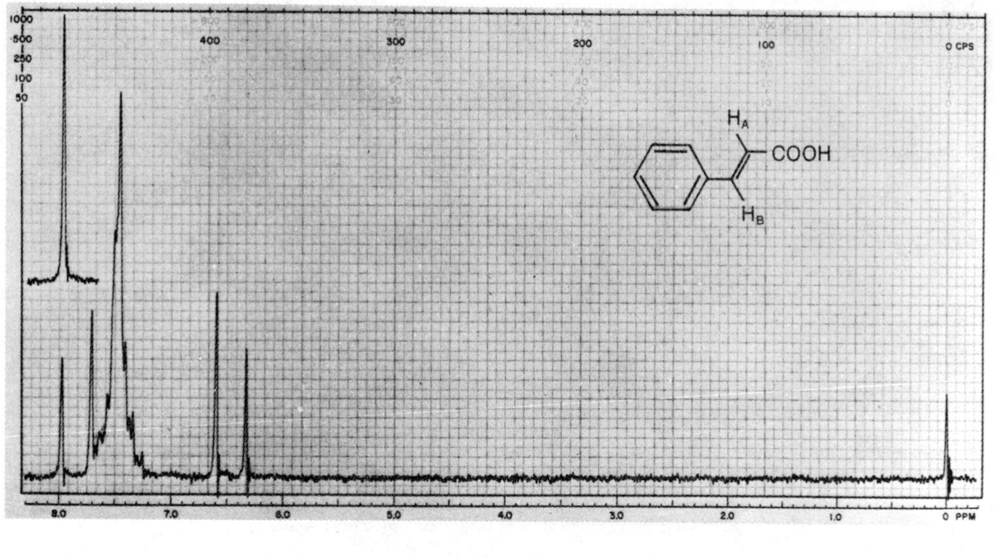

FIGURE 4–11 The NMR Spectrum of *trans*-Cinnamic Acid

The phenyl protons appear as a large group at 7.45 δ, and the acid proton is a singlet that appears off-scale at 13.21 δ. The two vinyl protons H_A and H_B split one another into two doublets, one centered at 7.83 δ to the left of the phenyl resonances, and the other at 6.46 δ to the right of the phenyl resonances. The proton H_B, attached to the carbon bearing the phenyl ring, is assigned the larger chemical shift since it should be in a deshielding area of the anisotropic field generated by the π electrons of the aromatic ring. The coupling constant J_{AB} can be determined easily. The splitting extends slightly more than 3 chart divisions (5 Hz each), which is equal to about 16 Hz – a common value for *trans* proton-proton coupling across a double bond. The *cis* isomer would exhibit a smaller splitting.

Molecules which have a symmetry element (plane or axis of symmetry) passing through the C=C double bond do not show any *cis* or *trans* splitting, since the protons H_A and H_B are magnetically equivalent. An example of each type can be seen in *cis*- and *trans*-stilbene, respectively. In each compound, the vinyl protons give rise to only a single *unsplit* resonance peak.

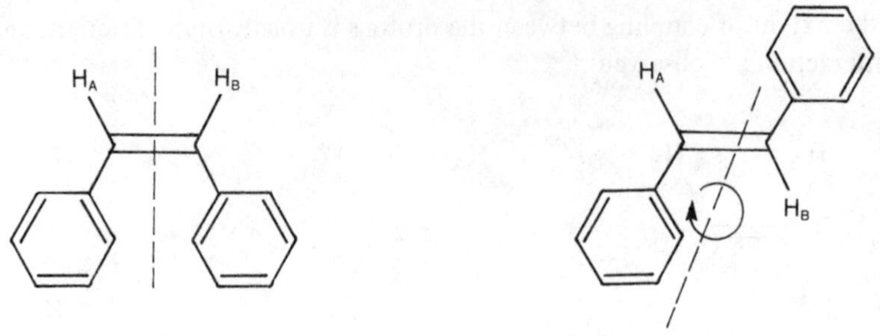

Plane of Symmetry
cis-stilbene

Twofold Axis of Symmetry
trans-stilbene

Vinyl acetate gives an nmr spectrum typical of a compound having three hydrogens substituted on a double bond. Each of these protons has a different chemical shift and a different coupling constant to each of the other protons. This spectrum, shown in Figure 4–12, is not unlike that of styrene oxide (Figure 4-9). The graphical analysis of the vinyl portion is given in Figure 4–13. Notice that J_{BC}(*trans*) is larger than J_{AC}(*cis*), and that J_{AB}(geminal) is very small, the usual situation for vinyl compounds.

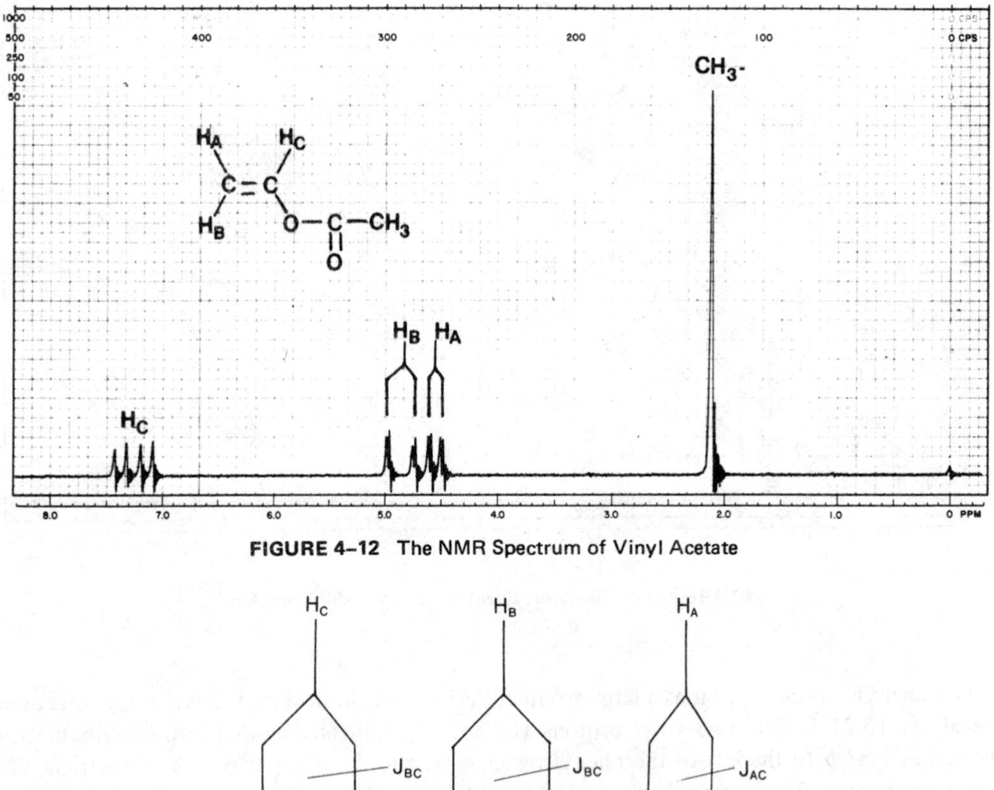

FIGURE 4–12 The NMR Spectrum of Vinyl Acetate

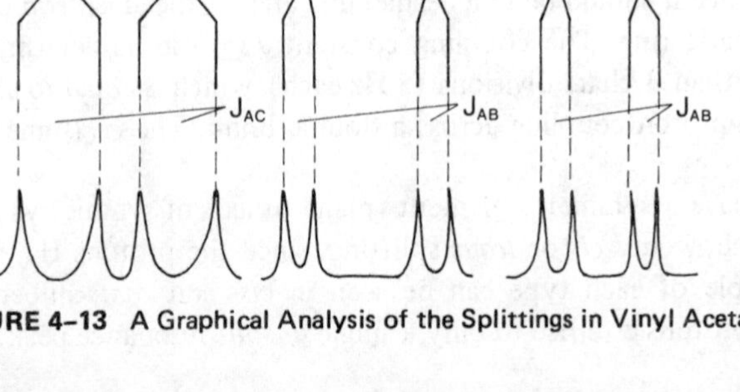

FIGURE 4–13 A Graphical Analysis of the Splittings in Vinyl Acetate

Again, as in the last section, we close with a note of caution. There exist many cases in which the splitting patterns of apparently simple alkenes cannot be explained either by this simple graphical approach or by use of the $n + 1$ rule.

4.5 MECHANISMS OF COUPLING IN ALKENES; ALLYLIC COUPLING

The mechanism of *cis* and *trans* coupling in alkenes is no different from that of any other vicinal coupling, and that for the terminal methylene protons is just a case of geminal coupling. All three types have been discussed previously and are illustrated in Figure 4–14.

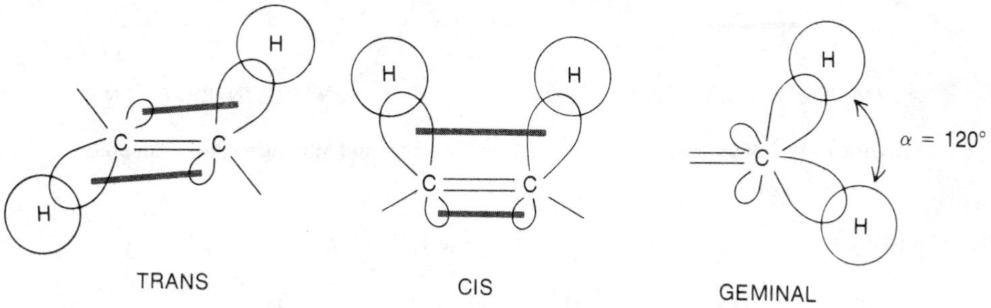

FIGURE 4–14 The Types of Coupling Present in Alkenes

In explaining the relative magnitudes of the coupling constants, notice that the two CH bonds are parallel for *trans* coupling, while they are tilted away from each other in *cis* coupling. Note also that the HCH angle for geminal coupling is nearly 120°, a virtual minimum in the graph of Figure 4–7. In addition to these three types of coupling, alkenes often show small couplings between protons substituted on carbons α to the double bond and those on the opposite end of the double bond:

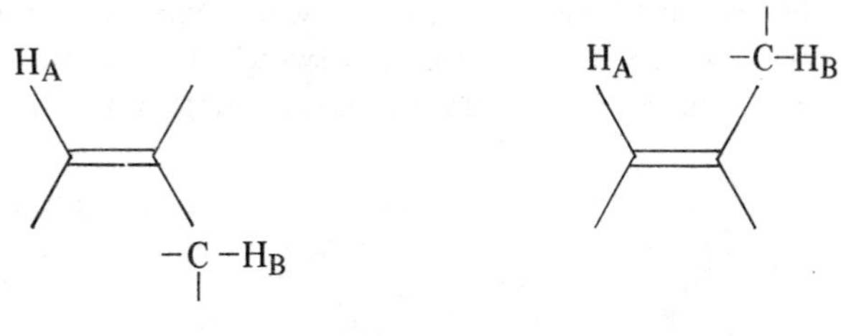

$J \sim 0$–3 Hz $J \sim 0$–3 Hz

This is called *allylic coupling*. The π electrons of the double bond apparently help to transmit the spin information from one nucleus to the other, as shown in Figure 4–15.

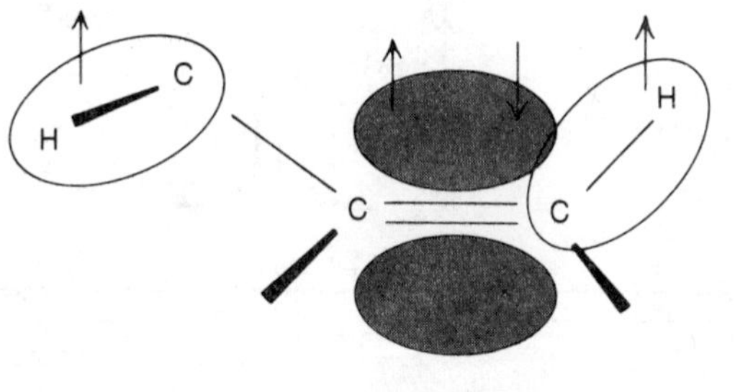

FIGURE 4–15 Allylic Coupling

When all nuclei are co-planar, there is no interaction of the allylic CH bond orbital with the π system, and $J = 0$ Hz. However, when the allylic CH bond is perpendicular to the C=C plane (as is the π bond), the interaction assumes the maximum value, $J = 3$ Hz. These geometries are illustrated in Figure 4–16.

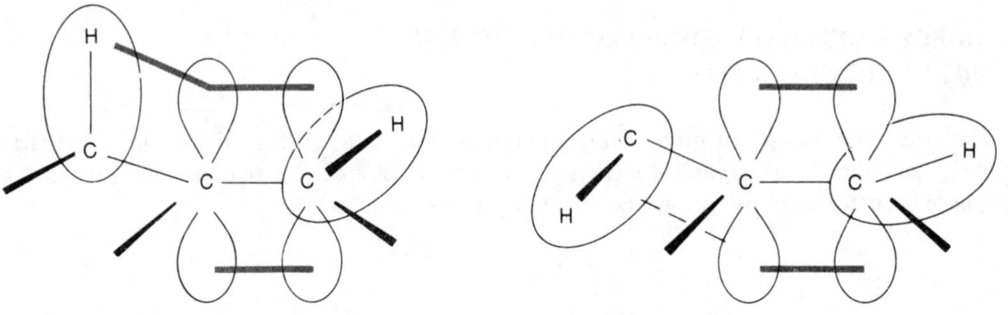

FIGURE 4–16 Geometric Arrangements That Maximize and Minimize Allylic Coupling

Allylic splitting is observed in compounds such as the following.

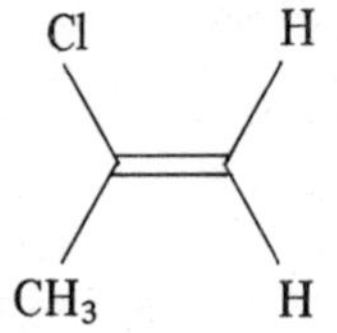

1

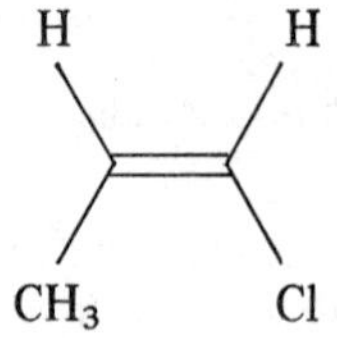

2

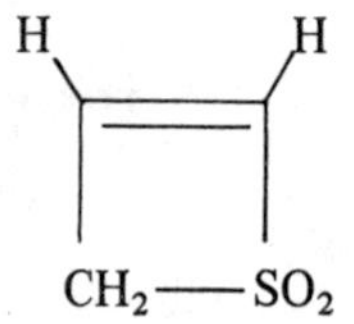

3

The nmr spectrum of the last of these is given in Figure 4–17. See if you can assign the peaks and explain the splittings in this compound. A graphical analysis (without labels) of this spectrum is given in Figure 4–18. Since the spectrum itself is rather crowded when determined under the usual conditions, the peaks have been redetermined at greater sweep width in the traces which are found above the baseline. This expansion of the spectrum allows for a better observation of the splittings.

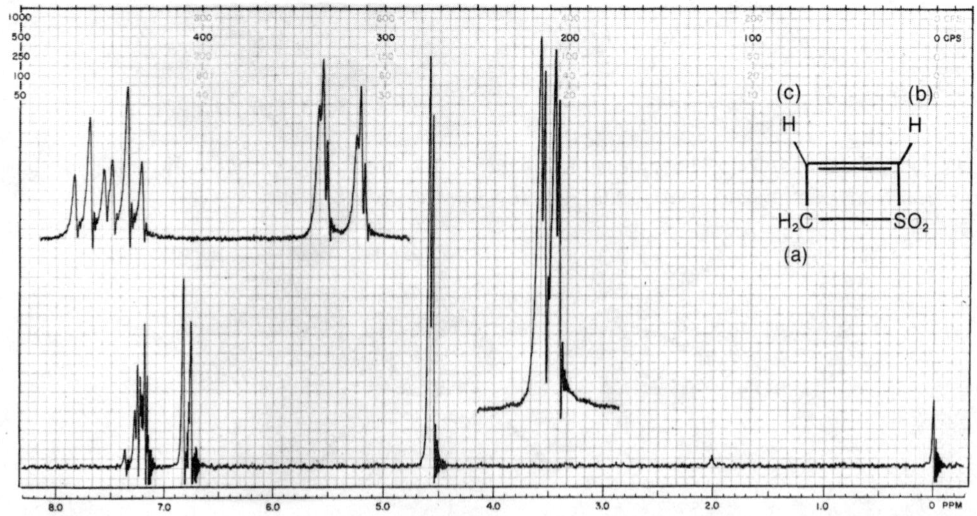

FIGURE 4–17 The NMR Spectrum of Compound 3, Page 000.

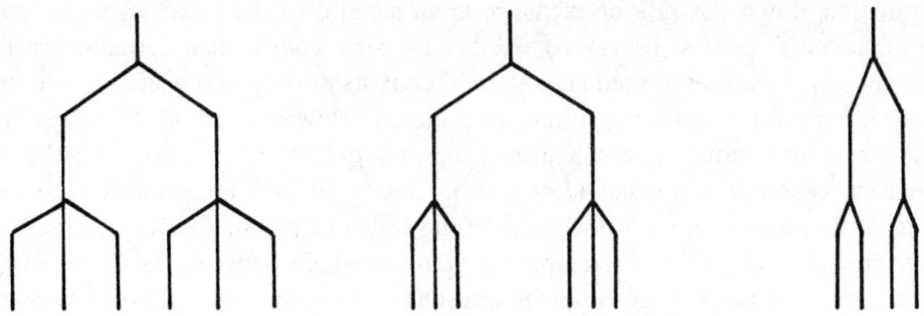

FIGURE 4–18 A Graphical Analysis of the NMR Spectrum of Compound 3 (Figure 4–17). No Labels are Given.

The reader is cautioned that in many molecules J is small or vanishingly small for the allylic coupling of type H_A–C–C=C–H_B mentioned above, and often *no* spin-spin splitting is observed. It is difficult to generalize, but this is particularly true for molecules in which the allylic carbon bearing H_A has free rotation. Rigid systems of the correct geometry are the most common systems in which to observe allylic coupling.

4.6 PROTONS ON OXYGEN: ALCOHOLS

Under the usual conditions of determining the nmr spectrum of an alcohol, and for most alcohols, no coupling is observed between the hydroxyl hydrogen and hydrogens on the carbon atom to which the hydroxyl group is attached (–CH–OH). As in geminal coupling, however, there actually is coupling between these hydrogens, but, owing to other factors, spin-spin splitting often is not observed. Whether or not spin-spin splitting involving the hydroxyl hydrogen is observed in a given alcohol depends on several factors; among these are temperature, purity of the sample, and the solvent used. These factors are all related to the rate at which hydroxyl protons exchange with one another (or the solvent) in the solution. Under normal conditions, the rate of exchange of protons between alcohol molecules occurs at a rate faster than the nmr spectrometer can respond.

$$R{-}OH_a + R'{-}OH_b \rightleftharpoons R{-}OH_b + R'{-}OH_a$$

It requires about 10^{-2} to 10^{-3} second for an nmr transitional event to occur and be recorded. At room temperature, a typical pure liquid alcohol sample undergoes intermolecular proton exchange at a rate of about 10^5 protons/second. This means that the average residence time of a single proton on the oxygen atom of a given alcohol is about 10^{-5} second before it exchanges. This is a much shorter time than is required for a spin transition, and leads to complications. For instance, a proton just starting to have resonance may suddenly begin to exchange; or one that is in between two molecules may begin to have resonance and then suddenly affix itself to a molecule. Since the nmr spectrometer cannot respond rapidly to these situations, the net result, as far as it is concerned, is that the proton is unattached more frequently than it is attached to oxygen, and the spin interaction between the hydroxyl proton and any other proton in the molecule is effectively *decoupled*. *Rapid chemical exchange decouples spin interactions*, and the nmr spectrometer records only the *average environment* which it senses for the exchanging proton. The hydroxyl proton, for instance, often exchanges between alcohol molecules so rapidly that it "sees" all the possible spin orientations of hydrogens attached to the α carbon as a single time-averaged spin configuration. Similarly, the α hydrogens see so many different protons on the hydroxyl oxygen (some with spin +1/2, some with spin −1/2) that the spin configuration they sense is an average or intermediate value between +1/2 and −1/2, that is, *zero*. In effect, the nmr spectrometer behaves like a camera with a slow shutter speed in photographing a fast event. Events which are faster than the click of the shutter mechanism are either blurred or averaged.

If one can slow down the rate of exchange in an alcohol to the point where it approaches the "time scale of the nmr" (*i.e.*, $< 10^2$ to 10^3 exchanges per second), then coupling can be observed. For instance, the nmr spectrum of methanol at 25°C consists of only two peaks, both unsplit singlets, integrating for one proton and three protons, respectively. However, at −40°C the spectrum changes dramatically. The one-proton OH resonance becomes a quartet (J = 5 Hz) and the three-proton methyl resonance becomes a doublet (J = 5 Hz). Clearly, at −40°C chemical exchange has been slowed down to the point where it is now on the time scale of the nmr spectrometer, and coupling to the hydroxyl proton is observed. At temperatures intermediate between 25°C and −40°C one sees transitional spectra. Figure 4–19 gives representations of the nmr spectrum of methanol at various temperatures between 25°C and −40°C.

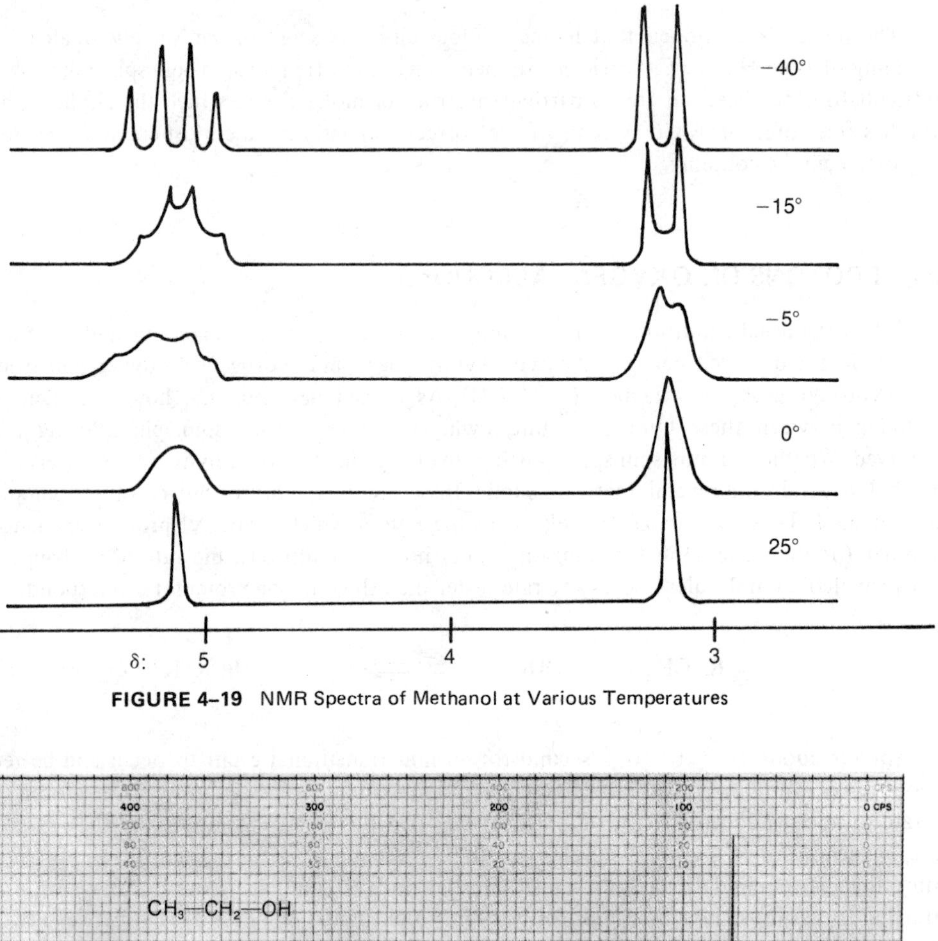

FIGURE 4–19 NMR Spectra of Methanol at Various Temperatures

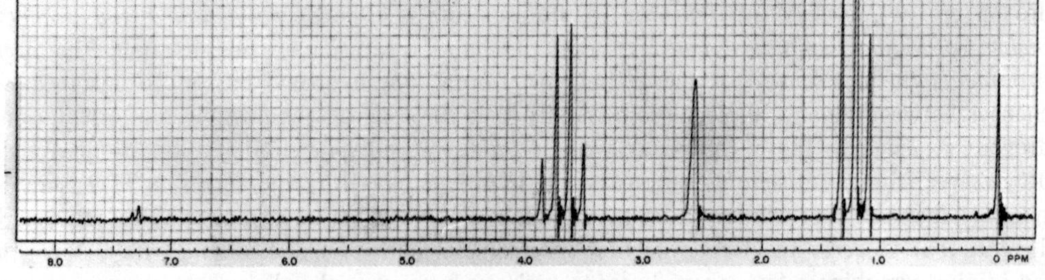

FIGURE 4–20 The NMR Spectrum of an Ordinary Ethanol Sample

The room temperature spectrum of an ordinary sample of ethanol (Figure 4–20) shows no coupling of the hydroxyl proton to the methylene protons. Thus, the hydroxyl proton is seen as a singlet and the methylene protons (split by the methyl group) as a quartet. Apparently the rate of exchange in such a sample is faster than the nmr time scale, and decoupling of the hydroxyl and methylene protons is observed. However, if one purifies a sample of ethanol in such a way as to eliminate all traces of impurity (especially of acid and water, thereby slowing the proton exchange rate), the hydroxyl-methylene coupling can be observed in the form of increased complexity of the spin-spin splitting patterns. The hydroxyl absorption becomes a triplet, and the methylene absorptions are seen as an overlapping pair of quartets. The hydroxyl resonance is apparently split (just as the methyl group, but with different J) into a triplet by its two neighbors on the methylene carbon.

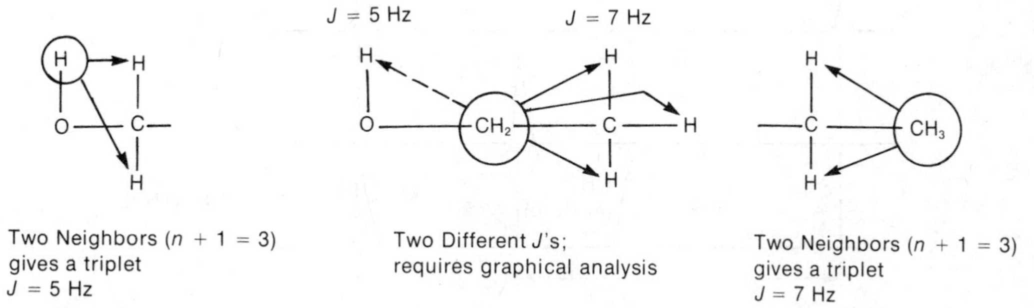

The coupling constant for hydroxyl-methylene interaction is found to be ~5 Hz. The methyl triplet is found to have a different coupling constant, $J \cong 7$ Hz, for the methylene-methyl coupling. The methylene protons *are not* split into a quintet by their four neighbors, since the coupling constants for hydroxyl-methylene and methyl-methylene are different. As noted in an earlier section, the $n + 1$ rule does not apply in such an instance; each interaction occurs independently of the other, and a graphical analysis is required to approximate the correct pattern. The analysis of the ultra-pure ethanol spectrum is shown in Figure 4–21.

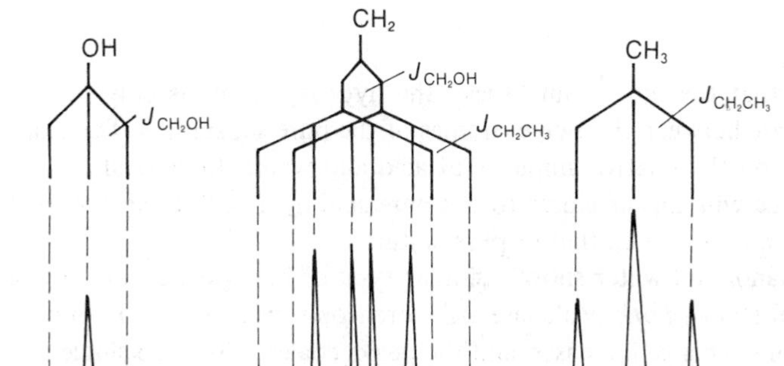

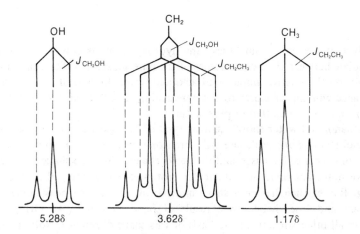

FIGURE 4–21 The Analysis and Appearance of the NMR Spectrum of an Ultrapure Ethanol Sample

If even a drop of acid is added to the ultrapure ethanol sample, the rate of proton exchange becomes so fast that the methylene and hydroxyl protons are decoupled, and the simpler spectrum is obtained.

4.7 ACID/WATER AND ALCOHOL/WATER MIXTURES

When two compounds, both of which contain an OH group, are mixed, one often observes only a single nmr absorption due to OH. For instance, consider first three spectra: (1) pure acetic acid, (2) pure water, and (3) a 1:1 mixture of acetic acid and water. The general appearance of these three spectra is indicated in Figure 4–22.

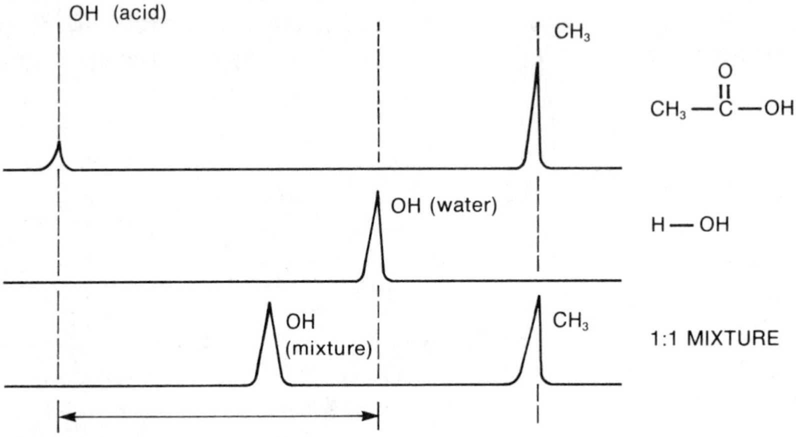

FIGURE 4–22 A Comparison of the Spectra of Acetic Acid, Water, and a 1:1 Mixture of the Two

Mixtures of acetic acid and water might be expected to show *three* peaks, since there are two distinct types of hydroxyl group in the solution – one on acetic acid and one on water. In addition, the methyl group on acetic acid should give an absorption peak. In actuality, however, only two peaks are seen in mixtures of these two reagents. The methyl peak occurs at its normal position in the mixture, but there is only a single hydroxyl peak *in between* the hydroxyl positions of the pure substances. Apparently, exchange of the type

$$CH_3COOH_a + H\text{-}OH_b \rightleftharpoons CH_3COOH_b + H\text{-}OH_a$$

occurs so rapidly that the nmr again "sees" the hydroxyl protons only in an average environment which is intermediate between the two extremes of the pure substances. The exact position of the OH resonance depends on the relative amounts of acid and water. In general, if there is more acid than water, the resonance will appear closer to the pure acid hydroxyl resonance. If one adds more water, the resonance will move closer to that of pure water.

Samples of ethanol and water show a similar type of behavior, except that at low concentration of water in ethanol (~1%) *both* peaks are still often observed. As the amount of water is increased, however, the rate of exchange increases, and the peaks coalesce into the single averaged peak.

In closing, we note that rapid intermolecular proton exchange often (but not always!) leads to *peak broadening*. Rather than having a sharp and narrow shape, the peak sometimes increases in width at the base and loses height as a result of rapid exchange. An OH peak can often be distinguished from all other singlets on the basis of this shape difference. Note the hydroxyl peak in Figure 4–20. Peak broadening is caused by factors which are rather complicated, and we will leave the explanation to more advanced texts. We note only that the phenomenon is *time dependent* and that the intermediate transitional stages of peak coalescence are sometimes seen in nmr spectra when the rate of exchange is neither slower nor faster than the nmr time scale, but instead is on roughly the same order of magnitude. These situations are illustrated in Figure 4–23.

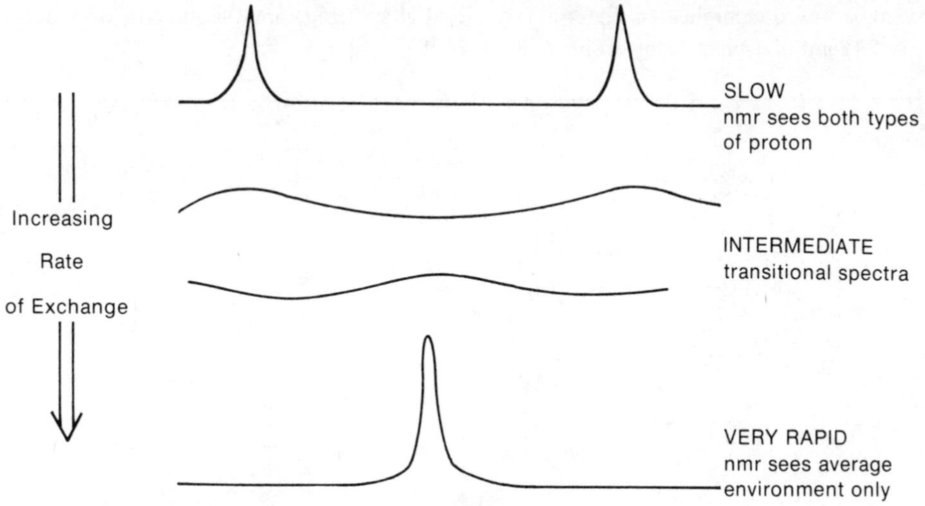

FIGURE 4–23 The Effect of the Rate of Exchange on the NMR Spectrum of a Hydroxylic Compound Dissolved in Water

Also, do not forget that when the spectrum of a pure alcohol is determined in an inert solvent (*e.g.*, CCl_4) the nmr absorption position is concentration dependent. You will recall that this is due to hydrogen-bonding differences. If not, now is a good time to reread pages 96–97.

4.8 PROTONS ON NITROGEN: AMINES

In simple amines, as in alcohols, intermolecular proton exchange is usually fast enough to decouple spin-spin interactions between protons on nitrogen and those on the α carbon atom. Under such conditions the amino hydrogens are usually seen as a sharp singlet (unsplit), and, in turn, the hydrogens on the α carbon are also unsplit by amino hydrogens. The rate of exchange can be slowed down by making the solution strongly acidic (pH < 1) and forcing the protonation equilibrium to favor the quaternary ammonium cation rather than the free amine.

$$R\text{-}CH_2\text{-}NH_2 + \underset{\substack{\text{excess}\\(\text{pH}<1)}}{H^+} \rightleftharpoons R\text{-}CH_2\text{-}NH_3^+$$

Under these conditions, the predominant species in solution is the protonated amine, and the rate of intermolecular proton exchange is slowed down. This often allows one to observe spin-spin coupling interactions that are decoupled and masked by exchange in the free amine.

In amides, which are less basic than amines, the rate of proton exchange is slow, and coupling is often observed between the protons on nitrogen and those on the α carbon of an alkyl substituent that is substituted on the same nitrogen.

$J_{HH} \sim 7$ Hz

Two examples of uncomplicated spectra (no NH–CH splitting) are the spectra of *n*-butyl amine (Figure 4–24) and of 1-phenylethylamine (Figure 4–25).

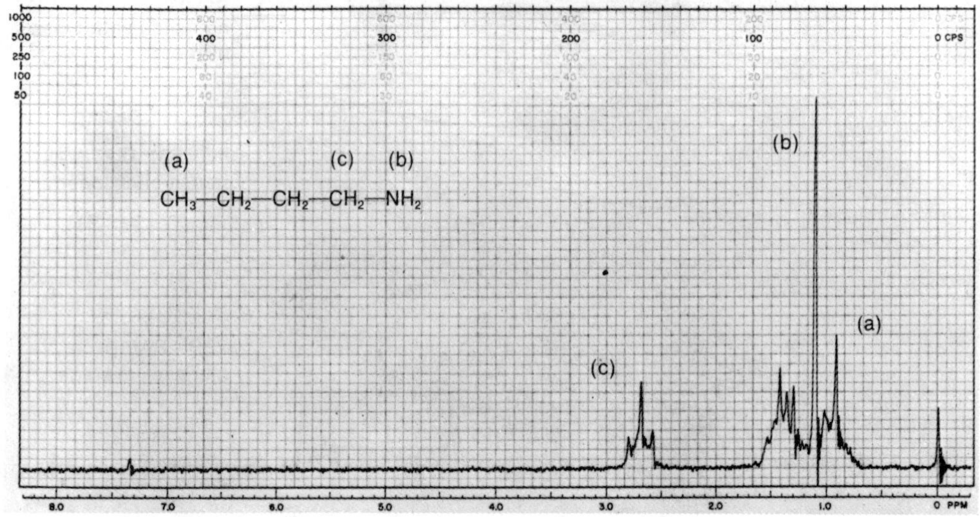

FIGURE 4–24 The NMR Spectrum of *n*-Butyl Amine

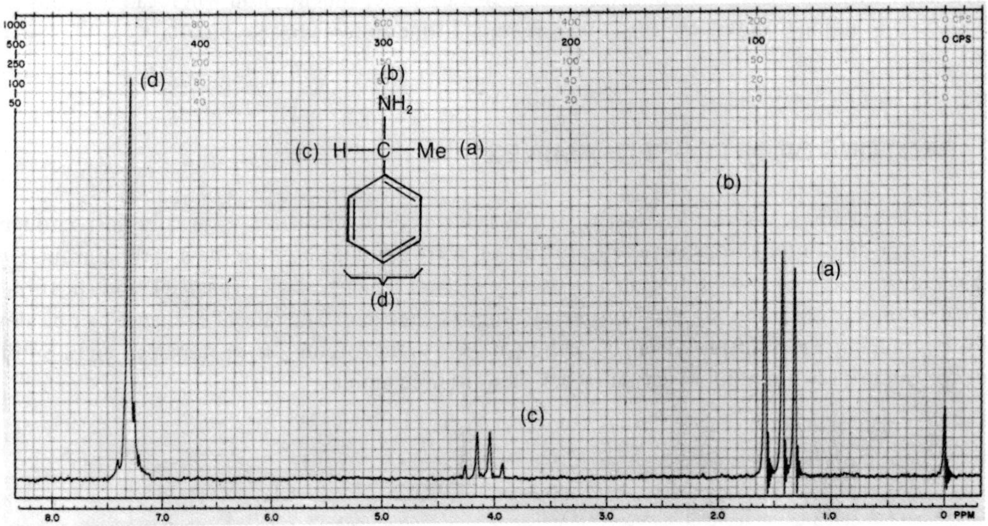

FIGURE 4–25 The NMR Spectrum of 1-Phenylethylamine

Unfortunately, the spectra of amines are not always as simple as these. There is another factor that can complicate the splitting patterns of both amines and amides: nitrogen itself has a nuclear spin. Its nuclear magnetic spin is unity ($I = 1$) and it can, therefore, adopt spin states +1, 0, and −1. Based on what we know so far of spin-spin coupling, we could predict the following possible types of interaction between H and N:

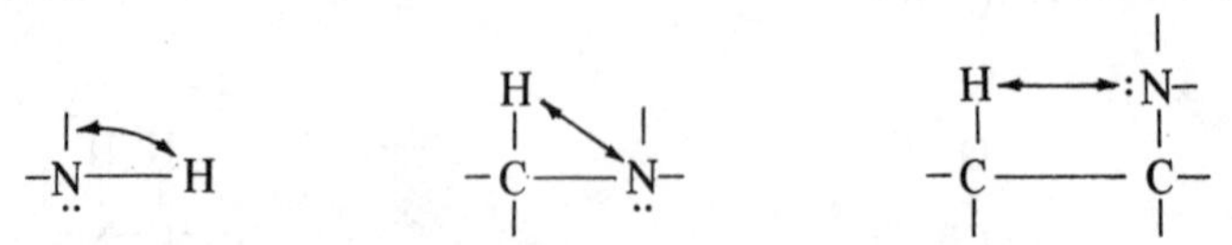

Direct Coupling — $J \sim 50$ Hz

Geminal Coupling — *Vicinal Coupling* — $J \sim$ negligible (*i.e.*, almost always zero)

Of these types of coupling, the geminal and vicinal types are very rarely seen, and we can dismiss them. Observation of direct coupling, however, is infrequent, but not unknown. Direct coupling is, of course, not observed if the hydrogen on the nitrogen is undergoing rapid exchange. The same conditions which decouple NH–CH or HO–CH proton-proton interactions will also decouple N–H nitrogen-proton interactions. In cases where direct coupling is observed, the coupling constant is found to be quite large: $J \sim 50$ Hz.

One of the cases in which both N–H and CH–NH proton-proton coupling can be observed is in the nmr spectrum of methylamine in aqueous hydrochloric acid solution (pH < 1). The species actually being observed in this medium is methylammonium chloride; *i.e.*, the hydrochloride salt of methylamine. This spectrum is simulated in Figure 4–26.

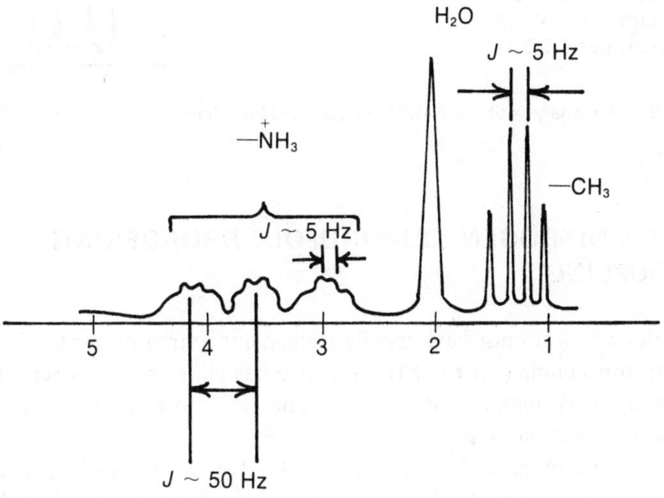

FIGURE 4–26 The NMR Spectrum of $CH_3NH_3^+$ in H_2O (pH < 1)

The peak at about 2.2 δ is due to water (of which there is plenty in aqueous hydrochloric acid solution!). The analysis of the remainder of the spectrum is completed in Figure 4–27.

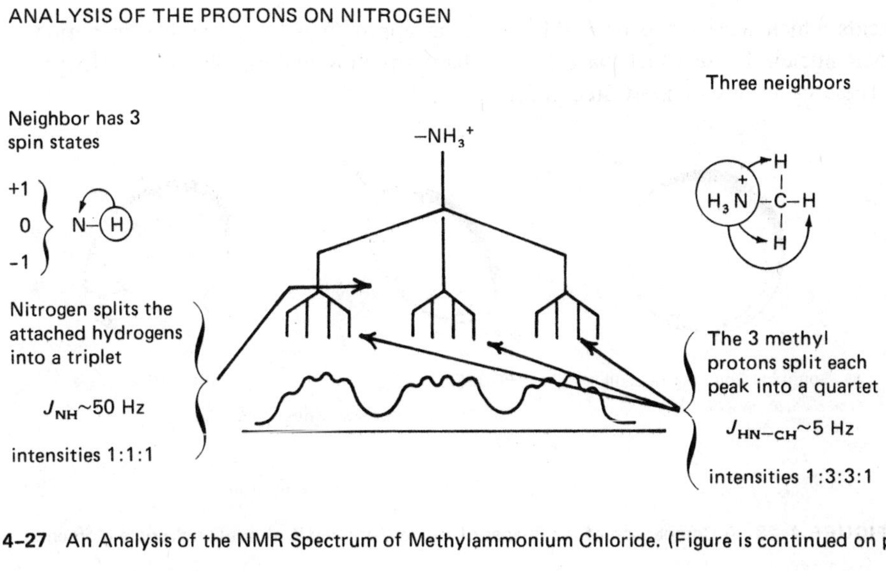

FIGURE 4–27 An Analysis of the NMR Spectrum of Methylammonium Chloride. (Figure is continued on page 134.)

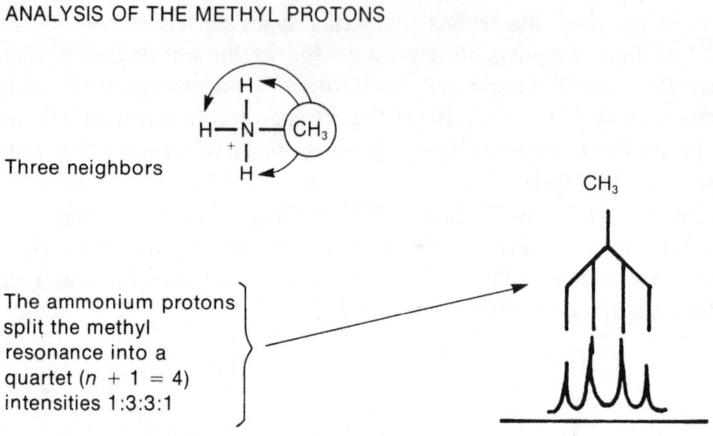

FIGURE 4–27 An Analysis of the NMR Spectrum of Methylammonium Chloride. (Continued)

4.9 PROTONS ON NITROGEN: QUADRUPOLE BROADENING AND DECOUPLING

Even in molecules which do not have rapidly exchanging amino protons, the *more usual* case is that the nitrogen-proton coupling in the NH resonance has either been decoupled or broadened by the large *electric quadrupole moment* of nitrogen. The clear observation of direct NH coupling as illustrated above for methylamine is uncommon.

We will not explain the origin of the electric quadrupole moment of nitrogen except in a cursory way. Nuclei which have either spin $I = 0$ or $I = 1/2$ have small electric quadrupole moments, while those elements for which $I > 1/2$ have relatively larger electric quadrupole moments. The electric quadrupole moments of the common nuclei are listed in order of increasing moment below:

$$\underbrace{{}^{1}H \ {}^{12}C \ {}^{16}O \ {}^{19}F}_{\text{Negligible}} \quad < {}^{2}H \ll {}^{14}N < \begin{matrix} {}^{35}Cl \\ {}^{37}Cl \end{matrix} \ll \begin{matrix} {}^{79}Br \\ {}^{81}Br \end{matrix} \approx {}^{127}I$$

Increasing Electric Quadrupole Moment (Absolute Values)

Elements which have $I = 0$ or $I = 1/2$ have an approximately spherical distribution of charge within their nuclei. Those which have $I > 1/2$ have an ellipsoidal distribution of charge within the nucleus. These situations are illustrated in Figure 4–28.

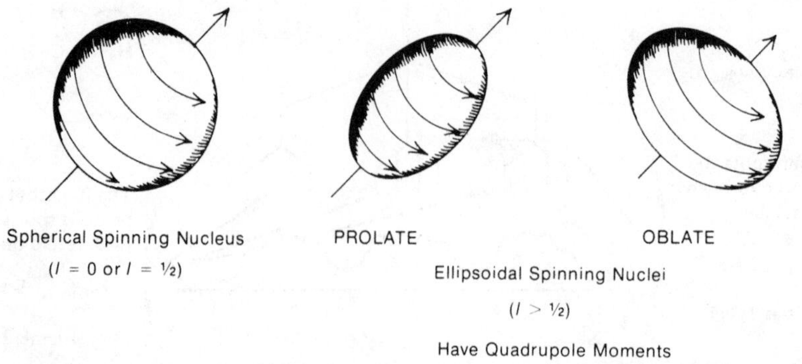

FIGURE 4–28 A Comparison of the Shapes of Nuclei With and Without Quadrupole Moments

Because of the non-spherical distribution of charge in ellipsoidal nuclei, they have a sort of "internal dipole moment," the *electric quadrupole moment*. Nuclei which have the non-spherical charge distributions (*i.e.*, have quadrupole moments) are much more sensitive both to interaction with the magnetic field of the nmr spectrometer and to magnetic and electric perturbations of the valence electrons and/or their environment. This means that they undergo transitions at a faster rate than spherically symmetric nuclei, and that they have very short lifetimes in their nuclear excited states. Thus, bromine ($I = 3/2$) rapidly undergoes transitions between all of its allowed states (+3/2, +1/2, -1/2, -3/2) while under the influence of a magnetic field. A hydrogen atom coupled to bromine would require 10^{-2} to 10^{-3} second to undergo an nmr transition. During this time, the typical bromine atom may have cycled its nucleus through all of the allowed states several times. Therefore, during its nmr transition from a +1/2 state to a -1/2 state, hydrogen will "see" only the average spin configuration of bromine – effectively, a single spin state – and will have been decoupled from it even though the coupling constant might be quite large for H-Br interaction. Chlorine, bromine, and iodine have large quadrupole moments and are effectively decoupled from interaction with the protons of most molecules under most conditions. Note, however, that fluorine (with $I = 1/2$) has essentially no quadrupole moment, and it *does* couple with protons in many molecules. Figure 4–29 illustrates quadrupole decoupling between hydrogen and bromine.

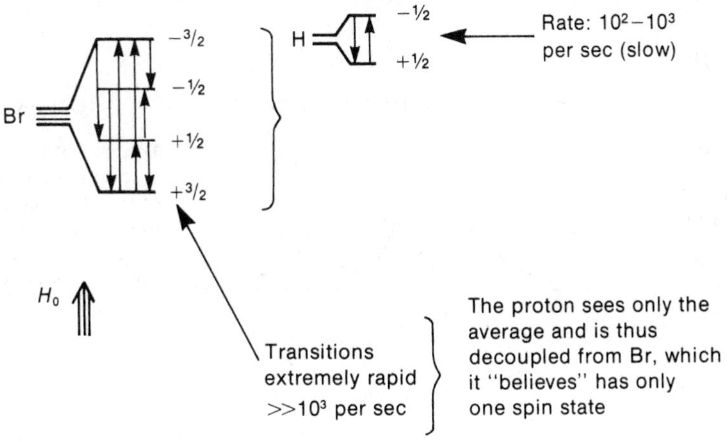

FIGURE 4–29 Decoupling of a Proton by Rapid Nuclear Spin Transitions Within a Bromine Atom

Now we return to nitrogen. Nitrogen has only a moderately sized quadrupole moment, and its transitions do not occur as rapidly as those in, say, bromine. Furthermore, the transitional rates and lifetimes of its excited spin states seem to vary slightly from one molecule to another. Solvent environment and temperature also seem to affect these rates. In many cases the transitional rates of nitrogen in a particular molecule are very close to the rate at which proton nmr absorption is taking place. In these cases, hydrogen is only partially decoupled from nitrogen. The proton is "less sure of what it sees," and the result is that the absorption peak *broadens*. This phenomenon is called *nuclear quadrupole broadening* of the nmr absorption peak. The three cases discussed here are summarized in Figure 4–30.

It should be carefully noted that not only must the rate of the spin transition in nitrogen be rapid (compared to that in hydrogen), but the lifetime of a +1 or 0 spin state must also be *short* to give the averaging of spins or spin decoupling of J_{NH}. If the rate of the upward transition were instantaneous; and if the nucleus stayed in the excited state for, say, 20 seconds; and if it then made an instantaneous transition downward, then all of the spin states of nitrogen would be represented in one molecule or another during the proton transitions. Spin-spin splitting would thus be observed. There would be three distinct types of molecules in the solution, and three absorptions would be seen since each proton in the three molecules represented would have a slightly different chemical shift owing to the different spin state of N in that molecule.

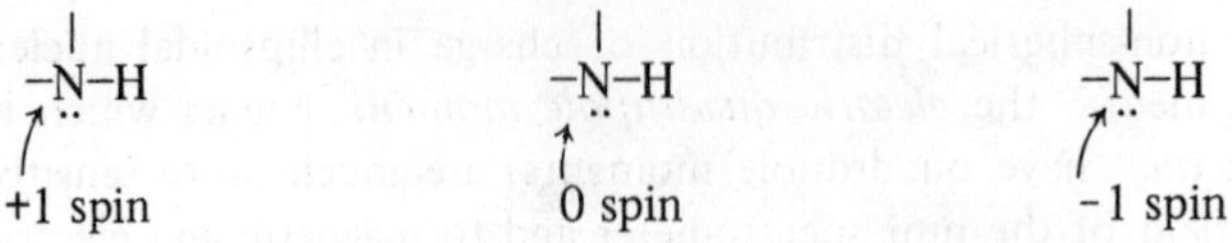
+1 spin
0 spin
−1 spin

Three Different Spin States of N in Three Different Molecules Would Give Rise to Three Absorption Peaks for Hydrogen

Of the three cases we have discussed, which are summarized in Figure 4–30, case three is the most common, followed by case two. Case one, discrete N–H spin-spin splitting, is not commonly observed. In Figure 4–31, the spectrum of the amide of propanoic acid clearly shows quadrupole broadening of the $-NH_2$ resonance. In some compounds (for instance, pyrroles and other heterocycles), the NH resonance can be subject to such a large broadening effect that it is difficult to distinguish the absorption from the baseline of the spectrum. The spectrum of pyrrole, shown in Figure 4–32, provides a good example.

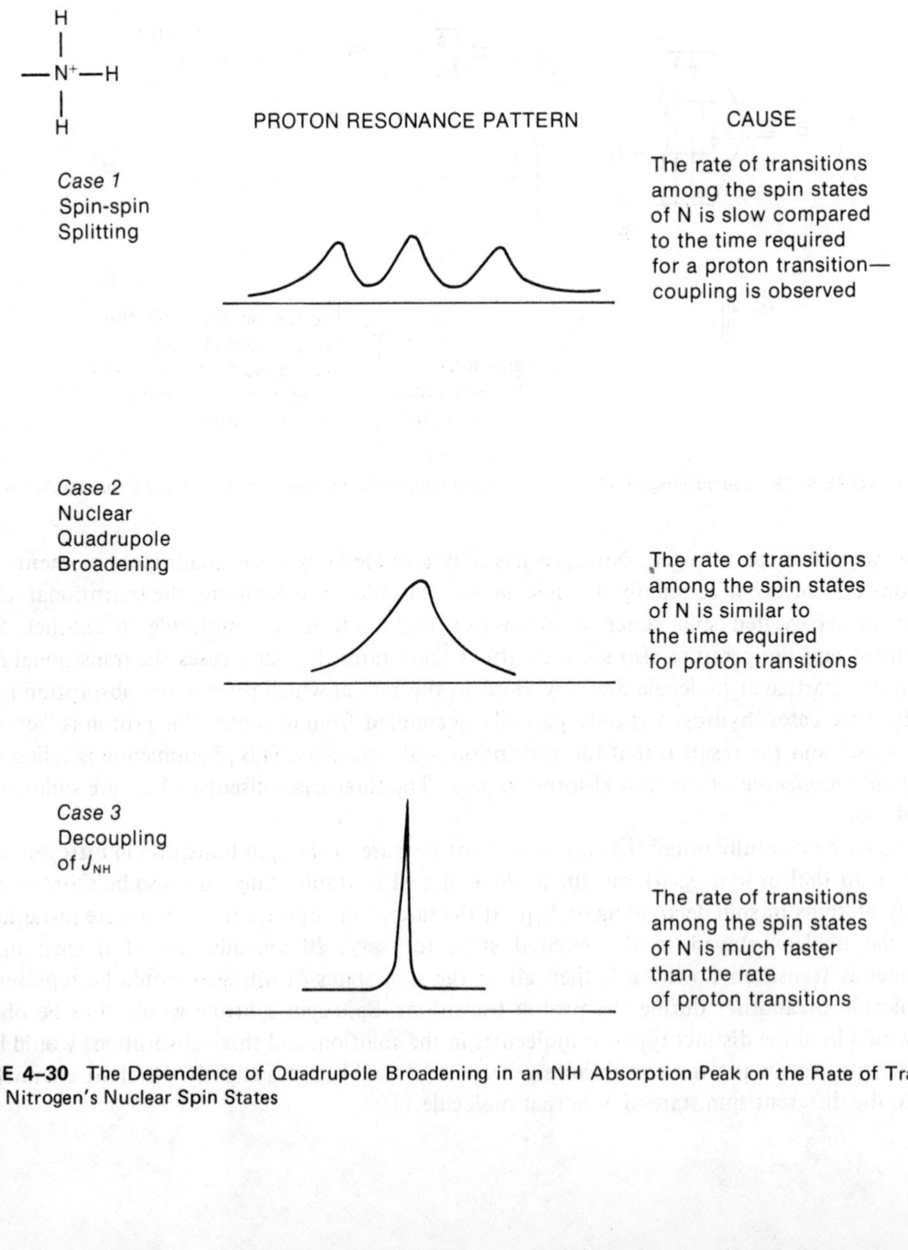
PROTON RESONANCE PATTERN
CAUSE
Case 1
Spin-spin
Splitting
The rate of transitions among the spin states of N is slow compared to the time required for a proton transition—coupling is observed
Case 2
Nuclear
Quadrupole
Broadening
The rate of transitions among the spin states of N is similar to the time required for proton transitions
Case 3
Decoupling
of J_{NH}
The rate of transitions among the spin states of N is much faster than the rate of proton transitions

FIGURE 4–30 The Dependence of Quadrupole Broadening in an NH Absorption Peak on the Rate of Transitions Among Nitrogen's Nuclear Spin States

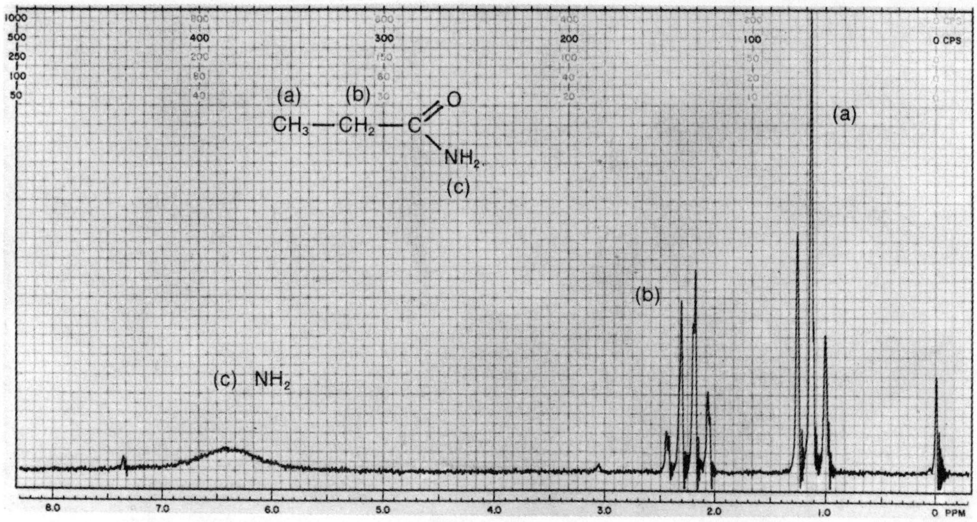

FIGURE 4-31 The NMR Spectrum of Propionamide

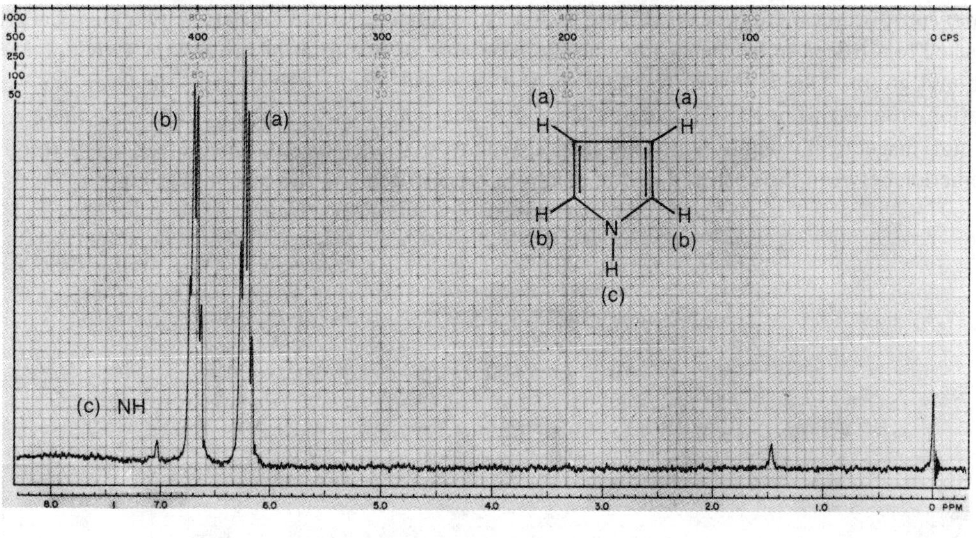

FIGURE 4–32 The NMR Spectrum of Pyrrole

4.10 AMIDES

Quadrupole broadening usually affects only the proton (or protons) attached directly to nitrogen. For instance, in *N*-acetyl-2-phenylethylamine (Figure 4–33) there is obviously no exchange of the NH proton since it couples with the protons of the adjacent methylene group; the protons in group B are split into a quartet by the two protons of group A and the amino hydrogen C (three neighbors). Yet, the NH resonance at ~ 6.50 δ is broadened and does not show the CH_2–NH coupling which should make it a triplet. The coupling has been obscured by nuclear quadrupole broadening.

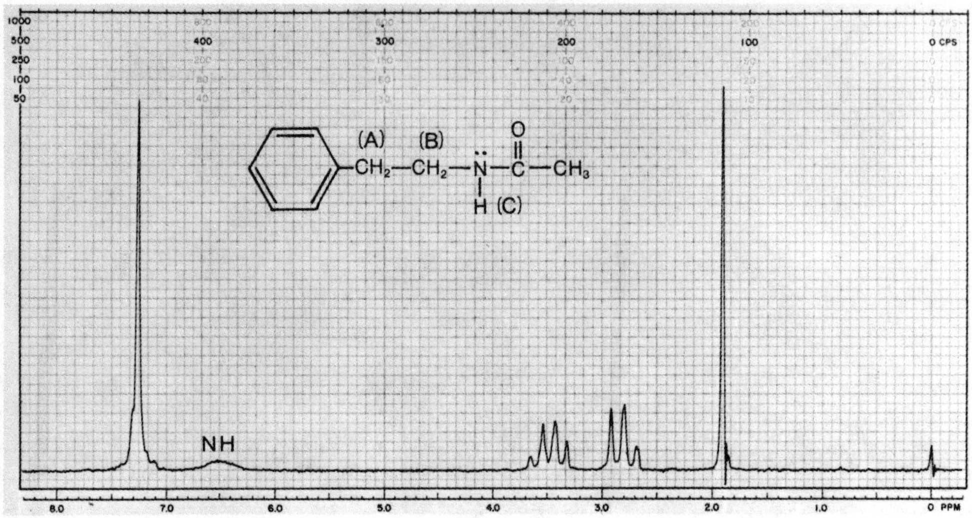

FIGURE 4–33 The NMR Spectrum of *N*-Acetyl-2-phenylethylamine

While discussing amides, one should also note that groups attached to an amide nitrogen are often found to be magnetically non-equivalent. For instance, the nmr spectrum of *N,N*-dimethylformamide shows *two* distinct methyl peaks (Figure 4–34). Normally one would expect the two identical groups attached to nitrogen to be chemically equivalent because of free rotation around the C–N bond to the carbonyl group. This is indeed the case on the chemical reaction time scale. However, the rate of rotation around this bond is apparently slowed to the point that on the nmr time scale these groups are not magnetically equivalent. One may speculate that this circumstance arises from resonance interaction between the unshared pair of electrons on nitrogen and the carbonyl group:

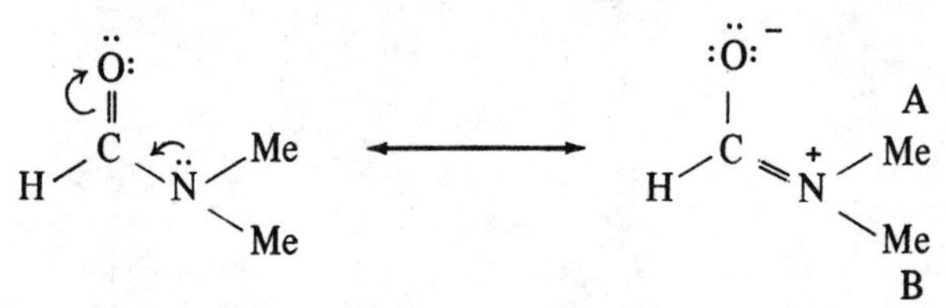

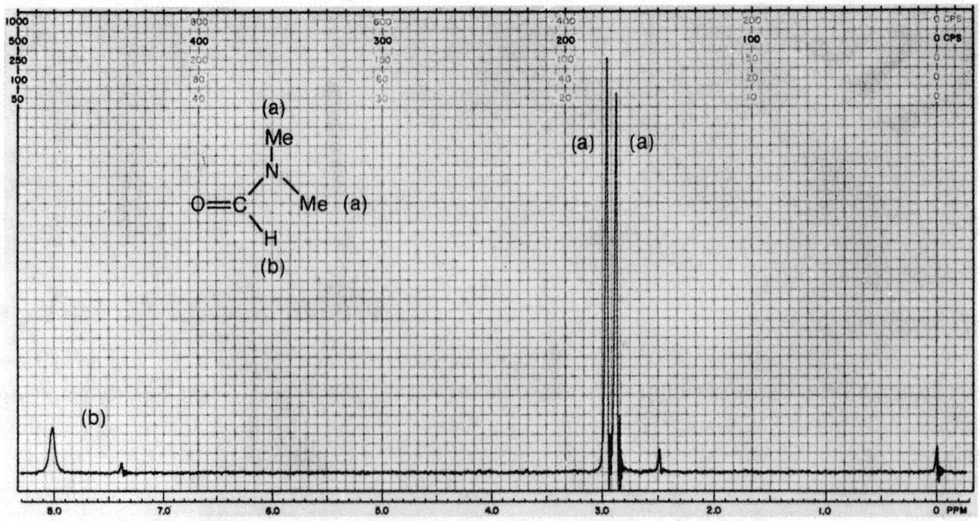

FIGURE 4–34 The NMR Spectrum of *N,N*-Dimethylformamide

This resonance requires that the molecule adopt a planar geometry, and it thus interferes with free rotation. If the free rotation is slowed to the point where it is slower than the time required for an nmr transition, then the nmr spectrometer will "see" two different methyl groups. One of these will be on the same side of the C=N bond as the carbonyl group; the other will be on the opposite side. Thus they will be in magnetically different environments, and they will have slightly different chemical shifts.

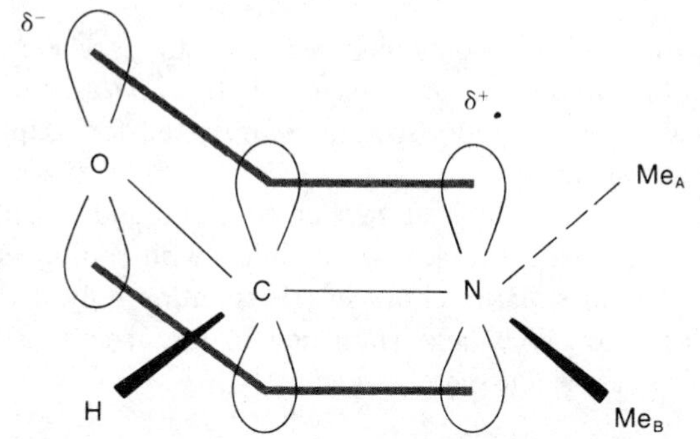

If one successively raises the temperature of the sample and redetermines the spectrum, the two peaks first broaden (80–100°C); then they merge to a single broad peak (~120°C), and finally they give a sharp singlet (150°C). The increase of temperature apparently speeds up the rate of rotation to the point where the nmr spectrometer records an "average" methyl group. That is, the methyl groups are changing environments so rapidly that during the period of time required for the nmr excitation of one of the methyl protons, that proton is simultaneously experiencing all of its possible conformational positions. The changes in the appearance of the methyl resonances of *N,N*-dimethylformamide with temperature are illustrated in Figure 4–35.

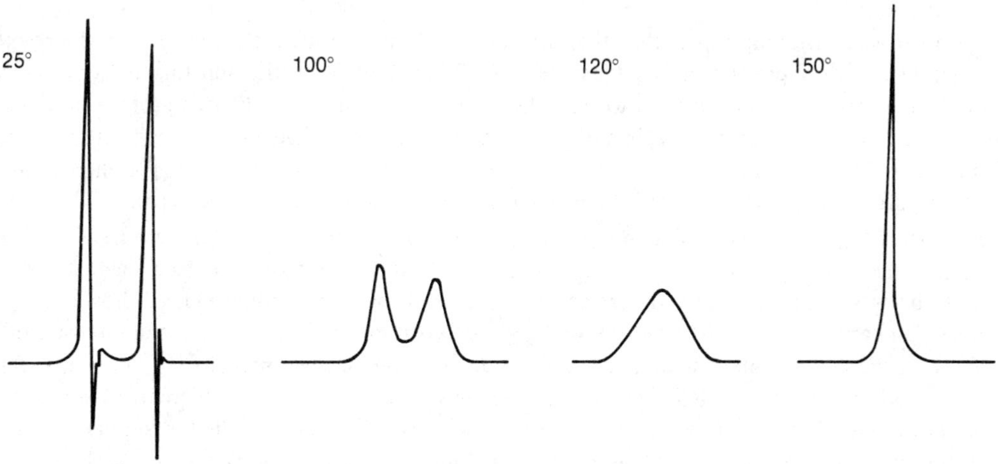

FIGURE 4–35 The Appearance of the Methyl Resonances of *N, N*-Dimethylformamide with Increasing Temperature

4.11 SPIN-SPIN SPLITTING BETWEEN NEARLY EQUIVALENT NUCLEI; ANOTHER FAILURE OF THE N + 1 RULE – SECOND ORDER SPECTRA

Previously, the failure of the $n + 1$ rule owing to magnetic non-equivalence within a group was discussed (Section 4.3). If either (1) the protons within a group (usually those bound to the same carbon) have different chemical shifts or (2) each of the protons within a group (even if they have the same chemical shift) has a different degree of coupling (J) to an adjacent group of protons, then the $n + 1$ rule fails. Spectra which can be interpreted by using the $n + 1$ rule or a simple graphical analysis are said to be *first order spectra*. There are cases, however, in which neither of these approaches suffices to explain the splitting patterns observed. In these latter cases a mathematical analysis must be carried out, usually by computer, in order to explain the spectrum. Obviously, this again is a case in which the $n + 1$ rule fails. Spectra which require this more advanced analysis are said to be *second order spectra*.

Second order spectra are most commonly observed in situations where the difference in chemical shift between two groups of protons is similar in magnitude (in Hz) to the coupling constant J (also in Hz) which links them. That is, second order spectra are observed for coupling between nuclei which are *nearly equivalent*, but not identical.

To examine this case, let us first look at two protons H_A and H_B placed on adjacent carbon atoms. Using the $n + 1$ rule, we expect to see two doublets with components of equal intensity. In actuality, we see two doublets of equal intensity in this situation only if the difference in chemical shift ($\Delta\nu$) between protons A and B is large compared to the magnitude of the coupling constant (J_{AB}) which links them. This case is illustrated in Figure 4–36.

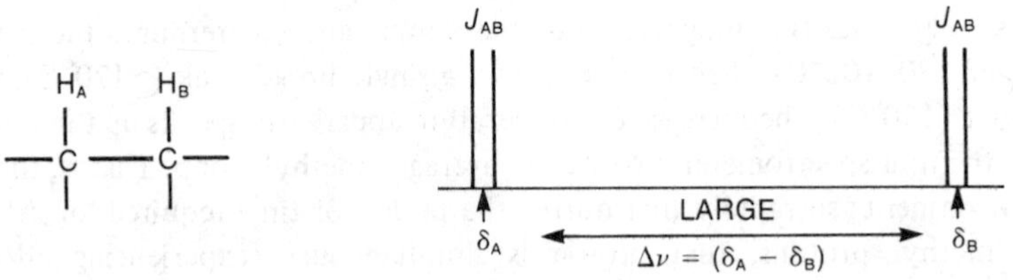

FIGURE 4–36 The Case in Which First Order Coupling (N + 1 Rule) Applies ($\Delta\nu$ Large)

Another way of saying this is that the ratio of the chemical shift difference ($\Delta\nu$) to the coupling constant (J), both expressed in Hz, must be large. When $\Delta\nu/J > 10$, the splitting pattern typically approximates first order splitting. However, when the chemical shifts of the two protons move closer together and $\Delta\nu/J$ approaches unity, one begins to see second order changes in the splitting pattern.

Figure 4–37 shows how the splitting pattern for the two-proton system H_AH_B changes as the chemical shifts of H_A and H_B come closer together and the ratio $\Delta\nu/J$ becomes smaller. The figure is to scale, with $J_{AB} = 7$ Hz. When $\delta_{H_A} = \delta_{H_B}$ (that is, when the protons H_A and H_B have the same chemical shift), then $\Delta\nu/J = 0$, and no splitting is observed; both protons give rise to a single absorption peak. In between this extreme and that of the simple first order spectrum which follows the $n + 1$ rule, subtle changes in the splitting pattern take place. Most obvious is the decrease in intensity of the outer peaks of the doublets with a corresponding increase in the intensity of the inner peaks. However, other changes which are not as obvious occur as well. Mathematical analysis by theoreticians has shown that while the chemical shifts of H_A and H_B in the simple first order spectrum correspond to the center point of each doublet, a more complex situation holds in the second order cases – the chemical shifts of H_A and H_B are found to be closer to the inner peaks than to the outer peaks. The actual position of δ_A and δ_B must be calculated. The difference in chemical shift must be determined from the line positions (in Hz) of the individual peak components of the group using the equation:

$$(\delta_A - \delta_B) = \sqrt{(\delta_1 - \delta_4)(\delta_2 - \delta_3)}$$

where δ_1 is the position (in Hz downfield from TMS) of the first line of the group and δ_2, δ_3, and δ_4 are the second, third, and fourth lines, respectively (see Figure 4–38). The chemical shifts of H_A and H_B are then displaced ½($\delta_A - \delta_B$) to either side of the center of the group.

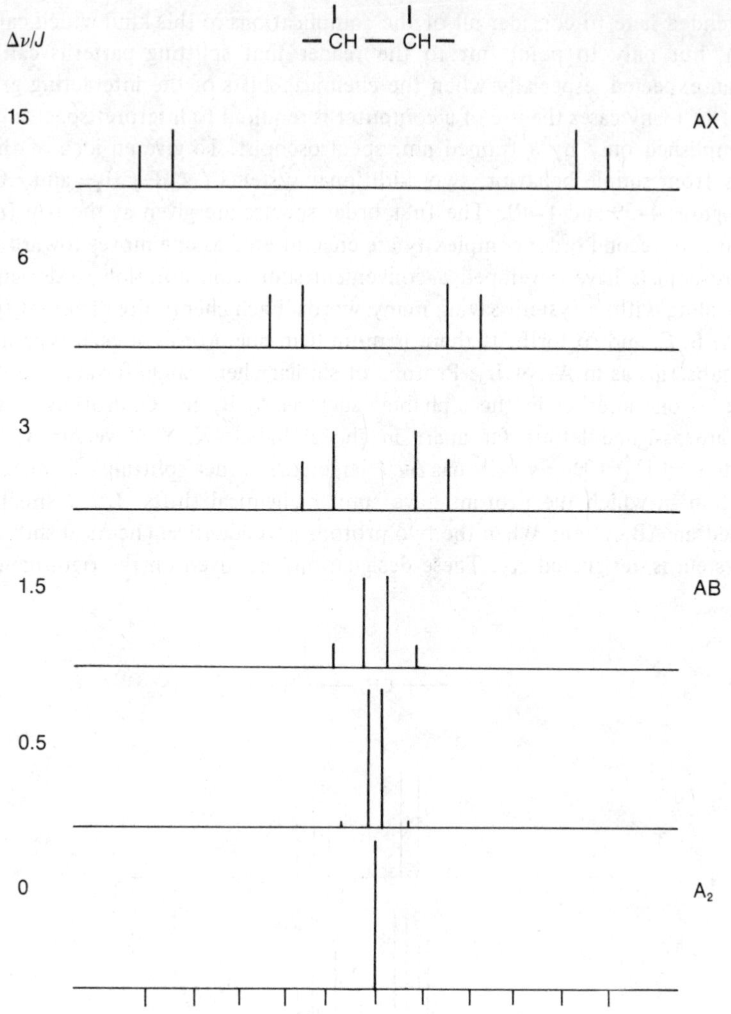

FIGURE 4–37 The Splitting Patterns of a Two-Proton System H_AH_B for Various Ratios of $\Delta\nu/J$

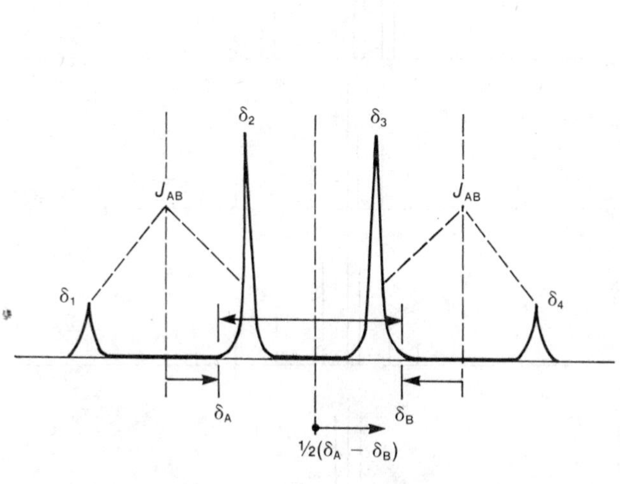

FIGURE 4–38 The Relationships Between the Chemical Shifts, Line Positions, and Coupling Constant in a Two-Proton Spectrum Which Exhibits Second Order Splitting Effects

It is not intended here to consider all of the complications of this kind which can occur for each type of system, but only to point out to the reader that splitting patterns can often be more complicated than expected, especially when the chemical shifts of the interacting groups of protons are very similar. In many cases the use of a computer is required to interpret spectra of this type, and it can be accomplished only by a trained nmr spectroscopist. To give an idea of the magnitude of these variations from simple behavior, two additional systems ($-CH-CH_2-$ and $-CH_2-CH_2-$) are illustrated in Figures 4–39 and 4–40. The first order spectra are given at the top ($\Delta\nu/J > 10$), while increasing amounts of second order complexity are encountered as one moves toward the bottom.

NMR spectroscopists have developed a convenient shorthand notation to designate the type of system one is dealing with, a system saving many words. Each chemically different type of proton is given a letter, A, B, C, and so forth. If there is more than one proton of each type in a group, this is indicated by a subscript as in A_2 or B_3. Protons of similar chemical shift values are assigned letters which are close to one another in the alphabet, such as A, B, and C. Protons of widely different chemical shift are assigned letters far apart in the alphabet: X, Y, Z versus A, B, C. Thus, the two-proton system $-CH_A-CH_B-$ which has $\Delta\nu/J$ large (first order splitting) is called an AX system, while the situation in which the protons have similar chemical shifts ($\Delta\nu/J$ small, second order splitting) is called an AB system. When the two protons have identical chemical shifts and give rise to a singlet, the system is designated A_2. These designations are given on the right-hand side of Figure 4–37.

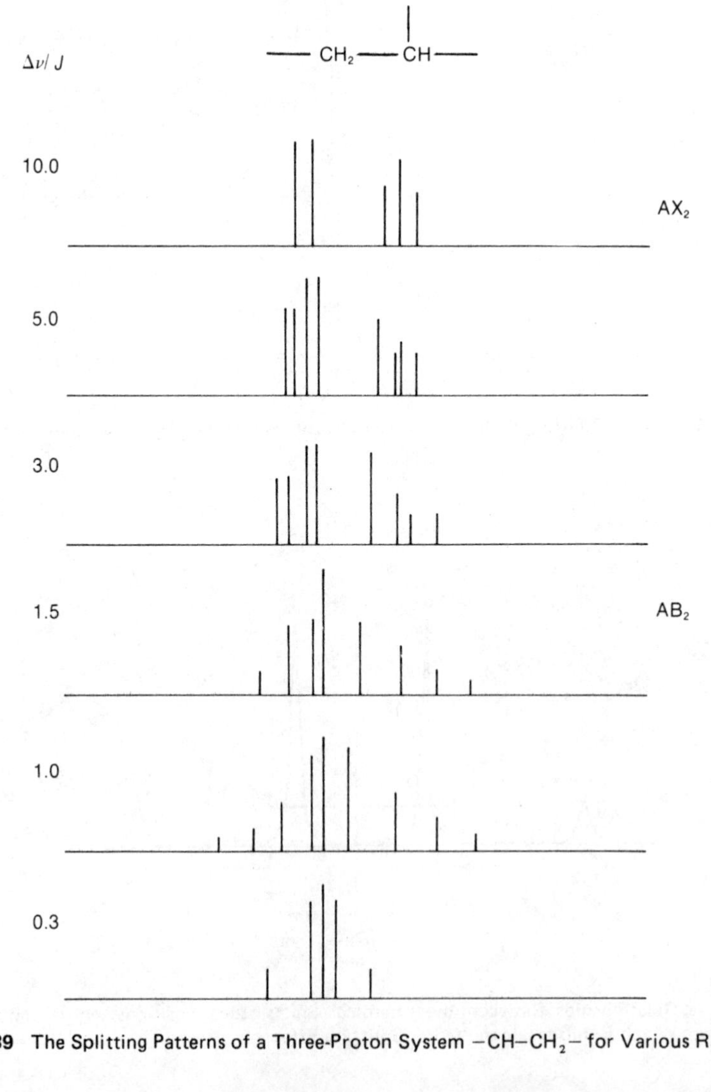

FIGURE 4–39 The Splitting Patterns of a Three-Proton System $-CH-CH_2-$ for Various Ratios of $\Delta\nu/J$

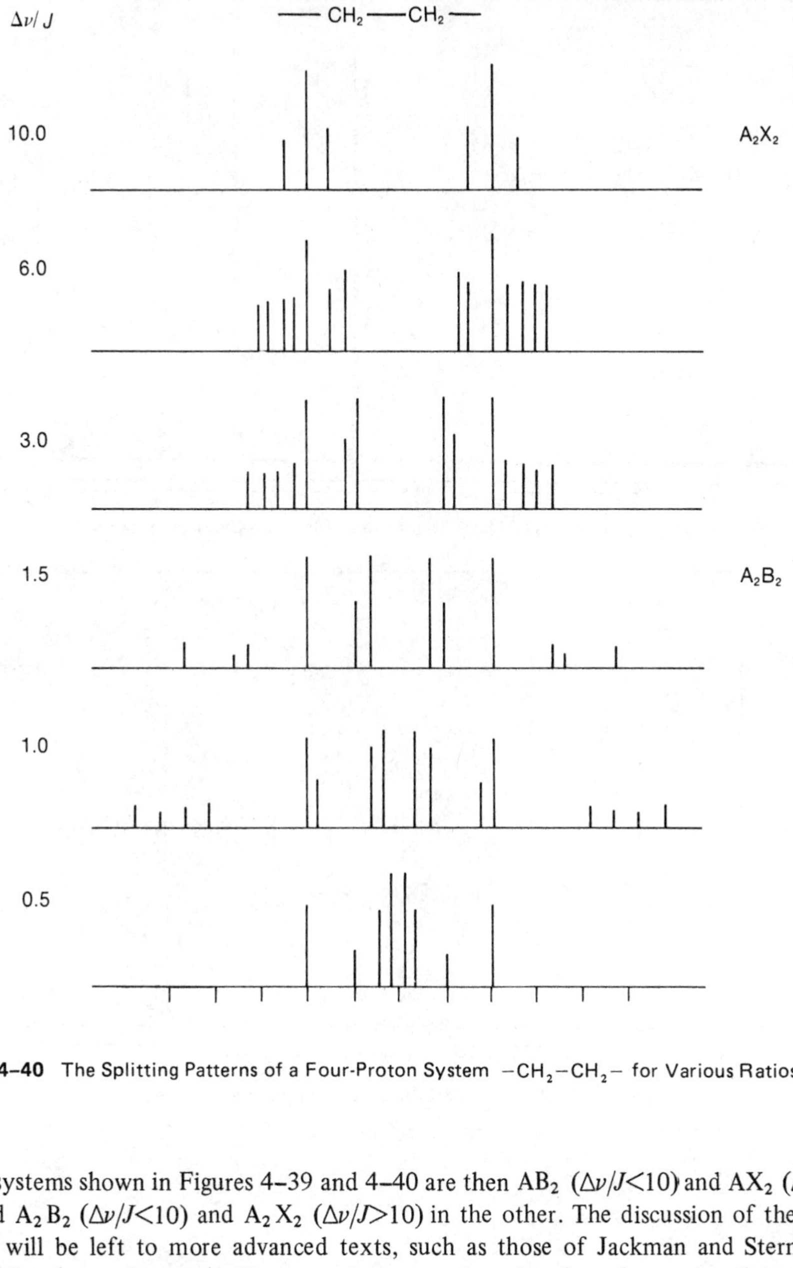

FIGURE 4–40 The Splitting Patterns of a Four-Proton System $-CH_2-CH_2-$ for Various Ratios of $\Delta\nu/J$

The two systems shown in Figures 4–39 and 4–40 are then AB_2 ($\Delta\nu/J<10$) and AX_2 ($\Delta\nu/J>10$) in one case, and A_2B_2 ($\Delta\nu/J<10$) and A_2X_2 ($\Delta\nu/J>10$) in the other. The discussion of these types of spin systems will be left to more advanced texts, such as those of Jackman and Sternhell and of Wiberg and Nist (see references). However, it is necessary to alert the reader that occasionally situations will be encountered which cannot possibly be interpreted with only a limited knowledge. On the following pages (Figures 4–41 through 4–46) the nmr spectra of some molecules of the A_2B_2 type are shown. The student will find it convenient to examine these spectra and compare them to the expected patterns shown in Figure 4–40. The patterns in Figure 4–40 were calculated from theory by use of a computer.

Finally, it should be mentioned that in two earlier sections the spectra of another commonly encountered type of system were introduced. Both the nmr spectra of styrene oxide (Figure 4–9) and of vinyl acetate (Figure 4–12) contain examples of an AMX system, where M, being in the middle of the alphabet, indicates a chemical shift intermediate between those of A and X. That is, all three chemical shifts (A, M, and X) are separated widely. Trimethylene oxide (Figure 4–8) provides an example of an A_2X_4 system, and *trans*-cinnamic acid (Figure 4–11) gives an example of an AX system.

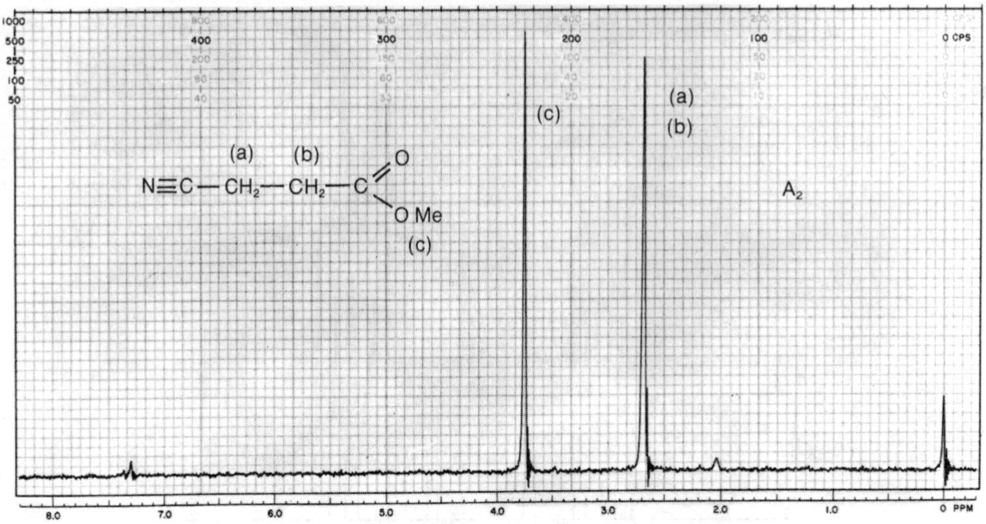

FIGURE 4–41 The NMR Spectrum of Methyl β-Cyanopropionate

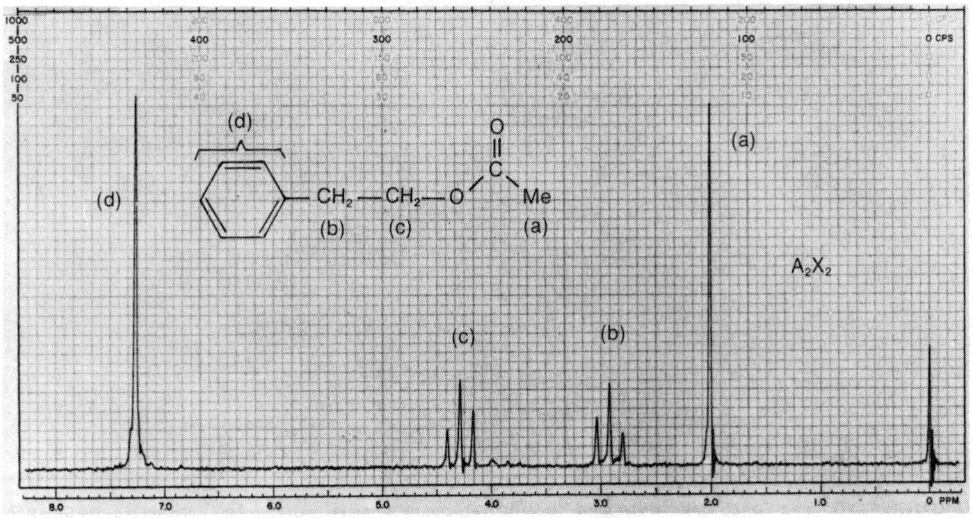

FIGURE 4–42 The NMR Spectrum of Phenylethyl Acetate

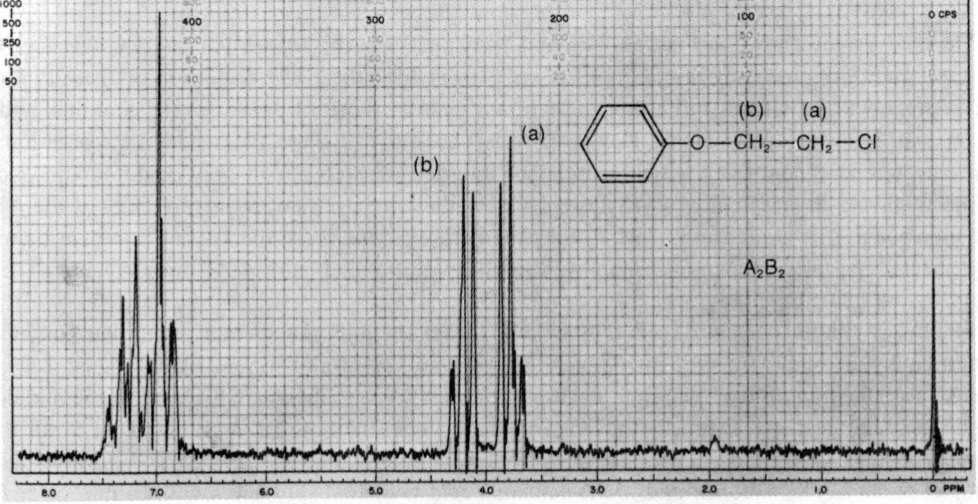

FIGURE 4–43 The NMR Spectrum of β-Chlorophenetole

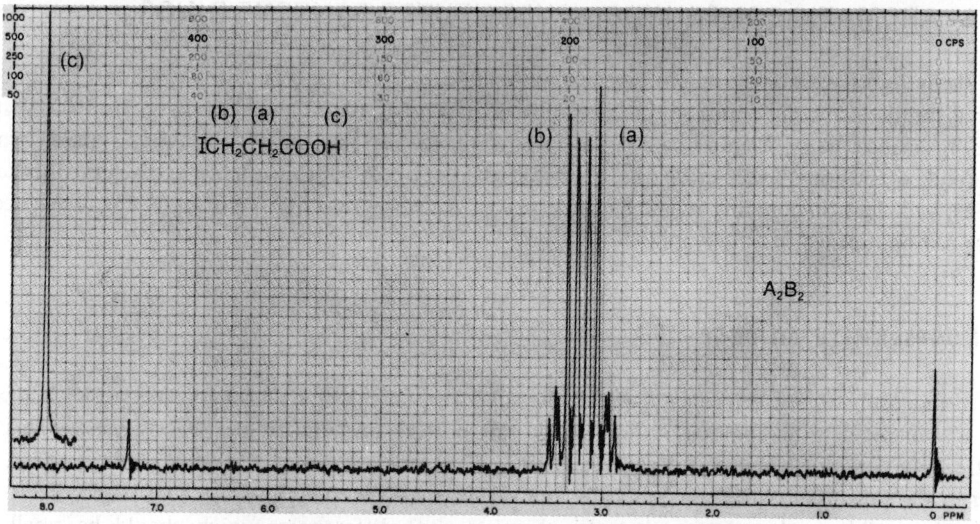

FIGURE 4–44 The NMR Spectrum of β-Iodopropionic Acid

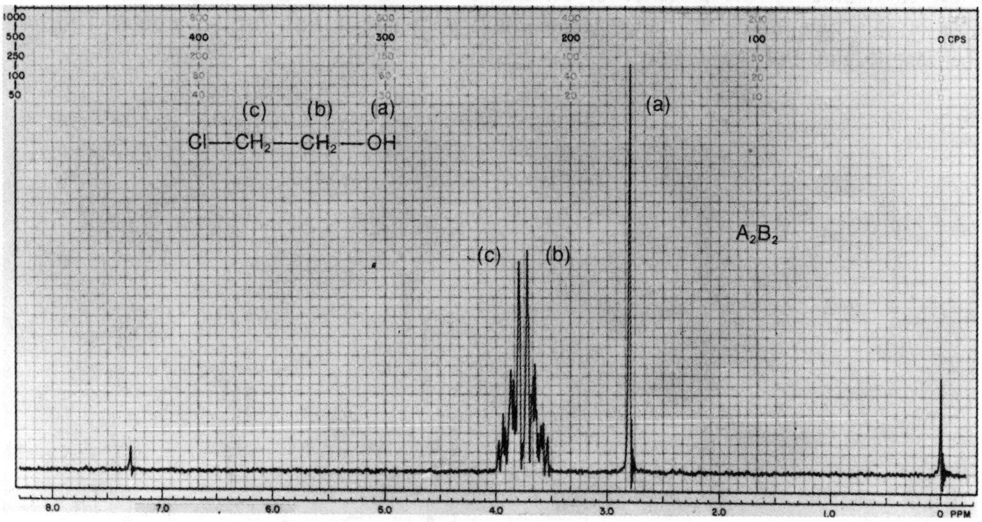

FIGURE 4–45 The NMR Spectrum of 2-Chloroethanol

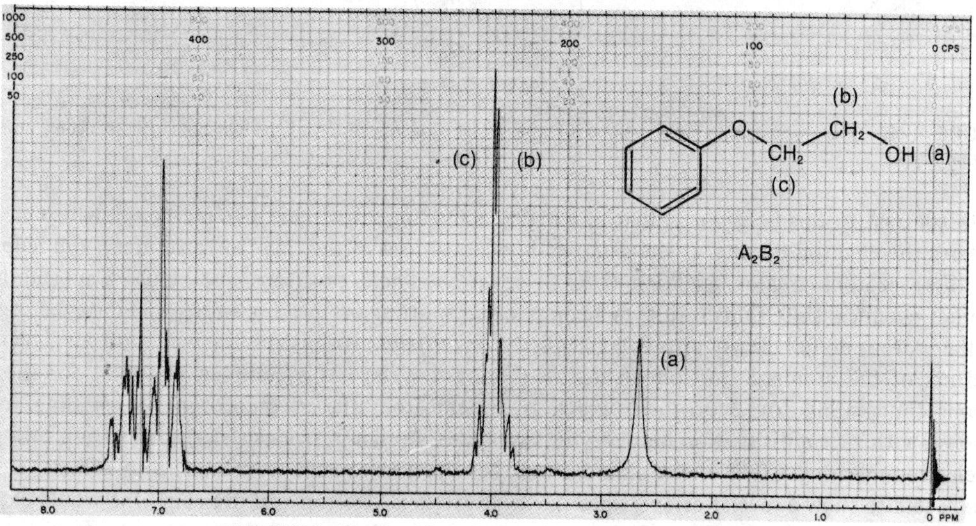

FIGURE 4–46 The NMR Spectrum of 2-Phenoxyethanol

4.12 AROMATIC COMPOUNDS – SUBSTITUTED BENZENE RINGS

Phenyl rings are so common in organic compounds that it is important to know a few facts about nmr absorptions in compounds which are aromatic. In general, the ring protons of an aromatic system have resonance near 7.2 δ; however, electron-withdrawing ring substituents (*e.g.*, nitro, cyano, carboxyl, or carbonyl) will move the resonance of these protons downfield (1,3,5-trinitrobenzene: 9.5 δ), and electron-donating ring substituents (*e.g.*, methoxy or amino) will move the resonance of these protons upfield (*p*-ethoxyaniline: 6.7 δ).

A. MONOSUBSTITUTED RINGS

In monosubstituted benzenes in which the substituent is neither a strongly electron-withdrawing nor a strongly electron-donating group, all the ring protons give rise to a *single resonance*. This is a particularly common occurrence in alkyl substituted benzenes. Although the protons *ortho*, *meta*, and *para* to the substituent are not chemically equivalent, they generally give rise to a single absorption peak. A possible explanation is that the chemical shift differences, which should be small, are somehow eliminated by the presence of the ring current which tends to equalize them. A good example of this type of circumstance is seen in the nmr spectrum of the aromatic portion of ethylbenzene (Figure 4–47A).

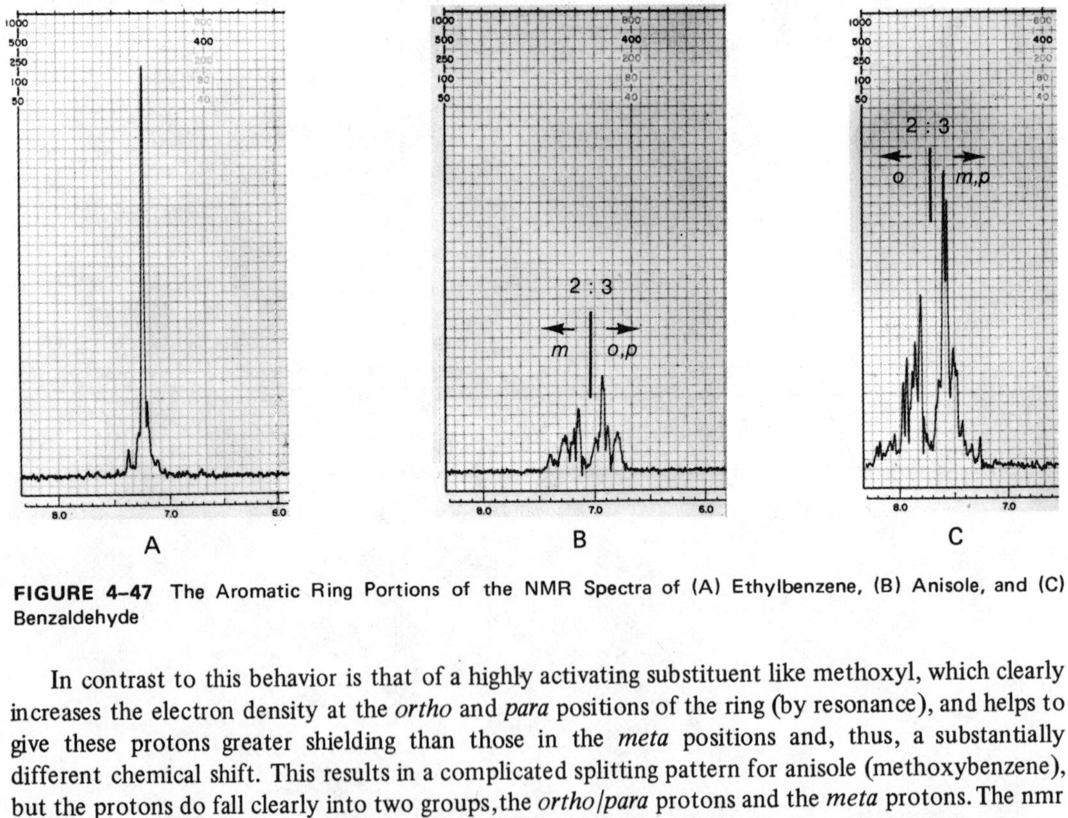

FIGURE 4–47 The Aromatic Ring Portions of the NMR Spectra of (A) Ethylbenzene, (B) Anisole, and (C) Benzaldehyde

In contrast to this behavior is that of a highly activating substituent like methoxyl, which clearly increases the electron density at the *ortho* and *para* positions of the ring (by resonance), and helps to give these protons greater shielding than those in the *meta* positions and, thus, a substantially different chemical shift. This results in a complicated splitting pattern for anisole (methoxybenzene), but the protons do fall clearly into two groups, the *ortho*/*para* protons and the *meta* protons. The nmr spectrum of the aromatic portion of anisole (Figure 4–47B) has a complex multiplet for the *o,p* protons (integrating for 3 protons) which is upfield from the *meta* protons (integrating for 2 protons), with a clear distinction between the two types. Aniline provides a similar spectrum owing to the electron-releasing effect of the amino group. Nitrobenzene (aside from anisotropy effects) would be expected to show a reverse effect, since the nitro group is electron-withdrawing. One would expect that the nitro group would act to decrease the electron density around the *ortho* and *para* positions, thus deshielding the *ortho* and *para* hydrogens, and providing a pattern which was exactly the reverse of the one shown for anisole (3:2 ratio, downfield:upfield). The reader should convince himself or herself of this by drawing out resonance structures. Nevertheless, the actual nmr spectrum of nitrobenzene does not have the appearance one would predict based on resonance structures. Instead, the *ortho* protons are much more deshielded than the *meta* and *para* protons. This effect is due to the magnetic anisotropy of the nitro group. A similar effect is observed when a substituent group bonds a carbonyl directly to the ring (Figure 4–48). Once again, the ring protons fall into two groups, with the *ortho* protons downfield from the *meta*/*para* protons. Benzaldehyde (Figure 4–47C) and acetophenone both show this effect in their nmr spectra. A similar effect is *sometimes* observed when a carbon-carbon double bond is attached to the ring.

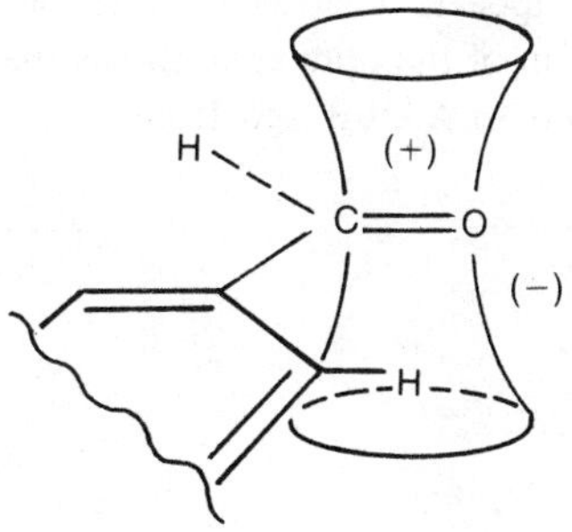

FIGURE 4–48 Anisotropic Deshielding of the *ortho* Protons of Benzaldehyde

B. PARA-DISUBSTITUTED RINGS

Of the possible substitution patterns of a benzene ring, only a few are easily recognized. One of these is the *para*-disubstituted benzene ring. Examine *p*-chloroaniline as a first example. It is illustrated in Figure 4-49A.

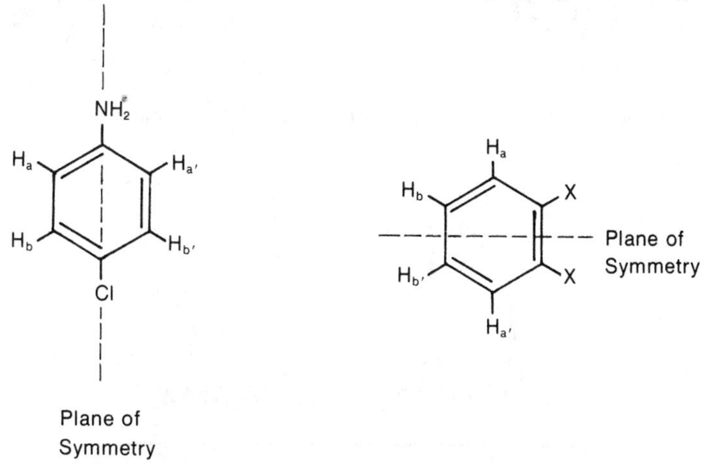

FIGURE 4–49 The Planes of Symmetry Present in (A) a *para*-Disubstituted Benzene Ring (*p*-Chloroaniline) and (B) a Symmetric *ortho*-Disubstituted Benzene Ring

Since this compound has a plane of symmetry (passing through the amino and chloro groups), one would expect the protons H_a and $H_{a'}$ (both *ortho* to NH_2) to have the same chemical shift. Similarly, the protons H_b and $H_{b'}$ should also have the same chemical shift. This is found to be the case. Both sides of the ring should then have identical splitting patterns. On this assumption one is tempted to look at each side of the ring separately. If one does this, a pattern is expected in which proton H_a is split by proton H_b into a doublet and proton H_b is split by proton H_a into a second doublet. Examination of the nmr spectrum of *p*-chloroaniline (Figure 4–50A) shows (crudely) just such a four-line pattern for the ring protons. In fact, a *para*-disubstituted ring is easily recognized by this pattern. *However*, the four lines *do not* correspond to a simple splitting pattern. This is because although the two protons H_a and $H_{a'}$ are chemically equivalent (same chemical shift), they are not magnetically equivalent. Protons H_a and $H_{a'}$ interact with one another and have a finite coupling constant $J_{aa'}$. Similarly, H_b and $H_{b'}$ also interact with each other and have coupling constant $J_{bb'}$. More importantly, H_a does not interact equally with H_b (*ortho* to H_a) and with $H_{b'}$ (*para* to H_a); that is, $J_{H_aH_b} \neq J_{H_aH_{b'}}$. If H_b and $H_{b'}$ are coupled differently to H_a, they cannot be magnetically equivalent. Turning the argument around, neither can H_a and $H_{a'}$ be magnetically equivalent because they are coupled differently to H_b, and also to $H_{b'}$. This suggests that the situation is more complicated than it might at first appear. A closer look at the pattern in Figure 4–50A shows that this is indeed the case. An expansion of the ppm scale shows that this pattern actually resembles four triplets (Figure 4–51). The pattern is an AA′BB′ spectrum.

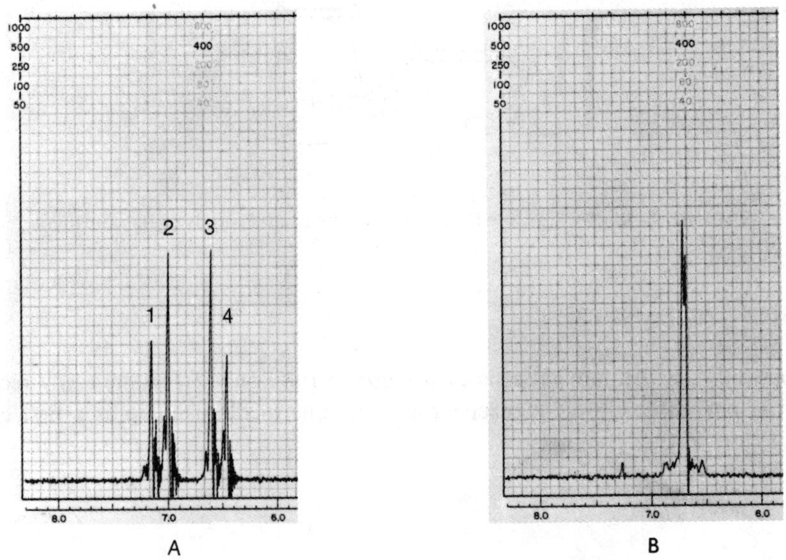

FIGURE 4–50 The Aromatic Ring Portions of the NMR Spectra of (A) *p*-Chloroaniline and (B) *p*-Methoxyaniline

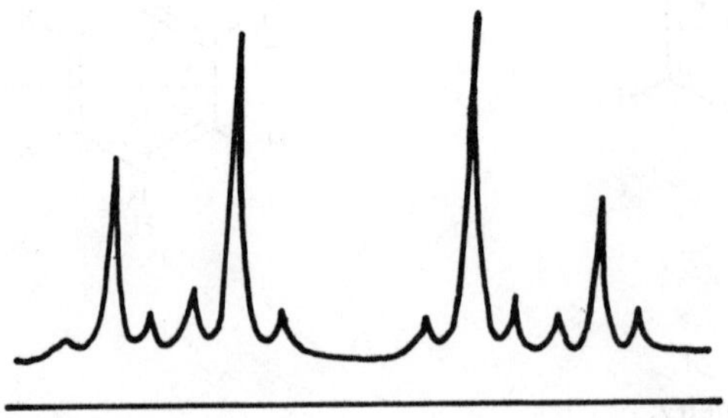

FIGURE 4–51 The Expanded *p*-Disubstituted Benzene Pattern

We will leave an analysis of this pattern to more advanced texts. Note, however, that a crude four-line spectrum is a characteristic pattern for a *para*-disubstituted ring. It is also characteristic for an *ortho*-disubstituted ring of the type shown in Figure 4–49B, where *both ortho* substituents are identical, leading to a plane of symmetry. The *ortho*-disubstituted case is relatively uncommon. As the chemical shifts of H_a and H_b approach one another, the *p*-disubstituted pattern becomes similar to that of *p*-methoxyaniline (Figure 4–50B). The inner peaks move closer together, and the outer ones become smaller or even disappear. Ultimately, when H_a and H_b approach each other in chemical shift closely enough, the outer peaks disappear and the two inner peaks merge into a *singlet*. Hence, a single aromatic resonance integrating for *four* protons could easily represent a *para*-disubstituted ring, but the substituents would obviously be either identical or very similar in nature. Note that a monosubstituted ring with alkyl groups would also give a singlet, but that this singlet would integrate for *five* protons.

C. OTHER SUBSTITUTION

Other modes of ring substitution can often lead to more complicated splitting patterns than those of the cases mentioned above. In aromatic rings, coupling usually extends beyond the adjacent carbon atoms. In fact, *ortho*, *meta*, and *para* protons can all interact, although the last interaction (*para*) is not usually observed. Typical *J* values for each of these interactions are given below.

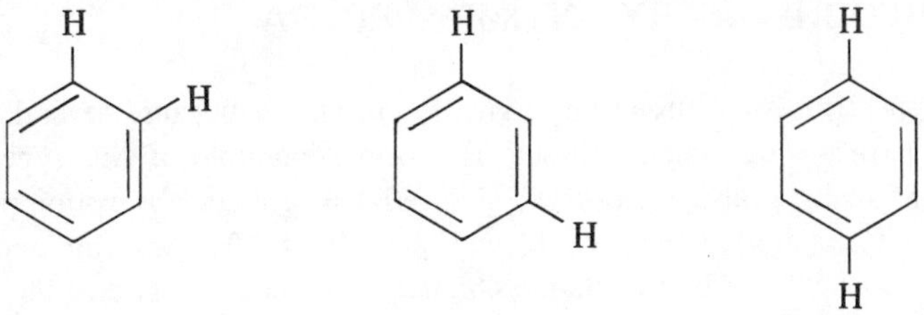

Ortho	*Meta*	*Para*
J = 7–10 Hz	J = 2–3 Hz	J = ∿1 Hz

The trisubstituted compound 2,4-dinitroanisole shows all of the types of interactions mentioned. The analysis is shown in Figure 4–53, following the nmr spectrum (Figure 4–52). In this example, as is usual, the coupling between the *para* protons is essentially zero. (The peak designated by the arrow is due to an impurity, $CHCl_3$, in the $CDCl_3$ solvent.)

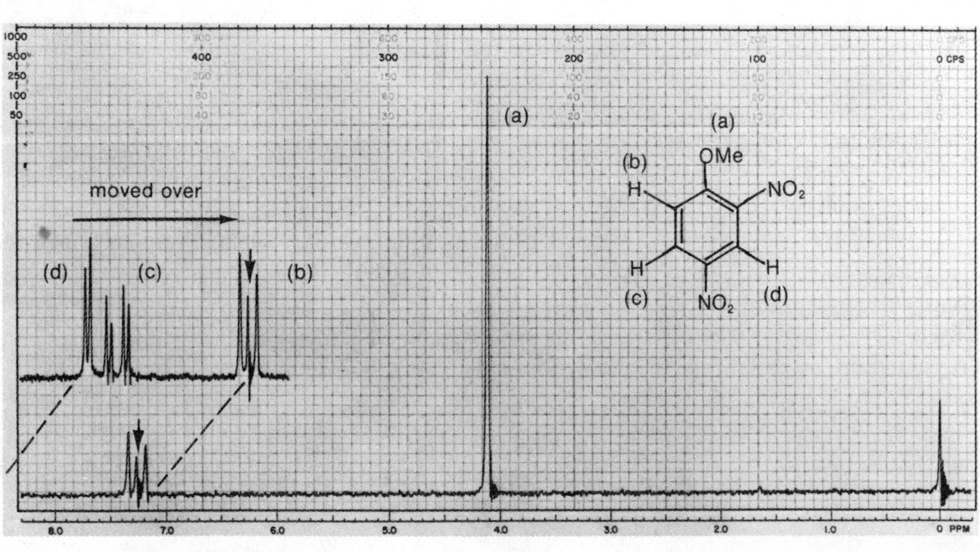

FIGURE 4–52 The NMR Spectrum of 2,4-Dinitroanisole Taken in $CDCl_3$ Solvent

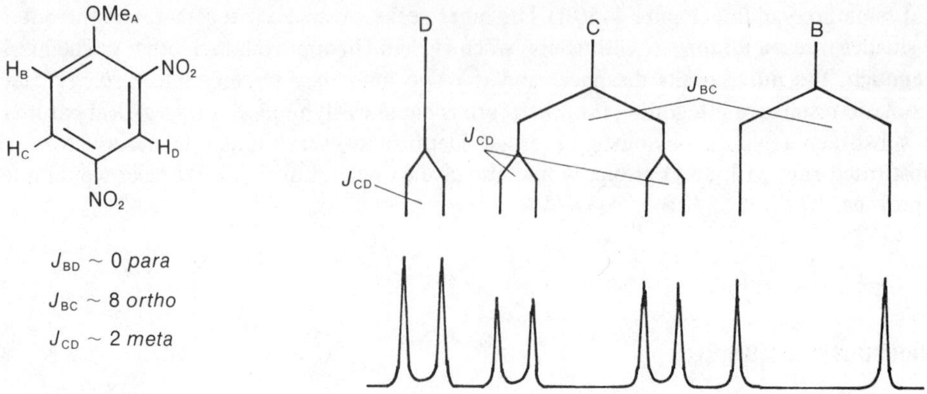

FIGURE 4–53 An Analysis of the Splitting Pattern in the NMR Spectrum of 2,4-Dinitroanisole

4.13 A CLOSER LOOK AT THE EFFECTS OF CONFORMATION AND STEREOCHEMISTRY ON NMR SPECTRA

Often the nmr spectrum of a molecule is greatly simplified because several of its protons are chemically and magnetically equivalent. Ethane is a prime example of this type of molecule, the chemically and magnetically equivalent protons giving rise to a single absorption peak. The magnetic equivalence of the protons arises from a conformational effect– free rotation about the C–C single bond. If ethane were locked into one of its possible staggered conformations, the situation might be quite different. Although all the protons would be in an identical environment magnetically, there would be two different coupling constants between the protons on adjacent carbons. Taking a look at H_A, the topmost proton in the front of the Newman projection of ethane, it should have two different types of vicinal interaction with the protons on the carbon in the rear:

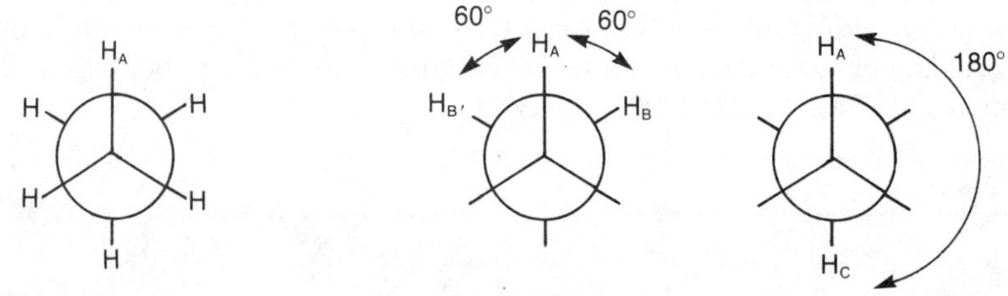

Ethane
Staggered Conformation

Two of the protons in the rear are related to this proton by a 60° dihedral angle, and the third by a dihedral angle of 180°. As we saw in Section 4.1, the magnitude of the coupling constant is dependent on this dihedral angle. At an interproton dihedral angle of 60°, the coupling constant J_{AB} should have a magnitude of about 4–5 Hz, while at an angle of 180° J_{AC} should be in the vicinity of 10–12 Hz. This is the type of situation to which one applies graphical analysis, and we would expect on this basis to see a four-line pattern for the resonance of H_A.

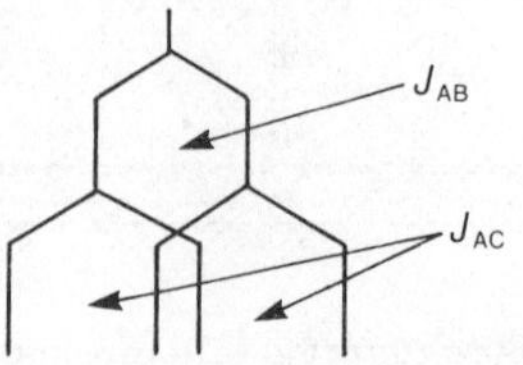

However, there are at least two reasons why this does not happen. First, there is free rotation in ethane, which averages all the possible dihedral values to a single average value. Second, even though coupling exists, it need not necessarily lead to spin-spin splitting (review geminal coupling, Section 4.2). Even if ethane were locked into the staggered conformation, it is likely that no spin-spin splitting would be observed. All protons are chemically and magnetically equivalent, as well as equivalent by symmetry, and thus would have the same chemical shift. In this situation, just as for methylene (geminal) protons, even though coupling existed, it would not be observed.

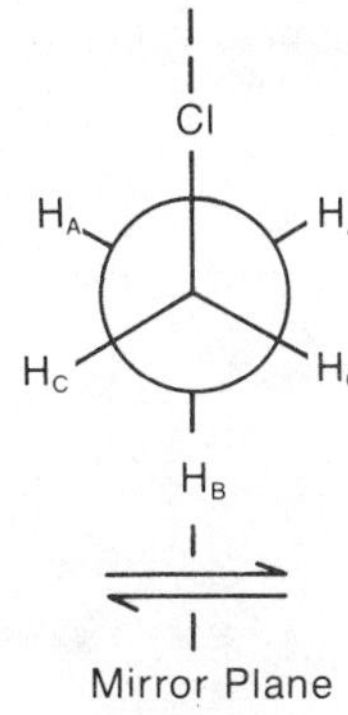

The next step is to make the situation somewhat less symmetric. Substitution of a chlorine atom for one proton does this and makes a good example. In chloroethane not all the protons are equivalent by symmetry, nor are they magnetically equivalent. The methylene protons have a larger chemical shift value than the methyl protons. To start with, "freeze" chloroethane in its staggered conformation (see diagram above). Clearly, the protons labeled H_A, H_B, and H_C are in different magnetic environments. Protons $H_{A'}$ and $H_{C'}$ are equivalent to H_A and H_C, respectively. Protons H_C and $H_{C'}$ are on the carbon atom bearing the chlorine atom and should have the greatest chemical shift. On the other carbon atom, protons H_A and $H_{A'}$ are closer to chlorine than is proton H_B, and they should show an intermediate chemical shift value. That is, all three types of protons (H_A, $H_{A'}$), H_B, and (H_C, $H_{C'}$) are magnetically non-equivalent and should have different chemical shift values. Additionally, protons H_A and $H_{A'}$, and protons H_C and $H_{C'}$, have a special relationship to one another. Since there is a mirror plane of symmetry in this conformation of chloroethane, protons H_A and $H_{A'}$ (as well as H_C and $H_{C'}$) are in *mirror image environments*. Protons which are magnetically equivalent and exist in mirror image environments are called enantiomeric or, better, *enantiotropic* protons. One proton is in a left-hand environment, while the other is in a right-hand environment. They are, nevertheless, in magnetically equivalent environments. Such protons cannot be distinguished from one another with ordinary nmr measurements except under special conditions. To have different chemical shifts, they must be placed in a dissymmetric or chiral environment. We will not consider this in any detail here, but one of the ways in which this can be accomplished is through the use of an optically active nmr solvent.

If we were to determine the nmr spectrum of chloroethane while it was locked in the staggered conformation, there would be six parameters determining the appearance of the spectrum: three different chemical shift values and three different coupling constants. These are given below:

Chemical Shifts	*Coupling Constants*
1. $\delta_{H_A} = \delta_{H_{A'}}$ (2 protons)	1. J_{AC} ($= J_{A'C'}$) vicinal, 60°
2. δ_{H_B} (1 proton)	2. J_{BC} ($= J_{BC'}$) vicinal, 60°
3. $\delta_{H_C} = \delta_{H_{C'}}$ (2 protons)	3. J_{AB} ($= J_{A'B}$) geminal, 109°

While the vicinal and geminal couplings would not be observed in staggered ethane because of the magnetic equivalence of all the protons involved, in staggered chloroethane the three different types of protons should have substantially different chemical shifts and all three coupling interactions should lead to observable spin-spin splitting.

In actuality, of course, chloroethane is not locked in its staggered conformation, and rotation around the C–C bond occurs at a rate much faster than the time required for an nmr transition. This rotation eliminates the difference in chemical shift between H_A (and/or $H_{A'}$) and H_B since they interchange positions during the rotation. All three protons (H_A, $H_{A'}$ H_B) have the same chemical shift under these conditions. They will have a single chemical shift value somewhere *in between* the two different values expected for H_A and H_B in the fixed conformation (Figure 4–54). This single value will correspond to the average environment which these three protons experience.

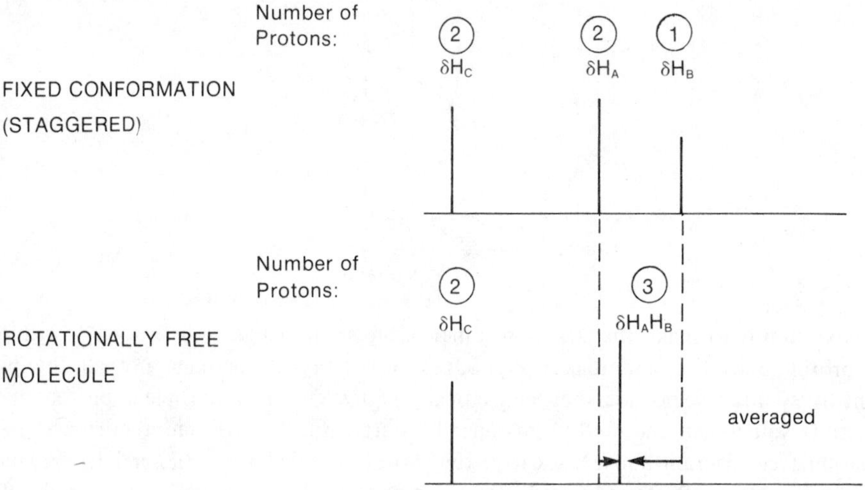

FIGURE 4–54 The Chemical Shift Positions of Chloroethane Protons

Similarly, since all three methyl protons of the rotating molecule effectively now have the same chemical shifts, J_{AB}, the geminal coupling in this group is not observed, and the vicinal coupling constants J_{AC} and J_{BC} are merged to a single averaged value intermediate between them in magnitude. Thus, we now have only three parameters to consider in the nmr spectrum: two chemical shift values (methyl and methylene protons) and one coupling constant $J_{CH_2-CH_3}$. Rotationally free chloroethane thus obeys the $n + 1$ rule and gives a quartet integrating for 2 protons and a triplet integrating for 3 protons. It should be realized, however, that unlike ethane or chloroethane, there could exist molecules that would not be rotationally free because of bulky substituents, or molecules that would have a preferred conformation for the same type of reason.

4.14 DIASTEREOMERIC PROTONS

Let us now turn to a new example involving stereochemistry and conformation: the nmr spectrum of 1,2-dibromo-1-phenylethane. Examination of the Newman projection diagrams of the three possible fixed staggered conformations shows that none of the protons are ever enantiomeric. There are no mirror image environments.

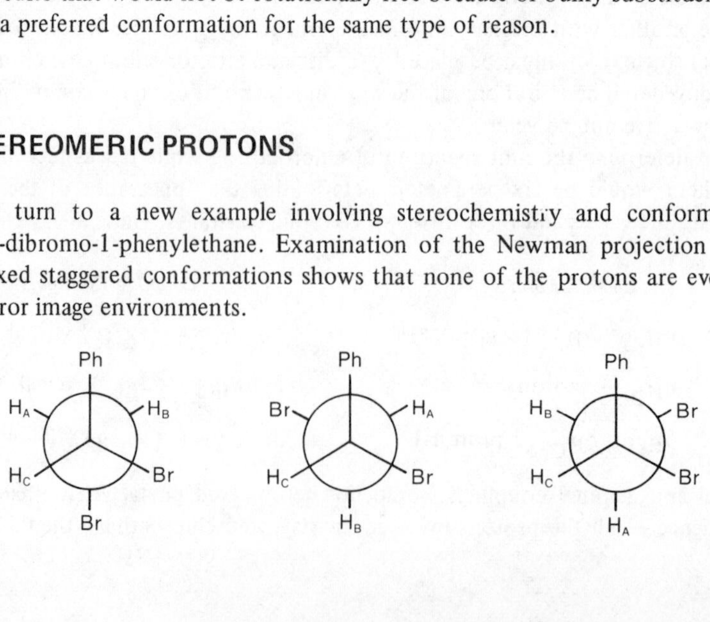

This examination also reveals that the methylene protons (H_A and H_B) are *never* magnetically equivalent. Because the front carbon is asymmetric (chiral), the left-hand side of the molecule is magnetically non-equivalent to the right-hand side. Protons such as H_A and H_B are said to be *diastereomeric* or *diastereotropic* protons. Protons of this type have different chemical shifts even in the rotationally free molecule. Furthermore, even in their rotationally averaged condition, they will couple differently to H_C, and will also couple to each other.

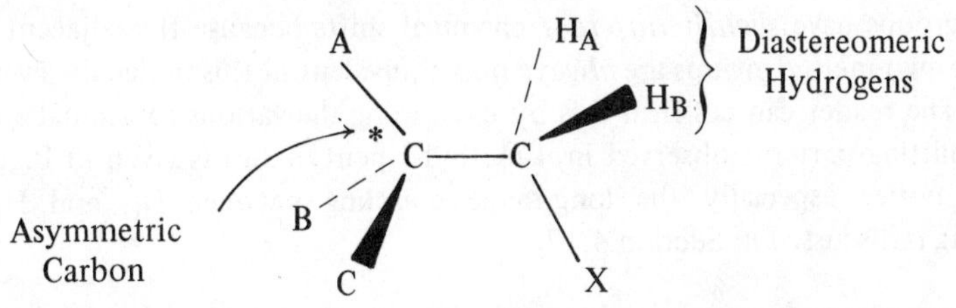

Thus, the possible spectral parameters of 1,2-dibromo-1-phenylethane are six in number:

1. δH_A	4. J_{AB} (geminal)
2. δH_B	5. J_{AC} (vicinal)
3. δH_C	6. J_{BC} (vicinal)

In the actual nmr spectrum of 1,2-dibromo-1-phenylethane, the geminal coupling is not observed in the spin-spin splitting pattern (this is a frequent occurrence), and the pattern shown in Figure 4–55 is seen for the methine (H_C) and methylene (H_A and H_B) protons.

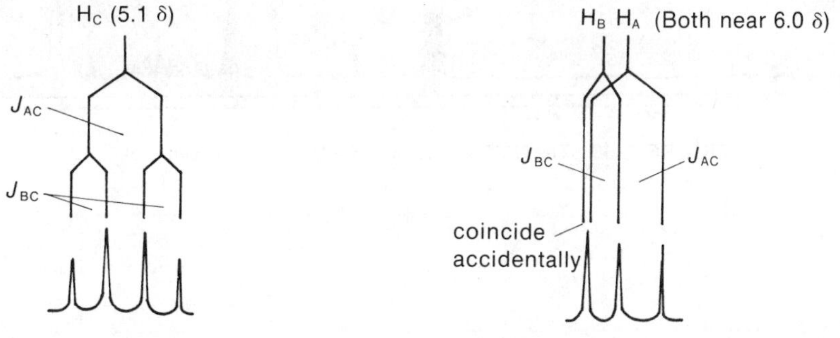

FIGURE 4–55 Appearance of the NMR Spectrum (Non-Aromatic Part) of $PhCHBrCH_2Br$.

The normal pattern for a CH–CH_2 system (AX_2) would be a 1:2:1 triplet integrating for one (CH) and a 1:1 doublet integrating for two (CH_2). Notice that the pattern for 1,2-dibromo-1-phenylethane would merge into this simple first order ($n + 1$) pattern if J_{AC} and J_{BC} were the same, and if H_A and H_B had the same chemical shift (see the graphical analysis in Figure 4–55).

In many compounds of the type expected to have diastereomeric hydrogens, the difference between the chemical shifts of the diastereomeric hydrogens, H_A and H_B, is so small that neither this difference nor the difference in coupling between H_A and H_B and some other proton, H_C, is easily detectible. In other compounds, however, δH_A and δH_B are quite different, leading to large differences in J_{AC} and J_{BC} as well. Then the $n + 1$ pattern is not observed, but instead, a more complicated splitting pattern emerges.

Occasionally, groups other than hydrogen will be diastereomeric. The alcohol 4-methyl-1-pentyn-3-ol is a molecule with diastereomeric *methyl groups*. Its spectrum is shown in Figure 4–56.

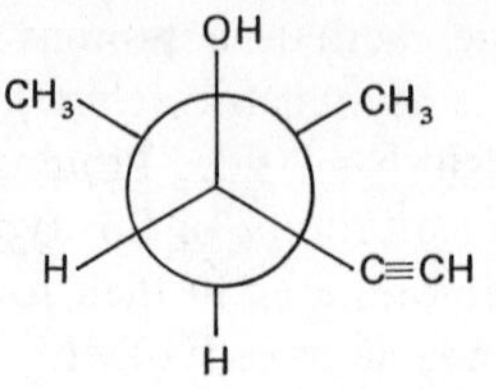

The two methyl groups have *slightly different* chemical shifts because the adjacent carbon atom is a chiral center. The two methyl groups are *always* non-equivalent in this molecule, even in the presence of free rotation. The reader can confirm this by examining the various rotational conformations. An analysis of the splitting patterns observed in 4-methyl-1-pentyn-3-ol is given in Figure 4–57. In this unusual scheme, notice especially the long-range coupling between H_D and H_F. This type of acetylenic coupling is discussed in Section 4.17.

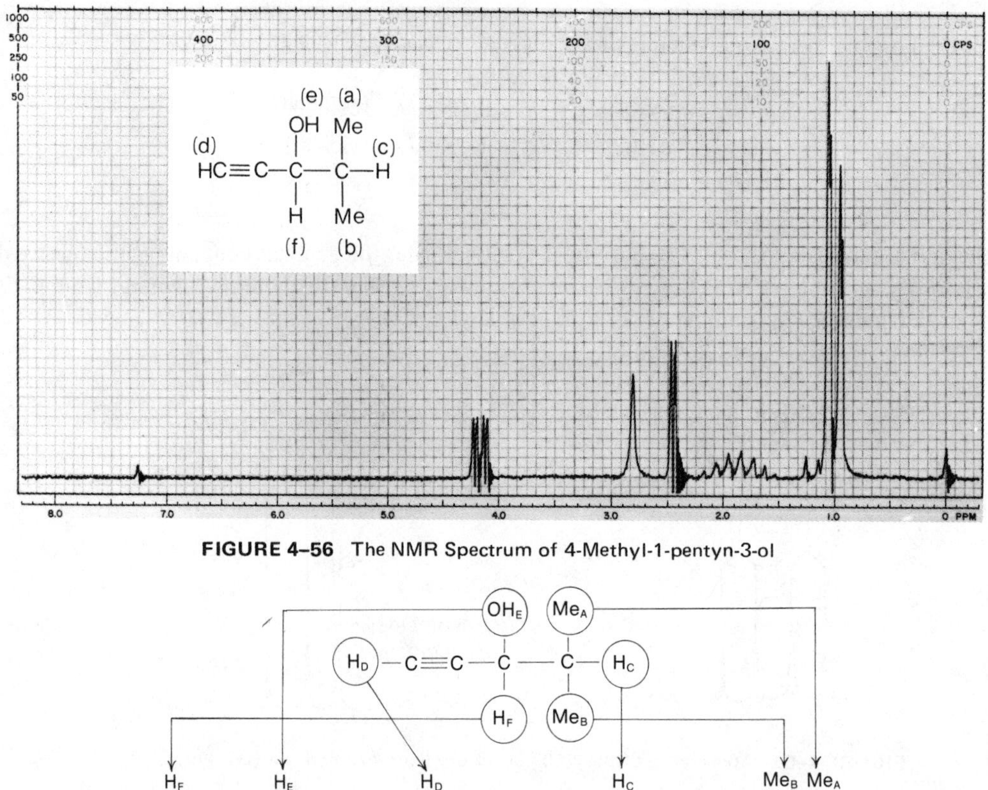

FIGURE 4–56 The NMR Spectrum of 4-Methyl-1-pentyn-3-ol

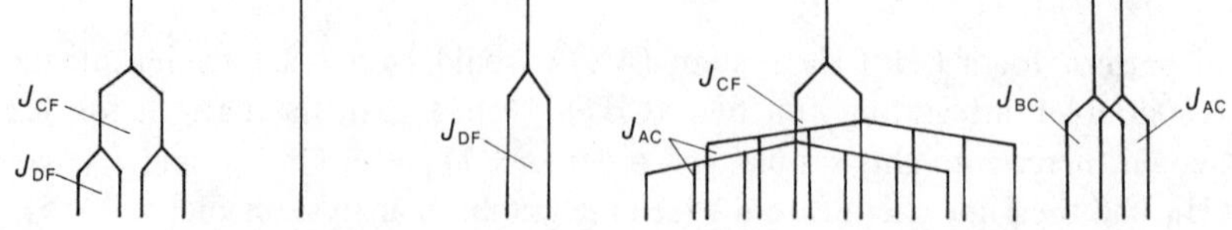

FIGURE 4-57 An Analysis of the Splitting in the NMR Spectrum of 4-Methyl-1-pentyn-3-ol (Figure 4–56)

4.15 IS THE N + 1 RULE REALLY EVER OBEYED?

In a linear chain the $n + 1$ rule is strictly obeyed only if the vicinal interproton coupling constants are the same for every successive pair of carbons.

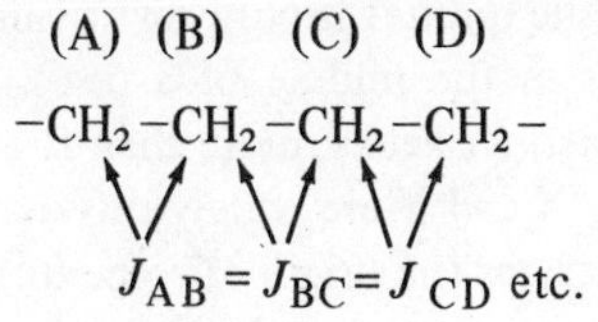

Condition for which the n + 1 Rule is Strictly Obeyed

Consider a three-carbon chain as an example. The protons on carbon B are split both by those on carbon A and by those on carbon C. If there is a total of four protons on carbons A and C, the $n + 1$ rule predicts a quintet. This occurs only if $J_{AB} = J_{BC}$. Graphically, this can be seen in Figure 4–58.

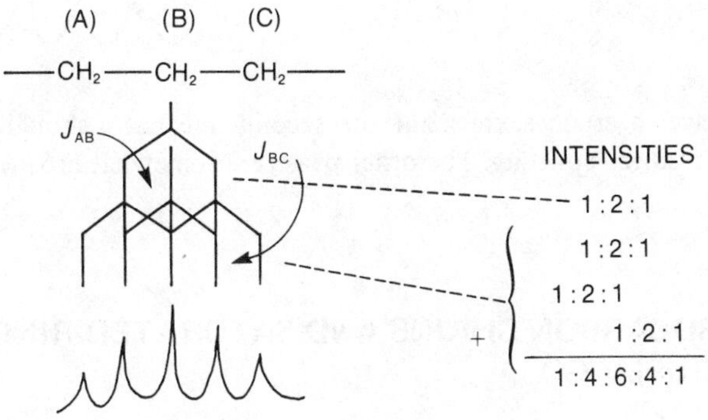

FIGURE 4–58 Construction of a Quintet for a Methylene Group with Four Neighbors All Coupled to the Same Extent

First, the protons on carbon B are split by those on carbon A, yielding a triplet (intensities 1:2:1). Each of these components of the triplet is then split into another triplet (1:2:1) by the protons on carbon C. At this stage it is seen that many of the lines from these second splittings *overlap* because they have the same spacings as the lines from the first splitting. Because of this coincidence, only five lines are observed. But it is seen easily that they arise in the fashion indicated by adding the intensities of each splitting (see Figure 4–58) to get a prediction of the intensities of the final five-line pattern. These agree with those predicted by using Pascal's triangle (Section 3.14). Thus it is seen that the $n + 1$ rule depends on a special condition. In many molecules J_{AB} is *slightly* different from J_{BC}. This leads to peak broadening in the multiplet, since the lines do not quite overlap. (Broadening occurs because the peak separation in Hz is too small in magnitude for the nmr instrument to be able to distinguish the separate peak components.) Non-equivalence of J_{AB} and J_{BC}, of course, happens most often when the chemical shifts of the protons on carbons A, B, and C are all quite different.

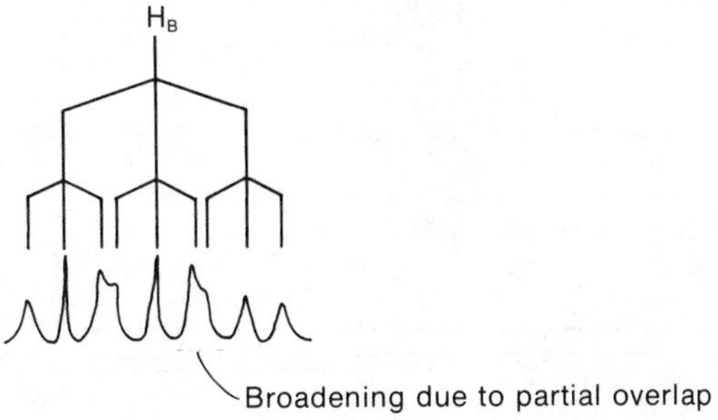

FIGURE 4–59 Loss of the Simple Quintet When $J_{AB} \neq J_{BC}$

Sometimes the perturbation of the quintet is only slight, and then either a shoulder is seen on the side of a peak, or a dip is obvious in the middle of a peak. At other times, when there is a large difference between J_{AB} and J_{BC}, distinct peaks, more than five in number, can be seen. In a chain of the type X–$CH_2CH_2CH_2$–Y, where X and Y are widely different in character, deviations of this type are most common. Figure 4–59 illustrates the origin of some of these deviations.

Of course, chains of any length can show this phenomenon, whether or not they consist solely of methylene groups. For instance, the spectrum of the protons in the second methylene group of propylbenzene is simulated below.

The splitting pattern gives a crude sextet, but the second line has a shoulder on the left, and the fourth line shows an unresolved splitting. The other peaks are somewhat broadened.

4.16 LONG HYDROCARBON CHAINS AND SATURATED RINGS; VIRTUAL COUPLING

In the nmr spectrum of a long hydrocarbon chain, most of the protons in the central portion of the chain appear as a single broadened absorption peak. These protons all have similar chemical shifts and similar coupling constants to hydrogens on adjacent carbons. However, the protons on the carbon next to a functional group often have a different chemical shift and are easily distinguished. The terminal methyl group of a chain is also easily distinguished owing to the fact that methyl group protons typically appear upfield from methylene protons. Decanol (Figure 4–60) illustrates this quite well. In addition, note that it can be seen that the terminal methyl group is split by the methylene group next to it into a *crude* triplet. This triplet is distorted and broadened by a coupling phenomenon called *virtual coupling*.

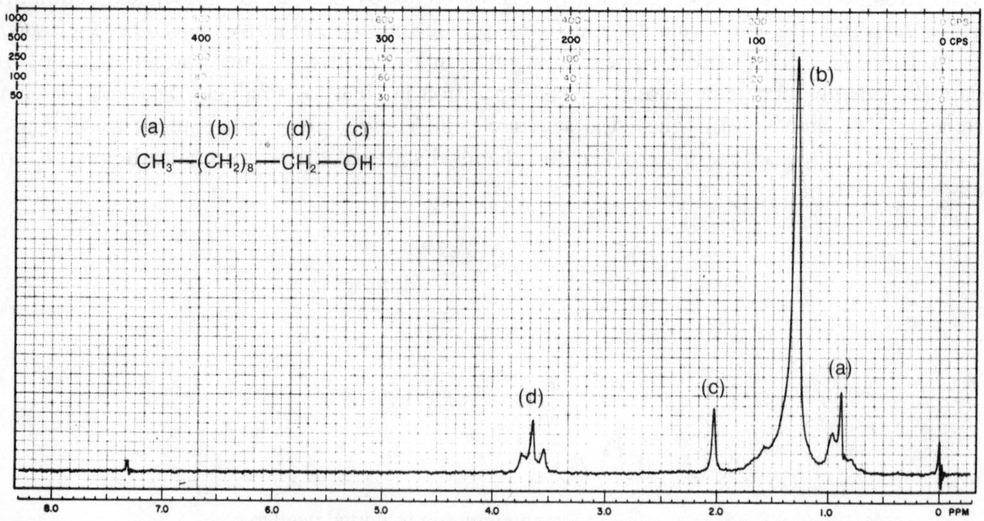

FIGURE 4–60 The NMR Spectrum of Decanol Showing Virtual Coupling

Normally, protons further than one carbon away do not couple. However, in a special case, protons on alternate carbons will appear to interact. They do not actually couple but, nevertheless, appear to do so. The figure below shows a system of this type.

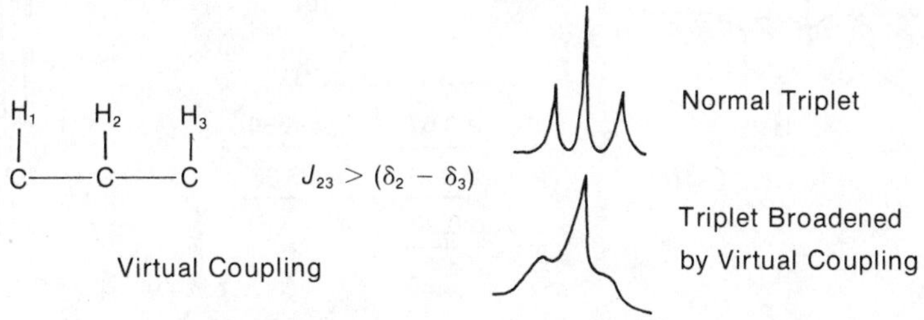

The vicinal protons H_1 and H_2 clearly couple, but H_1 *appears* to couple to H_3 as well, in spite of the fact that $J_{13} = 0$. This occurs because H_2 and H_3 have nearly the same chemical shift value, and because the coupling constant between them is larger than their chemical shift difference. In such a situation (H_2 and H_3 nearly equivalent), H_2 and H_3 behave as a unit and somewhat distort the expected triplet. In decanol, the same phenomenon broadens the triplet expected from the methylene protons attached to the hydroxyl-bearing carbon (*i.e.*, those attached to the functional group carbon).

The protons of an aliphatic ring, like those of the center portions of an aliphatic chain, also appear as an unresolvable "lump." 1-Ethynyl-1-cyclohexanol (Figure 4–61) shows just such a group of aliphatic ring protons.

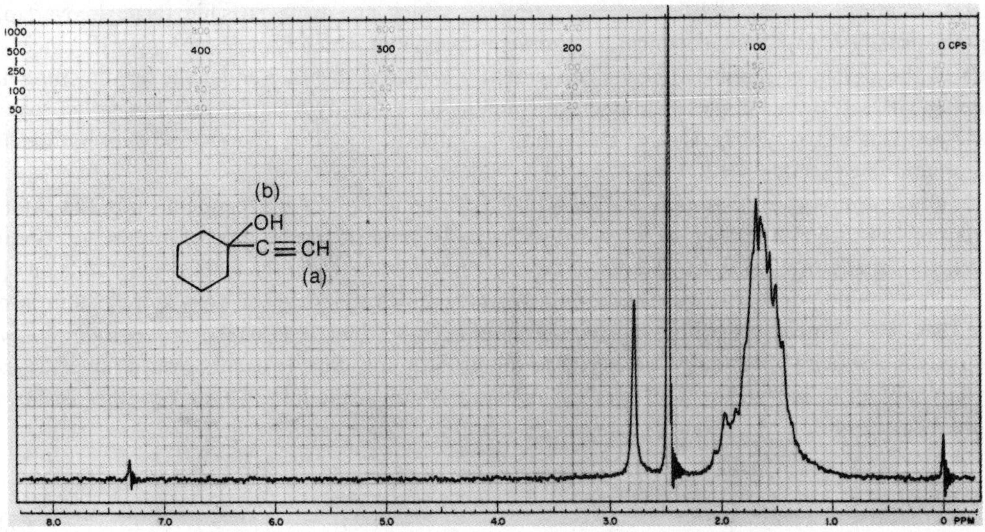

FIGURE 4–61 The NMR Spectrum of 1-Ethynyl-1-cyclohexanol

In large molecules like cholesterol (Figure 4–62), it is usually not possible to resolve the portion of the spectrum caused by the aliphatic ring and chain protons. However, it is usually possible to distinguish the sharp absorptions of the methyl groups or of those protons on the carbons that are attached to or part of a functional group.

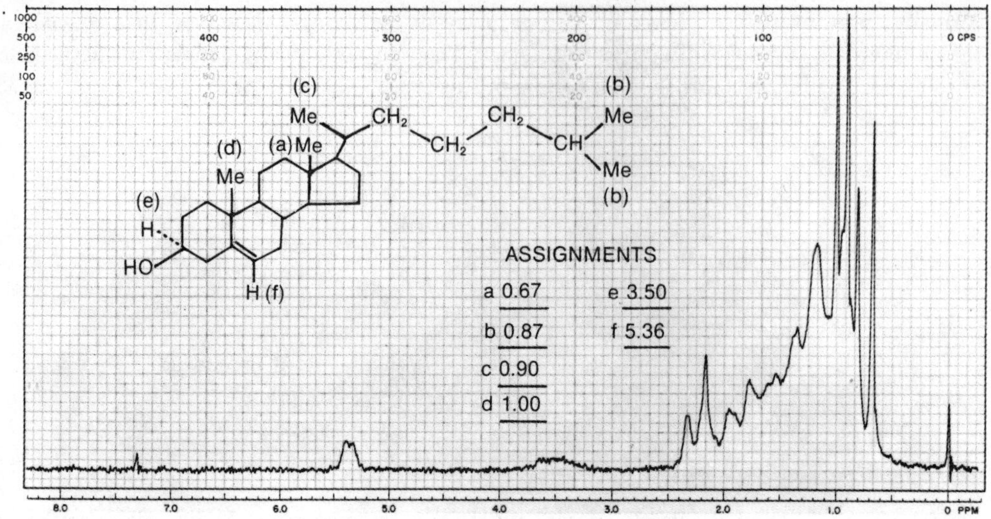

FIGURE 4-62 The NMR Spectrum of Cholesterol

4.17 LONG RANGE COUPLING

Normally, proton-proton coupling is observed between protons on adjacent atoms (if they are not equivalent) and only sometimes observed between protons on the same atom (if they are magnetically non-equivalent). These are called vicinal and geminal couplings, respectively. Coupling between protons which lie further apart than those in these two types of coupling will occur only under special circumstances. In addition, these couplings, which are made possible through overlapping orbitals, usually have a stereochemical requirement. Couplings which occur over longer distances than those of the vicinal or geminal type are conveniently collected together under the heading *long range coupling*. In fact, two examples of long range coupling have already been discussed in earlier sections: allylic coupling of the type H–C=C–CH (Section 4.5) and the coupling of *meta* and *para* hydrogens in an aromatic ring (Section 4.12C).

In allylic coupling in alkenes it was seen that the magnitude of J is dependent upon the extent of overlap of the carbon-hydrogen sigma bond with the pi bond. A similar type of interaction occurs in acetylenes. In addition, in some alkenes, coupling can occur between the C–H sigma bonds on either side of the double bond. This type of coupling is generally very small, or even non-existent in most molecules, but it may sometimes appear in an nmr spectrum. It is called *homoallylic coupling*. Homoallylic coupling is naturally weaker than allylic coupling since it occurs over a larger distance. Allylic coupling in acetylenes is illustrated in Figure 4–63, while homoallylic coupling is shown in Figure 4–64.

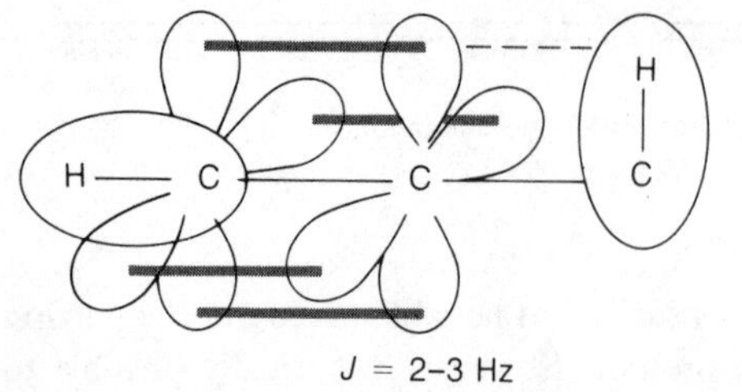

FIGURE 4–63 "Allylic" Coupling in Acetylenes

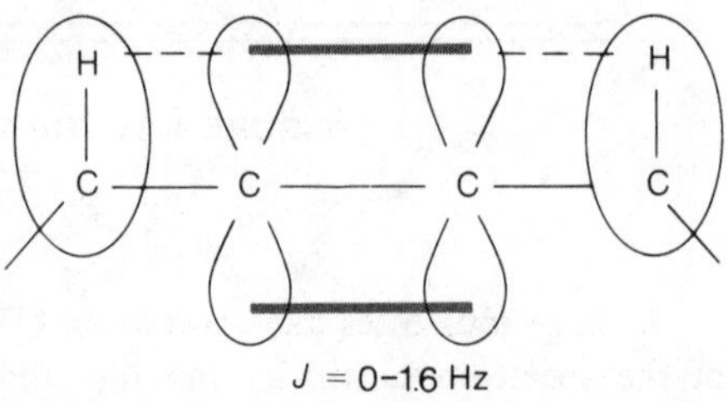

FIGURE 4–64 Homoallylic Coupling

The long range coupling that occurs in aromatic rings must also transmit the spin information through the pi system of the ring. This is illustrated in Figure 4–65 for the coupling between *meta* protons.

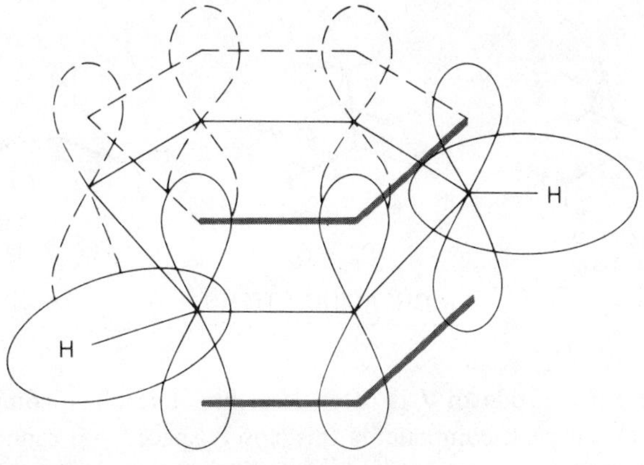

FIGURE 4–65 Coupling Between the *meta* Protons in an Aromatic Ring

Similar types of coupling exist in all of the aromatic heterocycles (furans, pyrroles, thiophenes, pyridines, etc.). In furan, for instance, there are couplings between *all* of the ring protons. In these compounds, the spin information is also transmitted through the pi system of the ring.

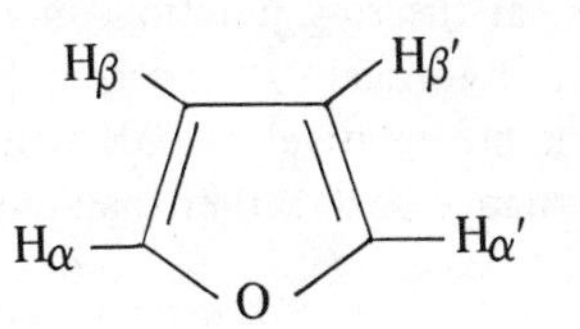

Furan

Couplings

$J_{\alpha\beta}$

$J_{\alpha\beta'}$

$J_{\alpha\alpha'}$

$J_{\beta\beta'}$

The magnitudes of the various types of proton-proton couplings for all the major heterocyclic systems are given in the tables of coupling constants in Appendix Three. Note that couplings of the allylic and homoallylic types seldom have a magnitude of more than 2 to 3 Hz. Couplings of the $J_{\alpha\beta}$ or $J_{\beta\beta'}$ type, either in furan or in other heterocycles, are *vicinal* couplings, however, and may have values up to 8 or 9 Hz. In practice, they are generally lower than this maximum value.

Long range couplings in compounds without pi systems are less common, but do occur in special cases. One special case of long range coupling in saturated systems occurs through an arrangement of atoms in the form of a W with hydrogens occupying the end positions. Two possible types of orbital overlap have been suggested to explain this type of coupling, but it is not currently known which method of transferring the spin information is correct. These two possible mechanisms are illustrated in Figure 4–66.

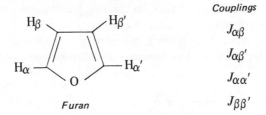

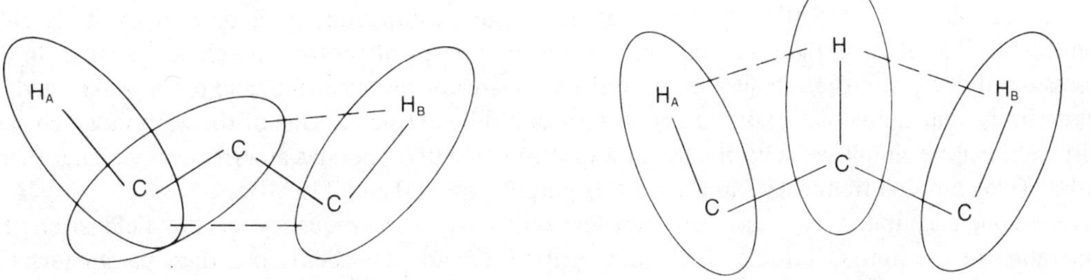

FIGURE 4–66 Two Possible Overlap Mechanisms Explaining the Transfer of Spin Information Between H_A and H_B When a Molecule Is Arranged in a W Pattern

The magnitude of J for W coupling is usually small except in highly strained ring systems. Apparently these rigid strained arrangements somehow have a favorable geometry for the overlaps involved.

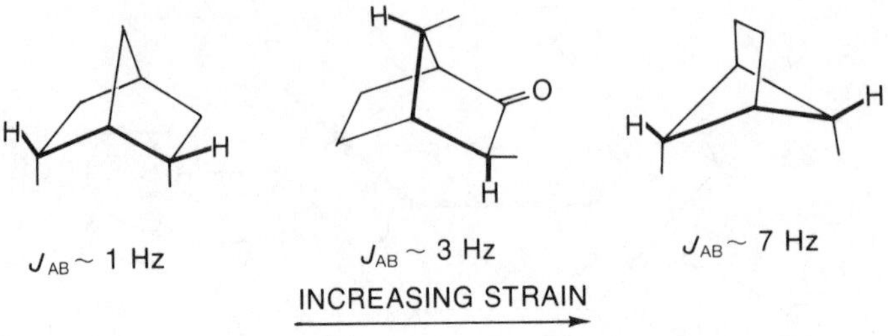

In other systems, the magnitude of J is often less than 1 Hz. The common nmr instrument is unable to distinguish between peak components this closely spaced – it cannot *resolve* them because the electronics are not sufficiently sensitive. Peaks which have spacings less than the resolving capabilities of the spectrometer are usually *broadened*. That is, two peaks very close to one another are seen as a single "fat" or broad peak. Their spacing is said to be beyond the *limits of resolution* of the spectrometer. Most W couplings are of this type; they give rise to peak broadening for the protons that are coupled rather than a discernible splitting. Small allylic couplings ($J < 1$ Hz) also give rise to peak broadening.

Angular methyl groups in steroids, or those at the ring junctions in a decalin system, often show peak broadening due to W coupling with several hydrogens of the ring. Several such interactions are illustrated in Figure 4–67. Since these systems are relatively unstrained, J_W is usually quite small. Hence, one sees a broadening of the methyl resonance peak rather than any splitting.

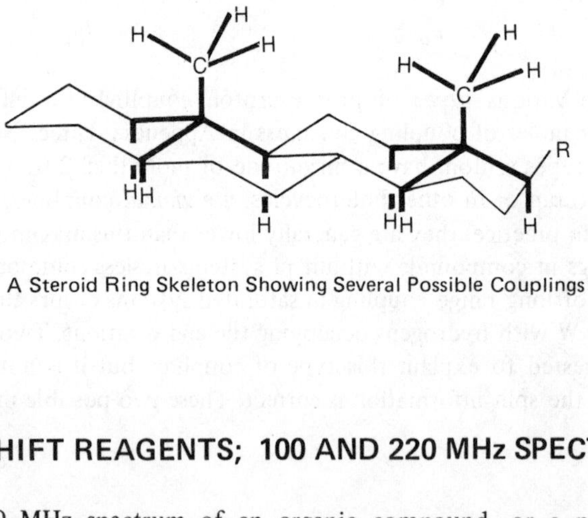

FIGURE 4–67 A Steroid Ring Skeleton Showing Several Possible Couplings of the W Type

4.18 CHEMICAL SHIFT REAGENTS; 100 AND 220 MHz SPECTRA

Occasionally the 60 MHz spectrum of an organic compound, or a portion of it, is almost undecipherable because the chemical shifts of several groups of protons are all very similar. In these cases all of the proton resonances occur in the same area of the spectrum, and often peaks overlap so extensively that individual peaks and splittings cannot be extracted. One of the ways in which such a situation can be simplified is by the use of a spectrometer that operates at a frequency higher than 60 MHz. Two common frequencies in use for this purpose are 100 and 220 MHz.

Although coupling constants are not dependent on the frequency or the field strength of operation of the nmr spectrometer, chemical shifts *in Hz* are dependent upon these parameters. (This was discussed previously in Section 3.15.) This circumstance can often be used to simplify an otherwise undecipherable spectrum. Suppose, for instance, that a compound contained three multiplets: a quartet and two triplets derived from groups of protons with very similar chemical shifts. At 60 MHz these peaks might overlap as illustrated in Figure 4-68 and simply give an unresolved envelope of absorption.

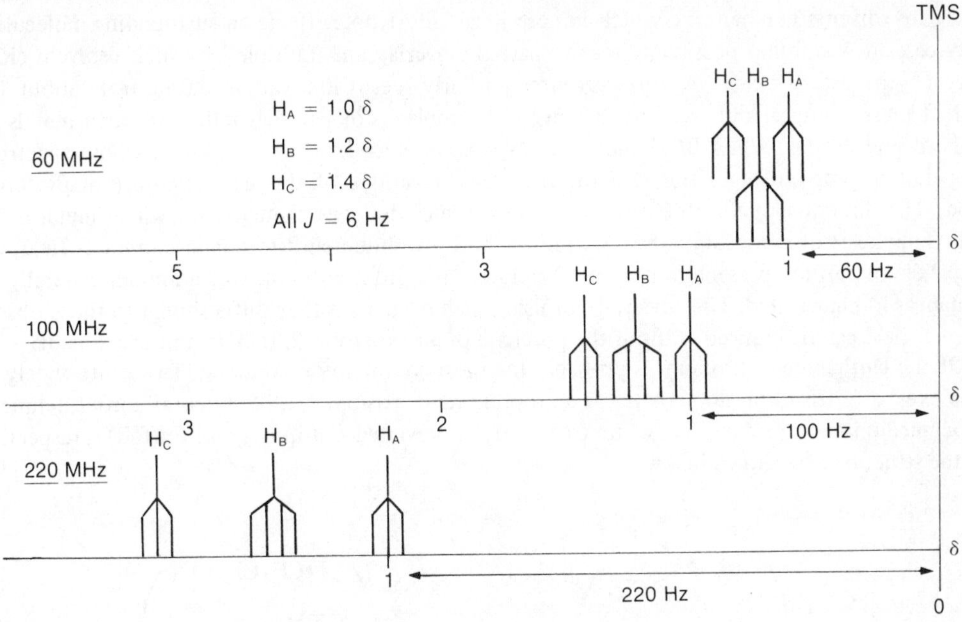

FIGURE 4–68 A Comparison of the Spectrum of a Compound With Overlapping Multiplets at 60 MHz with Spectra of the Same Compound Also Determined at 100 and 220 MHz. The Drawing is to Scale.

Figure 4–68 also shows the spectrum of the same compound redetermined at two higher field strengths. In redetermining the spectrum at higher field strengths, the coupling constants do not change, but the chemical shifts in Hz (not δ) of the proton groups (H_A,H_B, H_C) responsible for the multiplets do increase. It should be noted that at 220 MHz the individual multiplets are cleanly separated and resolved.

The fact that chemical shifts increase at higher field strength while coupling constants remain unchanged can often be used to advantage in distinguishing whether two unresolved peaks are caused by different protons (or groups) or whether they represent a splitting of a single proton's (or group's) resonance peak. For example, refer back to Figure 4–56, the nmr spectrum of 4-methyl-1-pentyn-3-ol. The original workers were not certain whether the small splittings of the methyl doublet were long range couplings or whether each of the two methyl groups had a different value of chemical shift. If long range coupling were involved, the appearance of the methyl resonances would have remained unchanged at higher field. The only change would have been that the entire group experienced a greater shift in Hz – all internal relationships within the peak grouping would have remained the same. However, if differences in chemical shift for the two methyl groups were involved, at higher field strength the shifts would increase and a larger splitting would have been found. At higher field, these workers found the splitting to increase, thereby confirming a chemical shift difference for the two diastereomeric methyl groups.

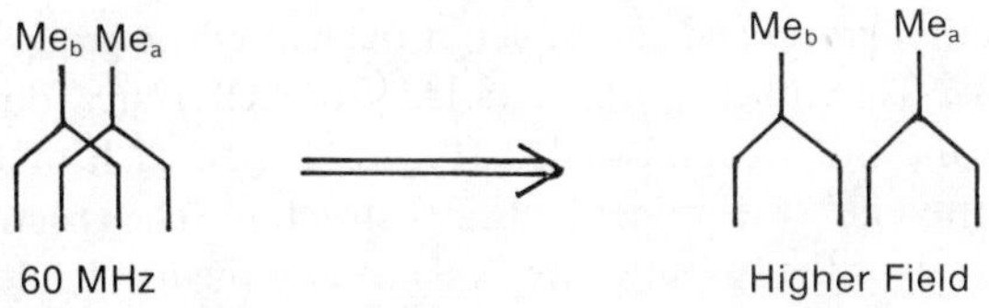

It has been known for some time that interactions between molecules and solvents, such as those due to hydrogen bonding, can cause large changes in the resonance positions of certain types of protons (*e.g.*, hydroxyl and amino). It has also been known that the resonance positions of some groups of protons can be greatly affected by changing from the usual nmr solvents such as CCl_4 and $CDCl_3$ to solvents like benzene which impose local anisotropic effects on surrounding molecules. In many cases it was found possible to resolve partially overlapping multiplets by such a solvent change. However, the use of *chemical shift reagents* is a fairly recent innovation, dating from about 1969. Most of these chemical shift reagents are organic complexes of paramagnetic rare earth metals from the lanthanide series. When these metal complexes are added to the compound whose spectrum is being determined, profound shifts in the resonance positions of the various groups of protons are noted. The direction of the shift (upfield or downfield) depends primarily on which metal is being used. Complexes of europium, erbium, thulium, and ytterbium shift resonances to lower field, while complexes of cerium, praseodymium, neodymium, samarium, terbium, and holmium generally shift resonances to higher field. The advantage of using such reagents is that shifts similar to those observed at higher field can be induced without the purchase of an expensive 220 MHz nmr instrument.

Of the lanthanides, europium is probably the most commonly used metal. Two of its widely used complexes are tris-(dipivalomethanato) europium and tris-6,6,7,7,8,8,8-heptafluoro-2,2-dimethyl-3,5-octanedionato europium. These are frequently abbreviated $Eu(dpm)_3$ and $Eu(fod)_3$, respectively, and the structures are shown below.

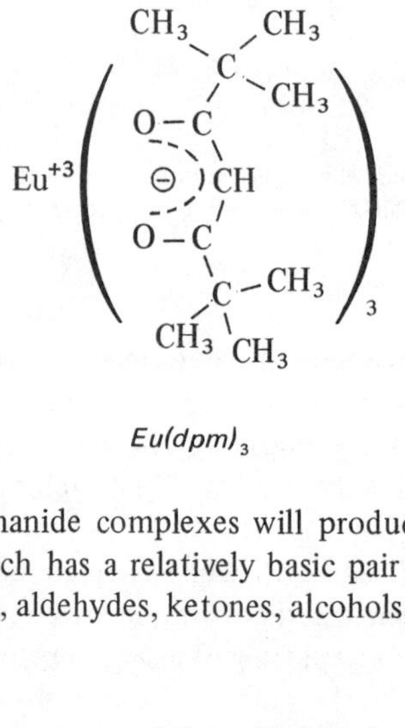
$Eu(dpm)_3$

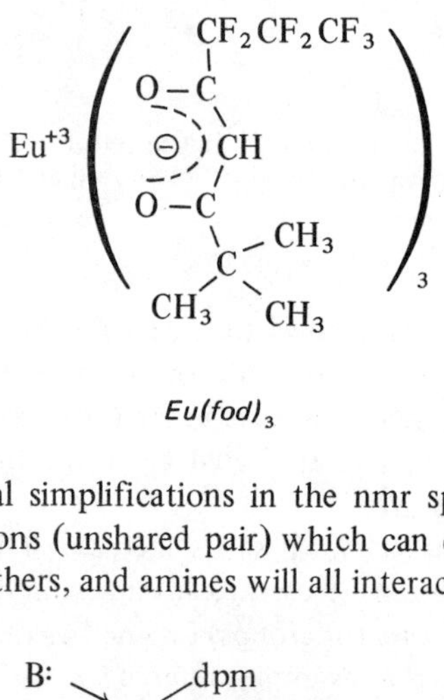
$Eu(fod)_3$

These lanthanide complexes will produce spectral simplifications in the nmr spectrum of any compound which has a relatively basic pair of electrons (unshared pair) which can coordinate with Eu^{+3}. Typically, aldehydes, ketones, alcohols, thiols, ethers, and amines will all interact:

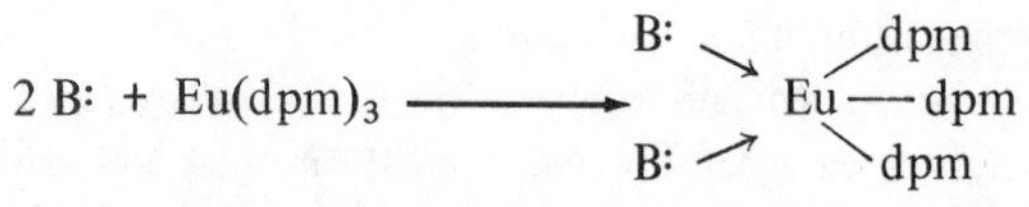

The amount of shift which a given group of protons will experience depends (1) on the distance separating the metal (Eu^{+3}) and that group of protons, and (2) on the concentration of the shift reagent in the solution. Because of the latter dependence, it is necessary when reporting a lanthanide-shifted spectrum to report the number of mole equivalents of shift reagent used or its molar concentration.

The distance factor is beautifully illustrated in the spectra of hexanol which are given in Figures 4–69 and 4–70. In the absence of shift reagent, the spectrum of Figure 4–69 is obtained. Only the triplet of the terminal methyl group and the triplet of the methylene group next to the hydroxyl are resolved in the spectrum. The other protons (aside from OH) are found together in a broad unresolved group. With the shift reagent added (Figure 4–70), each of the methylene groups is clearly separated and resolved into the proper multiplet structure. The spectrum is in every sense *first order* and, thus, simplified; all of the splittings are explained by the $n + 1$ rule.

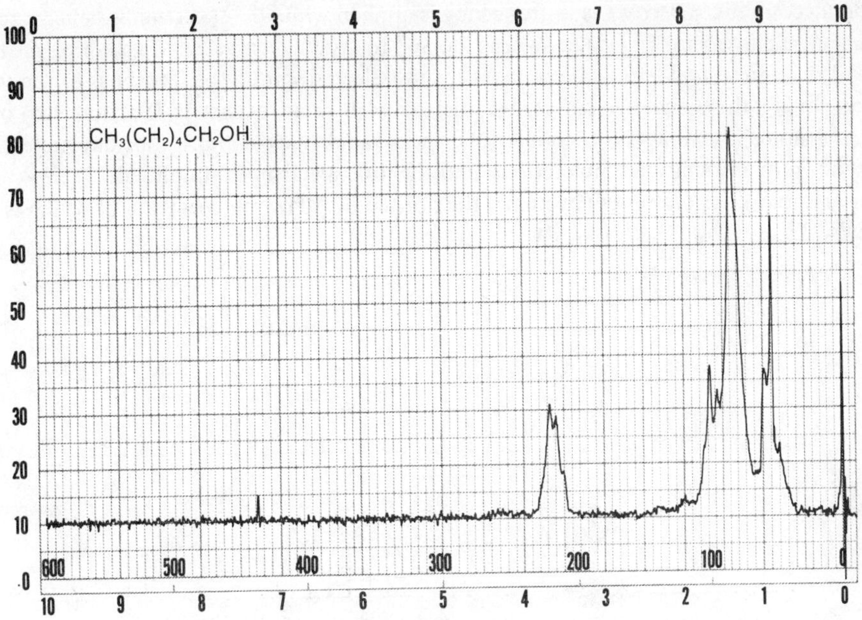

FIGURE 4–69 The Normal 60 MHz NMR Spectrum of Hexanol

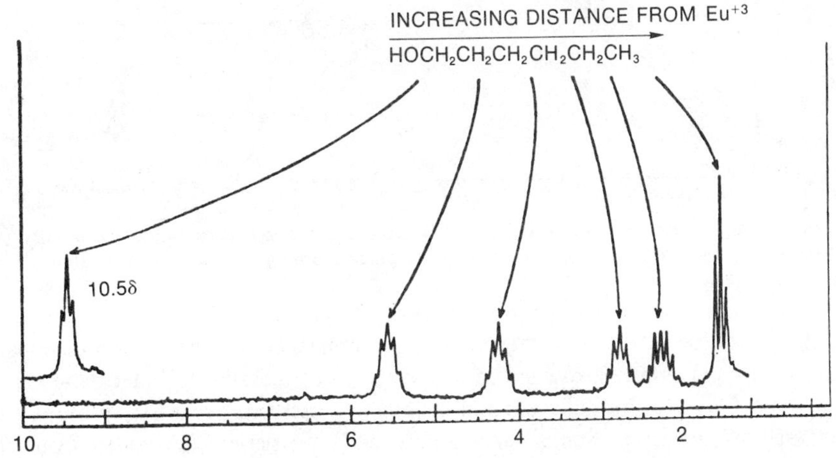

FIGURE 4–70 The 100 MHz NMR Spectrum of Hexanol with 0.29 Mole Equivalents of $Eu(dpm)_3$ Added [Reprinted with permission from J.K.M. Sanders and D.H. Williams, Chem. Commun., 422 (1970).]

One final consequence of the use of a shift reagent should be noted. It can be seen in Figure 4–70 that the multiplets are not as nicely resolved into sharp peaks as one usually expects. This is due to the fact that the shift reagents cause a small amount of line broadening. At high shift reagent concentrations this problem becomes serious, but at most useful concentrations the amount of broadening experienced is tolerable.

4.19 SPIN DECOUPLING METHODS; DOUBLE RESONANCE

Spin decoupling, sometimes called *double resonance*, is another technique used to simplify nmr spectra. It is useful in determining which spin-spin multiplets actually interact with one another, and also in locating absorptions that may be hidden or "buried" under other group absorptions.

In spin decoupling, a *second* radiofrequency is applied while the spectrum is being scanned in the usual fashion. This second rf frequency can be set to irradiate a selected group of protons in the molecule being measured. Irradiation causes the selected group of protons to undergo rapid transitions among their nuclear spin states. Just as with rapid exchange (Section 4.6) or with the nuclear quadrupole phenomenon (Section 4.9), these rapid nuclear transitions decouple the protons of this group from any spin-spin interactions, and the nmr spectrum is simplified.

As an example, consider propyl bromide, the normal 60 MHz nmr spectrum of which is illustrated in Figure 4–71A.

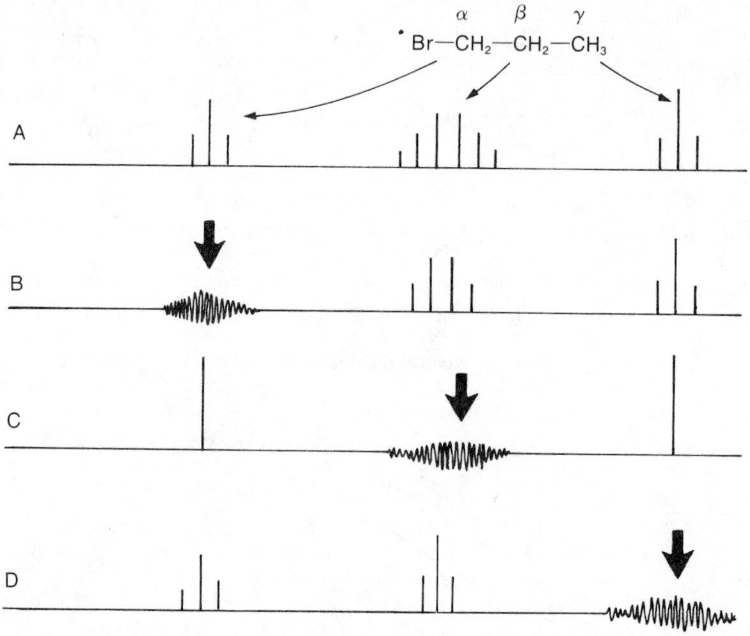

FIGURE 4–71 NMR Spectra of Propyl Bromide: (A) the normal spectrum at 60 MHz, (B) the effect of decoupling (irradiating) the α CH_2 protons, (C) the effect of decoupling the β CH_2 protons, and (D) the effect of decoupling the γ CH_3 protons.

Figure 4–71B shows the effect of redetermining the nmr spectrum with simultaneous irradiation of the α CH_2 protons (the large arrow indicates the point of irradiation). Under these conditions, the resonance of the α CH_2 protons collapses and cannot be observed. The α CH_2 protons are caused to cycle through all of their spin states very rapidly and are thereby decoupled from all spin-spin interactions with the adjacent β CH_2 protons. Therefore, the β CH_2 protons couple only with the three methyl protons on the other side, and their resonance is simplified to a quartet rather than the original sextet. The methyl resonance, which does not couple with the α CH_2 protons even under non-irradiative conditions, is left unchanged. It is split into a triplet by the two β CH_2 protons. Figures 4–71C and 4–71D show the effects of double irradiation (decoupling) of the β CH_2 and γ CH_3 protons, respectively.

The principal application of spin decoupling is as an aid in the interpretation of complicated spectra. For instance, the above experiments establish definitely the nature of the spin-spin interactions in propyl bromide. Spectrum B shows that the α CH_2 protons couple with the β CH_2 protons, but not with the γ CH_3 protons. The resonance of the latter protons is left unchanged by irradiation of the α CH_2 group, while the resonance of the β CH_2 protons, which do couple, is simplified. Spectrum C shows that the α CH_2 and γ CH_3 protons do not interact, but that the β CH_2 protons are responsible for the splitting of both groups in spectrum A. Similarly, spectrum D establishes that the γ CH_3 protons couple only with the β CH_2 protons. In other words, the decoupled spectra establish beyond a doubt which groups of protons interact to yield the multiplet structures in the propyl bromide spectrum. While these relationships may have seemed obvious for propyl bromide, a relatively simple compound, one can easily see that such experiments could be extremely useful in the spectrum of a complicated compound.

Often double resonance is useful in interpreting a spectrum that has extensive multiplet structure and is extremely difficult to interpret. By careful selection of the double resonance frequency, one can often simplify such a spectrum to the point where it may be interpreted easily. Allyl bromide (Figure 4–72) provides an illustrative example. The resonances of the double bond protons are very complex, principally owing to allylic couplings with the $-CH_2-$ group protons. However, if the allylic $-CH_2-$ protons are decoupled, the spectrum simplifies to the typical AMX pattern which one finds for many alkenes (see vinyl acetate, Figures 4–12 and 4–13). This AMX simplification is illustrated in the lower trace of Figure 4–72. One can easily see that, while it would have been difficult to determine J values for the various vinyl interactions in the original spectrum, it is relatively easy to do so in the decoupled spectrum which presents a much simpler splitting pattern.

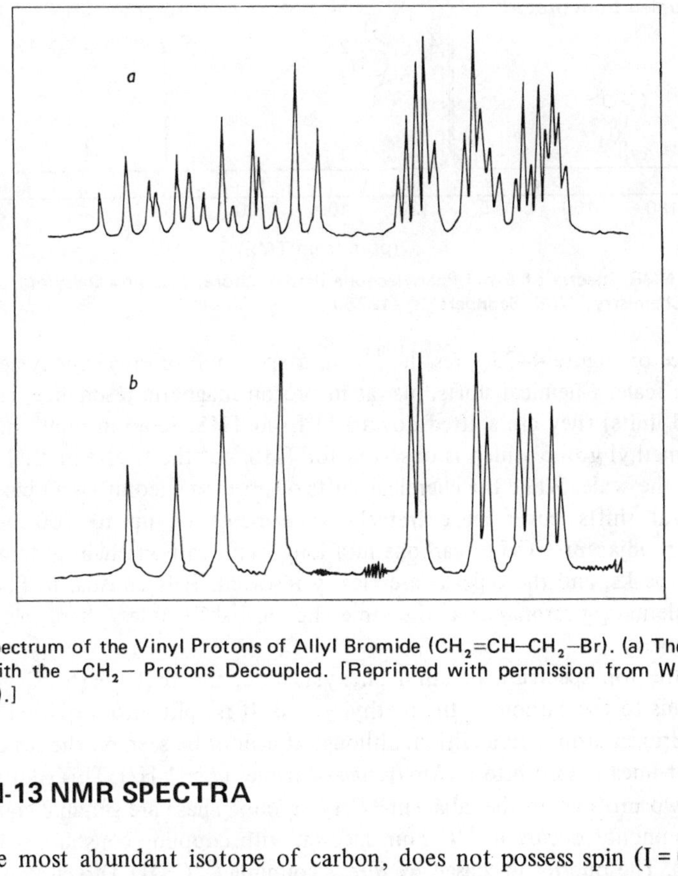

FIGURE 4–72 A Spectrum of the Vinyl Protons of Allyl Bromide (CH_2=CH–CH_2–Br). (a) The Normal Spectrum. (b) The Spectrum with the $-CH_2-$ Protons Decoupled. [Reprinted with permission from W. McFarlane, Chem. Britain, *5*, 143 (1969).]

4.20 CARBON-13 NMR SPECTRA

Carbon-12, the most abundant isotope of carbon, does not possess spin (I = 0); it has both an even atomic number and an even atomic weight (see Section 3.1). However, the second principal isotope of carbon, carbon-13, does have the nuclear spin property ($I = 1/2$). Unfortunately, carbon-13 atom resonances are not easy to observe owing to a combination of two factors. First, the natural abundance of ^{13}C is low; only 1.1% of all carbon atoms are ^{13}C. Second, the magnetic moment (μ) of ^{13}C is also low. For these two reasons, the resonances of ^{13}C atoms are about *6000 times* weaker than those of hydrogen. Nevertheless, by special instrumental techniques, which will not be discussed here, it is possible to observe ^{13}C nuclear magnetic resonance spectra.

Because of the nature of the instrumental techniques used, the only useful parameter derived from ^{13}C spectra is the chemical shift. Integrals are unreliable and are not necessarily related to the relative numbers of ^{13}C atoms. Finally, although hydrogens which are attached to ^{13}C atoms cause spin-spin splitting, spin-spin interaction between adjacent carbon atoms is rare. With its low natural abundance (0.011), the probability of finding a second ^{13}C atom next to a first one is extremely low.

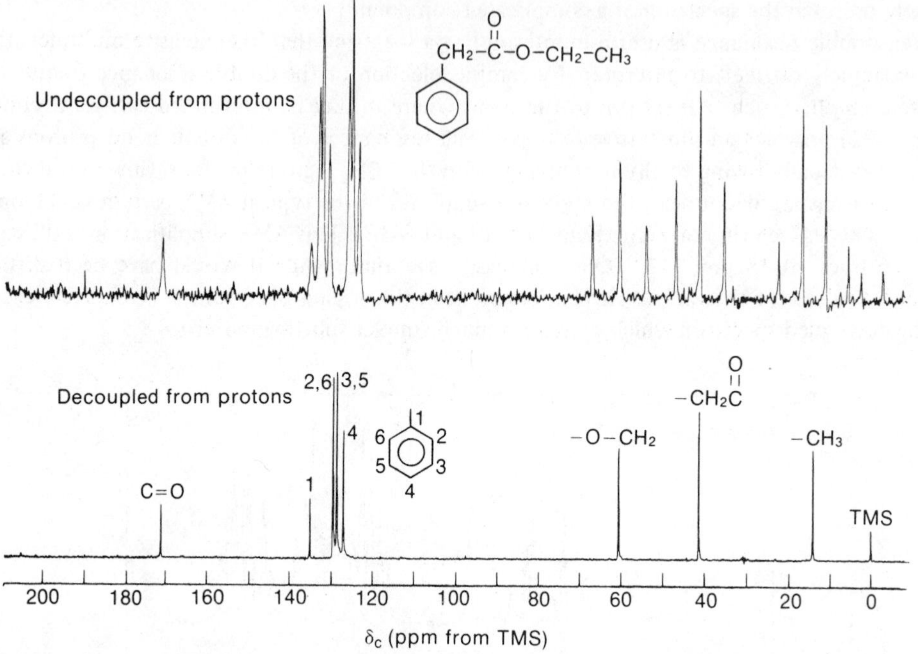

FIGURE 4–73 ^{13}C NMR Spectra of Ethyl Phenylacetate (From Moore, J.A., and Dalrymple, D.L., "Experimental Methods in Organic Chemistry," W.B. Saunders Co., 1976.)

The upper trace of Figure 4–73 gives the ^{13}C nmr spectrum of ethyl phenylacetate. Notice first the chemical shift scale. Chemical shifts, just as in proton magnetic resonance, are reported by the number of ppm (δ units) they are shifted downfield from TMS. Keep in mind, however, that it is a ^{13}C atom of the methyl group which is observed for TMS, not the twelve methyl hydrogens. Notice also the extent of the scale. While the chemical shifts of protons encompass a range of only about 20 ppm, ^{13}C chemical shifts cover an extremely wide range of up to 200 ppm! Under these circumstances, even adjacent $-CH_2-$ carbons in a long hydrocarbon chain generally have their own distinct resonance peaks, and these peaks are clearly resolved. It is unusual to find *any* two carbon atoms in a molecule having resonance at the same chemical shift unless these two carbon atoms are equivalent by symmetry.

Returning to the top spectrum of ethyl phenylacetate, the first quartet downfield from TMS (14.2 δ) corresponds to the carbon of the methyl group. It is split into a quartet ($J = 6$ Hz) by the three attached hydrogen atoms. In addition, although it cannot be seen on the scale of this spectrum, each of the quartet lines is split into a *closely spaced* triplet ($J \sim 1$ Hz). This additional fine splitting is caused by the two protons on the adjacent $-CH_2-$ group. These are geminal couplings ($H-C-^{13}C$) of a type that commonly occurs in ^{13}C nmr spectra, with coupling constants which are generally small ($J = 0$–2 Hz). The quartet is caused by *direct* coupling ($^{13}C-H$). Direct coupling constants are larger, usually about 100–200 Hz, and are more obvous on the scale in which the spectrum is presented.

There are two $-CH_2-$ groups in ethyl phenylacetate. The one corresponding to the ethyl $-CH_2-$ group is found further downfield (60.6 δ), as this carbon is deshielded by the attached oxygen. It is a triplet because of the two attached hydrogens. Again, although it is not seen in this unexpanded spectrum, each of the triplet peaks is finely split into a quartet by the three hydrogens on the adjacent methyl group. The benzyl $-CH_2-$ carbon is the intermediate triplet (41.4 δ). Furthest downfield is the carbonyl group carbon (171.1 δ). On the scale of presentation, it is a singlet (no directly attached hydrogens), but because of the adjacent benzyl$-CH_2-$ group, actually it is split finely into a triplet. The aromatic ring carbons remain, and they have resonances over the range from 127 δ to 136 δ.

Although the splittings in a simple molecule such as this one yield interesting structural information, namely the number of hydrogens attached to each carbon (as well as those adjacent if the spectrum is expanded), for large molecules the ^{13}C spectrum becomes very complex owing to these splittings, and they often overlap. It is customary, therefore, to decouple *all* the protons in the molecule by irradiating them all simultaneously with a broad spectrum of frequencies ("noise") in the proper range. This type of spectrum is said to be *noise decoupled*. The noise decoupled spectrum is much simpler, and for larger molecules, much easier to interpret. The noise decoupled spectrum of ethyl phenylacetate is presented in the lower trace of Figure 4–73.

In the noise decoupled ^{13}C spectrum, each peak represents a different carbon atom. If two carbons are represented by a single peak, they are then equivalent by symmetry. Thus the *ortho* carbons of the phenyl ring give a single peak and the *meta* carbons give another single peak in the lower trace of Figure 4–73. Also, just as for proton spectra, the chemical shift of each carbon indicates both its type and its structural environment. In fact, a correlation table can be presented for ^{13}C chemical shift ranges. Figure 4–74 gives the typical ranges for the various types of carbon atoms.

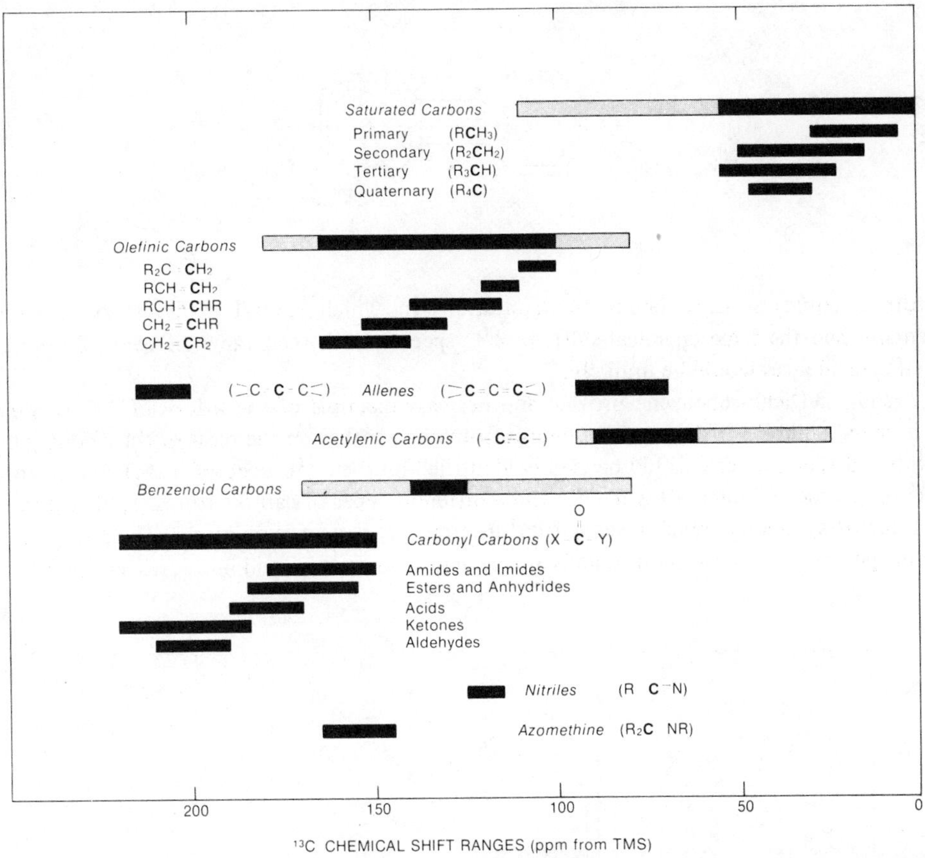

FIGURE 4–74 ^{13}C Chemical Shift Ranges (ppm from TMS). Extended Ranges Sometimes Occur When Polar Substituents are Attached to the Carbons. These Extended Ranges Are Indicated by the Lightly Shaded Areas. (From Moore, J.A., and Dalrymple, D.L., "Experimental Methods in Organic Chemistry," W.B. Saunders Co., 1976.)

Electronegativity, hybridization, and anisotropy effects all affect ^{13}C chemical shifts just as they do for protons, but in a more complex fashion. These factors will not be discussed in any detail here, but note that the $-CH_2-$ group carbon attached to oxygen in ethyl phenylacetate has a larger chemical shift than the benzyl carbon. Note also that the carbonyl carbon appears relatively far downfield, probably owing to an anisotropy effect.

One of the most powerful features of ^{13}C spectra is that they allow one to determine the number of different carbons in a molecule and thus allow a recognition of equivalences or symmetry elements. For instance, cyclooctatetraene readily dimerizes to a solid. On the basis of theory, two possible structures can be proposed for this dimer:

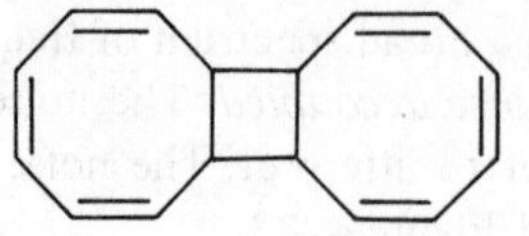

1

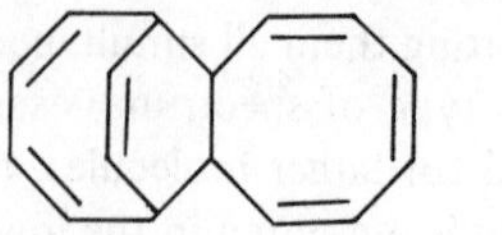

2

The proton decoupled ^{13}C spectrum of the dimer shows only *four* sharp lines. This effectively proves structure 1 which, having two planes of symmetry, fits rather well.

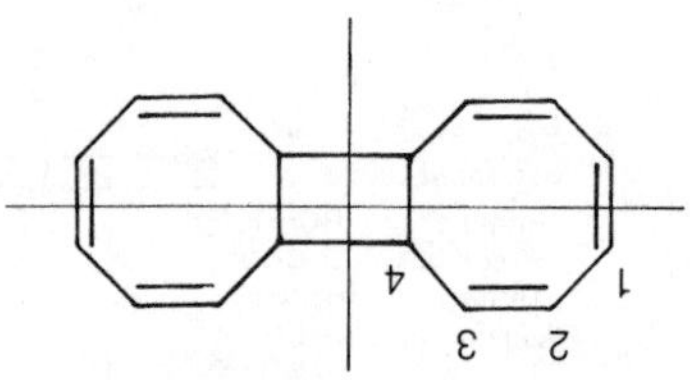

Structure 2 could fit only under the conditions in which several *different types* of carbons accidentally had the same chemical shift. In ^{13}C spectra, with their range of over 200 ppm, such a series of coincidences would be unlikely.

1,2- and 1,3-Dichlorobenzene provide another easy example of the utility of ^{13}C spectra. While these isomers could be difficult to distinguish from one another on the basis of their boiling points or their infrared spectra, each would be clearly identified by their ^{13}C spectra. 1,2-Dichlorobenzene has a plane of symmetry which gives it only three different types of carbon atoms. 1,3-Dichlorobenzene has a plane of symmetry which gives it four different types of carbon atoms. The proton decoupled ^{13}C nmr spectra of these two compounds are given in Figures 4–75 and 4–76, respectively.

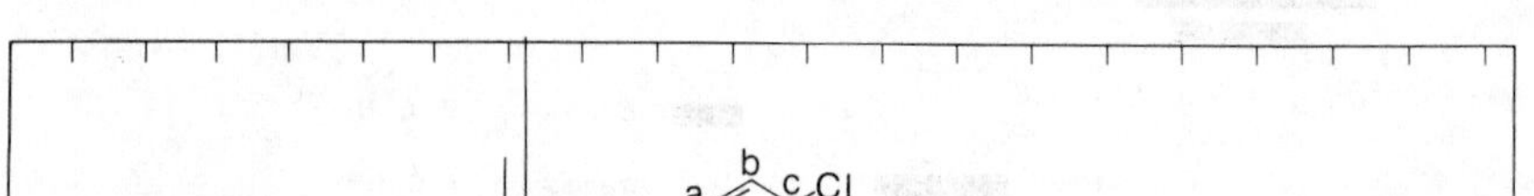

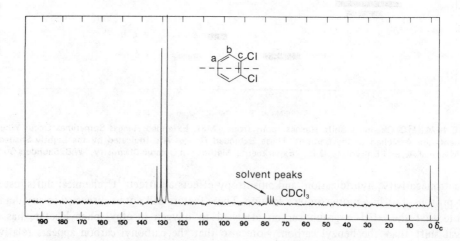

FIGURE 4–75 The Proton Decoupled ^{13}C NMR Spectrum of 1,2-Dichlorobenzene. (From L.F. Johnson and W.C. Jankowski, "Carbon-13 NMR Spectra: A Collection of Assigned, Coded, and Indexed Spectra," Wiley-Interscience, © 1972. Reprinted by permission of John Wiley and Sons, Inc., New York).

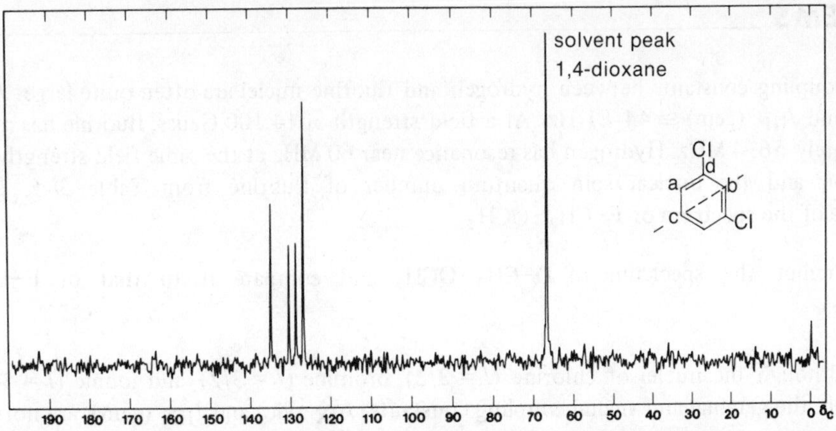

FIGURE 4–76 The Proton Decoupled ^{13}C NMR Spectrum of 1,3-Dichlorobenzene (From L.F. Johnson and W.C. Jankowski, "Carbon-13 NMR Spectra: A Collection of Assigned, Coded, and Indexed Spectra," Wiley-Interscience, © 1972. Reprinted by permission of John Wiley and Sons, Inc., New York.)

Other applications of ^{13}C nmr spectroscopy involve isotopic tracer or labeling studies, the study of reaction mechanisms, and conformational analysis. The interested reader is referred to more advanced texts.

REFERENCES

GENERAL

J. R. Dyer, "Applications of Absorption Spectroscopy of Organic Compounds," Prentice-Hall, Englewood Cliffs, N.J., 1965.

C. J. Pouchert and J. R. Campbell, "The Aldrich Library of NMR Spectra," 11 vols., Aldrich Chemical Company, Milwaukee, Wisconsin.

J. D. Roberts, "Nuclear Magnetic Resonance. Applications to Organic Chemistry," McGraw-Hill, New York, 1959.

R. M. Silverstein, G. C. Bassler, and T. C. Morrill, "Spectrometric Identification of Organic Compounds," 2nd ed., John Wiley, New York, 1974.

Varian Associates, "High Resolution NMR Spectra Catalogue," Varian Associates, Palo Alto, California, Vol. 1, 1962; Vol. 2, 1963.

ADVANCED

F. A. Bovey, "NMR Spectroscopy," Academic Press, New York, 1969.

L. M. Jackman and S. Sternhell, "Applications of Nuclear Magnetic Resonance Spectroscopy in Organic Chemistry," 2nd ed., Pergamon Press, London, 1969.

J. A. Pople, W. G. Schneider, and H. J. Bernstein, "High Resolution Nuclear Magnetic Resonance," McGraw-Hill, New York, 1959.

J. D. Roberts, "An Introduction to the Analysis of Spin-Spin Splitting in High Resolution Nuclear Magnetic Resonance Spectra," W. A. Benjamin, New York, 1962.

K. B. Wiberg and B. J. Nist, "The Interpretation of NMR Spectra," W. A. Benjamin, New York, 1962.

CARBON-13

L. F. Johnson and W. C. Jankowski, "Carbon-13 NMR Spectra," Wiley-Interscience, New York, 1972. A catalog of spectra.

G. C. Levy and G. L. Nelson, "Carbon-13 Nuclear Magnetic Resonance for Organic Chemists," Wiley-Interscience, New York, 1972.

J. B. Strothers, "Carbon-13 NMR Spectroscopy," McGraw-Hill, New York, 1959.

J. A. Moore and D. L. Dalrymple, "Experimental Methods in Organic Chemistry," 2nd ed., W. B. Saunders, Philadelphia, 1976, pp. 107–115.

FLUORINE-19

E. F. Mooney, "An Introduction to ^{19}F NMR Spectroscopy," Hayden/Sadtler, Sadtler Research Laboratories, Philadelphia, Pa., 1970.

PROBLEMS

1. Coupling constants between hydrogen and fluorine nuclei are often quite large: J_{HF} (vic) $\cong$ 3–25 Hz and J_{HF} (gem) $\cong$ 44–81 Hz. At a field strength of 14,100 Gauss, fluorine has resonance at approximately 56.4 MHz. Hydrogen has resonance near 60 MHz at the same field strength. Using this information and the nuclear spin quantum number of fluorine from Table 3–1, predict the appearance of the spectrum of $F{-}CH_2{-}OCH_3$.

2. Predict the spectrum of $D{-}CH_2{-}OCH_3$ and compare it to that of $F{-}CH_2{-}OCH_3$ (Problem 1).

3. Although the nuclei of chlorine (I = 3/2), bromine (I = 3/2), and iodine (I = 5/2) exhibit nuclear spin, the geminal and vicinal coupling constants, J_{HX} (vic) and J_{HX} (gem), are normally zero. These atoms are simply too large and diffuse to transmit spin information via their plethora of electrons. These halogens are completely decoupled from protons directly attached, or on adjacent carbon atoms, owing to strong electrical quadrupole moments. Predict the nmr spectrum of $Br{-}CH_2{-}OCH_3$ and compare it to the spectra of $F{-}CH_2{-}OCH_3$ (Problem 1) and $D{-}CH_2{-}OCH_3$ (Problem 2).

4. J_{HH} (gem) is typically 5 Hz in cyclopropyl compounds. Using the formula on page 118, calculate the approximate magnitude of J_{HD} (gem) for a cyclopropyl compound if one of these hydrogens were replaced by deuterium. (μ_D = 0.85738, μ_H = 2.79268).

5. Estimate the expected splitting (J in Hz) for the lettered protons in the following compounds; *i.e.*, give J_{ab}, J_{ac}, J_{bc}, etc.

a)

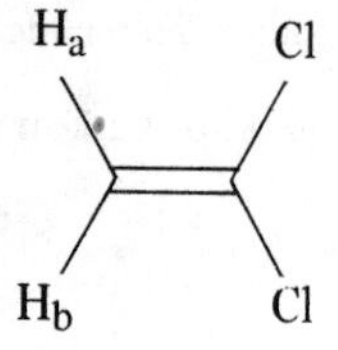

b)

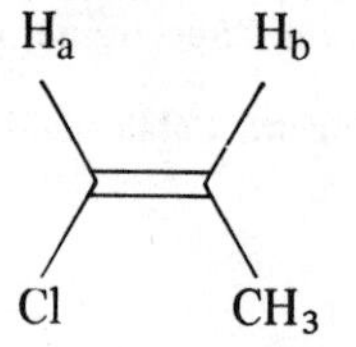

c)

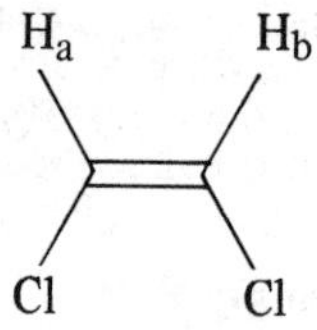

d)

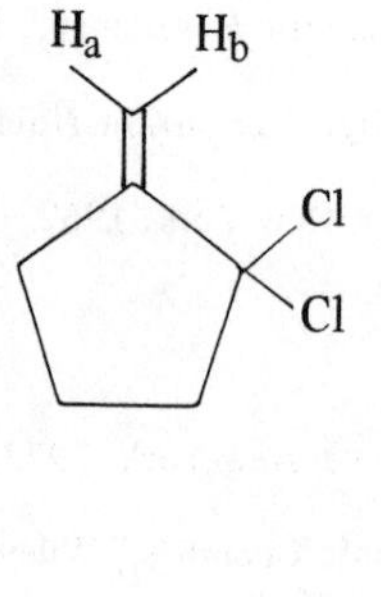

e)

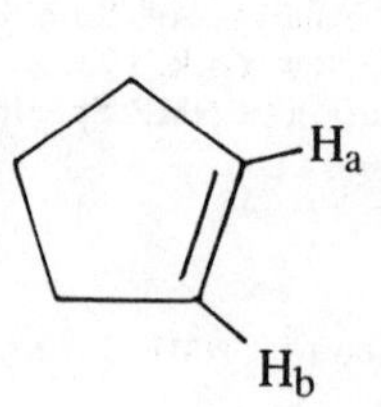

f)

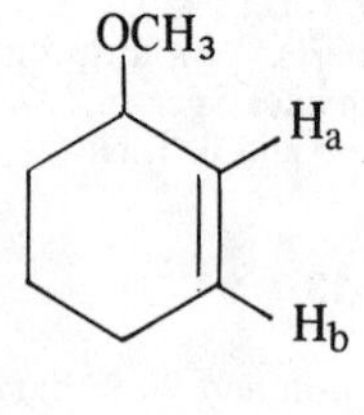

g)

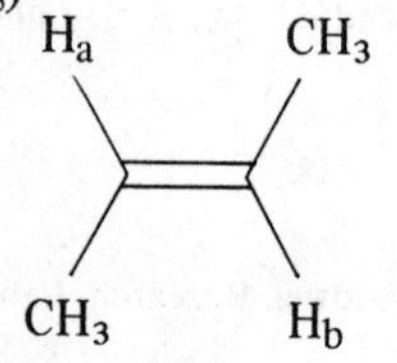

h)

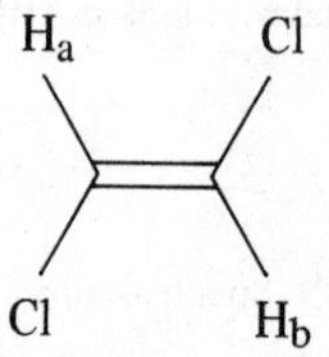

i)

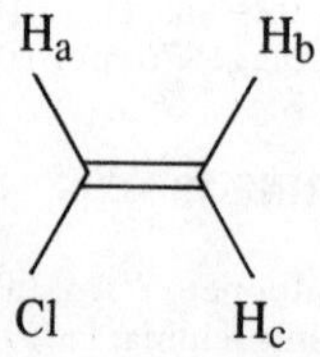

6. Analyze graphically the spectrum of methyl vinyl thioether given below. Assign the peaks and determine the approximate coupling constants for J_{BD}, J_{CD}, and J_{BC}.

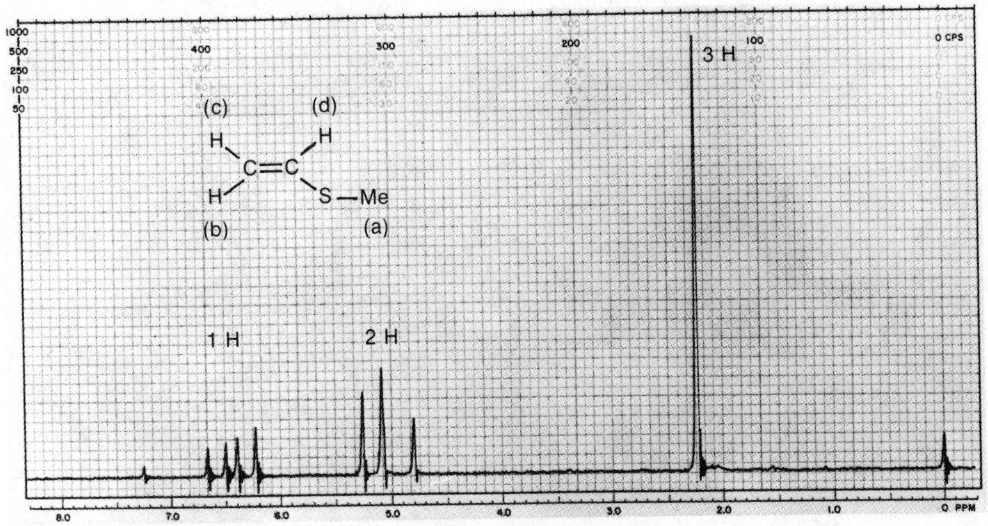

7. Give a full explanation of the spectrum of crotonaldehyde (C_4H_6O) which is given below. The peak at 7.35 δ is an impurity peak due to $CHCl_3$. The offset doublet has been moved upfield from 9.48 δ.

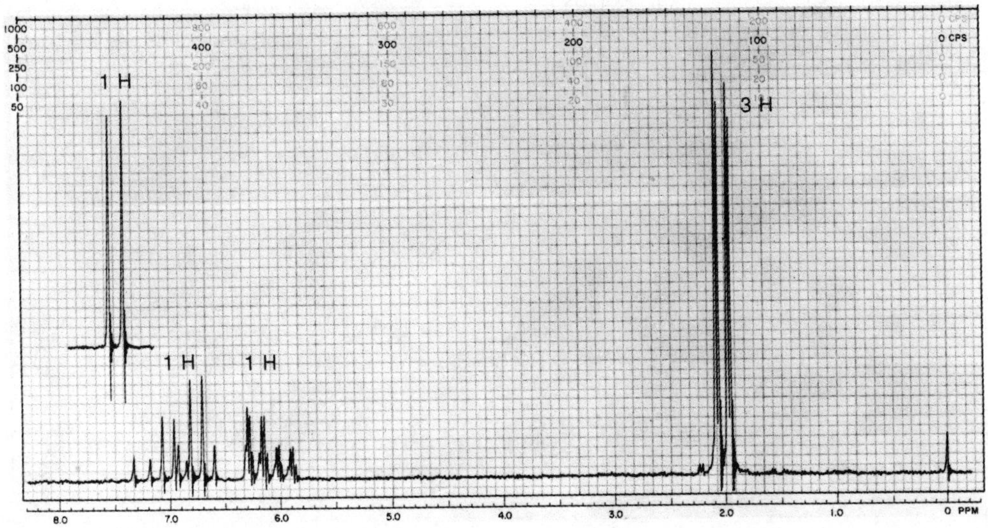

8. The following compound is a carboxylic acid, $C_4H_6O_2$. Deduce its structure and explain the spectrum. (Hint: There are two overlapping quartets centered near 7.10 δ.) The tall peak at the left of the spectrum has been moved upfield from 12.18 δ.

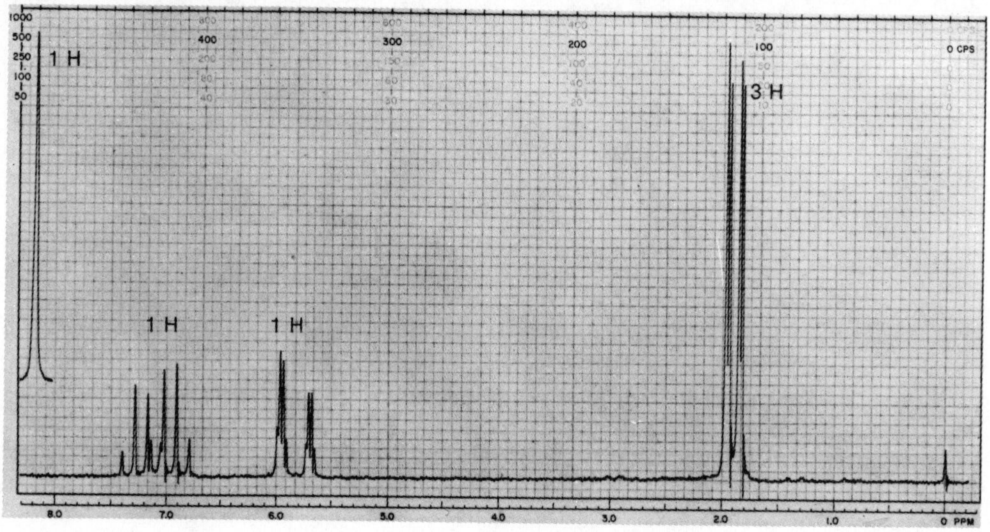

9. The following alkene with a formula C_7H_{12} is converted to methylcyclohexane on hydrogenation. Give its structure.

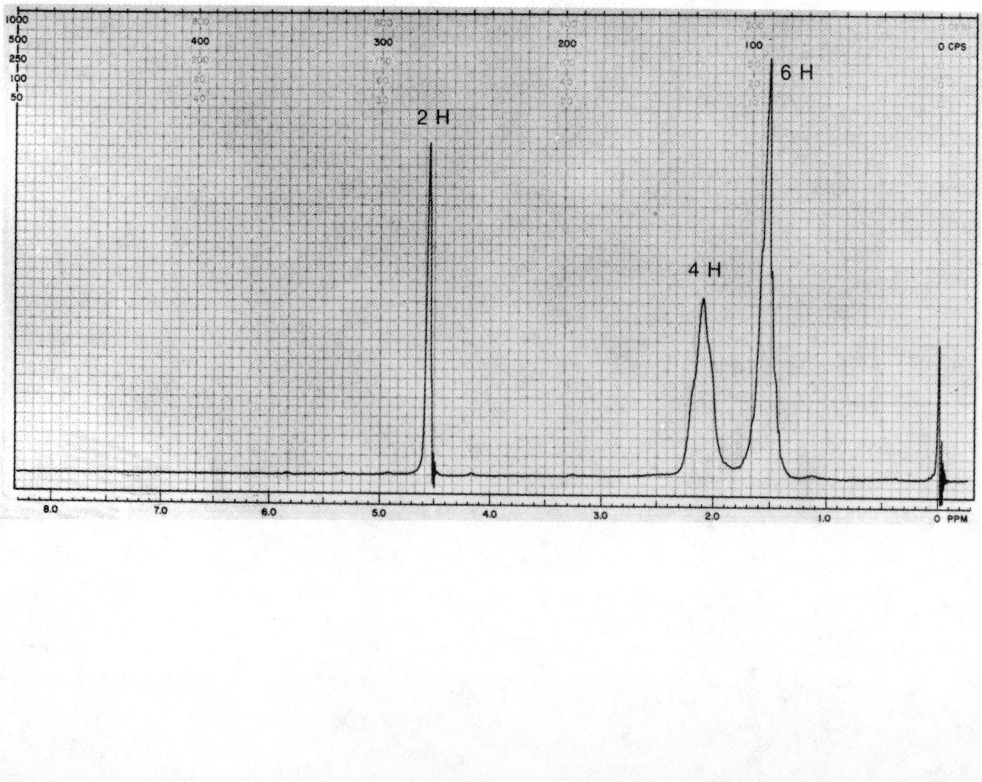

10. Deduce the structure of the following ester with a formula $C_8H_{12}O_4$, and assign the chemical shift values.

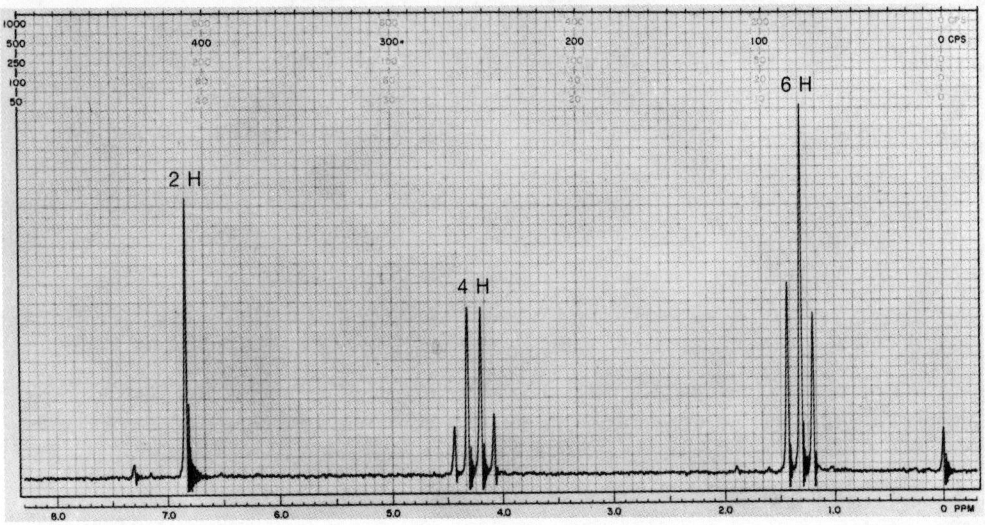

11. The spectra *A* and *B* of the two isomeric compounds β-chlorostyrene and *para*-chlorostyrene are given below. Assign a structure to each spectrum. In spectrum *B* a small peak belonging to the peaks to the right coincides with the large phenyl absorption. Explain each spectrum fully.

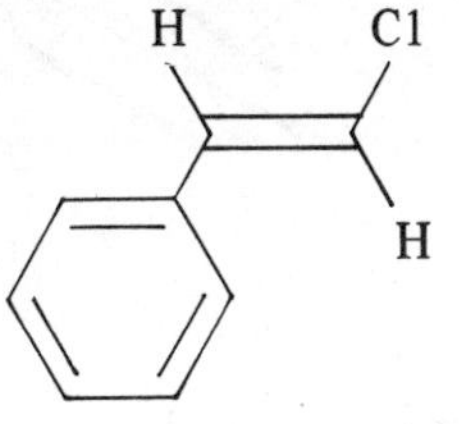

β-Chlorostyrene

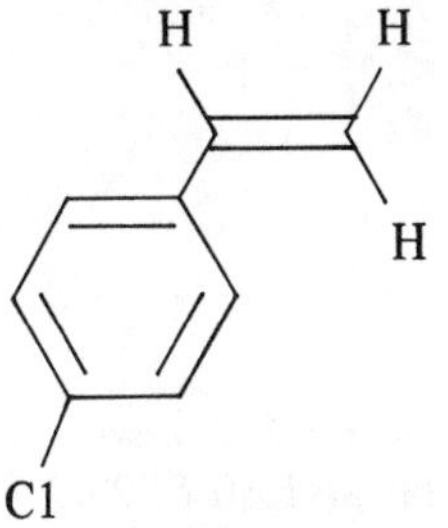

p-Chlorostyrene

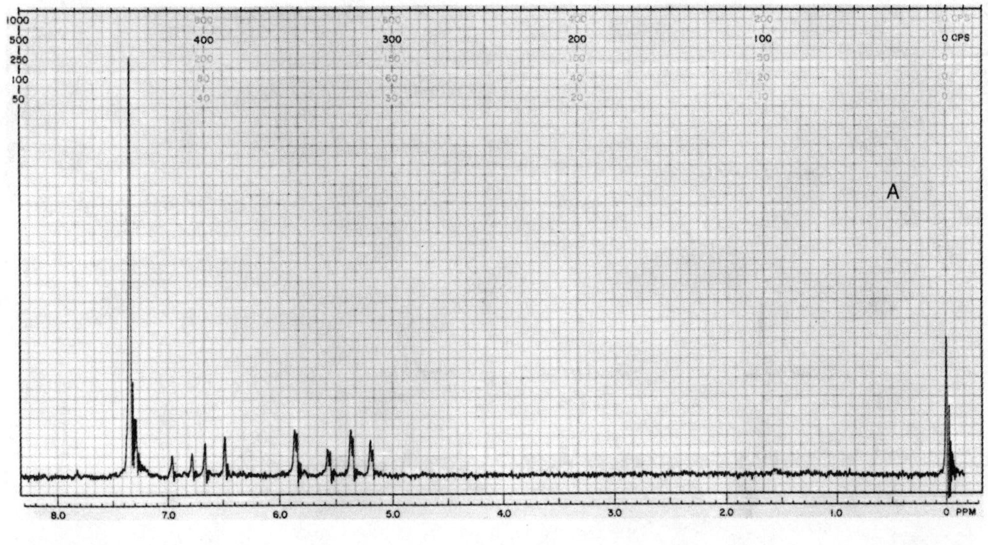

11. (Cont'd)

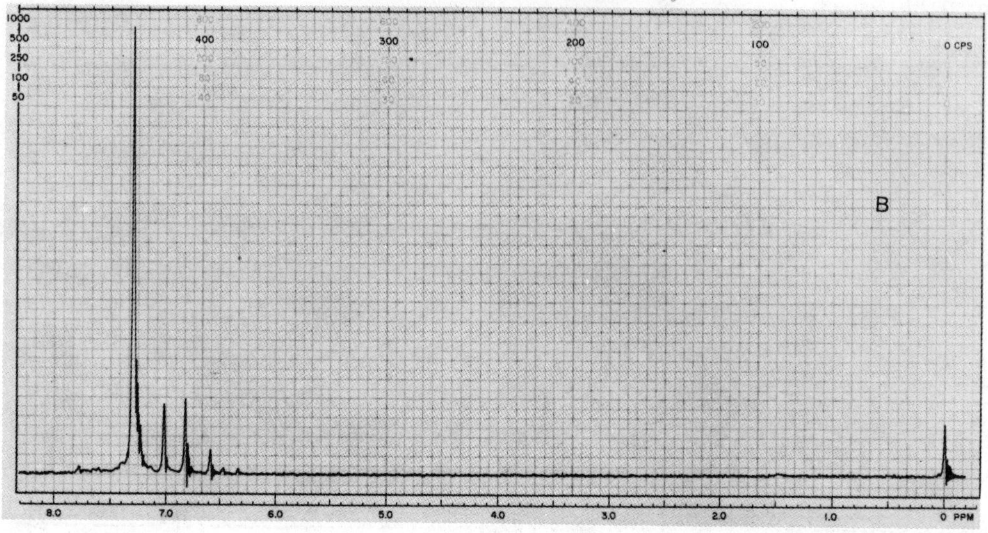

12. In which of the following two compounds is one likely to see allylic coupling?

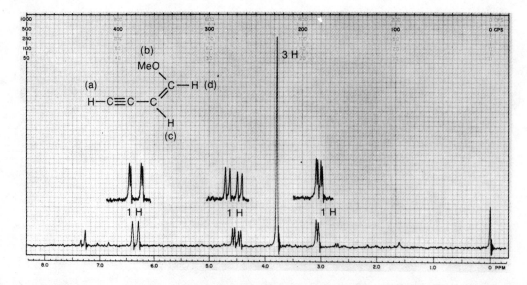

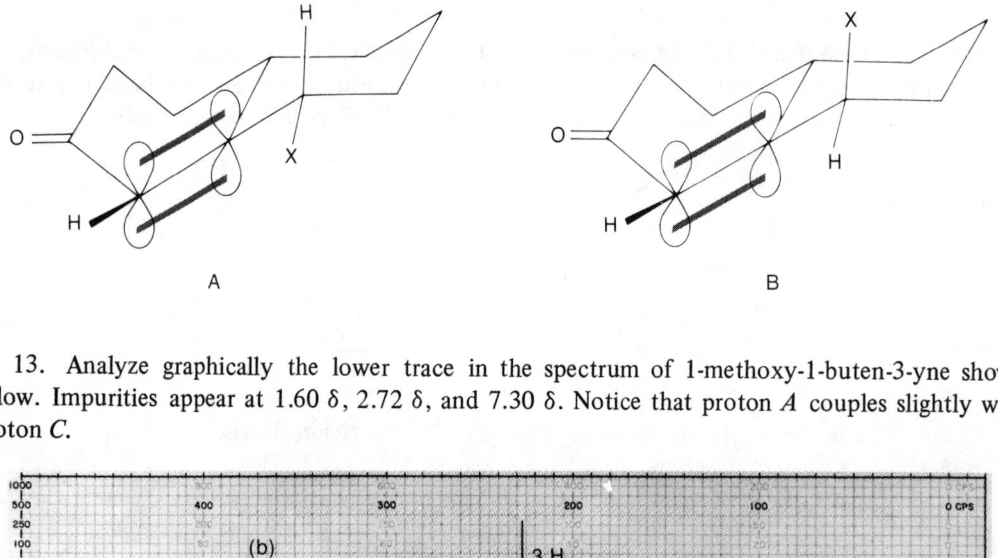

13. Analyze graphically the lower trace in the spectrum of 1-methoxy-1-buten-3-yne shown below. Impurities appear at 1.60 δ, 2.72 δ, and 7.30 δ. Notice that proton *A* couples slightly with proton *C*.

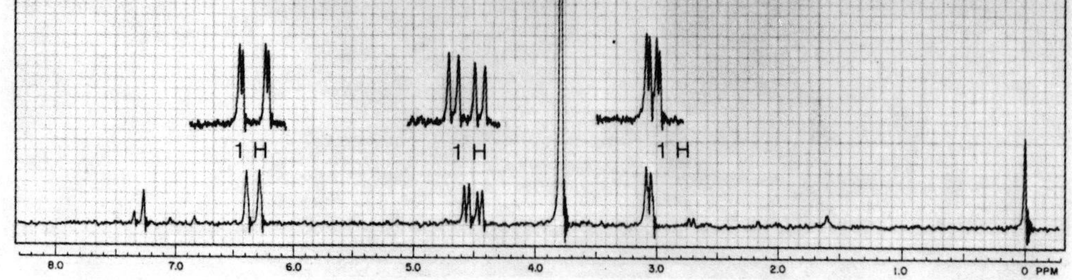

14. A naturally occurring amino acid with a formula $C_3H_7NO_2$ gives the nmr spectrum shown below when determined in deuterium oxide solvent. Determine the structure of this amino acid and comment on the peak at about 4.9 δ.

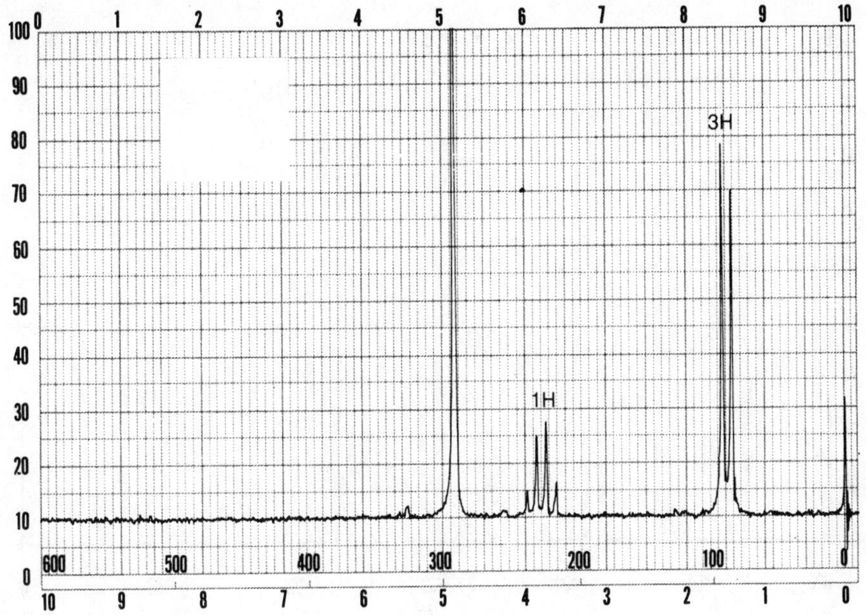

15. A compound with a formula C_4H_9NO has the nmr spectrum shown below. Give the structure of this compound and interpret the spectrum.

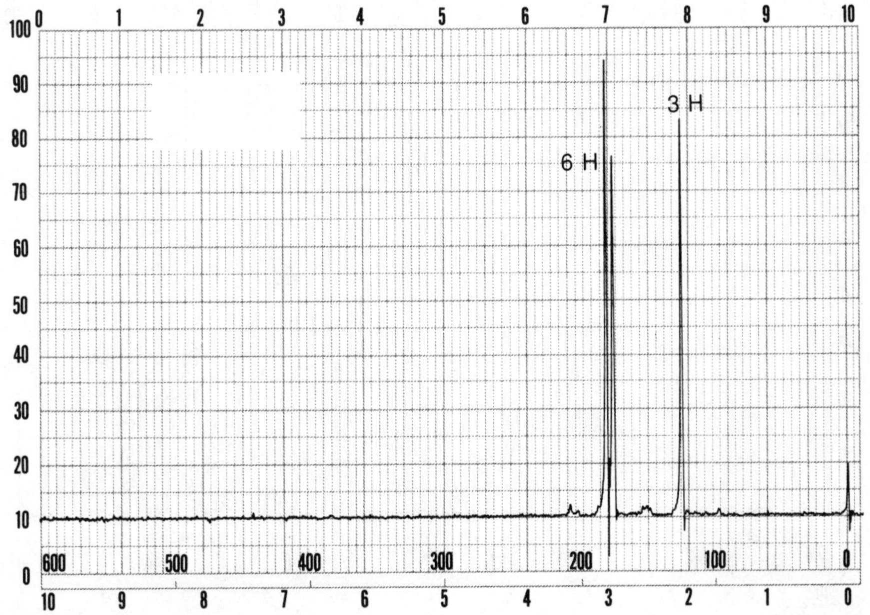

16. An aromatic compound with a formula $C_8H_8O_2$ shows weak infrared absorption at about 3000 cm^{-1} (3.33 μ), 2850 cm^{-1} (3.51 μ), and 2750 cm^{-1} (3.64 μ) and strong absorption at 1680 cm^{-1} (5.95 μ), 1260 cm^{-1} (7.94 μ), 1030 cm^{-1} (9.71 μ), and 840 cm^{-1} (11.90 μ). From these ir data and the nmr spectrum given below, determine the structure of the compound.

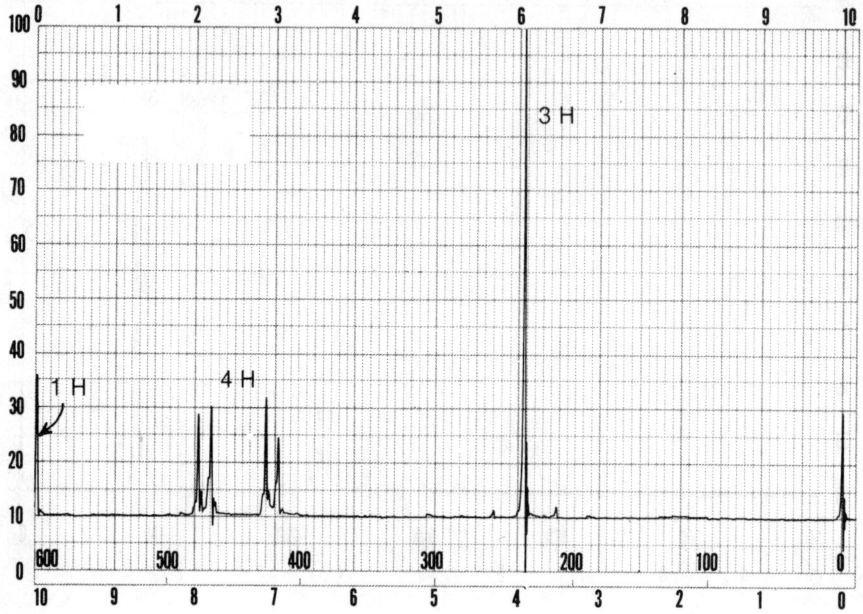

17. Phenacetin ($C_{10}H_{13}NO_2$) is sometimes mixed with aspirin and sold as an over-the-counter analgesic drug. Phenacetin contains an amide linkage and has the nmr spectrum shown below. Give the structure of this compound.

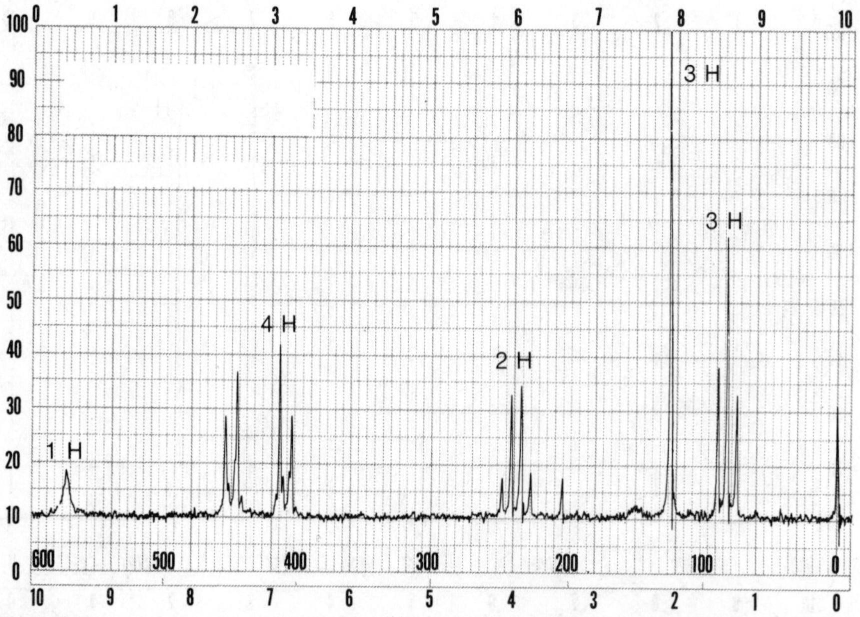

18. The fragrant natural product anethole ($C_{10}H_{12}O$) is obtained from anise by steam distillation. The nmr spectrum of the purified material is shown below. Deduce the structure of anethole.

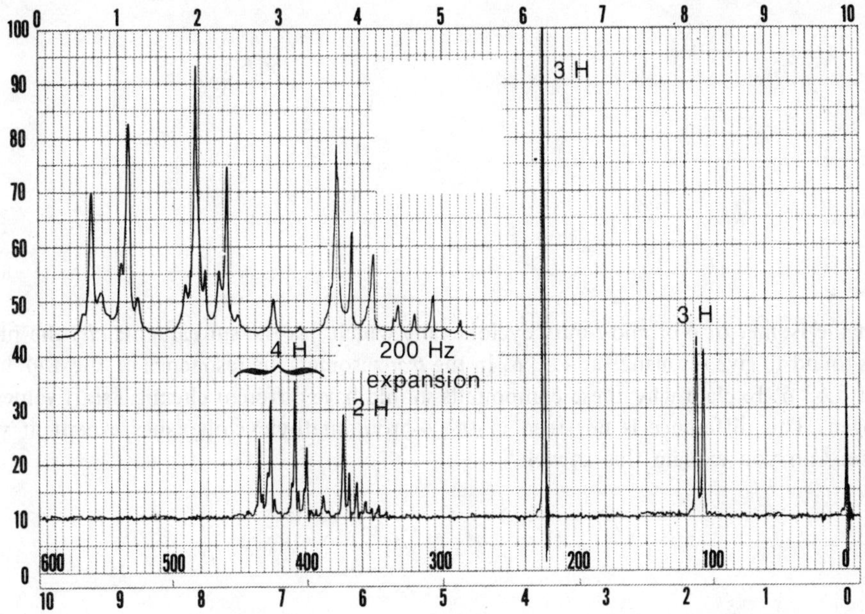

19. An aromatic compound with a formula $C_6H_4N_2O_5$ gives strong infrared absorption at about 3300 cm^{-1} (broad); 1520 cm^{-1}, and 1350 cm^{-1}. Determine the functional groups present and the substitution pattern on the ring. Explain the splitting patterns. The functional group indicated in the infrared spectrum at 3300 cm^{-1} does not appear in this nmr spectrum.

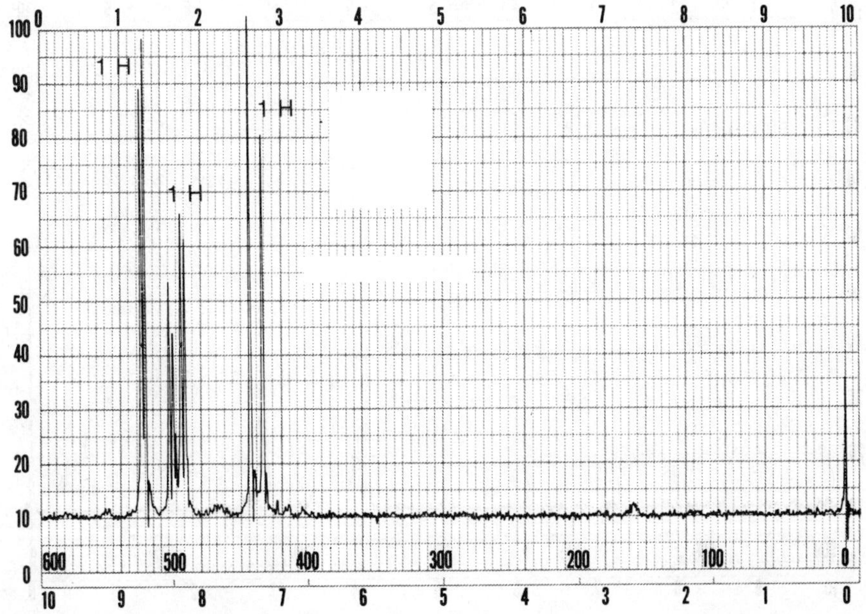

20. At room temperature the nmr spectrum of cyclohexane shows only a single resonance peak. As the temperature of the sample is lowered, the sharp single peak broadens until at −66.7°C it begins to split into two peaks, both broad. As the temperature is lowered further to −100°C, each of the two broad bands begins to give a splitting pattern of its own. Explain the origin of these two families of bands.

21. In the isomer of 4-bromo-*t*-butylcyclohexane which has the substituents *cis*, the proton on carbon-4 is found to give resonance at 4.33 δ. In the *trans* isomer, the resonance of the C-4 hydrogen is found at 3.63 δ. Explain why these compounds should have different chemical shift values for the C-4 hydrogen. This difference is not seen in the 4-bromomethylcyclohexanes except at very low temperature. Can you explain this as well?

22. The spectrum of 2,2,2-trifluoroethanol is shown below. A trace of acid has been added to decouple the hydroxyl hydrogen. Explain the spectrum completely and give an approximate value for J_{HF}.

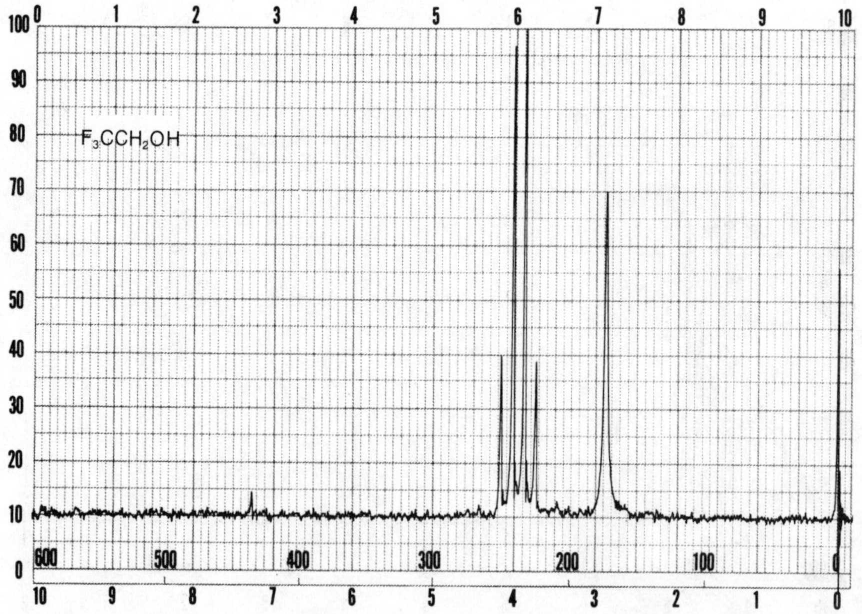

23. The following compound with a formula $C_5H_8O_2$ is a cyclic ester (lactone). Give its structure.

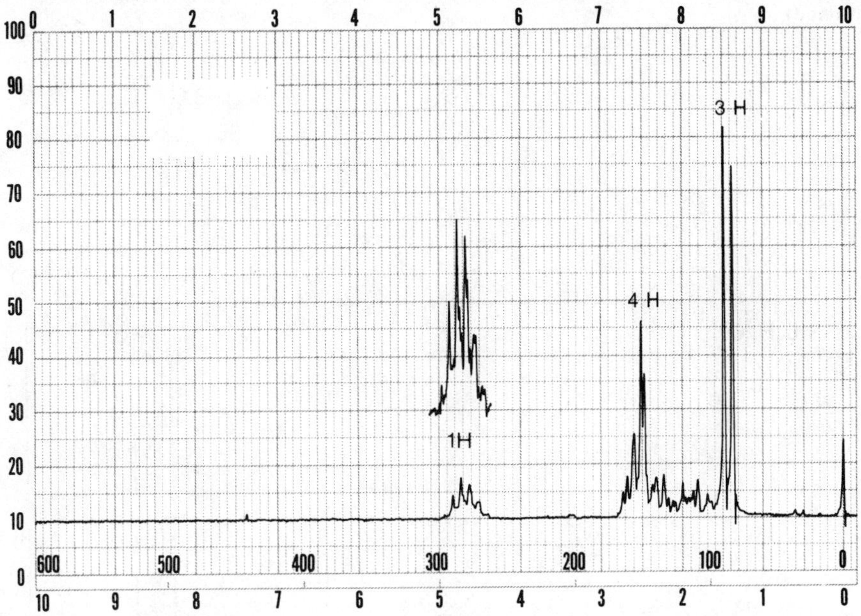

24. Below are the noise-decoupled carbon-13 nmr spectra of the three other isomers of *n*-butyl alcohol ($C_4H_{10}O$). Identify the alcohol responsible for each spectrum and, using the correlation chart in Figure 4–74, assign the peaks in each spectrum to an appropriate carbon atom (or appropriate carbon atoms) of your chosen structure. (Note: The peaks in an nmr spectrum are usually labeled a,b,c. . . in terms of increasing chemical shift. Hence, the corresponding carbon atoms should be labeled a,b,c. . . .) (Spectra from L.F. Johnson and W.C. Jankowski, "Carbon-13 NMR Spectra: A Collection of Assigned, Coded, and Indexed Spectra," Wiley-Interscience, © 1972. Reprinted by permission of John Wiley and Sons, Inc., New York.)

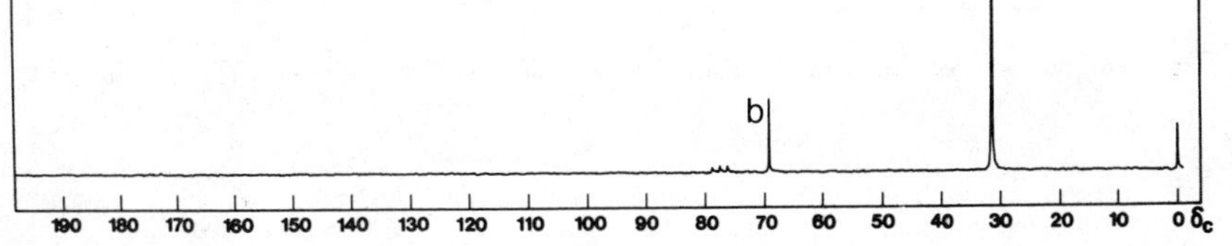

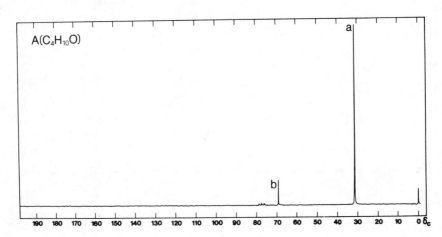

24. (Cont'd)

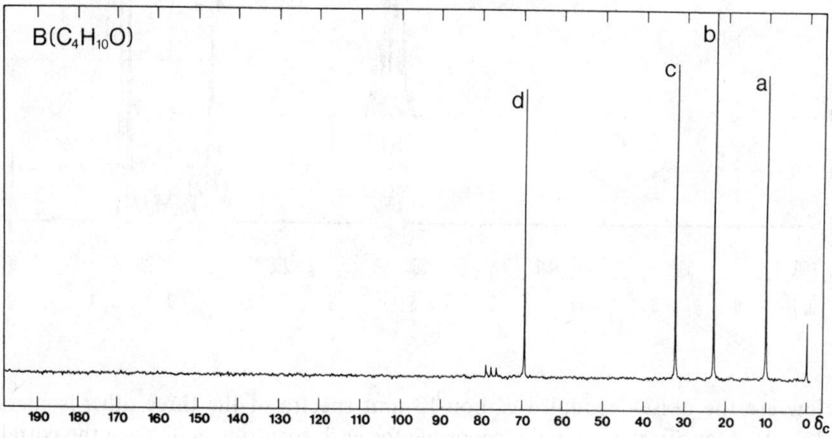
B($C_4H_{10}O$)
d
c
b
a
190 180 170 160 150 140 130 120 110 100 90 80 70 60 50 40 30 20 10 0 δ_c

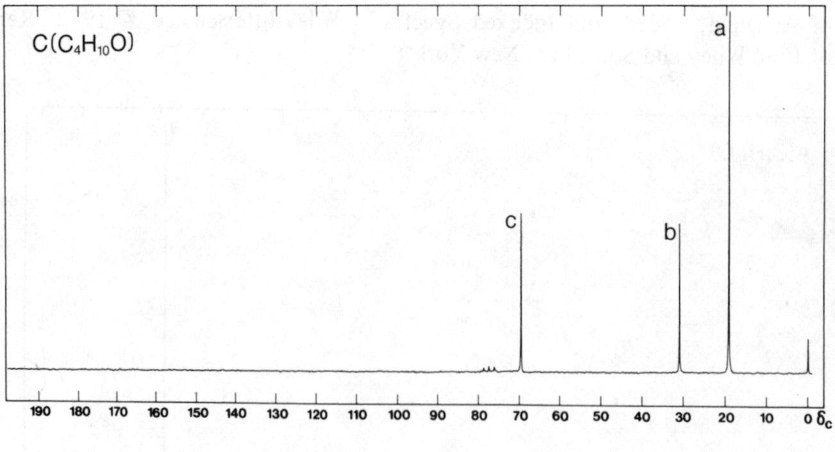
C($C_4H_{10}O$)
c
b
a
190 180 170 160 150 140 130 120 110 100 90 80 70 60 50 40 30 20 10 0 δ_c

25. Spectra D and E are the noise-decoupled carbon-13 nmr spectra of two isomers of formula C_7H_8O. One compound is an aromatic ether, and the other is a methyl substituted phenol. Identify the structure of the compound to which each spectrum corresponds, and assign all the peaks in each of the spectra to the appropriate carbon atoms of your structures. (Spectra from L.F. Johnson and W.C. Jankowski, "Carbon-13 NMR Spectra: A Collection of Assigned, Coded, and Indexed Spectra." Wiley-Interscience, © 1972. Reprinted by permission of John Wiley and Sons, Inc., New York.)

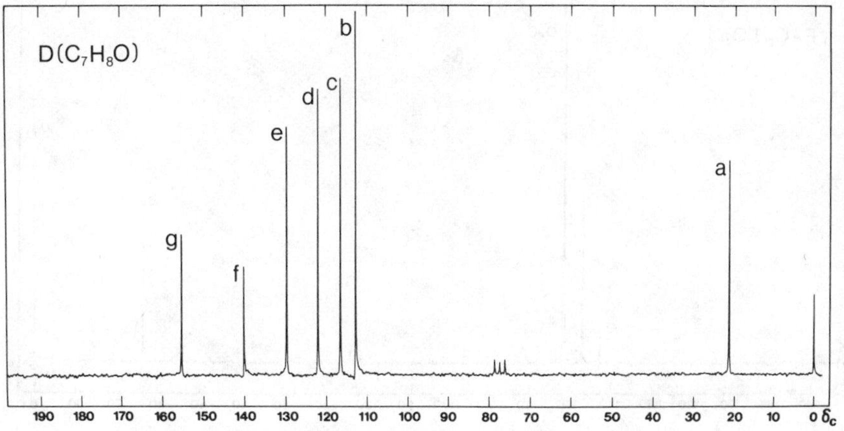

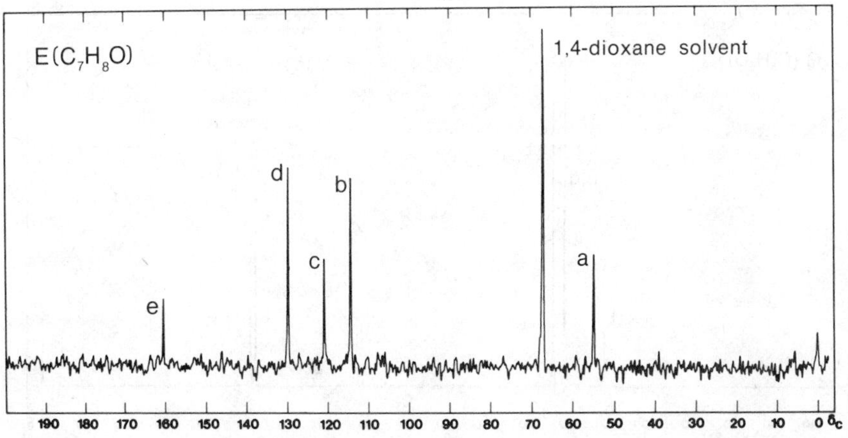

26. Spectra F and G are the noise-decoupled carbon-13 nmr spectra of two isomeric compounds of formula C_8H_8O. Both are aromatic compounds. One is a ketone and one is an epoxide. Give the structure of the compound responsible for each spectrum, and assign the peaks to specific carbon atoms. (Spectra from L.F. Johnson and W.C. Jankowski, "Carbon-13 NMR Spectra: A Collection of Assigned, Coded, and Indexed Spectra," Wiley-Interscience, © 1972. Reprinted by permission of John Wiley and Sons, Inc., New York.)

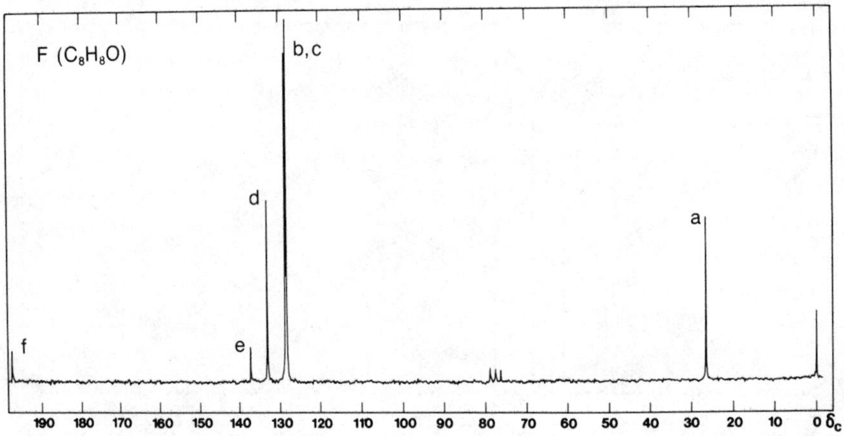

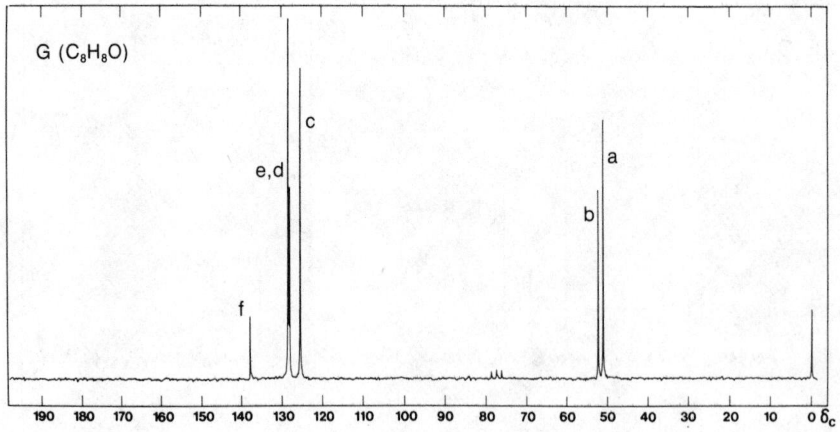

CHAPTER 5

ULTRAVIOLET SPECTROSCOPY

Most organic molecules and functional groups are transparent in the portions of the electromagnetic spectrum which we call the ultraviolet (uv) and visible (vis) regions, that is, the regions whose wavelengths range from 190 nm to 800 nm. Consequently, absorption spectroscopy which deals with this range of wavelengths is of limited utility. However, in some cases useful information can be derived from these regions of the spectrum, and this information, when combined with the detail provided by infrared and nuclear magnetic resonance spectra, can lead to valuable structural proposals.

5.1 THE NATURE OF ELECTRONIC EXCITATIONS

When continuous radiation passes through a transparent material, a portion of the radiation may be absorbed. If this occurs, the residual radiation, when it is passed through a prism, will yield a spectrum with gaps in it. This is called an absorption spectrum. As a result of energy absorption, atoms or molecules pass from a state of low energy (the initital or ground state) to a state of higher energy (the excited state). This excitation process, which is quantized, is shown in Figure 5–1. The electromagnetic radiation which is absorbed has energy which is exactly equal to the energy difference between the excited and the ground states.

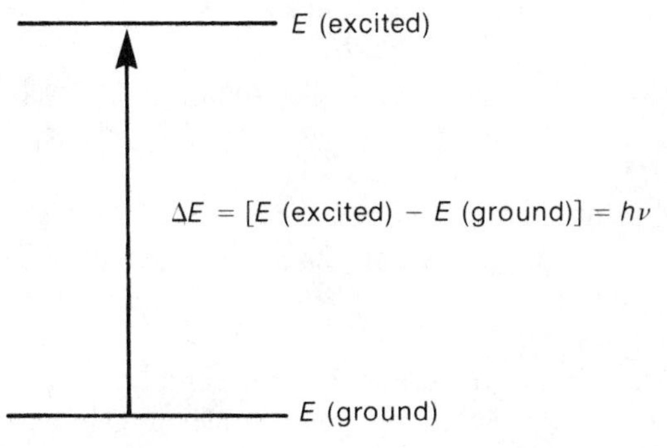

FIGURE 5–1 The Excitation Process

In the case of ultraviolet and visible spectroscopy, the transitions which result in the absorption of electromagnetic radiation in this region of the spectrum are transitions between *electronic* energy levels. The energy differences between the electronic levels in most molecules vary from 30 to 150 kcal/mole.

For most molecules, the lowest energy occupied molecular orbitals are the σ-orbitals, those which correspond to σ-bonds. The π-orbitals lie at somewhat higher energy levels, and those orbitals which hold unshared electron pairs, the non-bonding or *n*-orbitals, lie at even higher energies. The unoccupied or anti-bonding orbitals (π^* and σ^*) are the orbitals of highest energy. A typical progression of electronic energy levels is illustrated in Figure 5–2A.

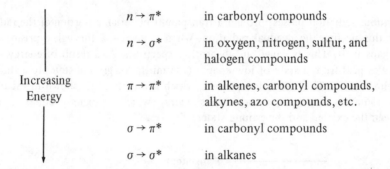

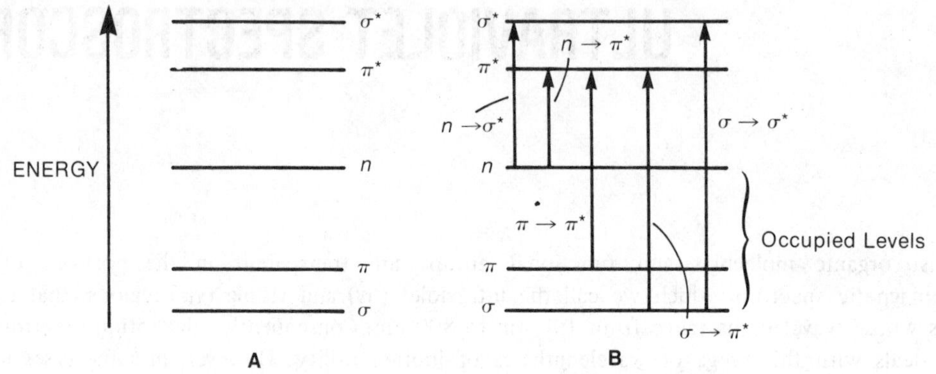

FIGURE 5–2 Electronic Energy Levels and Transitions

In all compounds other than alkanes, the electrons may undergo several possible transitions of different energies. Some of the most important transitions are:

Increasing Energy	$n \rightarrow \pi^*$	in carbonyl compounds
	$n \rightarrow \sigma^*$	in oxygen, nitrogen, sulfur, and halogen compounds
	$\pi \rightarrow \pi^*$	in alkenes, carbonyl compounds, alkynes, azo compounds, etc.
	$\sigma \rightarrow \pi^*$	in carbonyl compounds
	$\sigma \rightarrow \sigma^*$	in alkanes

These transitions are illustrated in Figure 5–2B. Electronic energy levels in aromatic molecules are more complicated than this simple illustration describes. The electronic transitions of aromatic compounds are described in Section 5.17.

Clearly, the energy required to bring about transitions from the highest occupied energy level in the ground state to the lowest unoccupied energy level is less than the energy required to bring about a transition from a lower occupied energy level. For many purposes it is the transition of lowest energy which is the most important.

Not all of the transitions which would at first sight appear to be possible are in fact observed. Certain restrictions, called *selection rules*, must be considered. One important selection rule states that transitions which involve a change in the spin quantum number of an electron during the transition are not allowed to take place, or are *forbidden*. Other selection rules deal with the numbers of electrons which may be excited at one time, with symmetry properties of the molecule and of the electronic states, and with other factors which need not be discussed here. Transitions that are formally forbidden by the selection rules will often not be observed. However, theoretical treatments are only rather approximate, and in certain cases "forbidden" transitions are in fact observed, although the intensity of light absorption tends to be much lower than for transitions which are allowed by the selection rules. The $n \rightarrow \pi^*$ transition is the most common type of "forbidden" transition.

5.2 THE ORIGIN OF UV BAND STRUCTURE

For an atom which absorbs in the ultraviolet, the absorption spectrum often consists of very sharp lines of absorption, as would be expected for a quantized process occurring between two discrete energy levels. For molecules, however, the uv absorption usually occurs over a wide range of wavelengths. This occurs because molecules (as opposed to atoms) normally have many excited modes of vibration and rotation at room temperature. In fact, the vibration of molecules cannot be completely "frozen out" even at absolute zero. As a consequence, a collection of molecules will generally have its members in many states of vibrational and rotational excitation. The energy levels for these states are quite closely spaced, corresponding to considerably smaller energy differences than for electronic levels. The rotational and vibrational levels are thus "superimposed" on the electronic levels. A molecule may therefore undergo electronic and vibrational/rotational excitation simultaneously, as shown in Figure 5–3.

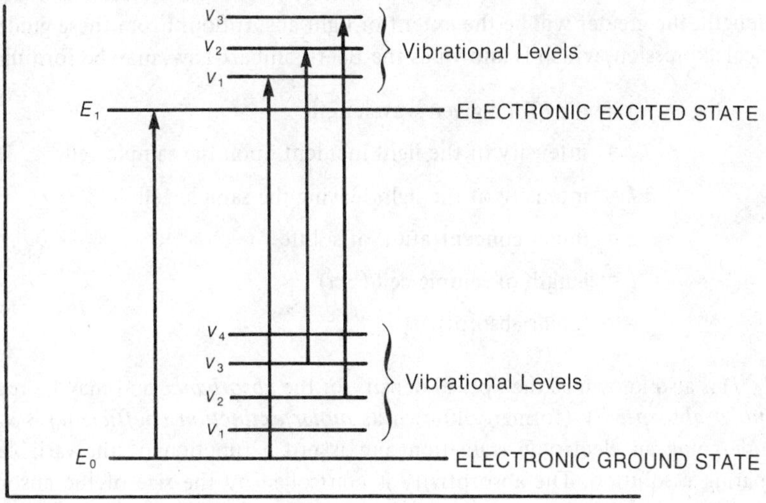

FIGURE 5–3 Electronic Transitions with Vibrational Transitions Superimposed (Rotational levels, which are very closely spaced within the vibrational levels, are omitted for clarity.)

Because there are so many possible transitions, each differing from the others by only a slight amount, each electronic transition consists of a vast number of lines very closely spaced. The spacing of these lines is so small that the spectrophotometer cannot resolve them. Rather, the instrument traces an "envelope" over the entire pattern. What is observed from these types of combined transitions is that the uv spectrum of a molecule usually consists of a *band* of absorption centered near the wavelength of the major transition, as shown in Figure 5–4A.

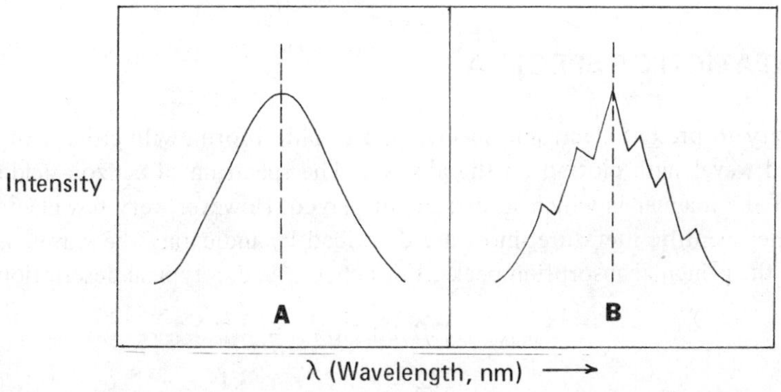

FIGURE 5–4 A UV Absorption Band at Room Temperature (A) and at Lowered Temperature (B)

If the temperature is lowered, rotational and collisional energy transfer processes are slowed down or stopped in a collection of molecules. Under these conditions, a uv band usually begins to develop "structure," the transitions corresponding to pure electronic-vibrational transitions. An example of how a molecule with a uv band like that in Figure 5–4A might develop fine structure at lower temperature is illustrated in Figure 5–4B. Many larger molecules, which often have high energy rotational levels, often show very characteristic vibrational fine structure in their uv spectra, even at room temperature.

5.3 PRINCIPLES OF ABSORPTION SPECTROSCOPY

The greater the number of molecules capable of absorbing light of a given wavelength, the greater will be the extent of light absorption. Furthermore, the more effective a molecule is in absorbing light of a given wavelength, the greater will be the extent of light absorption. From these guiding ideas, the following empirical expression, which is known as the Beer-Lambert Law, may be formulated:

$$\log (I_0/I) = \epsilon\, c\, l \text{ for a given wavelength}$$

I_0 = intensity of the light incident upon the sample cell

I = intensity of the light leaving the sample cell

c = molar concentration of solute

l = length of sample cell (cm)

ϵ = molar absorptivity

The term $\log (I_0/I)$ is also known as the optical density or the *absorbance* and may be represented by OD or A. The *molar absorptivity* (formerly known as *molar extinction coefficient*) is a property of the molecule undergoing an electronic transition and is not a function of the variable parameters involved in preparing a solution. The absorptivity is controlled by the size of the absorbing system and by the probability that the electronic transition will take place. The absorptivity may range in numerical size from zero to 10^6. Values above 10^4 are termed high intensity absorptions, while values below 10^3 are termed low intensity absorptions. Forbidden transitions have absorptivities in the range from 100 to 1000.

The Beer-Lambert Law is rigorously obeyed when a *single species* gives rise to the observed absorption. However, when different forms of the absorbing molecule are in equilibrium, when solute and solvent form complexes through some sort of association, when there is *thermal* equilibrium between the ground electronic state and a low-lying excited state, or when there are fluorescent compounds or compounds which are changed by irradiation, the law may not be obeyed.

5.4 PRESENTATION OF SPECTRA

It is customary to present electronic spectra in a graphical form, with either ϵ or $\log \epsilon$ plotted on the ordinate, and wavelength plotted on the abscissa. The spectrum of benzoic acid shown in Figure 5–5 is typical of the manner in which spectra are displayed. However, very few electronic spectra are reproduced in the scientific literature; most are described by indicating the wavelength maxima and absorptivities of the principal absorption peaks. For benzoic acid, a typical description might be:

$$\lambda_{max} = 230 \text{ nm};\ \log \epsilon = 4.2$$

272	3.1
282	2.9

The actual spectrum which corresponds to these data is shown in Figure 5–5.

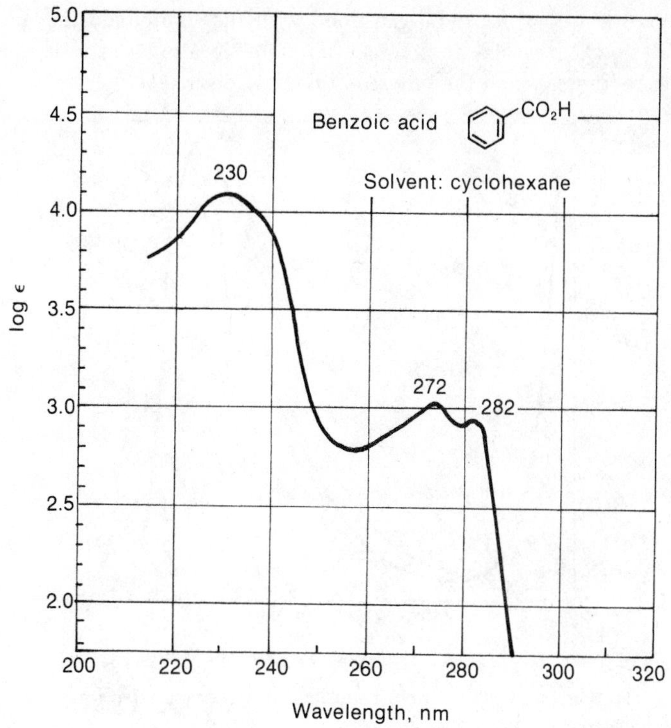

FIGURE 5–5 Ultraviolet Spectrum of Benzoic Acid in Cyclohexane (from R. A. Friedel and M. Orchin, "Ultraviolet Spectra of Aromatic Compounds," © 1951. Reprinted by permission of John Wiley and Sons, Inc., New York.)

5.5 SOLVENTS

The choice of solvent to be used in ultraviolet spectroscopy is quite important. The first criterion for a good solvent is that it should not absorb ultraviolet radiation in the same region as the substance whose spectrum is being determined. Usually solvents which do not contain conjugated systems are most suitable for this purpose, although they vary as to the shortest wavelength at which they remain transparent to ultraviolet radiation. Table 5–1 lists some common ultraviolet spectroscopy solvents and their "cut-off points" or minimum regions of transparency.

TABLE 5–1 Solvent Cut-Offs

Acetonitrile	190 nm	*n*-Hexane	201 nm
Chloroform	240	Methanol	205
Cyclohexane	195	Isooctane	195
1,4-Dioxane	215	Water	190
95% Ethanol	205	Trimethyl Phosphate	210

Of the solvents listed in Table 5–1, water, 95% ethanol, and *n*-hexane are the most commonly used. Each of them is transparent in the regions of the ultraviolet spectrum where interesting absorption peaks from solute molecules are likely to occur.

A second criterion which must be considered is the effect of a solvent on the fine structure of an absorption band. Figure 5–6 illustrates the effect of polar and non-polar solvents on an absorption band. A non-polar solvent does not hydrogen bond with the solute, and the spectrum of the solute closely approximates what it would be in a gaseous state. In a polar solvent, the hydrogen bonding forms a solute-solvent complex, and the fine structure may disappear.

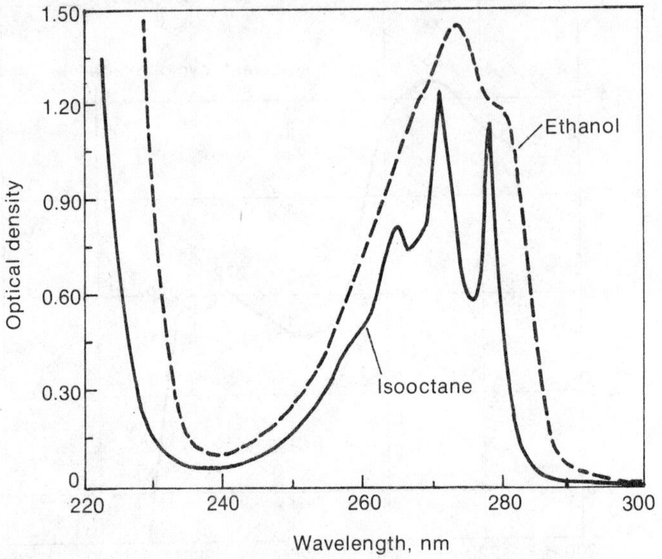

FIGURE 5–6 Ultraviolet Spectra of Phenol in Ethanol and in Isooctane [Reprinted with permission from N. D. Coggeshall and E. M. Lang. J. Am. Chem. Soc., *70*, 3288 (1948).]

A third property of a solvent which must be considered is the ability of a solvent to influence the wavelength of ultraviolet light which will be absorbed. Polar solvents may not form hydrogen bonds as readily with excited states as with ground states of polar molecules, and the energies of electronic transitions in these molecules will be increased by these polar solvents. Transitions of the $n \rightarrow \pi^*$ type are shifted to shorter wavelengths by polar solvents. On the other hand, in some cases the excited states may form stronger hydrogen bonds than the corresponding ground states. In such cases, a polar solvent would shift an absorption to longer wavelength, since the energy of the electronic transition would be decreased. Transitions of the $\pi \rightarrow \pi^*$ type are shifted to longer wavelengths by polar solvents. Table 5–2 illustrates a typical effect of a series of solvents on an electronic transition. This will be discussed further in Section 5.16.

TABLE 5–2 Solvent Shifts on the $n \rightarrow \pi^*$ Transition in Acetone

Solvent	H_2O	CH_3OH	C_2H_5OH	$CHCl_3$	C_6H_{14}
λ_{max}, nm	264.5	270	272	277	279

5.6 WHAT IS A CHROMOPHORE?

Although the absorption of ultraviolet radiation is brought about by the excitation of electrons from their ground states to their excited states, the nuclei which the electrons hold together in bonds play an important role in determining which wavelengths of radiation will be absorbed. The nuclei determine the strength with which the electrons are bound, and in this way influence the energy spacing between ground and excited states. Thus, the characteristic energy of a transition, and the wavelength of radiation absorbed, is a property of a group of atoms, rather than of electrons themselves. The group of atoms producing such an absorption is called a *chromophore*. As structural changes in this group of atoms are made, the exact energy and intensity of the absorption are expected to change accordingly. Very often, it is extremely difficult to predict from theory how the absorption will be changed as the structure of the chromophore is modified, and it is necessary to apply empirical working guides to predict such relationships.

For molecules, such as alkanes, which contain nothing but single bonds and which lack atoms with unshared electron pairs, the only electronic transitions possible are of the $\sigma \rightarrow \sigma^*$ type. These transitions are of such a high energy that they absorb ultraviolet energy at very short wavelengths – shorter than is experimentally accessible using typical spectrophotometers. This type of transition is illustrated in Figure 5–7. On the right of the figure, the excitation of the σ-bonding electron to the σ^*-antibonding orbital is depicted.

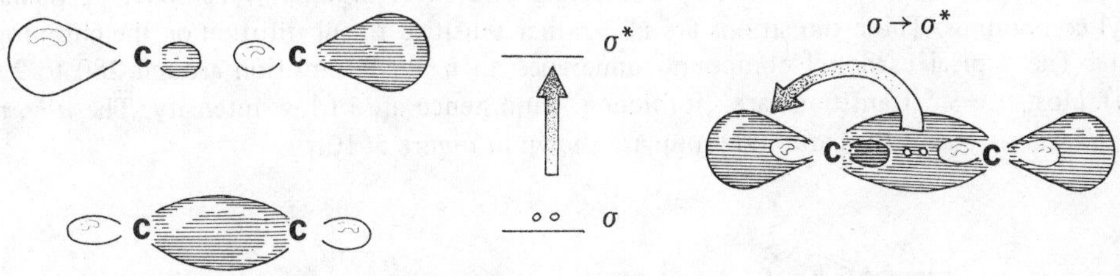

FIGURE 5–7 $\sigma \rightarrow \sigma^*$ Transition

In saturated molecules which contain atoms bearing unshared pairs of electrons, transitions of the $n \rightarrow \sigma^*$ type become important. These are also rather high energy transitions, but they do absorb radiation which lies within an experimentally accessible region. Alcohols and amines absorb in the range from 175 to 200 nm, while organic thiols and sulfides absorb between 200 and 220 nm. Most of the absorptions are below the cut-off points for the common solvents, and so they are not observed in solution spectra. An $n \rightarrow \sigma^*$ transition is illustrated in Figure 5–8. Again, on the right of the figure, the excitation of the non-bonding electron to the antibonding orbital is described.

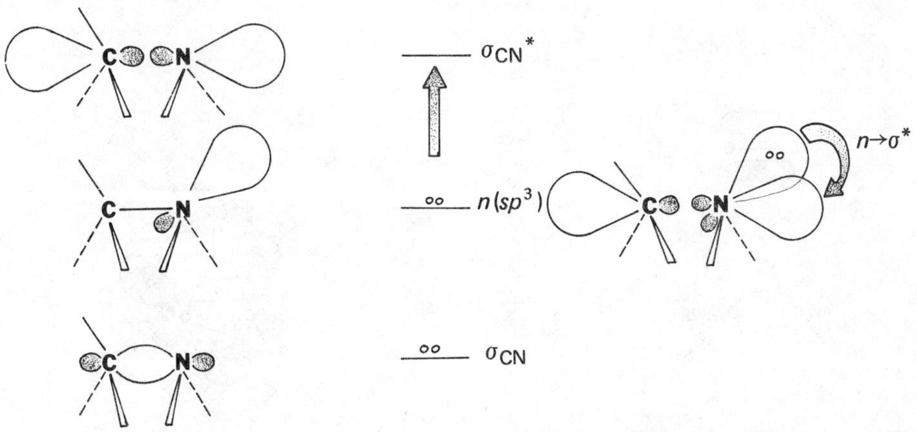

FIGURE 5–8 $n \rightarrow \sigma^*$ Transition

With unsaturated molecules, $\pi \rightarrow \pi^*$ transitions become possible. These transitions are of rather high energy as well, but their positions are sensitive to the presence of substitution, as will be seen later. Alkenes absorb around 175 nm, alkynes absorb around 170 nm, and carbonyl compounds absorb around 188 nm ($\epsilon = 900$). This type of transition is shown in Figure 5–9.

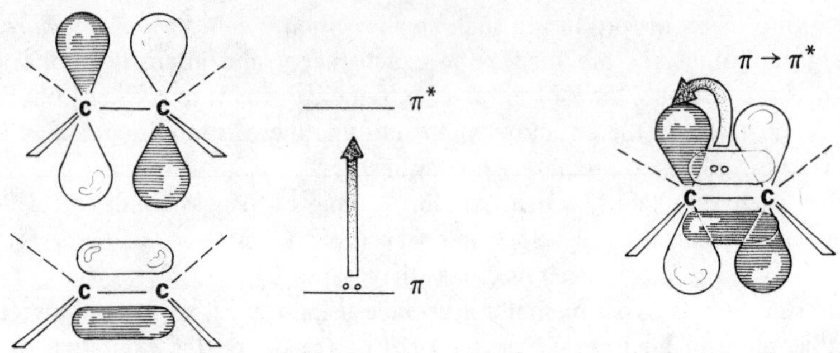

FIGURE 5–9 π → π* Transition

Unsaturated molecules which contain such atoms as oxygen or nitrogen may also undergo $n \rightarrow \pi^*$ transitions. These are perhaps the most interesting and most studied transitions, particularly in carbonyl compounds. These transitions are also rather sensitive to substitution on the chromophoric structure. The typical carbonyl compound undergoes an $n \rightarrow \pi^*$ transition around 280 to 290 nm (ϵ = 15). Most $n \rightarrow \pi^*$ transitions are "forbidden," and hence are of low intensity. The $n \rightarrow \pi^*$ and the $\pi \rightarrow \pi^*$ transitions of the carbonyl group are shown in Figure 5-10.†

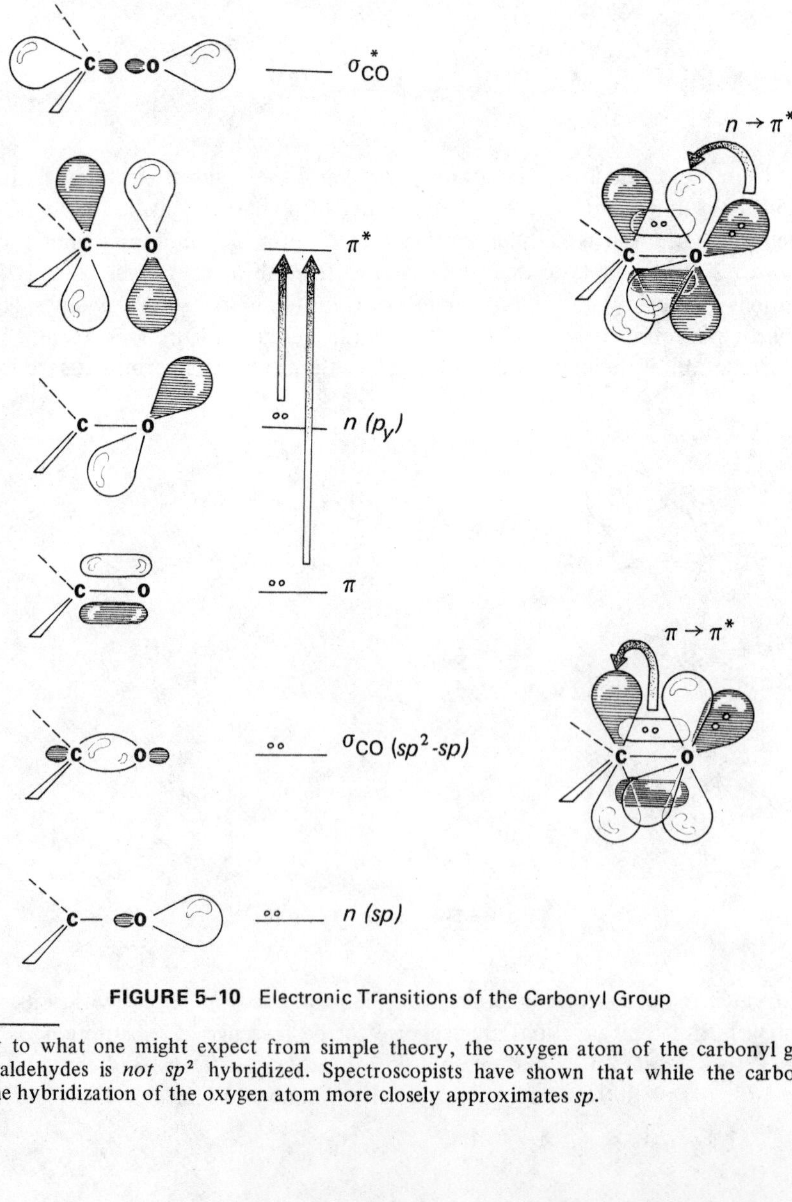

FIGURE 5–10 Electronic Transitions of the Carbonyl Group

†Contrary to what one might expect from simple theory, the oxygen atom of the carbonyl group in simple ketones and aldehydes is *not* sp^2 hybridized. Spectroscopists have shown that while the carbon atom is sp^2 hybridized, the hybridization of the oxygen atom more closely approximates sp.

Table 5–3 lists typical absorptions of simple isolated chromophores. It may be noticed that these *simple* chromophores nearly all absorb at approximately the same wavelength (160–210 nm).

TABLE 5-3 Typical Absorptions of Simple Isolated Chromophores

Class	Transition	λ_{max} (nm)	log ϵ	Class	Transition	λ_{max} (nm)	log ϵ
R–OH	$n \to \sigma^*$	180	2.5	$R–NO_2$	$n \to \pi^*$	271	<1.0
R–O–R	$n \to \sigma^*$	180	3.5	R–CHO	$\pi \to \pi^*$	190	2.0
$R–NH_2$	$n \to \sigma^*$	190	3.5		$n \to \pi^*$	290	1.0
R–SH	$n \to \sigma^*$	210	3.0	R_2CO	$\pi \to \pi^*$	180	3.0
$R_2C{=}CR_2$	$\pi \to \pi^*$	175	3.0		$n \to \pi^*$	280	1.5
R–C≡C–R	$\pi \to \pi^*$	170	3.0	RCOOH	$n \to \pi^*$	205	1.5
R–C≡N	$n \to \pi^*$	160	<1.0	RCOOR′	$n \to \pi^*$	205	1.5
R–N=N–R	$n \to \pi^*$	340	<1.0	$RCONH_2$	$n \to \pi^*$	210	1.5

The position and intensity of an absorption band of a chromophore may be modified by the attachment of substituent groups in place of hydrogen on the basic chromophore structure. The substituent groups may not give rise to the absorption of ultraviolet radiation themselves, but their presence will modify the absorption of the principal chromophore. Substituents which increase the intensity of the absorption, and possibly the wavelength, are called *auxochromes*. Typical auxochromes might include methyl, hydroxyl, alkoxy, halogen, and amino groups.

Other substituents may have any of four kinds of effects on the absorption:

1) *Bathochromic shift* (red shift) – shift to lower frequency or longer wavelength.
2) *Hypsochromic shift* (blue shift) – shift to higher frequency or shorter wavelength.
3) *Hyperchromic effect* – an increase in intensity.
4) *Hypochromic effect* – a decrease in intensity.

These types of effects are shown in Figure 5–11.

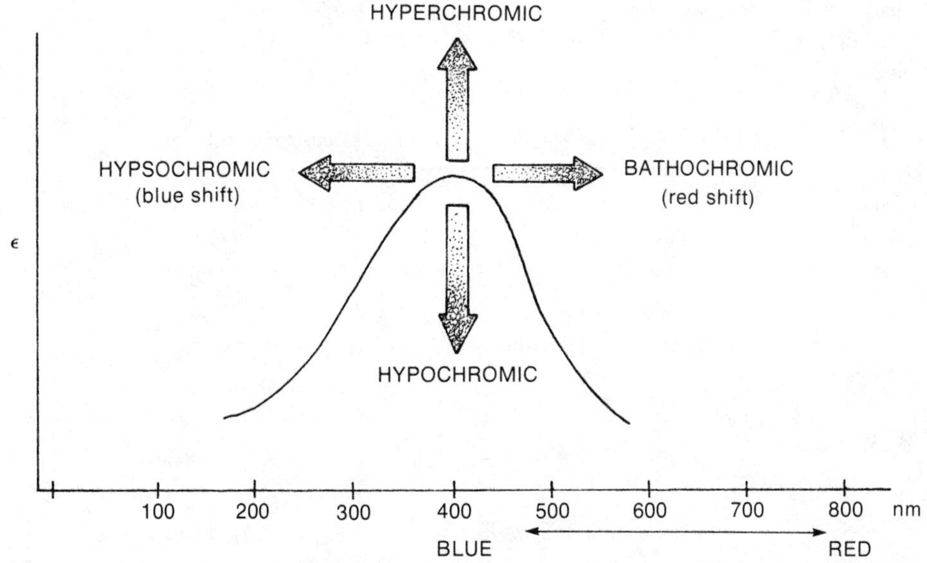

FIGURE 5–11 Substituent Effects

5.7 THE EFFECT OF CONJUGATION

One of the best ways to bring about a bathochromic shift is to increase the extent of conjugation in a double-bonded system. In the presence of conjugated double bonds, the electronic energy levels of a chromophore move closer together. As a result, the energy required to produce a transition from an occupied electronic energy level to an unoccupied level is decreased, and the wavelength of the light absorbed becomes longer. Figure 5–12 illustrates the bathochromic shift which is observed in a series of conjugated polyenes as the length of the conjugated chain is increased.

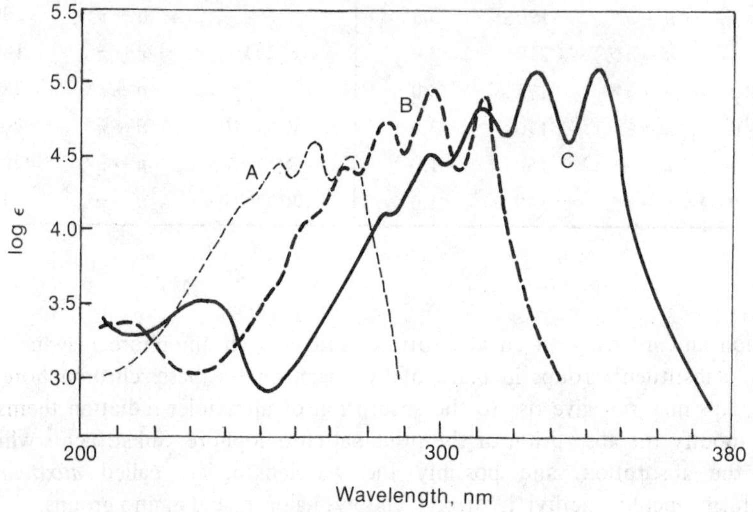

FIGURE 5–12 Ultraviolet Spectra of Dimethylpolyenes, $CH_3(CH{=}CH)_nCH_3$; A, $n = 3$; B, $n = 4$; C, $n = 5$ [Reprinted with permission from P. Nayler and M. C. Whiting, J. Chem. Soc., 3042 (1955).]

Conjugation of two chromophores not only results in a bathochromic shift, but it also increases the intensity of the absorption. These two effects are of prime importance in the use and interpretation of the electronic spectra of organic molecules, because conjugation shifts the selective light absorption of isolated chromophores from a region of the spectrum that is not readily accessible to a region that is easily studied with commercially available spectrophotometers. The exact position and intensity of the absorption band of the conjugated system can be correlated with the extent of conjugation in the system. Table 5–4 illustrates the effect of conjugation on some typical electronic transitions.

TABLE 5–4 The Effect of Conjugation on Electronic Transitions

Alkenes		λ_{max} (nm)	ε
Ethylene		175	15,000
1,3-Butadiene		217	21,000
1,3,5-Hexatriene		258	35,000
β-Carotene (11 double bonds)		465	125,000
Ketones			
Acetone	$\pi \to \pi^*$	189	900
	$n \to \pi^*$	280	12
3-Buten-2-one	$\pi \to \pi^*$	213	7,100
	$n \to \pi^*$	320	27

5.8 THE EFFECT OF CONJUGATION ON ALKENES

The bathochromic shift that results from an increase in the length of a conjugated system implies that an increase in conjugation decreases the energy required for electronic excitation. This is true and can be explained most easily by the use of molecular orbital theory. According to molecular orbital (MO) theory, the atomic *p* orbitals on each of the carbon atoms combine to make π molecular orbitals. For instance, in the case of ethylene, we have two atomic *p* orbitals, ϕ_1 and ϕ_2. From these two *p* orbitals, two π molecular orbitals, ψ_1 and $\psi_2{}^*$, are formed by taking linear combinations. The bonding orbital ψ_1 results from an addition of the wave functions of the two *p* orbitals, and the antibonding orbital $\psi_2{}^*$ results from the subtraction of these two wave functions. The new bonding orbital, a *molecular orbital*, has an energy lower than that of either of the original *p* orbitals; likewise, the antibonding orbital has an elevated energy. This is diagramatically illustrated in Figure 5–13.

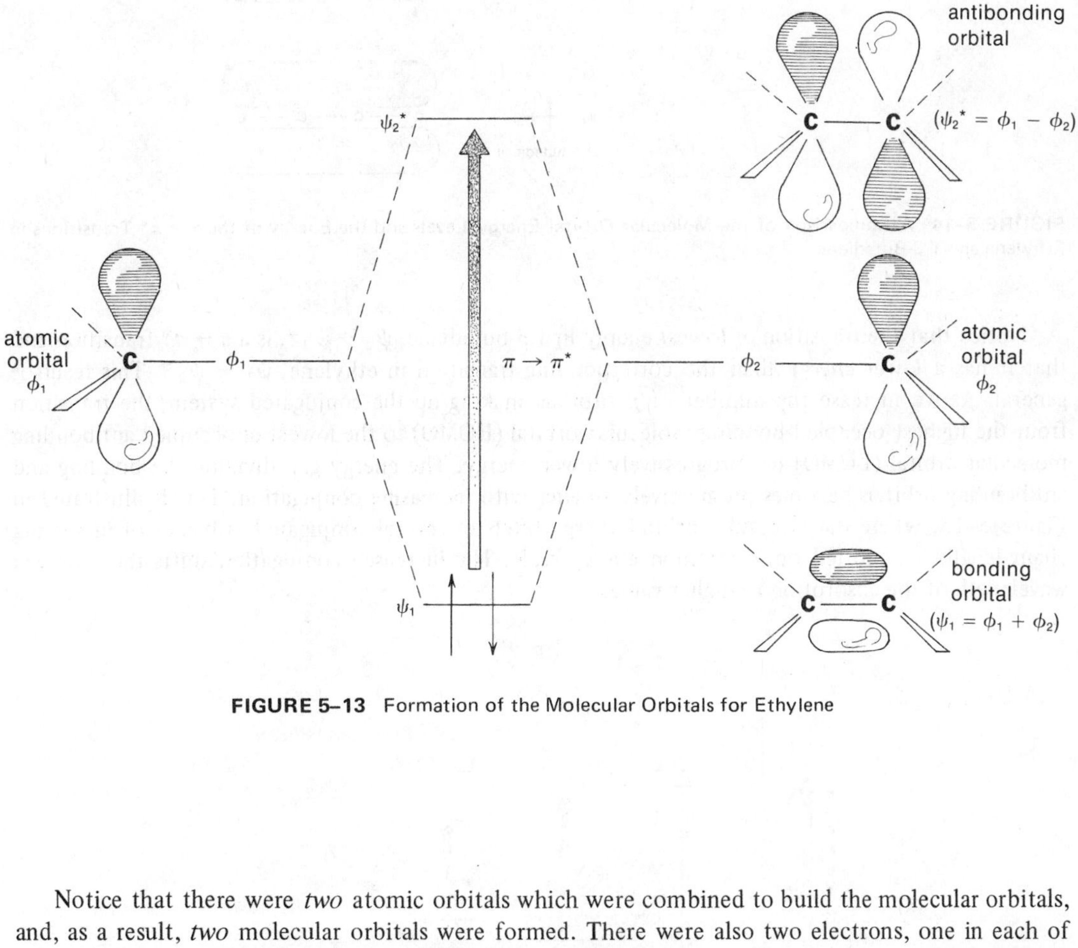

FIGURE 5–13 Formation of the Molecular Orbitals for Ethylene

Notice that there were *two* atomic orbitals which were combined to build the molecular orbitals, and, as a result, *two* molecular orbitals were formed. There were also two electrons, one in each of the atomic *p* orbitals. As a result of combination, the new π system will contain *two* electrons. Since we fill the lower energy orbitals first, these electrons will end up in ψ_1, the bonding orbital, and they constitute a new π bond. Electronic transition in this system is a $\pi \to \pi^*$ transition from ψ_1 to $\psi_2{}^*$.

Now, moving from this simple two-orbital case, consider 1,3-butadiene, which has *four* atomic *p* orbitals which form its π system of two conjugated double bonds. Since we had four atomic orbitals with which to build, *four* molecular orbitals will result. These are illustrated in Figure 5–14, where the orbitals of ethylene are represented on the same energy scale as the new orbitals for the sake of comparison.

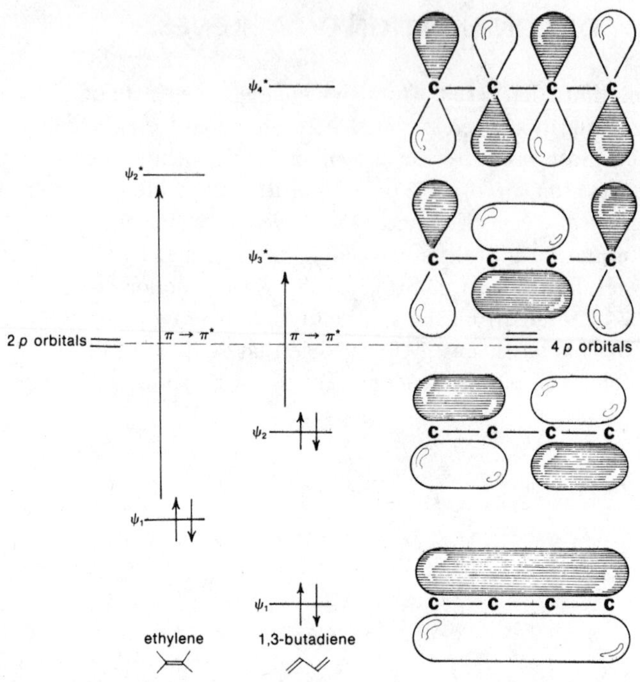

FIGURE 5–14 A Comparison of the Molecular Orbital Energy Levels and the Energy of the π → π* Transitions in Ethylene and 1,3-Butadiene

Notice that the transition of lowest energy in 1,3-butadiene, $\psi_2 \rightarrow \psi_3^*$, is a $\pi \rightarrow \pi^*$ transition and that it has a *lower energy* than the corresponding transition in ethylene, $\psi_1 \rightarrow \psi_2^*$. This result is general. As we increase the number of *p* orbitals making up the conjugated system, the transition from the highest occupied bonding molecular orbital (HOMO) to the lowest unoccupied antibonding molecular orbital (LUMO) has progressively lower energy. The energy gap dividing the bonding and antibonding orbitals becomes progressively smaller with increasing conjugation. This is illustrated in Figure 5–15, where the molecular orbital energy levels of several conjugated polyenes of increasing chain length are plotted on a common energy scale. The increased conjugation shifts the observed wavelength of the absorption to higher values.

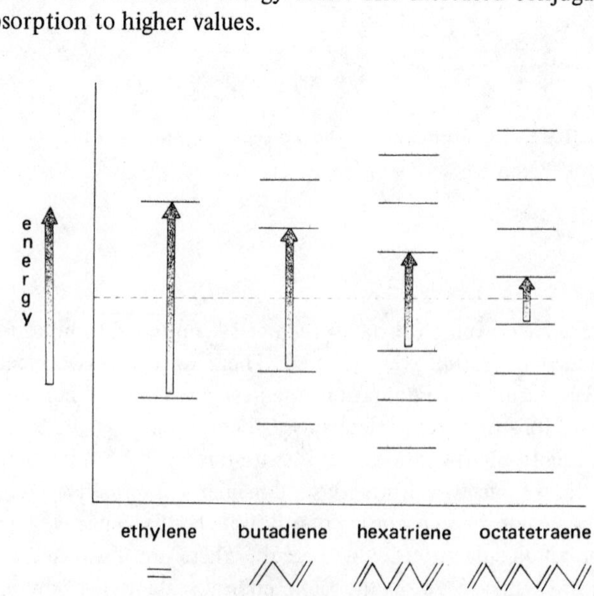

FIGURE 5–15 A Comparison of the π → π* Energy Gap in a Series of Polyenes of Increasing Chain Length

In a qualitatively similar fashion, many auxochromes exert their bathochromic shifts by means of an extension of the length of the conjugated system. Such bathochromes invariably possess a pair of unshared electrons on the atom attached to the double bond system. Resonance interaction of this lone pair with the double bond(s) increases the length of the conjugated system.

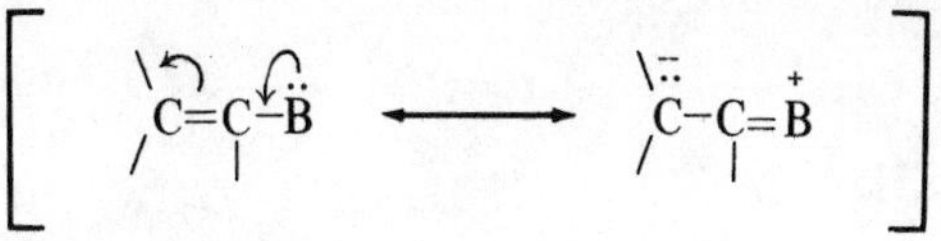

As a result of this interaction, as shown above, the non-bonded electrons become part of the π system of molecular orbitals, increasing its length by one extra orbital. This interaction is depicted in Figure 5–16 for ethylene and an unspecified atom B with an unshared electron pair. However, any of the typical auxochromic groups, –OH, –OR, –X, or $-NH_2$, could have been illustrated specifically.

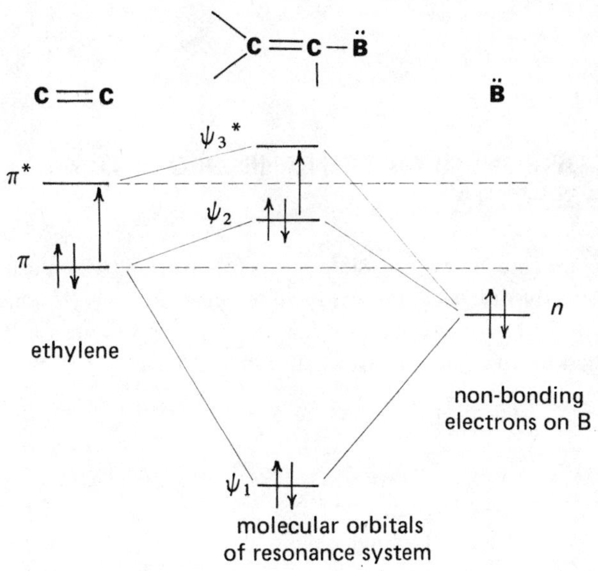

FIGURE 5–16 Energy Relationships of the New Molecular Orbitals and the Interacting π System and its Auxochrome

In the new system, the transition from the highest occupied orbital ψ_2 to the antibonding orbital $\psi_3{}^*$ always has lower energy than the transition $\pi \rightarrow \pi^*$ would have in the system without the interaction. This general result can be explained by MO theory, but it is beyond the scope of this text, and we will not pursue it further here.

In similar fashion, methyl groups also produce a bathochromic shift. However, as methyl groups do not have unshared electrons, the interaction is thought to result from overlap of the electrons in a C–H bond with the π system, as illustrated below.

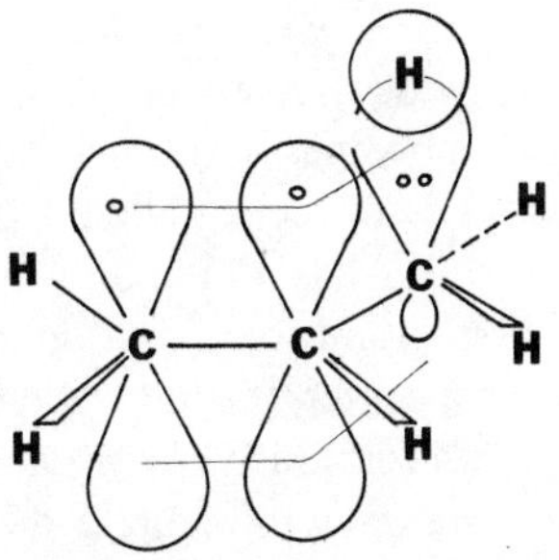

This type of interaction is often called "hyperconjugation" and may be represented by quasi-resonance structures as follows.

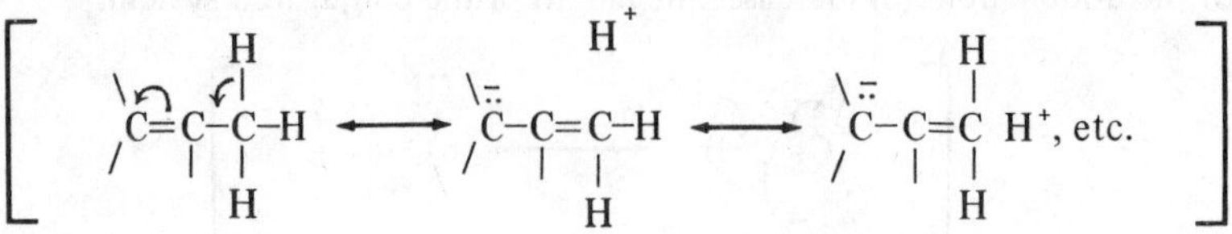

The net effect is an extension of the π system.

5.9 THE EFFECT OF CONFORMATION AND GEOMETRY ON POLYENE SPECTRA

In butadiene, there are actually two possible $\pi \rightarrow \pi^*$ transitions which can occur: $\psi_2 \rightarrow \psi_3{}^*$ and $\psi_2 \rightarrow \psi_4{}^*$. We have already discussed the easily observable $\psi_2 \rightarrow \psi_3{}^*$ transition. The $\psi_2 \rightarrow \psi_4{}^*$ transition is not often observed for two reasons. First, it lies near 175 nm for butadiene; and, second, it is a forbidden transition for the s-*trans* conformation of double bonds in butadiene.

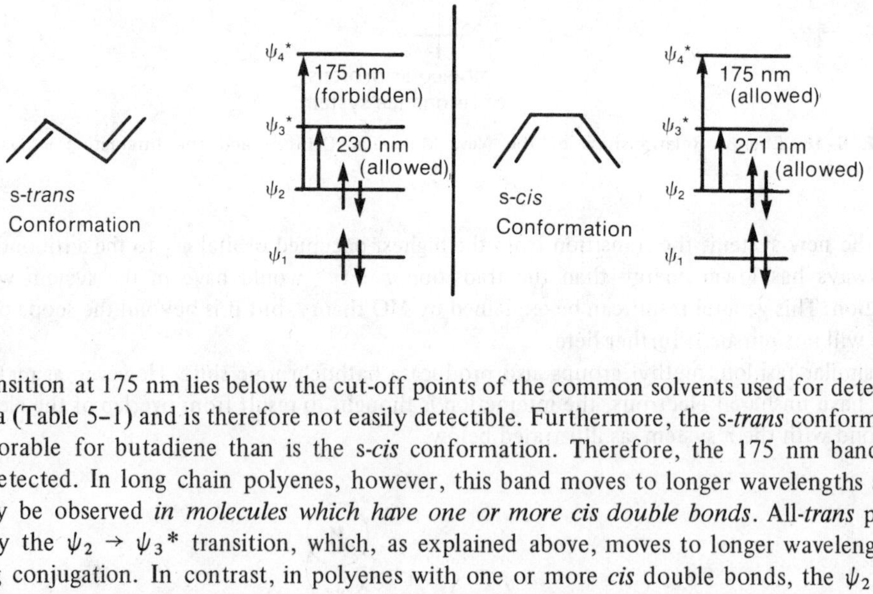

A transition at 175 nm lies below the cut-off points of the common solvents used for determining uv spectra (Table 5–1) and is therefore not easily detectible. Furthermore, the s-*trans* conformation is more favorable for butadiene than is the s-*cis* conformation. Therefore, the 175 nm band is not usually detected. In long chain polyenes, however, this band moves to longer wavelengths and can frequently be observed *in molecules which have one or more cis double bonds*. All-*trans* polyenes show only the $\psi_2 \rightarrow \psi_3{}^*$ transition, which, as explained above, moves to longer wavelength with increasing conjugation. In contrast, in polyenes with one or more *cis* double bonds, the $\psi_2 \rightarrow \psi_4{}^*$ transition *may* become allowed, giving rise to the so-called "*cis* band." This *cis* band occurs at shorter wavelength than the main transition, and it has a much lower intensity. This band does not occur in all molecules which have *cis* double bonds, and some caution must be used in structure assignment on this basis. Any molecule which retains a center of symmetry, even when possessing several *cis* double bonds, will not show the *cis* band. It is a forbidden transition in a system with a center of symmetry. Two examples illustrating the *cis* band are given in Figure 5–17 and 5–18.

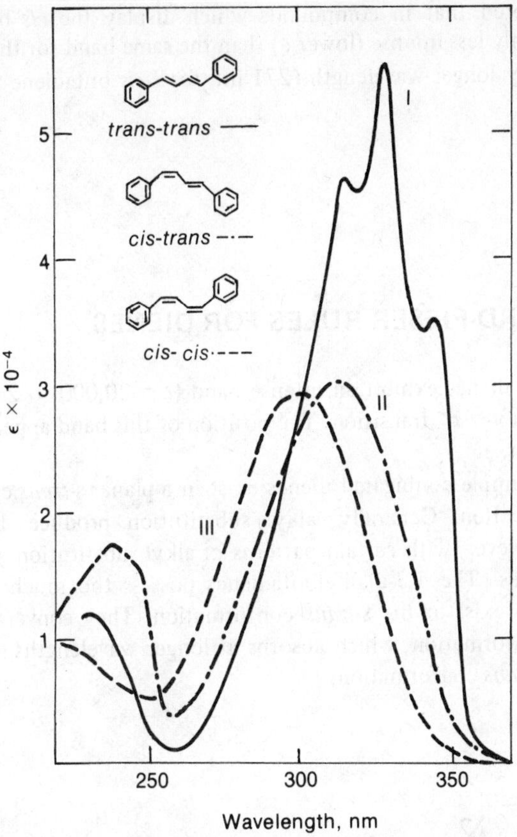

FIGURE 5–17 The spectra of (*I*) *trans-trans*, (*II*) *cis-trans*, and (*III*) *cis-cis*-1,4-diphenylbutadiene. [Reprinted by permission from J. H. Pinckard, B. Willie, and L. Zechmeister, J. Am. Chem. Soc., *70*, 1938 (1948).]

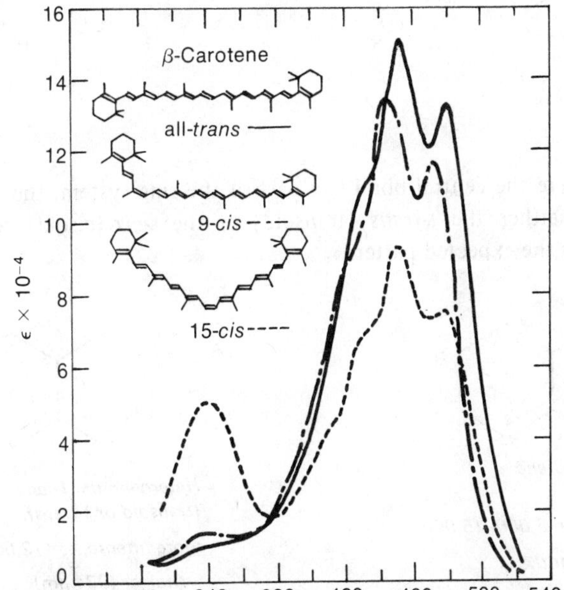

FIGURE 5–18 Spectra of isomeric β-carotenes, showing the development of the *cis* band. [Reprinted by permission from L. Zechmeister and A. Polgar, J. Am. Chem. Soc., *65*, 1523 (1943).]

It should also be noted that in compounds which display the *cis* band in their spectra, the $\psi_2 \rightarrow \psi_3{}^*$ band is not only less intense (lower ϵ) than the same band for the all-*trans* compound, but is also shifted to a slightly longer wavelength (271 nm for s-*cis* butadiene *versus* 230 nm for s-*trans* butadiene).

5.10 THE WOODWARD-FIESER RULES FOR DIENES

In general, conjugated dienes exhibit an intense band (ϵ = 20,000 to 26,000) in the region from 217 to 245 nm, owing to a $\pi \rightarrow \pi^*$ transition. The position of this band appears to be quite insensitive to the nature of the solvent.

Butadiene and many simple conjugated dienes exist in a planar s-*trans* conformation, as has been noted in the previous section. Generally, alkyl substitution produces bathochromic shifts and hyperchromic effects. However, with certain patterns of alkyl substitution, the wavelength increases, but the intensity decreases. The 1,3-dialkylbutadienes possess too much crowding between alkyl groups to permit them to exist in the s-*trans* conformation. They convert by rotation around the single bond to an s-*cis* conformation, which absorbs at longer wavelengths, but with lower intensity than the corresponding s-*trans* conformation.

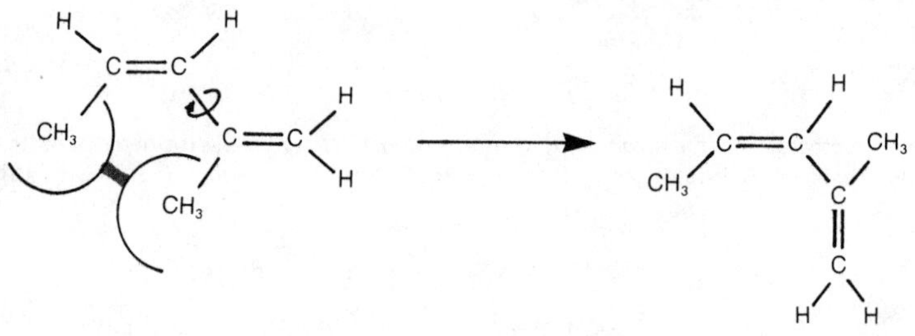

s-*trans*　　　s-*cis*

In cyclic dienes, where the central bond is a part of the ring system, the diene chromophore is usually held rigidly in either the s-*trans* (transoid) or the s-*cis* (cisoid) conformation. Typical absorption spectra follow the expected pattern:

Homoannular Diene (cisoid or s-cis)

less intense, ϵ = 5,000–15,000

λ longer (273 nm)

Heteroannular Diene (transoid or s-trans)

more intense, ϵ = 12,000–28,000

λ shorter (234 nm)

By studying a vast number of dienes of each type, Woodward and Fieser were able to devise an empirical correlation of structural variations which allow one to predict the wavelength at which an unknown conjugated diene will absorb. The rules are summarized in Table 5–5.

TABLE 5-5 Empirical Rules for Dienes

	Homoannular (cisoid)	Heteroannular (transoid)
Parent	λ = 253 nm	λ = 214 nm
Increments for:		
Double bond extending conjugation	30	30
Alkyl substituent or ring residue	5	5
Exocyclic double bond	5	5
Polar groupings:		
$-OCOCH_3$	0	0
$-OR$	6	6
$-Cl$, $-Br$	5	5
$-NR_2$	60	60

A few examples of how these rules are applied follow:

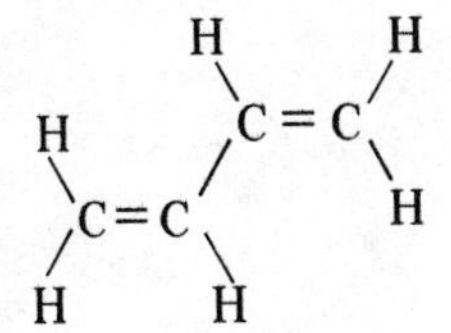

transoid: 214 nm
observed: 217 nm

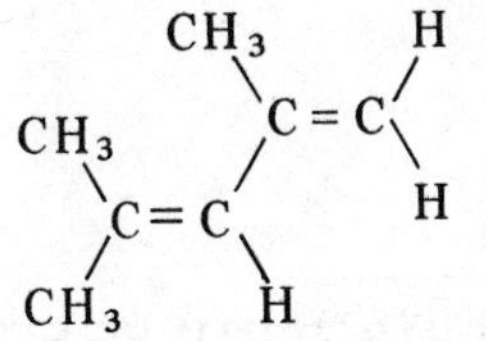

transoid: 214 nm
alkyl groups: 3 X 5 = 15
229 nm
observed: 228 nm

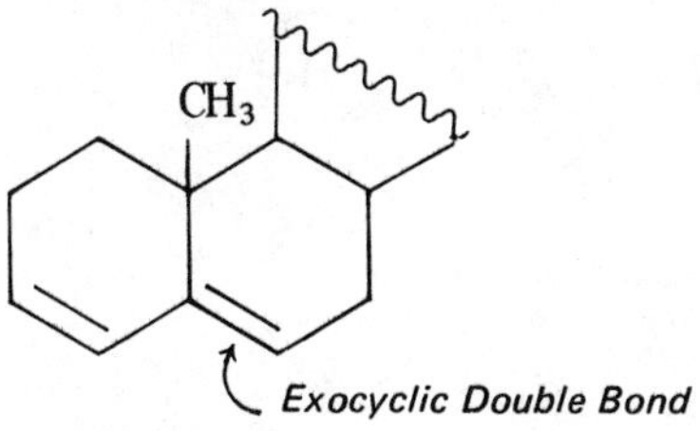

transoid: 214 nm
ring residues: 3 X 5 = 15
exocyclic double bond: 5
234 nm
observed: 235 nm

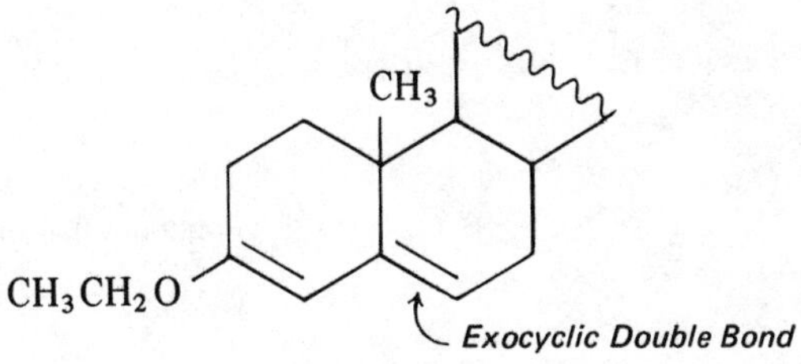

transoid: 214 nm
ring residues: 3 X 5 = 15
exocyclic double bond: 5
−OR: 6
240 nm
observed: 241 nm

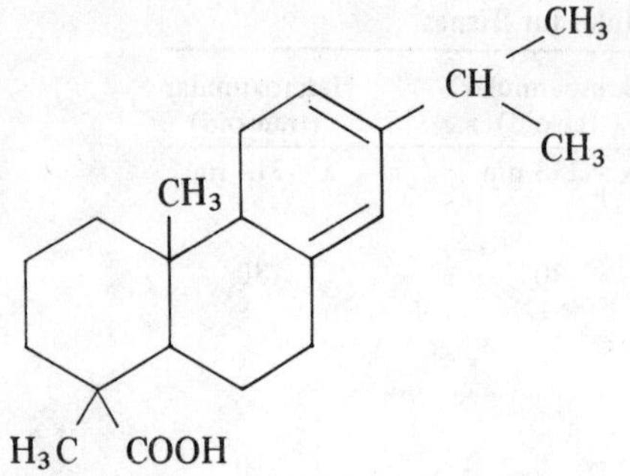

cisoid:	253 nm
alkyl substituent:	5
ring residues: 3 X 5 =	15
exocyclic double bond:	5
	278 nm
observed:	275 nm

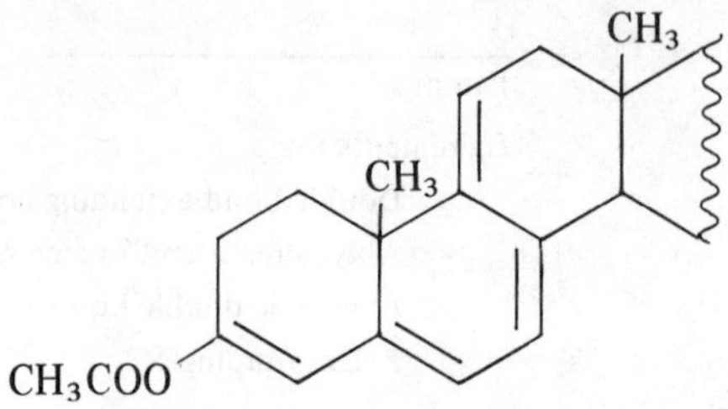

cisoid:	253 nm
ring residues: 5 X 5 =	25
double bond extending conjugation: 2 X 30 =	60
exocyclic double bonds: 3 X 5 =	15
CH_3COO–:	0
	353 nm
observed:	355 nm

5.11 THE FIESER-KUHN RULES FOR POLYENES

The Woodward-Fieser rules work well for polyenes with from one to four conjugated double bonds. However, for systems with a higher degree of conjugation they are less effective. Fieser and Kuhn have developed a simple set of empirical rules that work well for polyene systems such as those found in carotenoid pigments like β-carotene (found in carrots) and lycopene (found in tomatoes).

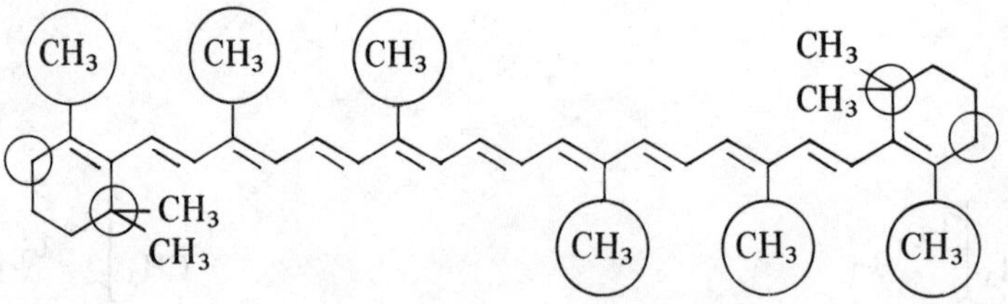

β-Carotene, 11 Double Bonds

λ_{max} = 452 nm (hexane), ε = 15.2 X 10⁴

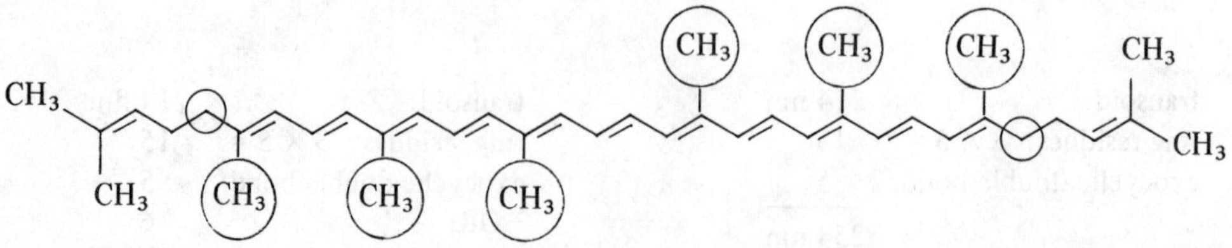

Lycopene, 13 Double Bonds (11 Conjugated)

λ_{max} = 474 nm (hexane), ε = 18.6 X 10⁴

These rules, given in Table 5–6, allow a calculation of both λ_{max} and ϵ_{max} for these systems.

TABLE 5–6 Empirical Rules for Polyenes

$$\lambda_{max}\ (\text{hexane}) = 114 + 5M + n(48.0 - 1.7n) - 16.5R_{endo} - 10R_{exo}$$

$$\epsilon_{max}\ (\text{hexane}) = 1.74 \times 10^4 n$$

where:

M	=	the number of alkyl substituents
n	=	the number of conjugated double bonds
R_{endo}	=	the number of rings with endocyclic double bonds
R_{exo}	=	the number of rings with exocyclic double bonds

As an example, consider *β*-carotene:

M = 10 (see circled positions in the structure shown above)
n = 11 conjugated double bonds
R_{endo} = 2
R_{exo} = 0

Therefore:

$$\begin{aligned}\lambda_{max} &= 114 + (5 \times 10) + 11(48.0 - 1.7 \times 11) - (16.5 \times 2) - (10 \times 0)\\ &= 114 + 50 + 11(48.0 - 18.7) - 33 - 0\\ &= 114 + 50 + 322.3 - 33\\ &= \boxed{453.3 \text{ CALCULATED}}\\ \epsilon_{max} &= 1.74 \times 10^4 \times 11\\ &= \boxed{19.1 \times 10^4 \text{ CALCULATED}}\end{aligned}$$

These figures compare well with the experimental data given above. The reader is encouraged to perform a similar calculation for lycopene and to check the calculated values against those given above. (Note that the double bonds at each end of the chain *are not* in conjugation with the others.)

5.12 CARBONYL COMPOUNDS; ENONES

As discussed earlier in Section 5.6, carbonyl compounds have two principal uv transitions, the allowed $\pi \rightarrow \pi^*$ transition and the forbidden $n \rightarrow \pi^*$ transition.

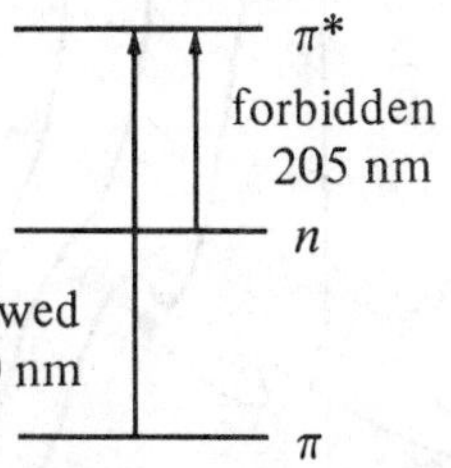

Of these two transitions, only the $n \rightarrow \pi^*$ transition, although it is weak (forbidden), is commonly observed above the usual cut-off point of solvents. Substitution of the carbonyl group by an auxochrome with a lone pair of electrons, such as $-NR_2$, $-OH$, $-OR$, $-NH_2$, or $-X$, as in amides, acids, esters, or acid chlorides, gives a pronounced hypsochromic effect on the $n \rightarrow \pi^*$ transition and a lesser bathochromic effect on the $\pi \rightarrow \pi^*$ transition. These latter bathochromic shifts are caused by resonance interaction similar to that discussed in Section 5.8. Seldom, however, are these effects large enough to bring the $\pi \rightarrow \pi^*$ band into the region above the solvent cut-off point. The hypsochromic effect on the $n \rightarrow \pi^*$ transition of an acetyl group is illustrated in Table 5-7.

TABLE 5-7 The Hypsocromic Effect of Lone Pair Auxochromes on the $n \rightarrow \pi^*$ Transition of a Carbonyl Group

	λ_{max}	ϵ_{max}	Solvent
$CH_3-C(=O)-H$	293 nm	12	Hexane
$CH_3-C(=O)-CH_3$	279	15	Hexane
$CH_3-C(=O)-Cl$	235	53	Hexane
$CH_3-C(=O)-NH_2$	214	–	Water
$CH_3-C(=O)-OCH_2CH_3$	204	60	Water
$CH_3-C(=O)-OH$	204	41	Ethanol

This shift is due primarily to the inductive effect of the oxygen, nitrogen, or halogen atoms. They withdraw electrons from the carbonyl carbon, causing the lone pair of electrons on oxygen to be held more firmly than would be the case in the absence of an inductive effect.

If the carbonyl group is a part of a conjugated system of double bonds, both the $n \rightarrow \pi^*$ and the $\pi \rightarrow \pi^*$ bands are shifted to longer wavelengths. However, the energy of the $n \rightarrow \pi^*$ transition does not decrease as rapidly as that of the $\pi \rightarrow \pi^*$ band, which is more intense. If the conjugated chain becomes long enough, the $n \rightarrow \pi^*$ band becomes "buried" under the more intense $\pi \rightarrow \pi^*$ band. This is illustrated in Figure 5-19 for a series of polyene aldehydes.

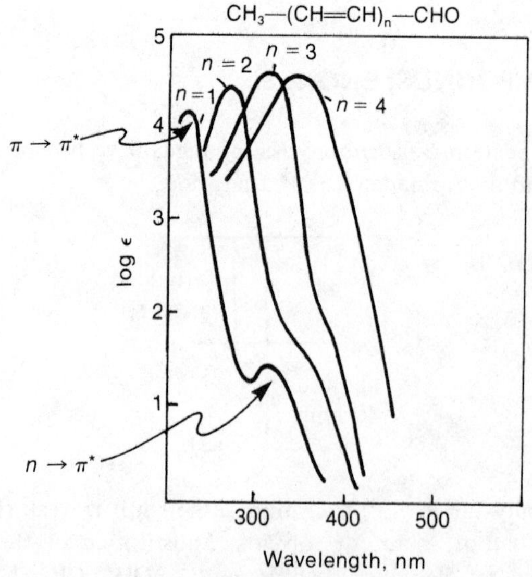

FIGURE 5-19 The Spectra of a Series of Polyene Aldehydes (From J. N. Murrell, "The Theory of the Electronic Spectra of Organic Molecules," © 1963. Reprinted by permission of Methuen and Co., Ltd., London.)

The molecular orbitals of a simple enone system, along with those of the non-interacting double bond and the carbonyl group, are shown in Figure 5–20.

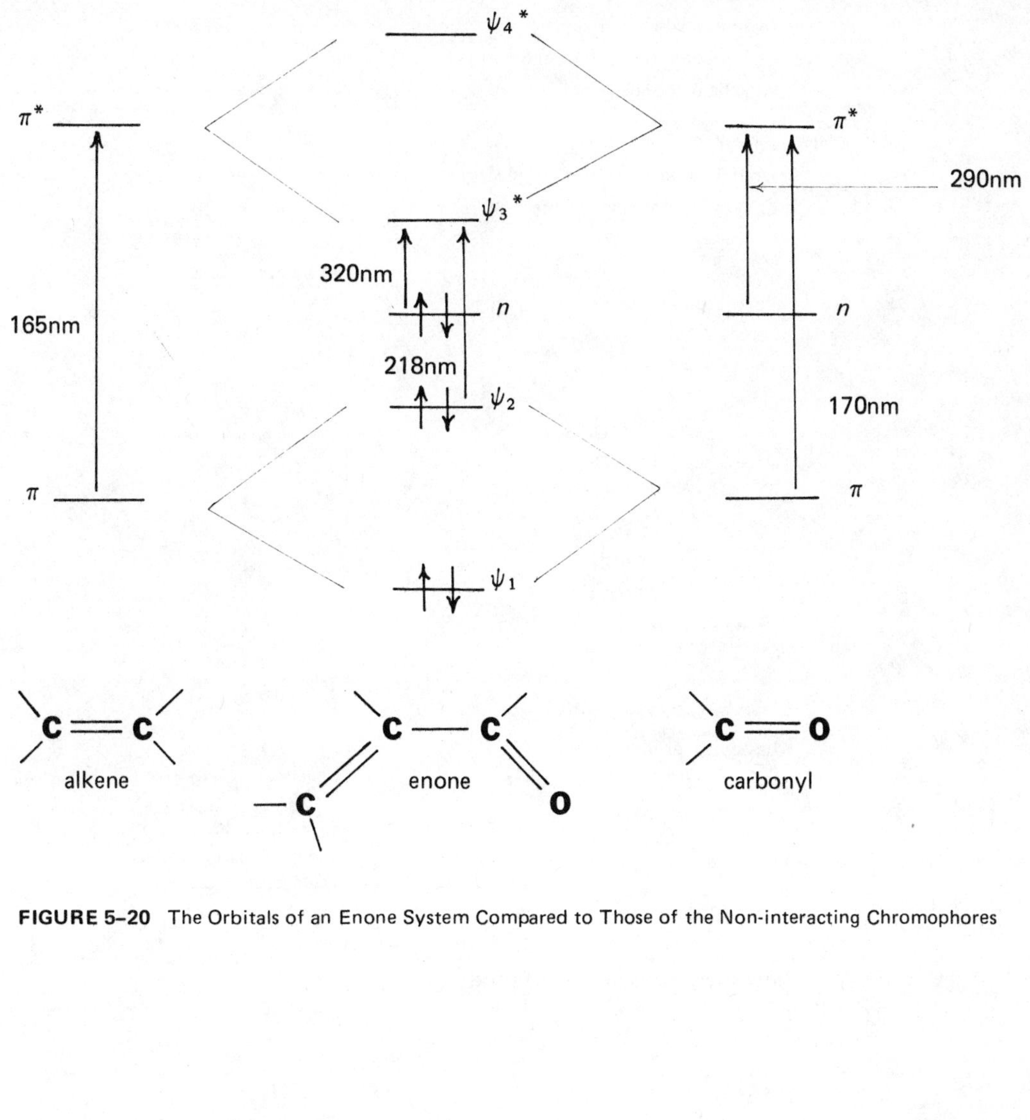

FIGURE 5–20 The Orbitals of an Enone System Compared to Those of the Non-interacting Chromophores

5.13 WOODWARD'S RULES FOR ENONES

The conjugation of a double bond with a carbonyl group leads to intense absorption (ϵ = 8,000 to 20,000) corresponding to a $\pi \rightarrow \pi^*$ transition of the carbonyl group. The absorption is found between 220 and 250 nm in simple enones. The $n \rightarrow \pi^*$ transition is much less intense (ϵ = 50 to 100), and it appears at 310–330 nm. While the $\pi \rightarrow \pi^*$ transition is affected in a predictable fashion by structural modifications of the chromophore, the $n \rightarrow \pi^*$ transition does not exhibit such predictable behavior. Woodward examined the ultraviolet spectra of a large number of enones, and he was able to devise a set of empirical rules which allow one to predict the wavelength at which the $\pi \rightarrow \pi^*$ transition will occur in an unknown enone. The rules are summarized in Table 5–8.

TABLE 5-8 Empirical Rules for Enones

β α β–C=C–C=O	δ γ β α δ–C=C–C=C–C=O		
Base Values:			
6-Membered ring or acyclic parent enone			= 215 nm
5-Membered ring parent enone			= 202 nm
Acyclic dienone			= 245 nm
Increments for:			
Double bond extending conjugation			30
Alkyl group or ring residue		α	10
		β	12
		γ and higher	18
Polar groupings:			
	–OH	α	35
		β	30
		δ	50
	$-OCOCH_3$	α, β, δ	6
	$-OCH_3$	α	35
		β	30
		γ	17
		δ	31
	–Cl	α	15
		β	12
	–Br	α	25
		β	30
	$-NR_2$	β	95
Exocyclic double bond			5
Homocyclic diene component			39
Solvent correction			variable
		λ_{max}^{EtOH} (calc)	= Total

A few examples of how these rules are applied follow:

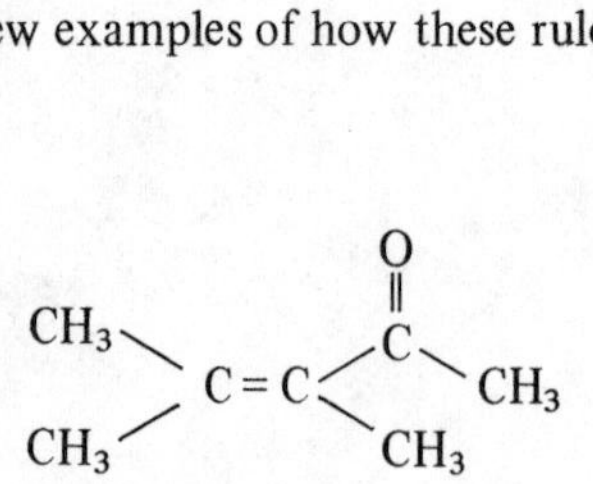

acyclic enone:	215 nm
$\alpha-CH_3$:	10
$\beta-CH_3$: 2 × 12 =	24
	249 nm
observed:	249 nm

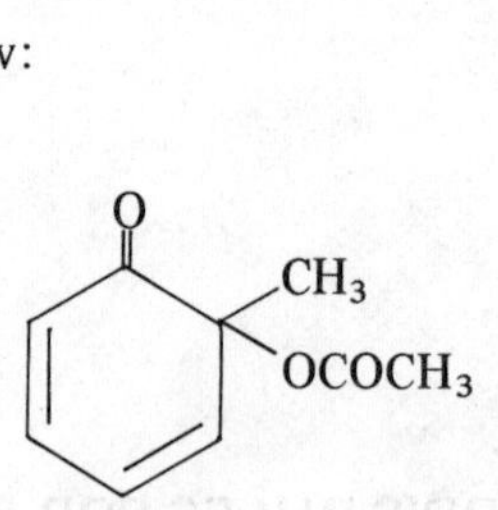

6-membered enone:	215 nm
double bond extending conjugation:	30
homocyclic diene:	39
δ-ring residue:	18
	302 nm
observed:	300 nm

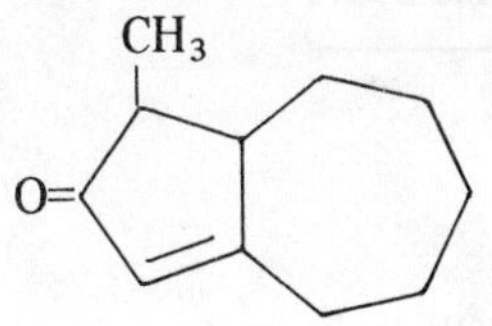

5-membered enone:	202 nm
β-ring residue: 2 × 12 =	24
exocyclic double bond:	5
	231 nm
observed:	226 nm

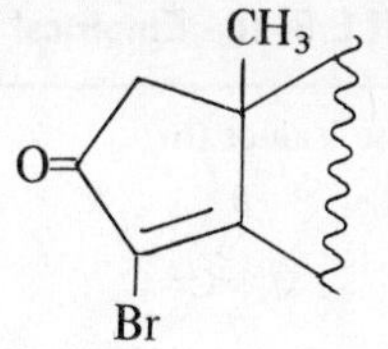

5-membered enone:	202 nm
α-Br:	25
β-ring residue: 2 × 12 =	24
exocyclic double bond:	5
	256 nm
observed:	251 nm

5.14 UNSATURATED ALDEHYDES

α,β-Unsaturated aldehydes generally follow the same rules as those given for enones (see previous section) except that their absorptions are displaced by about 5 to 8 nm toward shorter wavelength than those of the corresponding ketones. A table of empirical rules for unsaturated aldehydes is provided as Table 5–9.

TABLE 5–9 Empirical Rules for Unsaturated Aldehydes

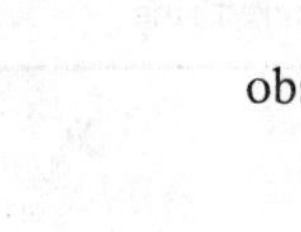

PARENT	208 nm
With α or β alkyl groups	220
With α, β or β,β alkyl groups	230
With α, β,β alkyl groups	242

5.15 NIELSEN'S RULES FOR α, β-UNSATURATED ACIDS AND ESTERS

A set of rules similar to those for enones has been developed for α,β-unsaturated acids and esters by Nielsen (Table 5–10).

TABLE 5–10 Empirical Rules for Unsaturated Acids and Esters

Base Values for:

with α or β alkyl group	208 nm
with α,β or β,β alkyl groups	217
with α,β,β alkyl groups	225
for an exocyclic α,β double bond	add 5 nm
for an endocyclic α,β double bond in a 5- or 7-membered ring	add 5 nm

Consider 2-cyclohexenoic and 2-cycloheptenoic acids as examples.

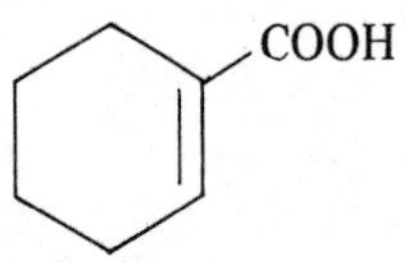

α,β-dialkyl	217 nm calc.
double bond is in a 6-membered ring, adds nothing	217 nm obs.

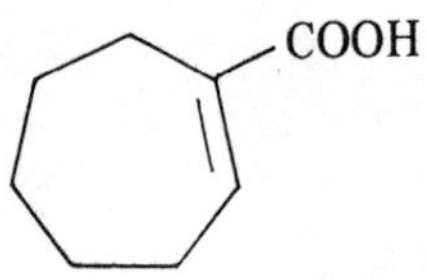

α,β-dialkyl	217 nm
double bond in a 7-membered ring	+ 5
	222 nm calc.
	222 nm obs.

5.16 SOLVENT SHIFTS – A MORE DETAILED EXAMINATION

As mentioned in Section 5.5, the position of an absorption band is often greatly affected by the nature of the solvent in which the spectrum is determined. In Table 5–2, the effect of solvent changes on the $n \rightarrow \pi^*$ transition of acetone was illustrated. In the current section, the origins of such solvent shifts in polyenes (alkenes) and in enones (ketones) will be examined in more detail.

Such solvent shifts are due to differences in the relative capabilities of solvents to solvate the ground state of a molecule as compared with their ability to solvate the excited state of the same molecule. Two general cases may be delineated.

1. In most $\pi \rightarrow \pi^*$ transitions, the ground state of the molecule is relatively non-polar, and the excited state is often more polar than the ground state. As a result, when a polar solvent is used, it interacts (stabilizes) more strongly with the excited state than with the ground state, and the transition is shifted to longer wavelength (bathochromic shift, lower energy). This is illustrated for a simple system (alkene) in Figure 5–21.

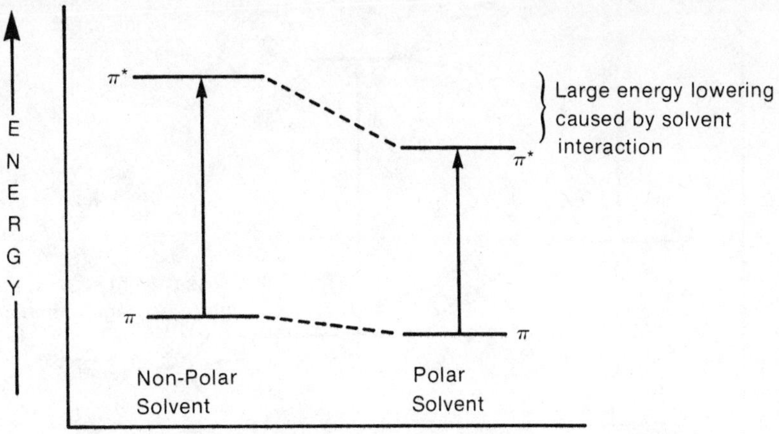

FIGURE 5–21 The Effect of a Polar Solvent on a $\pi \rightarrow \pi^*$ Transition

Ethylene affords an example of this effect. In the ground state, the π electron density in the double bond is more or less equally distributed between the two carbon atoms, and the nuclei of the carbon atoms are well shielded by their valence electrons. However, in the π^* excited state, one of these carbons (or both) becomes electron deficient after the electron has been promoted. This is shown below, where an asterisk is used to indicate the electron in the π^* orbital.

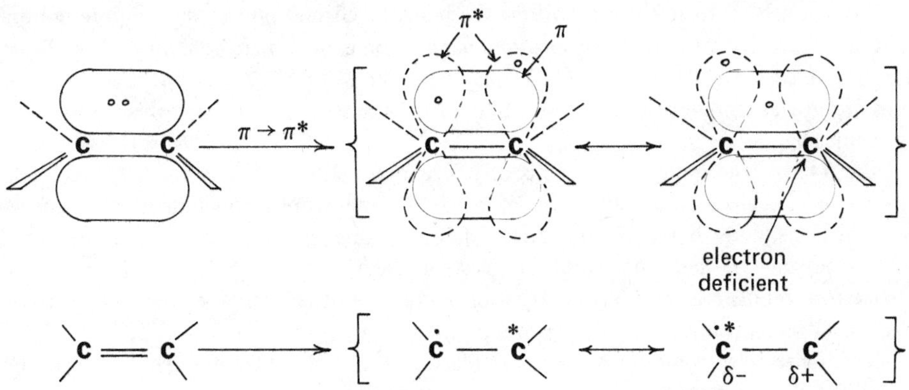

As a result, the excited state interacts more strongly with polar or hydrogen bonding solvents than does the ground state.

2. In most molecules exhibiting $n \rightarrow \pi^*$ transitions, the ground state is more polar than the excited state. In particular, hydrogen bonding solvents interact more strongly with unshared electron pairs in the ground state molecule than they do in the excited state molecule. As a result, an $n \rightarrow \pi^*$ transition will have its absorption maximum shifted to shorter wavelength (hypsochromic shift, higher energy) as the hydrogen bonding ability (polarity) of the solvent increases. This is illustrated in Figure 5–22 for a simple ketone. One should note than an $n \rightarrow \sigma^*$ transition would be affected in the same way as an $n \rightarrow \pi^*$ transition.

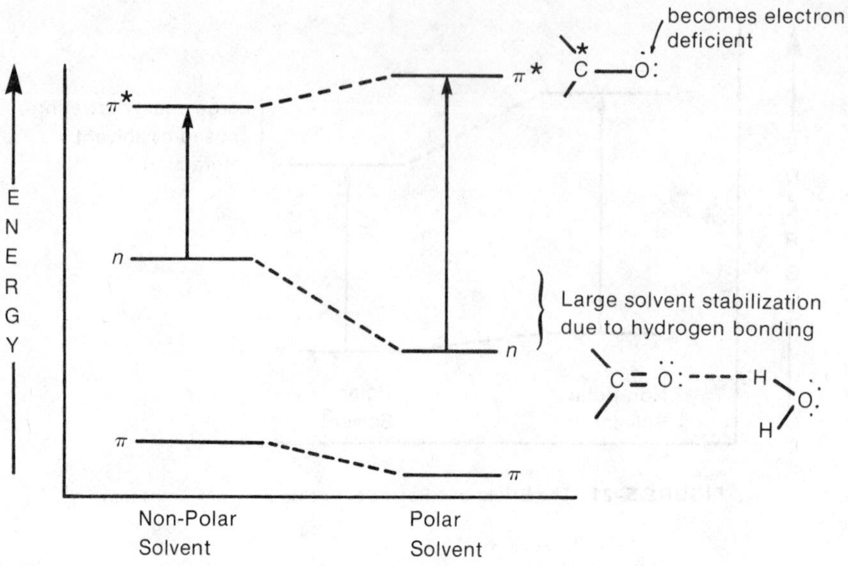

FIGURE 5–22 The Effect of a Polar Solvent on an $n \rightarrow \pi^*$ Transition

The utility of the observations above is that one may often identify a band as an $n \rightarrow \pi^*$ or a $\pi \rightarrow \pi^*$ band by observing the type of solvent shift which it exhibits.

5.17 AROMATIC COMPOUNDS

The absorptions due to transitions within the benzene chromophore can be quite complex. The ultraviolet spectrum contains three absorption bands, some of which contain a great deal of fine structure. The electronic transitions are basically of the $\pi \rightarrow \pi^*$ type, but the details of these transitions are not as simple as in the cases of the other classes of chromophores described in earlier sections of this chapter.

In Figure 5–23A, the molecular orbitals of benzene are shown. If one were to consider a simple explanation for the electronic transitions in benzene, one would conclude that there were four possible transitions, but that each transition had the same energy. One would predict that the ultraviolet spectrum of benzene would consist of one absorption peak. However, owing to electron-electron repulsions and symmetry considerations, the actual energy states from which electronic transitions occur are somewhat modified. The energy state levels of benzene are shown in Figure 5–23B. Three electronic transitions take place to these excited states. These transitions, which are also indicated in Figure 5–23B, are the so-called *primary* bands at 184 and 202 nm and the *secondary* (or fine-structure) band at 255 nm. The spectrum of benzene is shown in Figure 5–24. Of the primary bands, the 184 nm band (the *second primary band*) has a molar absorptivity of 47,000. It is an allowed transition. Nevertheless, this transition is not observed under usual experimental conditions, because absorptions at this wavelength are in the vacuum ultraviolet region of the spectrum, beyond the range of most commercial instruments. In polycyclic aromatic compounds, the second primary band is often shifted to longer wavelengths, in which case it can be observed under ordinary conditions. The 202 nm band is much less intense (ϵ = 7,400), and it corresponds to a forbidden transition. The secondary band is the least intense of the benzene bands (ϵ = 230). It also corresponds to a symmetry forbidden electronic transition. The secondary band, caused by interaction of the electronic energy levels with vibrational modes, appears with a great deal of fine structure. This fine structure is lost if the spectrum of benzene is determined in a polar solvent or if a single functional group is substituted onto the benzene ring. In such cases, the secondary band appears as a broad peak, lacking in any interesting detail.

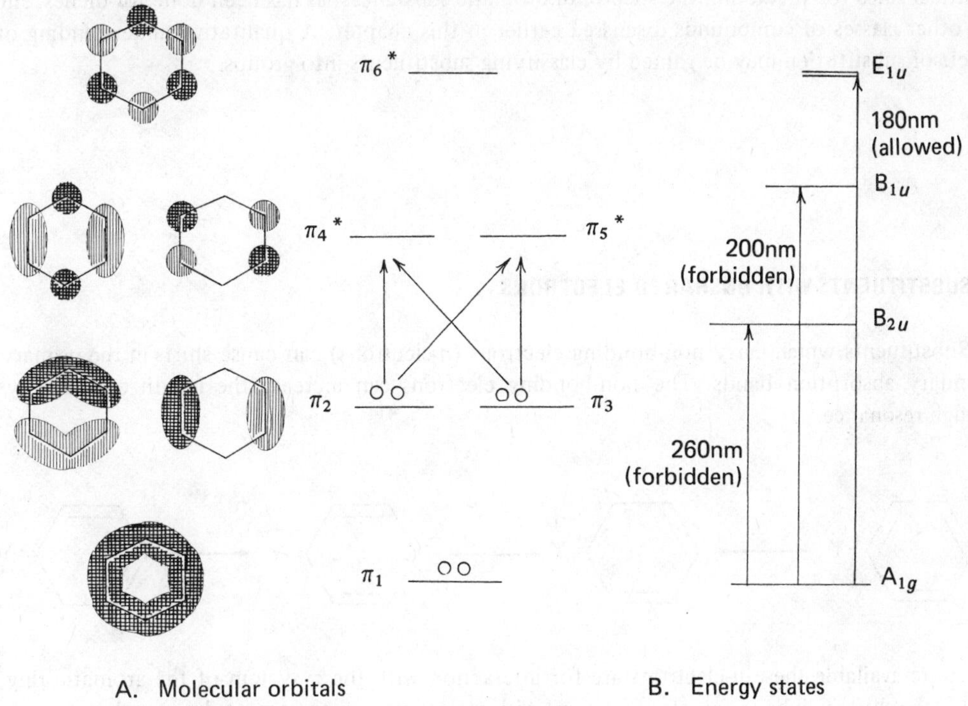

FIGURE 5–23 Molecular Orbitals and Energy States for Benzene

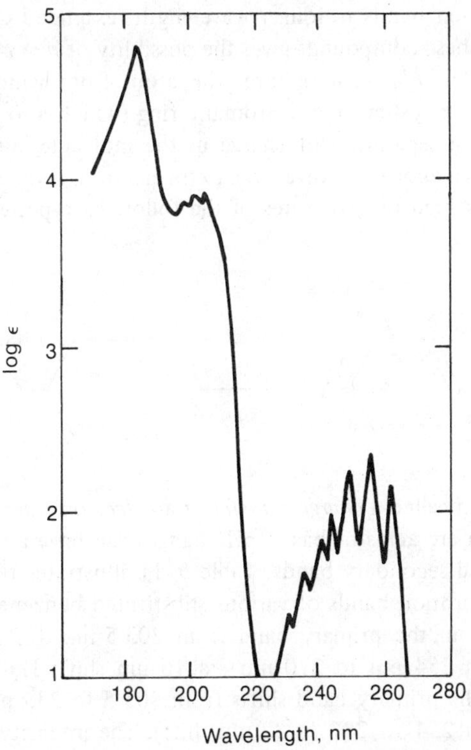

FIGURE 5–24 Ultraviolet Spectrum of Benzene [Reprinted by permission from J. Petruska, J. Chem. Phys., *34*, 1121 (1961)].

Substitution on the benzene ring is able to cause bathochromic and hyperchromic shifts. Unfortunately, these shifts are difficult to predict. As a consequence, it is not possible to formulate empirical rules for predicting the spectra of aromatic substances, as has been done for dienes, enones, and other classes of compounds discussed earlier in this chapter. A qualitative understanding of the effects of substitution may be gained by classifying substituents into groups.

A. SUBSTITUENTS WITH UNSHARED ELECTRONS

Substituents which carry non-bonding electrons (*n*-electrons) can cause shifts in the primary and secondary absorption bands. The non-bonding electrons can increase the length of the π system through resonance.

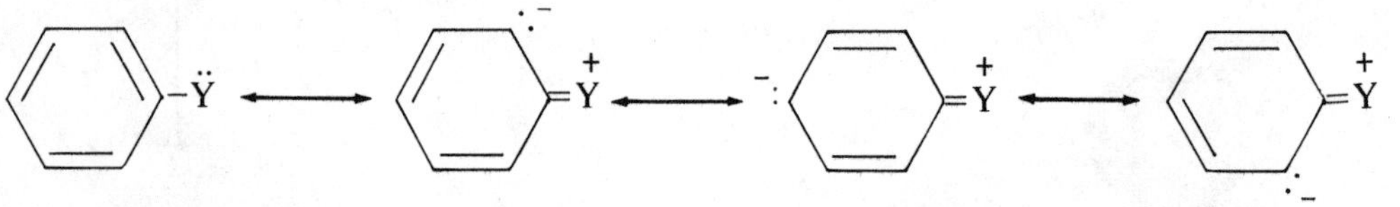

The more available these *n*-electrons are for interaction with the π system of the aromatic ring, the greater the shifts will be. Examples of groups with *n* electrons are the amino, hydroxyl, and methoxyl groups, as well as the halogens.

Interactions of this type between the *n* and π electrons usually cause shifts in the primary and secondary benzene absorption bands to longer wavelength (extended conjugation). In addition, the presence of *n*-electrons in these compounds gives the possiblity of $n \rightarrow \pi^*$ transitions. If an *n*-electron is excited into the extended π^* chromophore, the atom from which it was removed becomes electron-deficient, while the π system of the aromatic ring (which also includes atom Y) acquires an extra electron. This causes a separation of charge in the molecule and is generally represented as regular resonance as is shown above. However, the extra electron in the ring is actually in a π^* orbital and would be better represented by structures of the following type, with the asterisk representing the excited electron.

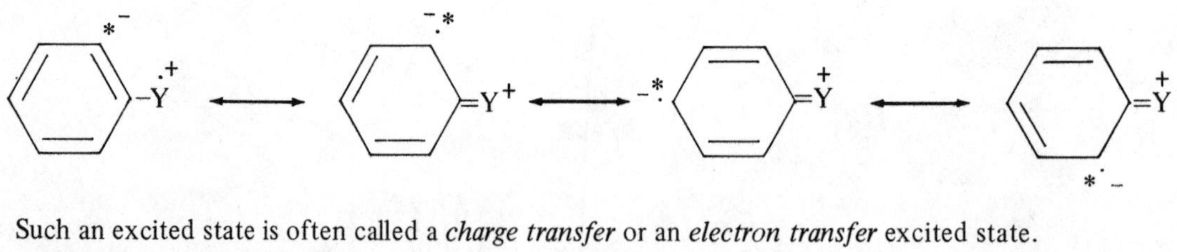

Such an excited state is often called a *charge transfer* or an *electron transfer* excited state.

With compounds which are acids or bases, pH changes can have a very significant effect on the positions of the primary and secondary bands. Table 5–11 illustrates the effects of changing the pH of the solution on the absorption bands of various substituted benzenes. In going from benzene to phenol, one notices a shift in the primary band from 203.5 nm to 210.5 nm – a 7 nm shift. The secondary band shifts from 254 nm to 270 nm – a 16 nm shift. However, in phenoxide ion, the conjugate base of phenol, the primary band shifts from 203.5 to 235 nm (a 31.5 nm shift), and the secondary band shifts from 254 to 287 (a 33 nm shift). The intensity of the secondary band also increases. In phenoxide ion, there are more *n*-electrons, and they are more available for interaction with the aromatic π system than in phenol.

TABLE 5–11 pH Effects on Absorption Bands

Substituent	PRIMARY λ (nm)	PRIMARY ε	SECONDARY λ (nm)	SECONDARY ε
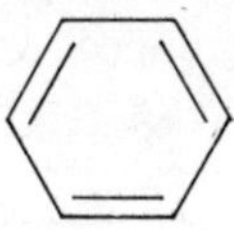 –H	203.5	7400	254	204
–OH	210.5	6200	270	1450
$-O^-$	235	9400	287	2600
$-NH_2$	230	8600	280	1430
$-NH_3^+$	203	7500	254	169
–COOH	230	11600	273	970
$-COO^-$	224	8700	268	560

A reverse case is illustrated in the comparison of aniline and anilinium ion. Aniline shows shifts similar to those of phenol. In going from benzene to aniline, the primary band shifts from 203.5 to 230 nm (a 26.5 nm shift), and the secondary band shifts from 254 to 280 nm (a 26 nm shift). However, in the case of anilinium ion, the conjugate acid of aniline, these large shifts are not observed. For anilinium ion, the primary band and the secondary band are not shifted at all. The quaternary nitrogen of anilinium ion has no unshared pairs of electrons to interact with the benzene π system. Consequently, the spectrum of anilinium ion is almost identical to that of benzene itself.

B. SUBSTITUENTS CAPABLE OF π-CONJUGATION

Substituents which are themselves chromophores usually contain π electrons. Interaction of the benzene ring electrons and the π electrons of the substituent, just as in the case of *n*-electrons, can also produce a new electron transfer band. At times, this new band may be so intense as to obscure the secondary band of the benzene system. Notice that the opposite polarity is induced; the ring becomes electron-deficient.

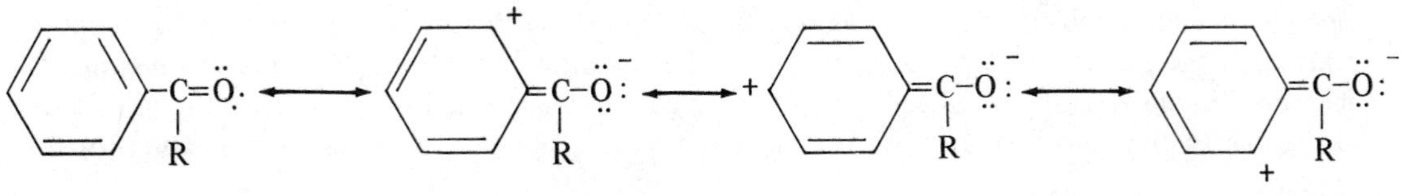

The effect of acidity or basicity of the solution on such a chromophoric substituent group can also be seen in Table 5–11. For benzoic acid, the primary and secondary bands are shifted substantially from those noted for benzene. However, the magnitudes of the shifts are somewhat smaller in the case of benzoate ion, the conjugate base of benzoic acid. The intensities of the peaks are lower than for benzoic acid, as well. Electron transfer of the sort shown above would be expected to be less likely when the functional group already bears a negative charge.

C. ELECTRON RELEASING AND ELECTRON WITHDRAWING EFFECTS

Finally substituents may have differing effects on the positions of absorption maxima, depending upon whether they are electron-releasing or electron-withdrawing. Any substituent, regardless of its influence on the electron distribution elsewhere in the aromatic molecule, will shift the primary absorption band to longer wavelength. Electron-withdrawing groups have essentially no effect on the position of the secondary absorption band, unless, of course, the electron-withdrawing group is also capable of acting as a chromophore. However, electron-releasing groups increase both the wavelength and the intensity of the secondary absorption band. These effects are summarized in Table 5–12, where electron-releasing and electron-withdrawing substituent groups are grouped together.

TABLE 5–12 Ultraviolet Maxima for Various Aromatic Compounds

Substituent		PRIMARY λ (nm)	PRIMARY ε	SECONDARY λ (nm)	SECONDARY ε
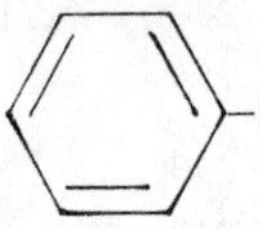	–H	203.5	7400	254	204
Electron-Releasing Substituents	$-CH_3$	206.5	7000	261	225
	–Cl	209.5	7400	263.5	190
	–Br	210	7900	261	192
	–OH	210.5	6200	270	1450
	$-OCH_3$	217	6400	269	1480
	$-NH_2$	230	8600	280	1430
Electron-Withdrawing Substituents	–CN	224	13000	271	1000
	–COOH	230	11600	273	970
	$-COCH_3$	245.5	9800		
	–CHO	249.5	11400		
	$-NO_2$	268.5	7800		

D. DISUBSTITUTED BENZENE DERIVATIVES

With disubstituted benzene derivatives, it is necessary to consider the effects of each of the two substituents. For *para*-disubstituted benzenes, two possiblities exist. If both groups are electron-releasing or if they are both electron-withdrawing, they exert effects which are similar to those observed for monosubstituted benzenes. The group with the stronger effect determines the extent of shifting of the primary absorption band. If one of the groups is electron-releasing while the other is electron-withdrawing, the magnitude of the shift of the primary band is greater than the sum of the shifts due to the individual groups. The enhanced shifting is due to resonance interactions of the type:

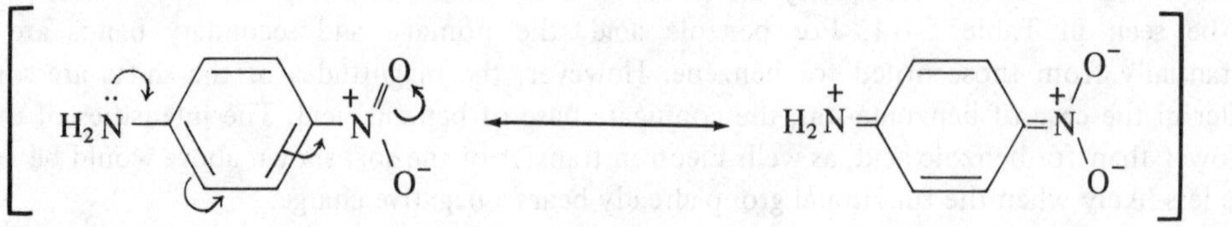

If the two groups of a disubstituted benzene derivative are either *ortho* or *meta* to each other, the magnitude of the observed shift is approximately equal to the sum of the shifts caused by each individual group. With substitution of these types, there is no opportunity for the kind of direct resonance interaction between substituent groups observed with *para* substituents. In the case of *ortho* substituents, the steric inability of both groups to achieve coplanarity inhibits resonance.

For the special case of substituted benzoyl derivatives, an empirical correlation of structure with the observed position of the primary absorption band has been developed; it is given in Table 5–13. It provides a means of estimating the position of the primary band for benzoyl derivatives within about 5 nm.

TABLE 5–13 Empirical Rules for Benzoyl Derivatives

Parent Chromophore:

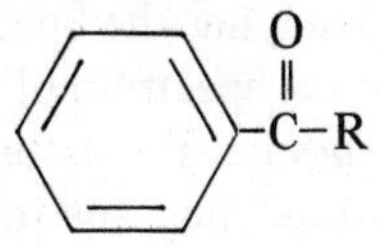

R = Alkyl or ring residue		246
R = H		250
R = OH or OAlkyl		230
Increment for each substituent:		
–Alkyl or ring residue	*o*-, *m*-	3
	p-	10
–OH, $-OCH_3$, or –OAlkyl	*o*-, *m*-	7
	p-	25
$-O^-$	*o*-	11
	m-	20
	p-	78
–Cl	*o*-, *m*-	0
	p-	10
–Br	*o*-, *m*-	2
	p-	15
$-NH_2$	*o*-, *m*-	13
	p-	58
$-NHCOCH_3$	*o*-, *m*	20
	p-	45
$-NHCH_3$	*p*-	73
$-N(CH_3)_2$	*o*-, *m*-	20
	p-	85

Two examples of how these rules are applied follow:

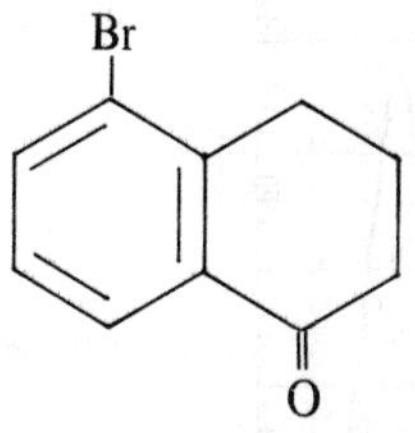

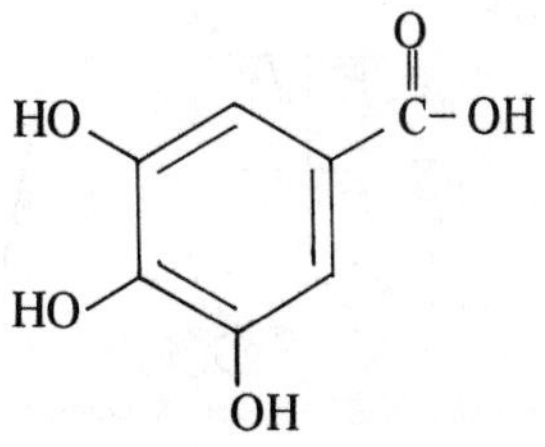

parent chromophore:	246 nm
o-ring residue:	3
m-Br:	2
	251 nm
observed:	253 nm

parent chromophore:	230 nm
m-OH: 2 X 7 =	14
p-OH:	25
	269 nm
observed:	270 nm

E. POLYNUCLEAR AROMATIC HYDROCARBONS AND HETEROCYCLIC COMPOUNDS

It is observed that the primary and the secondary bands in the spectra of polynuclear aromatic hydrocarbons are shifted to longer wavelength. In fact, even the second primary band, which appears at 184 nm for benzene, is shifted to a wavelength which lies within the range of most ultraviolet spectrophotometers. This band lies at 220 nm in the spectrum of naphthalene. As the extent of conjugation increases, the magnitude of the bathochromic shift also increases.

The ultraviolet spectra of the polynuclear aromatic hydrocarbons possess characteristic shapes and fine structure. Common practice in studying the spectra of substituted polynuclear aromatic derivatives is to compare their spectra with the spectrum of the unsubstituted hydrocarbon. Based on the similarity of peak shapes and fine structure, the nature of the chromophore can be identified. This technique involves the use of model compounds, and it is discussed further in Section 5.18.

The ultraviolet spectra of naphthalene and anthracene are shown in Figure 5-25. One should notice the characteristic shape and fine structure of each spectrum, as well as the effect of increased conjugation on the position of the absorption maxima.

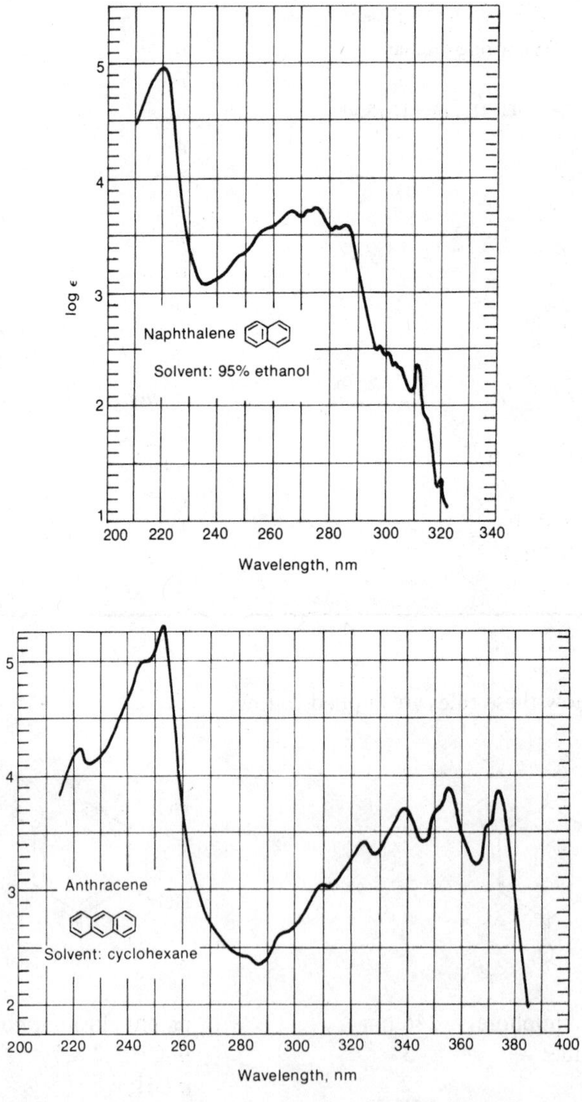

FIGURE 5-25 Ultraviolet Spectra of Naphthalene and Anthracene (From R. A. Friedel and M. Orchin, "Ultraviolet Spectra of Aromatic Compounds," © 1951. Reprinted by permission of John Wiley and Sons, Inc., New York).

Heterocyclic molecules have electronic transitions which include combinations of $\pi \rightarrow \pi^*$ and $n \rightarrow \pi^*$ transitions. The spectra can be rather complex, and an analysis of the transitions involved will be left to more advanced treatments. When studying derivatives of heterocyclic molecules, the commonly used method is to compare them to the spectra of the parent heterocyclic systems. The use of model compounds in this fashion is described further in Section 5.18.

The ultraviolet spectra of pyridine, quinoline, and isoquinoline are included in Figure 5–26. The reader may wish to compare the spectrum of pyridine with that of benzene (Figure 5–24) and the spectra of quinoline and isoquinoline with the spectrum of naphthalene (Figure 5–25).

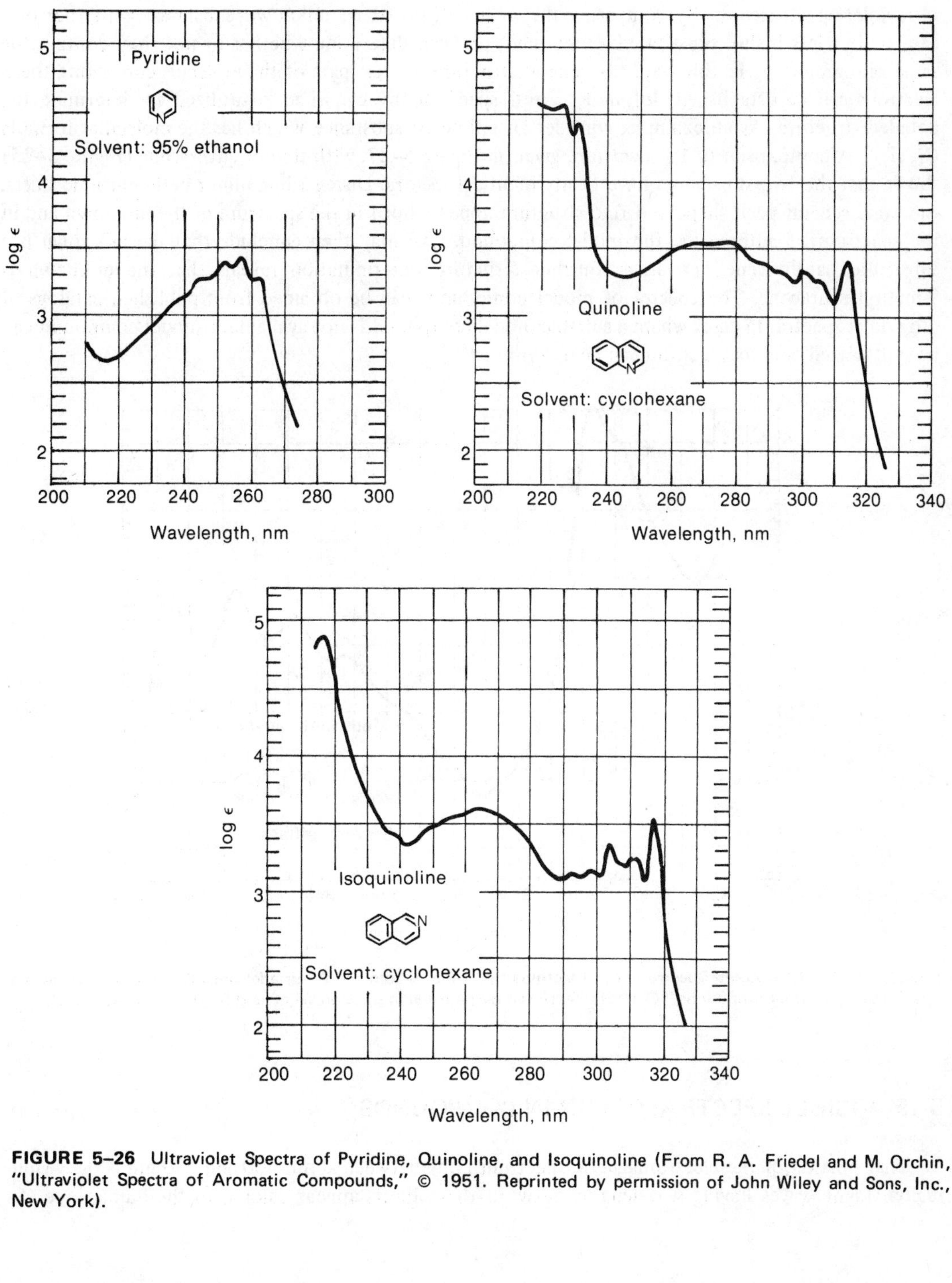

FIGURE 5–26 Ultraviolet Spectra of Pyridine, Quinoline, and Isoquinoline (From R. A. Friedel and M. Orchin, "Ultraviolet Spectra of Aromatic Compounds," © 1951. Reprinted by permission of John Wiley and Sons, Inc., New York).

5.18 MODEL COMPOUND STUDIES

Very often, the ultraviolet spectra of several members of a particular class of compounds are very similar. Unless one is thoroughly familiar with the spectroscopic properties of each member of the class of compounds, it is very difficult to distinguish the actual substitution patterns of individual molecules by their ultraviolet spectra. However, the gross nature of the chromophore of an unknown substance can be determined by this method. Then, based on the knowledge of the chromophore, one can employ the other spectroscopic techniques described in this book to elucidate the precise structure and substitution of the molecule.

This approach, using model compounds, is one of the best ways to use the technique of ultraviolet spectroscopy. By comparing the uv spectrum of an unknown substance with that of a similar, but less highly substituted, compound, one can determine whether or not they contain the same chromophore. In this way, the general structure of that part of the molecule containing the π electrons can be established. Infrared or nmr spectroscopy can then be utilized to determine the detailed structure. As an example, consider an unknown substance which has the molecular formula $C_{15}H_{12}$. A comparison of its spectrum, given in Figure 5–27, with that of anthracene (Figure 5–25) shows that the two substances have nearly identical spectra. Disregarding minor bathochromic shifts, the same general peak shape and fine structure appears both in the spectrum of the unknown and in the spectrum of anthracene, the model compound. One may then conclude that the unknown is a substituted anthracene derivative. Further structure determination reveals that the unknown is 9-methylanthracene. The spectra of model compounds can be obtained from published catalogs of ultraviolet spectra. In cases where a suitable model compound is not available, a model compound can be synthesized, and its spectrum can be determined.

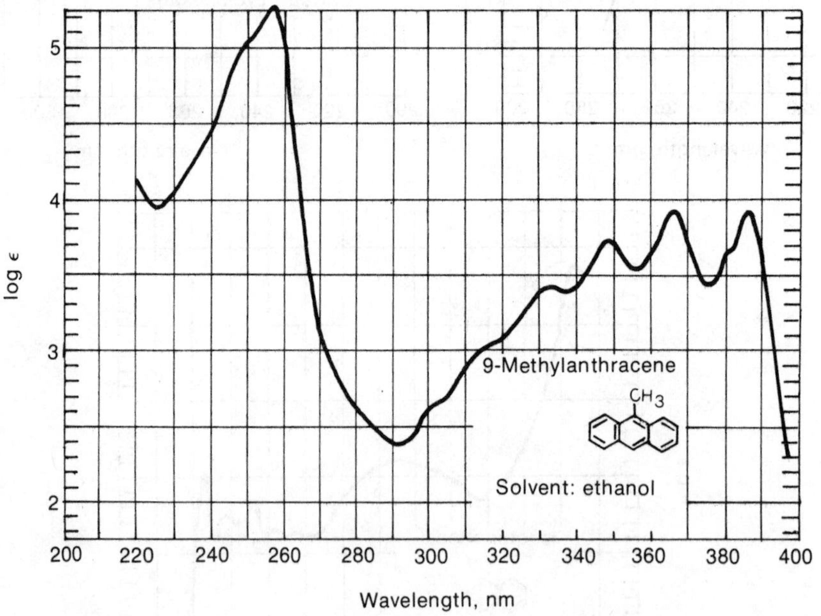

FIGURE 5–27 Ultraviolet Spectrum of 9-Methylanthracene (From R. A. Friedel and M. Orchin, "Ultraviolet Spectra of Aromatic Compounds," © 1951. Reprinted by permission of John Wiley and Sons, Inc., New York).

5.19 VISIBLE SPECTRA; COLOR IN COMPOUNDS

That portion of the electromagnetic spectrum lying between about 400 and 750 nm is the visible region. Light waves having wavelengths between these limits appear colored to the human eye. As anyone who has seen light diffracted by a prism or the diffraction effect of a rainbow will recall, one end of the visible spectrum is violet and the other is red. Light having wavelengths near 400 nm is violet, while that having wavelengths near 750 nm is red.

The phenomenon of color in compounds, however, is not as straightforward as the above discussion would suggest. If a substance absorbs visible light, it will appear to have a color; if not, it will appear white. However, compounds which absorb light in the visible region of the spectrum do not possess the color corresponding to the wavelength of the light which they absorb. Rather, there is an inverse relationship between the observed color and the color absorbed.

When we observe light *emitted* from a source, as from a lamp or an emission spectrum, we observe the color corresponding to the wavelength of light which is emitted. In such a case, a light source emitting violet light emits light at the high energy end of the visible spectrum. A light source emitting red light emits light at the low energy end of the spectrum.

However, when we observe the color of a particular object or of a substance, we do not observe that object or substance emitting light. Certainly the substance does not glow in the dark. Rather, we observe the light which is being *reflected*. The color which our eye perceives is not the color corresponding to the wavelength of the light absorbed, but rather its *complement*. When white light falls on an object, light of a particular wavelength will be absorbed. The remainder of the light will be reflected. The eye (and brain) registers all of the reflected light as the complementary color to that which was absorbed. In the case of transparent objects or solutions, the eye receives the light which is *transmitted*. Again, light of one particular wavelength is absorbed, and the remaining light passes through to reach the eye. As before, the eye registers this transmitted light as the complementary color to that which was absorbed. Table 5–14 illustrates the relationship between the wavelength of light absorbed by a substance and the color which is perceived by an observer.

TABLE 5-14 The Relationship Between the Color of Light Absorbed by a Compound and the Observed Color of the Compound

Color of Light Absorbed	Wavelength of Light Absorbed (nm)	Observed Color
violet	400	yellow
blue	450	orange
blue-green	500	red
yellow-green	530	red-violet
yellow	550	violet
orange-red	600	blue-green
red	700	green

Some familiar compounds may serve to underscore these relationships between the absorption spectrum and the observed color:

β-Carotene (pigment from carrots): λ_{max} = 452 nm, *orange*
Lycopene (pigment from tomatoes): λ_{max} = 474 nm, *red*
Cyanidin (blue pigment of cornflower): λ_{max} = 545 nm, *blue*
Chlorophyll *a* (leaf pigment): 2 peaks, λ_{max} = 420 and 680 nm, *green*

The general structural formulas of some colored substances are shown below. In each case, the reader should notice that these substances have highly extended conjugated systems of electrons. Such extensive conjugation shifts their electronic spectra to such long wavelengths that they absorb visible light and are colored.

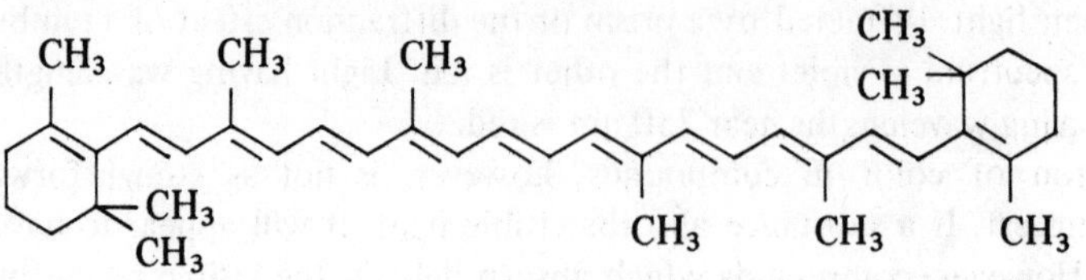

β-Carotene
(A carotenoid, which is a class of plant pigments)
λ_{max} = 452 nm

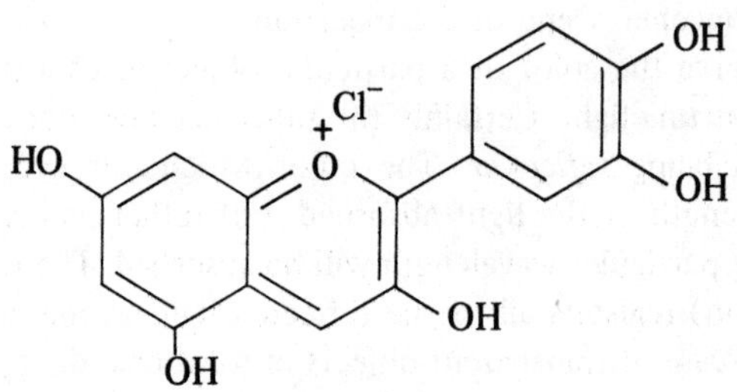

Cyanidin Chloride
(An anthocyanin, another class of plant pigments)
λ_{max} = 545 nm

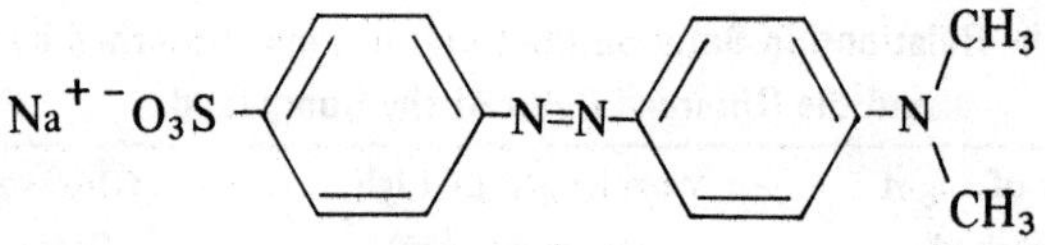

Methyl Orange
(An azo dye)
λ_{max} = 460 nm

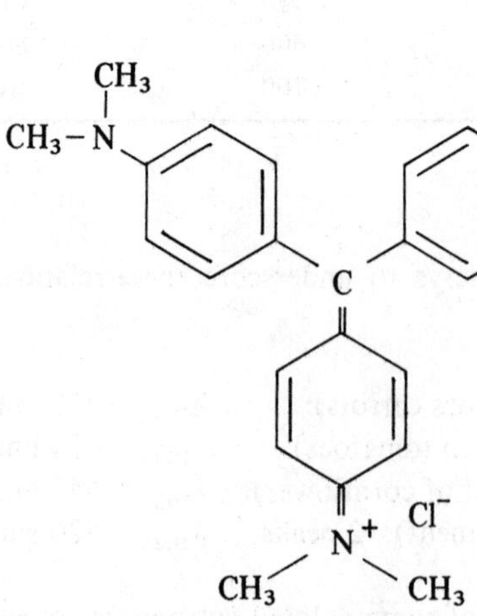

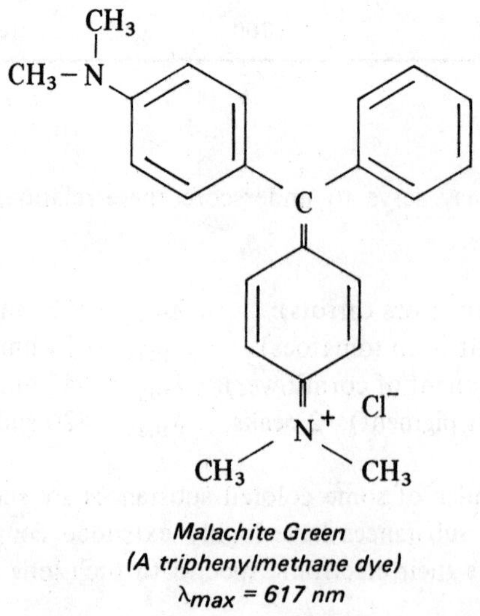

Malachite Green
(A triphenylmethane dye)
λ_{max} = 617 nm

5.20 WHAT DO YOU LOOK FOR IN AN ULTRAVIOLET SPECTRUM? A PRACTICAL GUIDE

When a uv spectrum is used by itself, it is often difficult to extract a great deal of information from it. However, several generalizations can be formulated which are a good guide to the use of uv data. These generalizations will, of course, be a good deal more meaningful when combined with infrared data, which can, for instance, definitely identify carbonyl groups, double bonds, aromatic systems, nitro groups, nitriles, enones, and other important chromophores. In the absence of infrared data, the following observations should be taken only as guidelines.

1. *A single band of low to medium intensity (ϵ = 100 to 10,000) at wavelengths less than 220 nm* usually indicates an $n \rightarrow \sigma^*$ transition. Amines, alcohols, ethers, and thiols are possibilities, provided that the non-bonded electrons are not included in a conjugated system. An exception to this generalization is that the $n \rightarrow \pi^*$ transition of cyano groups (−C≡N:) appears in this region. However, this is a weak transition (ϵ less than 100), and, of course, the cyano group is easily identified in the infrared. Do not neglect to look for N–H, O–H, C–O, or S–H bands in the infrared spectrum.

2. *A single band of low intensity (ϵ = 10 to 100) in the region 250–360 nm, with no major absorption at shorter wavelengths (200–250 nm),* usually indicates an $n \rightarrow \pi^*$ transition. Since the absorption does not occur at long wavelength, a simple, or unconjugated, chromophore is indicated, generally one which contains an O, N, or S atom. Examples of this might include C=O, C=N, N=N, $-NO_2$, −COOR, −COOH, or $-CONH_2$. Once again, the infrared spectrum should help a great deal.

3. *Two bands of medium intensity (ϵ = 1,000 to 10,000), both with λ_{max} above 200 nm,* generally indicate the presence of an aromatic system. If an aromatic system is present, there may be a good deal of fine structure in the longer wavelength band (in non-polar solvents only). Substitution on the aromatic rings will increase the molar absorptivity above 10,000, particularly if the substituent increases the length of the conjugated system.

In polynuclear aromatic substances, a third band will appear near 200 nm, a band which in simpler aromatics occurred below 200 nm where it could not be observed. Most polynuclear aromatics (and heterocyclic compounds) have very characteristic intensity and band shape (fine structure) patterns, and they may often be identified by comparison to spectra which are available in the literature. The texts by Jaffé and Orchin and by Scott, which are listed in the references at the end of this chapter, are good sources of spectra.

4. *Bands of high intensity (ϵ = 10,000 to 20,000) which appear above 210 nm* generally represent either an α, β-unsaturated ketone (check the infrared spectrum) or a diene or polyene. The longer the length of the conjugated system, the longer the observed wavelength will be. For dienes, the λ_{max} may be calculated using the Woodward-Fieser rules (Section 5.10), or, if there are more than four double bonds ($n > 4$), the Fieser-Kuhn rules (Section 5.11). Enones are discussed below.

5. *Simple ketones, acids, esters, amides, and other compounds containing both π systems and unshared electron pairs, will show two absorptions:* an $n \rightarrow \pi^*$ transition at longer wavelength (> 300 nm, low intensity) and a $\pi \rightarrow \pi^*$ transition at shorter wavelengths (< 250 nm, high intensity). With conjugation (enones), the λ_{max} of the $\pi \rightarrow \pi^*$ band moves to longer wavelengths and can be predicted by Woodward's rules (Section 5.13). The ϵ value usually rises above 10,000 with conjugation, and, as it is very intense, it may obscure or bury the weaker $n \rightarrow \pi^*$ transition.

For α, β-unsaturated esters and acids, Nielsen's rules (Section 5.15) may be used to predict the position of λ_{max} with increasing conjugation and substitution.

6. *Compounds which are highly colored* (have absorption in the visible region) are likely to contain a long-chain conjugated system or a polycyclic aromatic chromophore. Benzenoid compounds may be colored if they have enough conjugating substituents. For non-aromatic systems, usually a minimum of 4 to 5 conjugated chromophores are required to produce absorption in the visible region. However, some simple nitro, azo, nitroso, α-diketo, polybromo, and polyiodo

compounds may also exhibit color, as may many compounds with a quinoid structure.

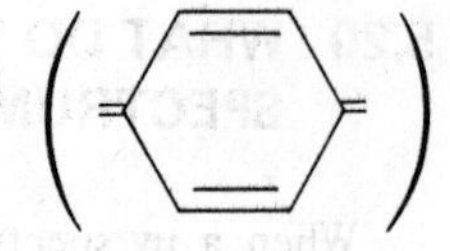

REFERENCES

J. R. Dyer, "Applications of Absorption Spectroscopy of Organic Compounds," Prentice-Hall, Inc., Englewood Cliffs, New Jersey, 1965.

R. A. Friedel and M. Orchin, "Ultraviolet Spectra of Aromatic Compounds," John Wiley and Sons, New York, 1951.

H. M. Hershenson, "Ultraviolet Absorption Spectra," Index for 1954–1957, Academic Press, New York, 1959.

H. H. Jaffé and M. Orchin, "Theory and Applications of Ultraviolet Spectroscopy," John Wiley and Sons, New York, 1964.

V. M. Parikh, "Absorption Spectroscopy of Organic Molecules," Addison-Wesley Publishing Company, Reading, Massachusetts, 1974. Chapter 2.

D. J. Pasto and C. R. Johnson, "Organic Structure Determination," Prentice-Hall, Inc., Englewood Cliffs, New Jersey, 1969. Chapter 3.

A. I. Scott, "Interpretation of the Ultraviolet Spectra of Natural Products," Pergamon Press, New York, 1964.

R. M. Silverstein, G. C. Bassler, and T. C. Morrill, "Spectrometric Identification of Organic Compounds," Third Edition, John Wiley and Sons, New York, 1974. Chapter 5.

E. S. Stern and T. C. J. Timmons, "Electronic Absorption Spectroscopy in Organic Chemistry," St. Martin's Press, New York, 1971.

PROBLEMS

1. The ultraviolet spectrum of benzonitrile shows a primary absorption band at 224 nm and a secondary band at 271 nm.

 a) If a solution of benzonitrile in water, whose concentration is 1×10^{-4} molar, is examined at a wavelength of 224 nm, the absorbance is determined to be 1.30. What is the molar absorptivity of this absorption band?

 b) If the same solution is examined at 271 nm, what will be the absorbance reading ($\epsilon = 1,000$)? What will be the intensity ratio, I_0/I?

2. Draw structural formulas which are consistent with the observations indicated:

 a) An acid, $C_7H_4O_2Cl_2$, shows a uv maximum at 242 nm.

 b) A ketone, $C_8H_{14}O$, shows a uv maximum at 248 nm.

 c) An aldehyde, $C_8H_{12}O$, absorbs in the uv with $\lambda_{max} = 244$ nm.

3. Predict the uv maximum for each of the following substances:

 a)

 $$CH{=}CH{-}\overset{\overset{\displaystyle O}{\|}}{C}{-}CH_3$$

 b)

 $$CH_2{=}\underset{\underset{\displaystyle CH_3}{|}}{C}{-}\overset{\overset{\displaystyle O}{\|}}{C}{-}CH_3$$

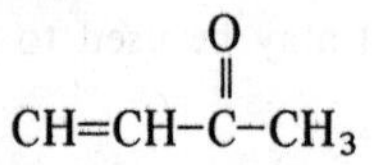

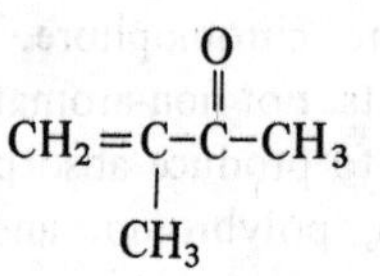

c)

$$CH_3-CH=CH-\overset{\overset{\displaystyle O}{\|}}{C}-CH_3$$

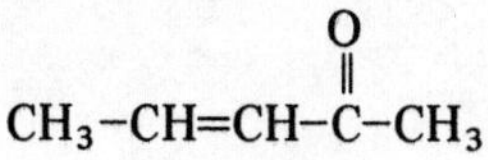

d)

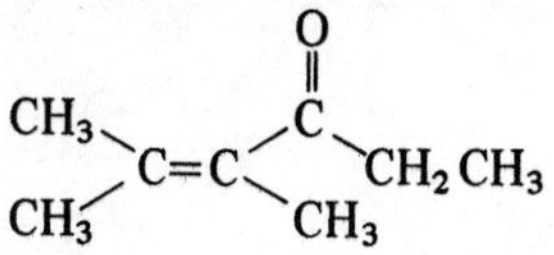

e)

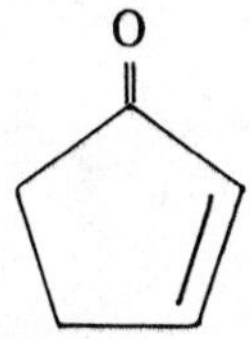

f)

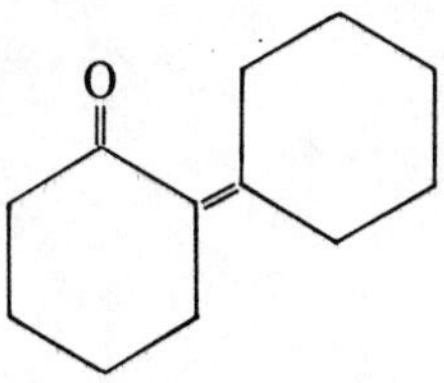

g)

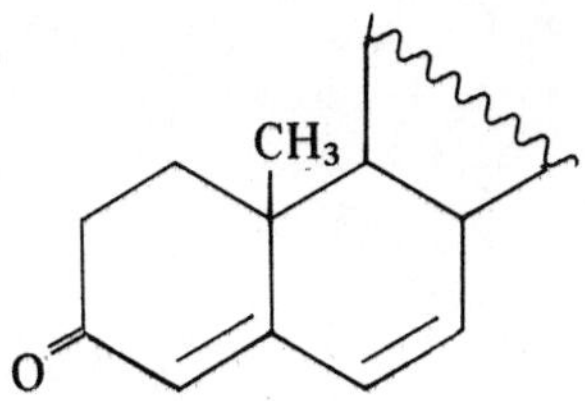

h)

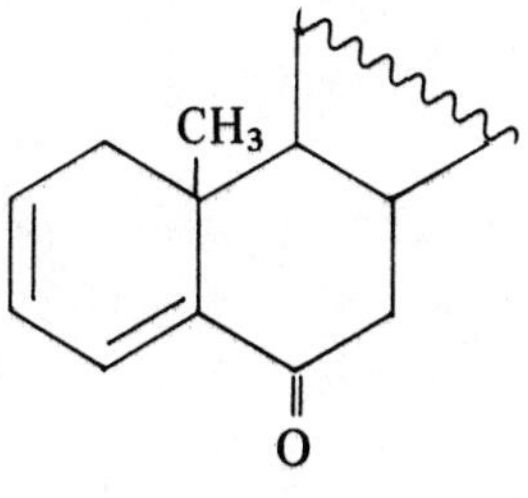

i)

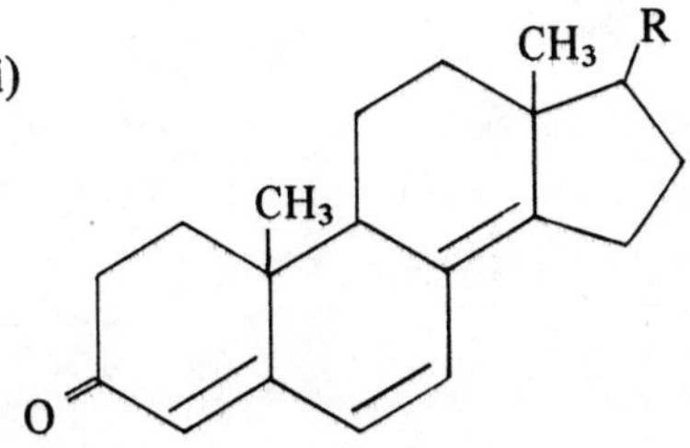

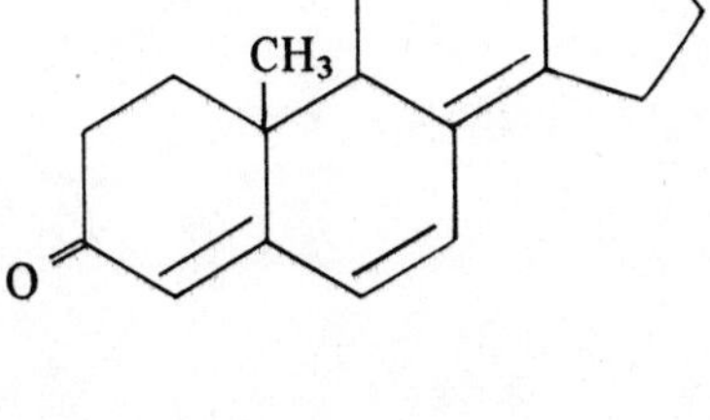

j)

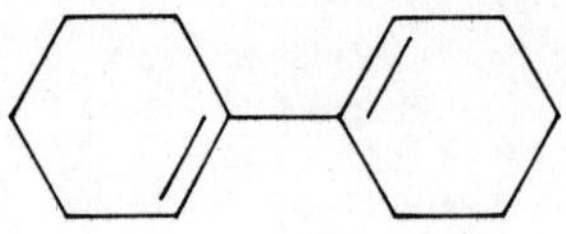

k)

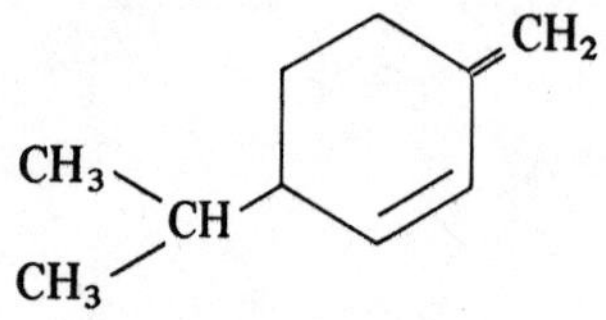

l)

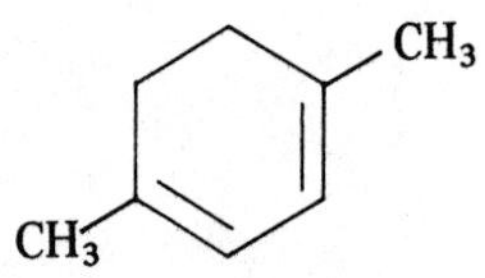

m)

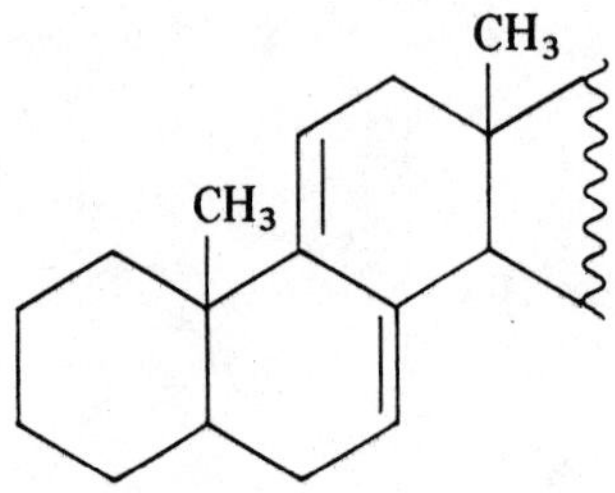

n)

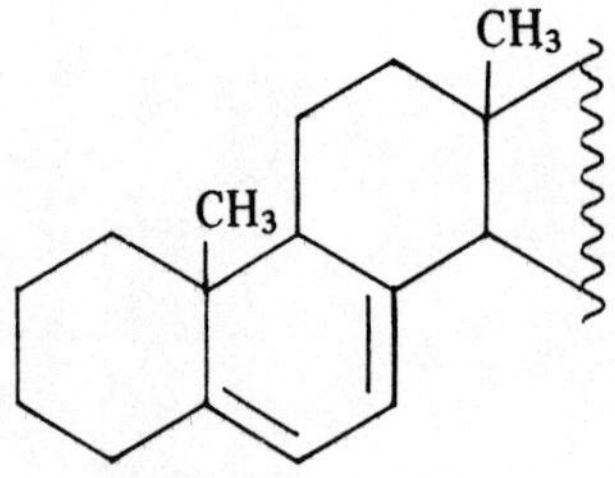

o)

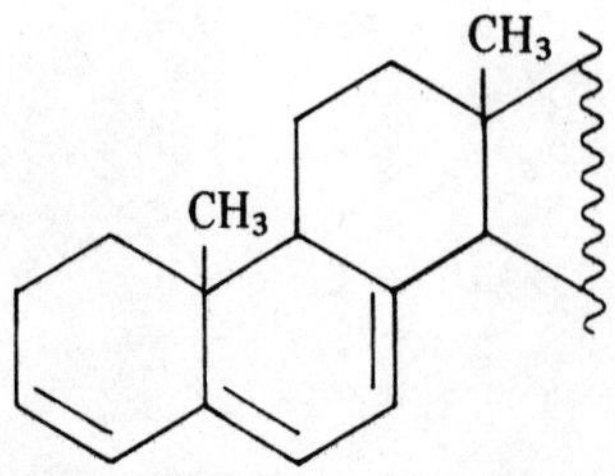

p)

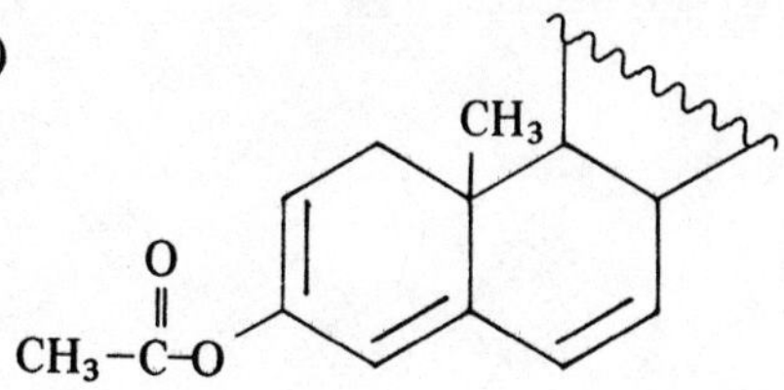
CH3
O
CH3–C–O

q)

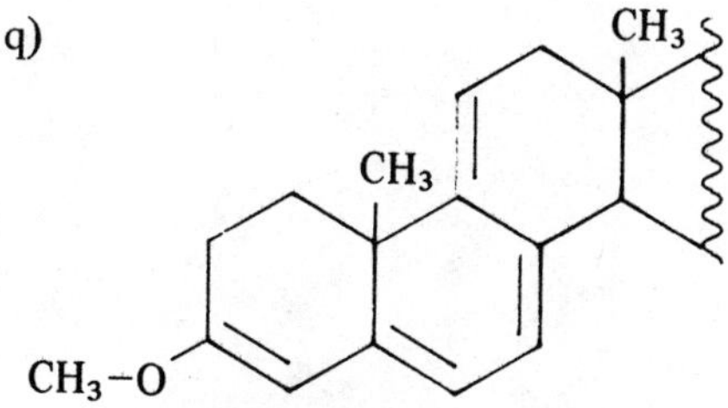
CH3
CH3
CH3–O

r)

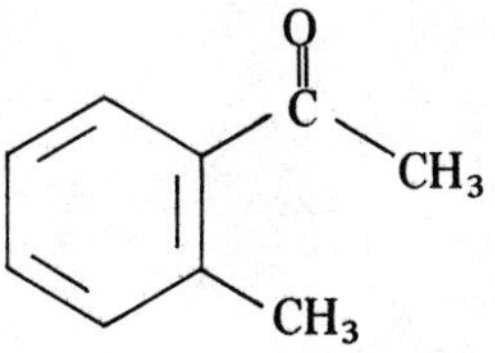
O
C
CH3
CH3

s)

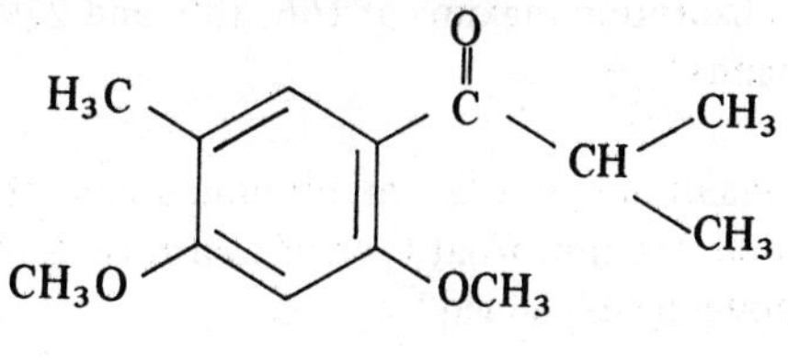
H3C
O
C
CH
CH3
CH3
CH3O
OCH3

t)

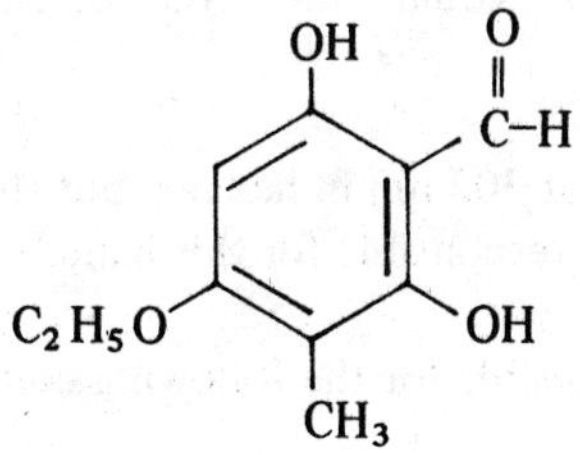
OH
O
C–H
C2H5O
OH
CH3

u)

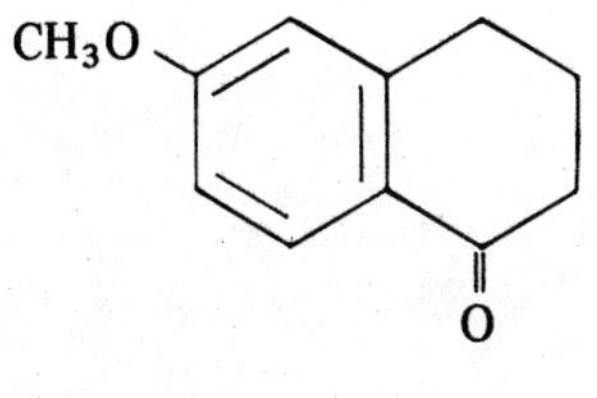
CH3O
O

v)

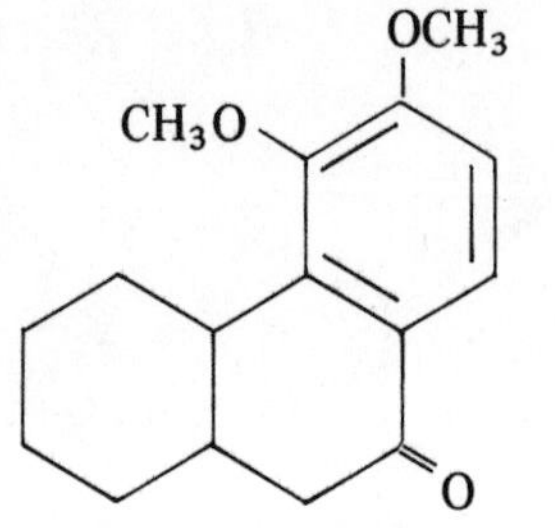

w)

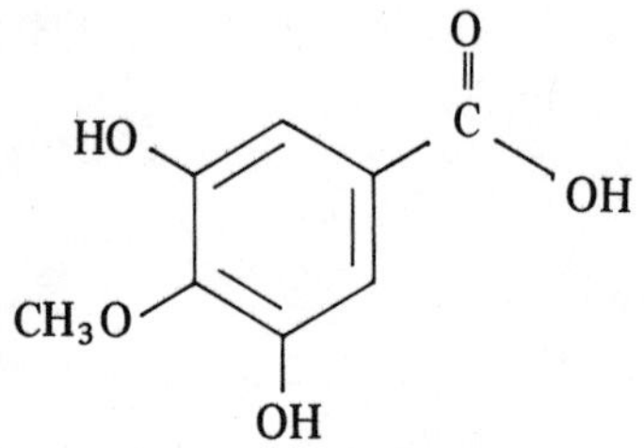

x)

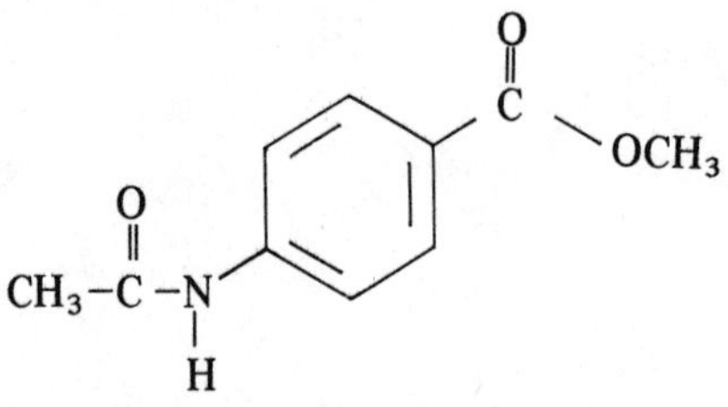

4. The uv spectrum of acetone shows absorption maxima at 166, 189, and 279 nm. What type of transition is responsible for each of these bands?

5. Chloromethane has an absorption maximum at 172 nm, bromomethane shows an absorption at 204 nm, and iodomethane shows a band at 258 nm. What type of transition is responsible for each band? How can the trend of absorptions shown be explained?

6. The uv spectrum of 3-buten-2-one in hexane shows an absorption maximum at 226 nm in hexane solution. Predict which direction the absorption would shift if the solvent were changed to ethanol.

7. If the uv spectrum of a substance shows a band at 305 nm in hexane, but the band is observed to shift to 307 nm in ethanol, what type of transition is responsible for this band?

8. What types of electronic transitions would be possible for the following compounds?

a) Cyclopentene

b) Acetaldehyde

c) Dimethyl ether

d) Methyl vinyl ether

e) Triethylamine

f) Cyclohexane

CHAPTER 6

MASS SPECTROMETRY

The principles which underlie mass spectrometry predate any of the other instrumental techniques described in this book. The fundamental principles date back to 1898. In 1911, J. J. Thomson used a mass spectrum to demonstrate the existence of neon-22 in a sample of neon-20, thereby establishing that elements could have isotopes. The earliest mass spectrometer, as we know it today, was built in 1918. However, the method of mass spectrometry has not come into common use until quite recently, when reasonably inexpensive and reliable instruments have become available. With the advent of commercial instruments which are capable of being maintained fairly easily, which are priced within the range of many industrial and academic laboratories, and which provide high resolution, the technique has become quite important in structure elucidation studies.

6.1 THE MASS SPECTROMETER

In its simplest form, the mass spectrometer performs three essential functions. First, molecules are subjected to bombardment by a stream of high-energy electrons, converting some of the molecules to ions. The ions are accelerated in an electric field. Second, the accelerated ions are separated according to their mass-to-charge ratio in a magnetic or electric field. Finally, the ions with a particular mass-to-charge ratio are detected by a device which is able to count the number of ions which strike it. The detector's output is amplified and fed to a recorder. The trace from the recorder is a *mass spectrum* – a graph of the number of particles detected as a function of mass-to-charge ratio.

When each function is examined in detail, one sees that the mass spectrometer is actually somewhat more complex than has been described above. Before the ions can be formed, some method must be found to introduce a stream of molecules into the *ionization chamber*, where the actual ionization takes place. A *sample inlet system* is used to provide this stream of molecules. Samples studied by mass spectrometry may be gases, liquids, or solids. Some means must be found to convert enough of the sample to the vapor state to obtain a stream of molecules flowing into the ionization chamber. With gases, of course, the substance is already vaporized, so a system such as the one depicted in Figure 6–1 can be used. This system is evacuated, such that the ionization chamber is at a lower pressure than the sample inlet system. The sample is introduced into a larger reservoir, from which the molecules of vapor can pass into the ionization chamber. To insure that there is a steady stream of molecules passing into the ionization chamber, the vapor passes through a small pinhole, called a *molecular leak*, before entering the chamber. The same system can be used for volatile liquids or solids. For less volatile materials, the system can be designed to fit within an oven, which can heat the sample to provide a greater vapor pressure. One must take care not to heat any sample to a temperature at which it might decompose.

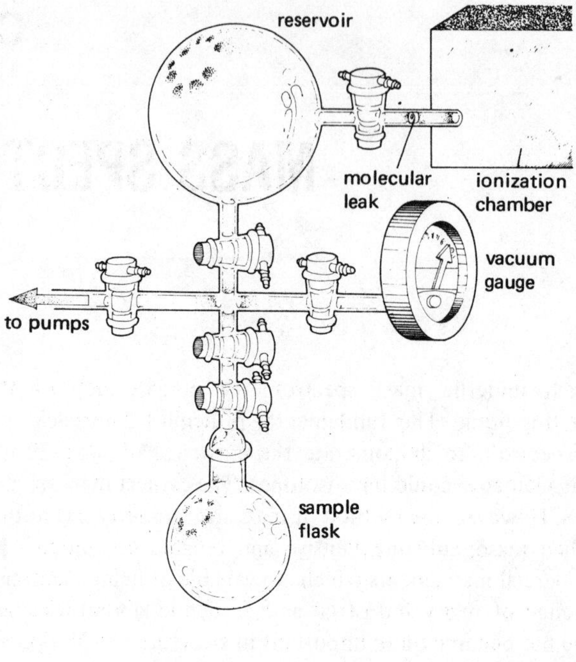

FIGURE 6–1 Sample Inlet System

With rather non-volatile solids, a direct probe method of introducing the sample may be used. This method is illustrated in Figure 6–2. The sample is placed on the tip of the probe, which is then inserted through a vacuum lock into the ionization chamber. The sample is placed very close to the ionizing beam of electrons. The probe is capable of being heated, thus causing vapor from the sample to be evolved in close proximity to the beam of electrons. A system such as this one can be used to study samples of molecules with vapor pressures lower than 10^{-9} mm Hg at room temperature.

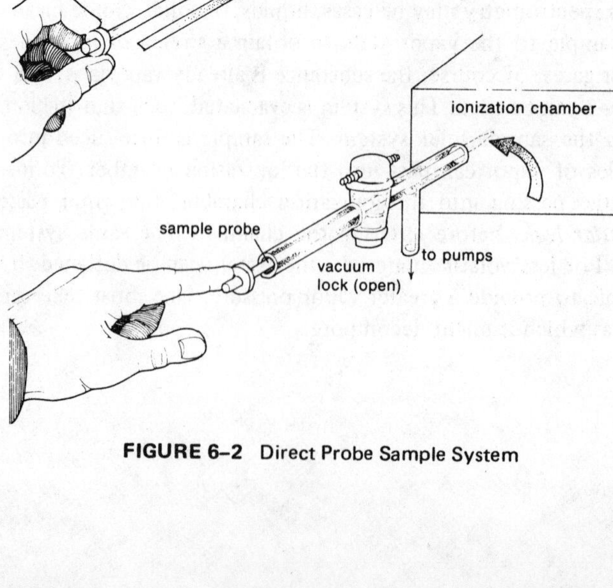

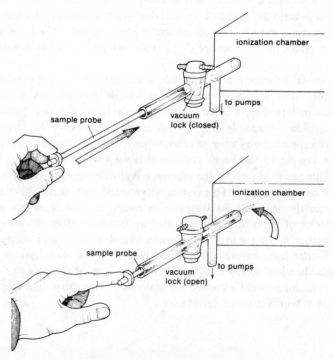

FIGURE 6–2 Direct Probe Sample System

Once the stream of sample molecules has entered the ionization chamber, it is bombarded by a beam of high-energy electrons. The molecules are converted to ions in this process. The ions are then accelerated in an electric field. A diagram of a typical ionization chamber is shown in Figure 6–3.

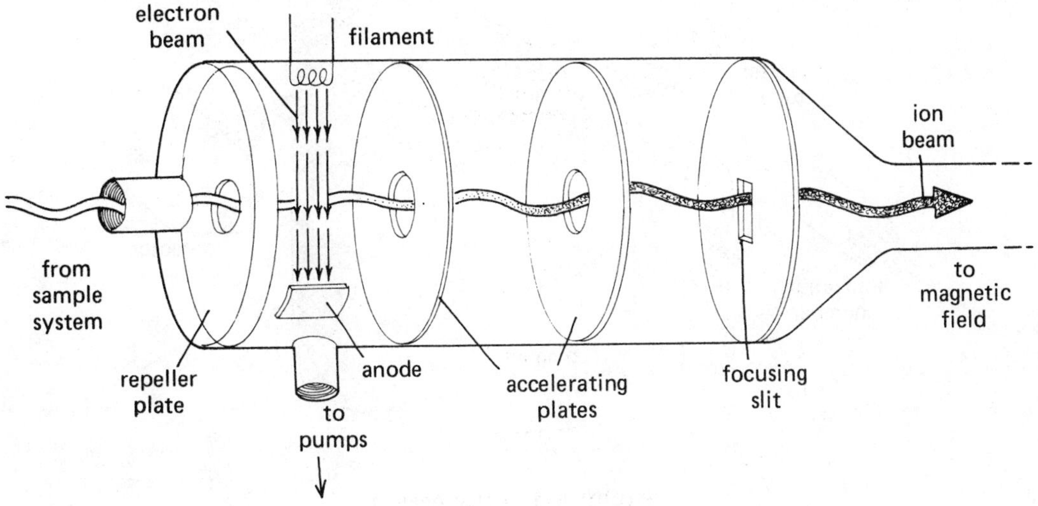

FIGURE 6–3 Ionization Chamber

In the ionization chamber, the beam of high-energy electrons is emitted from a heated *filament*. The filament is heated to several thousand degrees Celsius. In normal operation, the electrons have an energy of about 70 electron volts. These high-energy electrons strike the stream of molecules which has been admitted from the sample system, and ionize the molecules in the sample stream by removing electrons from them. The molecules are thus converted into positive ions. A *repeller plate*, which carries a positive electrical potential, directs the newly-created ions toward a series of *accelerating plates*. A large potential difference, ranging from 1 to 10 kilovolts, applied across these accelerating plates, produces a beam of rapidly traveling positive ions. The ions are directed into a uniform beam by one or more *focusing slits*.

Most of the sample molecules are not ionized at all. These are continuously drawn off by the vacuum pumps which are connected to the ionization chamber. Some of the molecules are converted to negative ions, through the absorption of electrons. These negative ions are absorbed by the repeller plate. A small proportion of the positive ions which are formed may have a charge greater than one (loss of more than one electron). These are accelerated in the same way as the singly-charged positive ions.

The energy required to remove an electron from an atom or molecule is its *ionization potential*. Most organic compounds have ionization potentials ranging between 8 and 15 electron volts. However, a beam of electrons does not create ions with high efficiency until the beam of electrons striking the stream of molecules has a potential of from 50 to 70 electron volts. In order to produce reproducible spectra, electrons of this energy range are used to ionize the sample.

From the ionization chamber, the beam of ions passes through a short field-free region. From there, the beam enters the *mass analyzer*, which is the region where the ions are separated according to their mass-to-charge ratio. A diagram of a simple mass analyzer is shown in Figure 6–4.

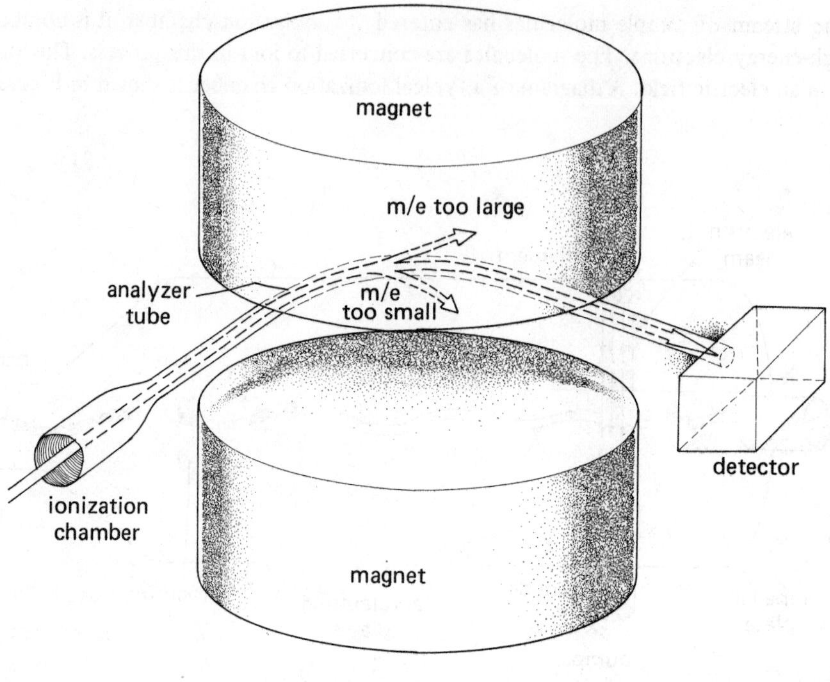

FIGURE 6–4 A Mass Analyzer

The kinetic energy of an accelerated ion is equal to:

$$\tfrac{1}{2}mv^2 = eV$$

where m is the mass of the ion, v is the velocity of the ion, e is the charge on the ion, and V is the potential difference of the ion accelerating plates. In the presence of a magnetic field, a charged particle will describe a curved flight path. The equation which yields the radius of curvature of this path is:

$$r = \frac{mv}{eH}$$

where r is the radius of curvature of the path and H is the strength of the magnetic field. If these two equations are combined to eliminate the velocity term, one obtains:

$$\boxed{\frac{m}{e} = \frac{H^2r^2}{2V}}$$

This is the important equation which governs the behavior of an ion in the mass analyzer portion of the mass spectrometer.

As can be seen from the equation, the larger the value of m/e, the larger will be the radius of the curved path. The analyzer tube of the instrument is constructed to have a fixed radius of curvature. A particle with the correct m/e ratio will be able to negotiate the curved analyzer tube and reach the detector. Particles having m/e ratios which are either too large or too small will strike the sides of the analyzer tube and will not reach the detector. Of course, the method would not be very interesting if ions of only one particular mass could be detected. Therefore, either the accelerating voltage or the magnetic field strength is continuously varied in order that all of the ions produced in the ionization chamber can be detected. The record produced from the detector system is in the form of a plot of the numbers of ions versus their value of m/e.

An important factor to be considered in a mass spectrometer is that of *resolution*. Resolution is defined according to the relationship:

$$R = \frac{M}{\Delta M}$$

where R is the resolution, M is the mass of the particle, and ΔM is the difference in mass between a particle of mass M and the particle of next higher mass which can be resolved by the instrument. Low resolution instruments have R values ranging as high as 2000. For some applications, resolutions which are some five to ten times that amount are required.

To obtain higher resolutions, modifications of the basic instrument design described in Figure 6–4 are used. Since the particles leaving the ionization chamber do not all have precisely the same velocity, a *double-focusing mass spectrometer* may be used. In such an instrument, the beam of ions passes through an electric field region before entering the magnetic field. The particles describe a curved path in each of these regions. In the presence of an electric field, the particles all travel at the same velocity, so the resolution of the magnetic field region improves.

In order to provide a relatively compact instrument which provides high resolution, a recent modification has been to replace the magnetic field region with a quadrupole system. In a *quadrupole mass spectrometer*, which is described in Figure 6–5, a set of four solid rods is arranged parallel to the direction of the ion beam. The rods should be hyperbolic in cross-section, although cylindrical rods may be used. A direct-current voltage and a radiofrequency are applied to the rods, with the result that an oscillating electrostatic field is generated in the region between the rods. Depending upon the ratio of the radiofrequency amplitude to the direct current voltage, ions acquire an oscillation in this electrostatic field. Ions of the incorrect mass-to-charge ratio (too small or too large) undergo an unstable oscillation. The amplitude of the oscillation continues to increase until the particle strikes one of the rods. Ions of the correct mass-to-charge ratio undergo a stable oscillation of constant amplitude. These ions do not strike the rods, but rather pass through the analyzer to reach the detector. The resolution of the system can be varied by varying the ratio of radiofrequency amplitude to DC voltage. Resolutions as high as 10,000 may be obtained with this type of mass analyzer.

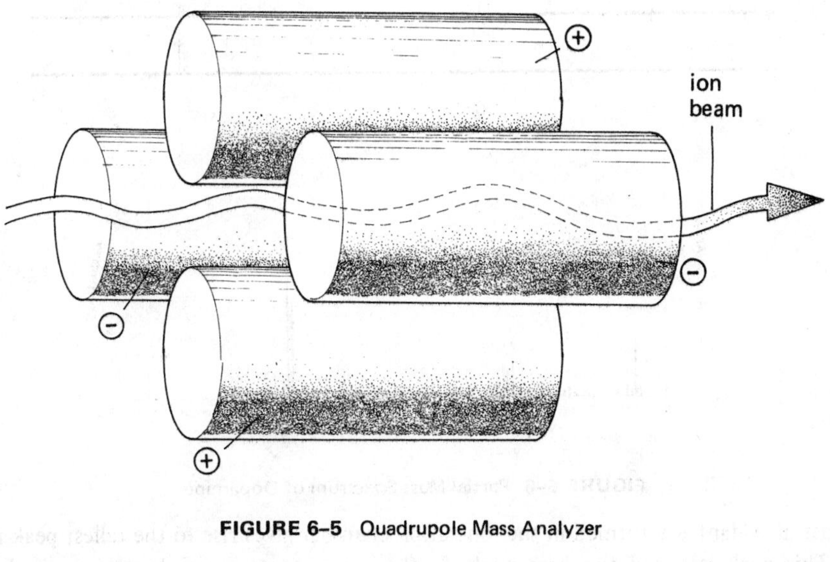

FIGURE 6–5 Quadrupole Mass Analyzer

The *detector* of most instruments consists of a counter which produces a current that is proportional to the number of ions which strike it. Through the use of electron multiplier circuits, this current can be measured so accurately that the current caused from just one ion striking the detector can be measured. The signal from the detector is fed to a *recorder*, which produces the mass spectrum itself. In the recorder, the electron current from the detector is fed to a series of five galvanometers of varying sensitivity. The galvanometers frequently range in sensitivity in the ratio: 1, 1/3, 1/10, 1/30, 1/100. A light beam striking mirrors attached to the galvanometers is reflected onto a piece of photosensitive paper. In this way, a mass spectrum with five simultaneous traces, each at a different sensitivity, is produced. By using the five traces, it is possible to record the weakest peaks while still being able to keep the strongest peaks on scale.

6.2 THE MASS SPECTRUM

The mass spectrum is shown as a plot of ion abundance versus *m*/*e* ratio. A portion of a typical mass spectrum is shown in Figure 6–6. The spectrum shown is that of dopamine, a substance which acts as a neurotransmitter in the central nervous system. Because a spectrum of this sort may be somewhat difficult to use, frequently one sees the information from a spectrum converted into the form of a bar graph. Such a graph is also shown beneath the spectrum in Figure 6–6. Commonly, mass spectral results are presented in tabular form, as can be seen in Table 6–1.

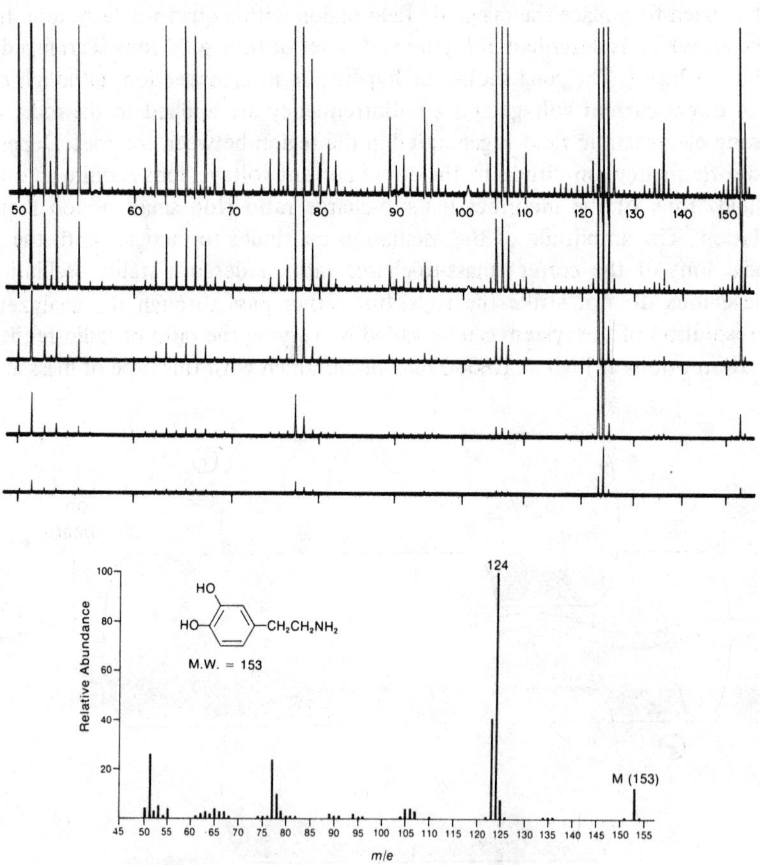

FIGURE 6–6 Partial Mass Spectrum of Dopamine

The most abundant ion formed in the ionization chamber gives rise to the tallest peak in the mass spectrum. This peak is called the *base peak*. In the mass spectrum of dopamine, the base peak is indicated at a *m*/*e* value of 124. The relative abundances of all of the other peaks in the spectrum are reported as percentages of the abundance of the base peak.

TABLE 6-1 Partial Mass Spectrum of Dopamine

m/e	Relative Abundance	*m/e*	Relative Abundance	*m/e*	Relative Abundance
50	4.00	76	1.48	114	0.05
50.5	0.05	77	24.29	115	0.19
51	25.71	78	10.48	116	0.24
51.5	0.19	79	2.71	117	0.24
52	3.00	80	0.81	118	0.14
52.5	0.62	81	1.05	119	0.19
53	5.43	82	0.67	120	0.14
53.5	0.19	83	0.14	121	0.24
54	1.00	84	0.10	122	0.71
55	4.00	85	0.10	123	41.43
56	0.43	86	0.14	124	100.00 (base peak)
56.5	0.05 (metastable peak)	87	0.14	125	7.62
57	0.33	88	0.19	126	0.71
58	0.10	89	1.57	127	0.10
58.5	0.05	89.7	0.10 (metastable peak)	128	0.10
59	0.05	90	0.57	129	0.10
59.5	0.05	90.7	0.10 (metastable peak)	131	0.05
60	0.10	91	0.76	132	0.19
60.5	0.05	92	0.43	133	0.14
61	0.52	93	0.43	134	0.52
61.5	0.10	94	1.76	135	0.52
62	1.57	95	1.43	136	1.48
63	3.29	96	0.52	137	0.33
64	1.57	97	0.14	138	0.10
65	3.57	98	0.05	139	0.10
65.5	0.05	99	0.05	141	0.19
66	3.14	100.6	0.19 (metastable peak)	142	0.05
66.5	0.14	101	0.10	143	0.05
67	2.86	102	0.14	144	0.05
67.5	0.10	103	0.24	145	0.05
68	0.67	104	0.76	146	0.05
69	0.43	105	4.29	147	0.05
70	0.24	106	4.29	148	0.10
71	0.19	107	3.29	149	0.24
72	0.05	108	0.43	150	0.33
73	0.14	109	0.48	151	1.00
74	0.67	110	0.86	152	0.38
74.5	0.05	111	0.10	153	13.33 (molecular ion)
75	1.00	112	0.05	154	1.48
75.5	0.14	113	0.05	155	0.19

As was mentioned previously, the beam of electrons in the ionization chamber converts some of the sample molecules into positive ions. Some of these types of ions are of sufficient importance to warrant a further examination. The simple removal of an electron from a molecule yields an ion whose weight is the actual molecular weight of the original molecule. This ion is the *molecular ion*, which is frequently symbolized by M^{+}. The value of *m/e* at which the molecular ion appears on the mass spectrum, assuming that the ion has only one electron missing, gives the molecular weight of the original molecule. If one is able to identify the molecular ion peak in the mass spectrum, one is able to use the spectrum to determine the molecular weight of an unknown substance. Ignoring heavy isotopes for the moment, the molecular ion peak is the heaviest peak in the mass spectrum. The molecular ion peak has been indicated in the graphed presentation of the mass spectrum in Figure 6–6. Strictly speaking, the molecular ion is a *radical-cation* since it contains an unpaired electron as well as a positive charge.

Molecules which occur in nature do not occur as isotopically pure species. Virtually all atoms have heavier isotopes which occur in varying natural abundances. Hydrogen occurs largely as ^{1}H, but a few per cent of hydrogen atoms are the isotope ^{2}H. Further, carbon normally occurs as ^{12}C, but a few per cent of carbon atoms are the heavier isotope, ^{13}C. With the possible exception of fluorine and a few other elements, most other elements have a certain percentage of heavier isotopes occurring naturally. Peaks caused by ions bearing these heavier isotopes are also found in the mass spectrum. The relative abundances of these isotopic peaks are proportional to the abundances of the isotopes in nature. Most often, the isotopes occur at one or two mass units above the mass of the "normal" atom. Therefore, besides looking for the molecular ion (M^+) peak, one also attempts to locate the *M+1* and *M + 2* peaks. As will be demonstrated in Section 6.4, the relative abundances of these M + 1 and M + 2 peaks can be used to determine the molecular formula of the substance being studied. In Figure 6–6, the isotopic peaks are the low intensity peaks at *m/e* values which are higher than the molecular ion peak.

We have seen that the beam of electrons in the ionization chamber can produce the molecular ion. This beam is also sufficiently powerful to break some of the bonds in the molecule, thus producing a series of molecular fragments. Those fragments which are positively charged are also accelerated in the ionization chamber, sent through the analyzer, detected, and recorded on the mass spectrum. These *fragment ions* appear at *m/e* values corresponding to their individual masses. Very often a fragment ion, rather than the parent ion, will be the most abundant ion produced in the mass spectrum. A second means of producing fragment ions occurs if the molecular ion, once it is formed, is so unstable that it disintegrates before it is able to pass into the accelerating region of the ionization chamber. Lifetimes which are less than 10^{-6} second are typical in this type of fragmentation. Those fragments which are charged then appear as fragment ions in the mass spectrum. A great deal of structural information about a substance can be determined from an examination of the fragmentation pattern in the mass spectrum. An examination of some fragmentation patterns for common classes of compounds will be presented in Section 6.5.

Ions having lifetimes of the order of 10^{-6} second will be accelerated in the ionization chamber before they have an opportunity to disintegrate. These ions may disintegrate into fragments while they are passing into the analyzer region of the mass spectrometer. The fragment ions formed at this point will have considerably lower energy than normal ions, since the uncharged portion of the original ion will carry away some of the energy which the ion received as it was accelerated. As a result, the fragment ion produced in the analyzer will follow an abnormal flight path on its way to the detector. This ion will appear at an *m/e* ratio which depends upon its mass as well as the mass of the original ion from which it was formed. Such an ion will give rise to what is termed a ***metastable ion peak*** in the mass spectrum. Metastable ion peaks usually are broad peaks, and frequently they may appear at nonintegral values of *m/e*. The equation which relates the position of the metastable ion peak in the mass spectrum to the mass of the original ion is:

$$m^* = \frac{(m_2)^2}{m_1}$$

where m^* is the apparent mass of the metastable ion in the mass spectrum, m_1 is the mass of the original ion from which the fragment was formed, and m_2 is the mass of the new fragment ion. Metastable ion peaks may be useful in some applications, since the presence of such a peak definitely links two peaks together. Metastable ion peaks can be used to prove a proposed fragmentation pattern or to aid in the solution of structure proof problems.

6.3 MOLECULAR WEIGHT DETERMINATION

In Section 6.1, it was shown that when a beam of high-energy electrons impinges upon a stream of sample molecules, the ionization of electrons from the molecules takes place. The resulting ions, called *molecular ions*, are then accelerated, sent through a magnetic field, and detected. If these molecular ions have a liftetime of at least 10^{-5} seconds, they will reach the detector without breaking into fragments. It then remains to observe the *m/e* ratio which corresponds to the molecular ion in order to determine the molecular weight of the sample molecule.

In practice, molecular weight determination is not quite as easy as the preceding paragraph would suggest. First, one must understand that the value of the mass of any ion accelerated in a mass spectrometer is its true mass, and not its molecular weight using chemical atomic weights. The chemical scale of atomic weights is based on weighted averages of the weights of all of the isotopes of a given element. However, the mass spectrometer is capable of distinguishing between masses of particles bearing the most common isotopes of the elements and particles bearing heavier isotopes. Consequently, the masses which are observed for molecular ions are the masses of the molecules in which every atom is present as its most common isotope.

In the second place, molecules subjected to bombardment by electrons may break apart into fragment ions. As a consequence of this fragmentation, mass spectra can be quite complex, with peaks appearing at a variety of *m/e* ratios. One must be quite careful to be certain that the suspected peak is indeed that of the molecular ion, rather than of a fragment ion. This becomes particularly crucial when the abundance of the molecular ion is low, as in those cases where the molecular ion is rather unstable and fragments easily.

The most important consideration is that the masses of the ions detected in the mass spectrum be measured accurately. An error of only one mass unit in the assignment of mass spectral peaks can render a structure determination impossible.

One method which can be used to confirm that a particular peak corresponds to a molecular ion is to vary the energy of the ionizing electron beam. If the energy of the beam is lowered, the tendency of the molecular ion to fragment is lessened. As a result, the intensity of the molecular ion peak should increase with decreasing electron potential, while the intensities of the fragment ion peaks should decrease.

For a molecular ion peak, certain facts must apply. First, the peak must correspond to the ion of highest mass in the spectrum. Of course, isotopic peaks occur at even higher masses, but these peaks are usually of much lower intensity than the molecular ion peak. At the sample pressures used in most spectral studies, the probability that ions and molecules may collide to form heavier particles is quite low. Second, the ion must have an odd number of electrons. When a molecule is ionized by an electron beam, it loses one electron to become a radical-cation. The charge on such an ion is *one*, thus making it an ion with an odd number of electrons. Third, the ion must be capable of forming the important fragment ions in the spectrum, particularly the fragments of relatively high mass, by loss of logical neutral fragments. These fragmentation processes will be explained in detail in Section 6.5.

The observed abundance of the suspected molecular ion must correspond to expectations based on the assumed molecule structure. Highly branched substances undergo fragmentation very easily. It thus would be unlikely to observe an intense molecular ion peak for a highly branched molecule. The lifetimes of molecular ions vary according to the following generalized sequence:

aromatic compounds > conjugated alkenes > alicyclic compounds >
organic sulfides > unbranched hydrocarbons > mercaptans>
ketones > amines > esters > ethers > carboxylic acids >
branched hydrocarbons > alcohols

Another rule which is sometimes used to verify that a given peak corresponds to the molecular ion is the so-called "nitrogen rule." This rule states that if a compound has an even number of nitrogen atoms (or no nitrogen atoms), its molecular ion will appear at an even mass value. On the other hand, a molecule with an odd number of nitrogen atoms will form a molecular ion with an odd mass. This rule stems from the fact that nitrogen, though it has an even mass, has an odd-numbered valence. Consequently, an extra hydrogen atom is included as a part of the molecule, giving the molecule an odd mass. To illustrate, consider ethylamine, $C_2H_5NH_2$. This substance has one nitrogen atom, and its mass is an odd number (45). On the other hand, ethylenediamine, $H_2N{-}CH_2CH_2{-}NH_2$, has two nitrogen atoms, and its mass is an even number (60).

One must be careful when studying molecules containing chlorine or bromine atoms, since these elements have two commonly occurring isotopes. Chlorine has isotopes of 35 (relative abundance = 75.77%) and 37 (relative abundance = 24.23%); bromine has isotopes of 79 (relative abundance = 50.5%) and 81 (relative abundance = 49.5%). Special caution must be taken when these elements are present not to confuse the molecular ion peak with a peak corresponding to the molecular ion with a heavier halogen isotope present.

In most cases which one is likely to encounter in mass spectrometry, the molecular ion can be observed in the mass spectrum. Once one has been able to identify that peak in the spectrum, the problem of molecular weight determination is solved. However, with molecules which form unstable molecular ions, the molecular ion peak may not be observed. Molecular ions whose lifetimes are less than 10^{-5} seconds break up into fragments before they can be accelerated. The only peaks which are observed in such cases are those due to fragment ions. One is obliged to deduce the molecular weight of the substance from the fragmentation pattern. This deduction is based on known patterns of fragmentation for certain classes of compounds. The fragmentation patterns are discussed in some detail in Section 6.5, but it is possible to provide some idea of the general approach used. Alcohols undergo dehydration very easily. Consequently, the molecular ion loses water (mass = 18) as a neutral fragment before it can be accelerated. In order to determine the mass of an alcohol molecular ion, one locates the heaviest fragment and keeps in mind that it may be necessary to add 18 to its mass. Acetates also undergo loss of acetic acid (mass = 60) easily. If acetic acid is lost, the weight of the molecular ion will be 60 mass units higher than the mass of the heaviest fragment.

The conjugate acids of oxygen and nitrogen compounds are reasonably stable. Spectra run with sample pressures higher than 0.5 mm Hg may show peaks due to ion-molecule collisions. In such collisions, a hydrogen atom is transferred from a molecule to an ion. The resulting ion is then accelerated. Since oxygen compounds form fairly stable oxonium ions and nitrogen compounds form ammonium ions, ion-molecule collisions form peaks in the mass spectrum which appear at one mass unit *higher* than the mass of the molecular ion. The formation of ion-molecule products may be helpful at times in the determination of the molecular weight of an oxygen or nitrogen compound.

Recently introduced techniques, such as *field ionization* and *chemical ionization*, are useful in studying unstable molecular ions. These methods are discussed in detail in Section 6.6.

Perhaps the most important application of high resolution mass spectrometers is in determining very precise molecular weights of substances. One is normally accustomed to thinking of atoms as having integral atomic masses; for example, H = 1, C = 12, and O = 16. However, if one is able to determine atomic masses with sufficient precision, one finds that this is not true. In 1923, Aston discovered that every isotopic mass is characterized by a small "mass defect." The mass of each atom actually differs from a whole mass number by an amount known as the *nuclear packing fraction*. The actual masses of some atoms are given in Table 6–2.

Depending upon the atoms which are contained within a molecule, it is possible for particles of the same nominal mass to have slightly different measured masses when precise mass determinations are possible. To illustrate, a molecule whose molecular weight is 60 could be C_3H_8O, $C_2H_8N_2$, $C_2H_4O_2$, or CH_4N_2O. These species have the following precise masses:

C_3H_8O	60.05754
$C_2H_8N_2$	60.06884
$C_2H_4O_2$	60.02112
CH_4N_2O	60.03242

TABLE 6-2 Precise Masses of Some Common Elements

Element	Atomic Weight	Nuclide	Mass
Hydrogen	1.00797	^{1}H	1.00783
		^{2}H	2.01410
Carbon	12.01115	^{12}C	12.0000
		^{13}C	13.00336
Nitrogen	14.0067	^{14}N	14.0031
		^{15}N	15.0001
Oxygen	15.9994	^{16}O	15.9949
		^{17}O	16.9991
		^{18}O	17.9992
Fluorine	18.9984	^{19}F	18.9984
Silicon	28.086	^{28}Si	27.9769
		^{29}Si	28.9765
		^{30}Si	29.9738
Phosphorus	30.974	^{31}P	30.9738
Sulfur	32.064	^{32}S	31.9721
		^{33}S	32.9715
		^{34}S	33.9679
Chlorine	35.453	^{35}Cl	34.9689
		^{37}Cl	36.9659
Bromine	79.909	^{79}Br	78.9183
		^{81}Br	80.9163
Iodine	126.904	^{127}I	126.9045

The observation of a molecular ion with a mass of 60.058 would establish that the unknown molecule was C_3H_8O. An instrument with a resolution of about 5320 would be required to distinguish among these peaks. This is well within the capability of modern mass spectrometers. Modern instruments can attain resolutions greater than one part in 20,000. A precise molecular weight determination can also be used to provide information about the molecular formula of a substance under study.

It is interesting to compare the precision of molecular weight determinations by mass spectrometry with the chemical methods described in Chapter 1, Section 1.2. The chemical methods could give results which were accurate to only two or three significant figures (± 0.1 to 1%). Molecular weights by mass spectrometry can be determined with an accuracy of about ± 0.005%. Clearly, mass spectrometry is much more precise than chemical methods of molecular weight determination.

6.4 MOLECULAR FORMULAS FROM ISOTOPE RATIO DATA

In the previous section, a method of determining molecular formulas using data available from high resolution mass spectrometers was described. Another method which can be used to determine molecular formulas is by examining the relative intensities of the peaks due to the molecular ion and related ions which bear one or more heavy isotopes. The advantage of this latter method is that it is much less costly. A high resolution instrument is not required. Unfortunately, the isotopic peaks may be difficult to locate in the mass spectrum. Furthermore, this method is useless when the molecular ion peak is very weak or when it does not appear. The results obtained by this method may at times be somewhat ambiguous.

To illustrate how one may determine a molecular formula from a comparison of the intensities of mass spectral peaks of the molecular ion and ions bearing heavier isotopes, let us use the example of ethane. Ethane, C_2H_6, has a molecular weight of 30 when the most common isotopes of carbon and hydrogen are present in the molecule. For ethane, the molecular ion peak would appear at a position in the spectrum corresponding to a mass of 30. However, occasionally, in a sample of ethane, one may observe a molecule in which one of the carbon atoms is a heavy isotope of carbon, ^{13}C. This particular molecule would appear in the mass spectrum at a mass of 31. The relative abundance of ^{13}C in nature is 1.08% of the ^{12}C atoms. In all the tremendous number of molecules in a sample of ethane gas, either of the carbon atoms of ethane will be a ^{13}C atom 1.08% of the time. Since there are two carbon atoms in ethane, a molecule of mass 31 will appear (2 × 1.08) or 2.16% of the time. One would expect to observe a peak of mass 31 with an intensity of 2.16% of the molecular ion peak intensity. This mass 31 peak is called the M + 1 peak, since its mass is one unit higher than that of the molecular ion.

The observant reader may notice that a particle of mass 31 could be formed in another manner. If one of the hydrogen atoms of ethane were replaced by a deuterium atom, ^2H, the molecule would also have a mass of 31. The natural abundance of deuterium is only 0.016% of the abundance of ^1H atoms. The intensity of the M + 1 peak would be only (6 × 0.016) or 0.096% of the intensity of the molecular ion peak, if one considered only contributions due to deuterium. When these contributions are added to those of ^{13}C, the observed intensity of the M + 1 peak of 2.26% of the intensity of the molecular ion peak is obtained.

A peak of mass 32 could be formed if both of the carbon atoms were replaced by ^{13}C atoms simultaneously. The probability that a molecule of the formula $^{13}C_2H_6$ would appear in a natural sample of ethane is (1.08 × 1.08)/100 or 0.01%. A peak which appears at two mass units higher than the mass of the molecular ion peak is called the M + 2 peak. The intensity of the M + 2 peak of ethane would be only 0.01% of the intensity of the molecular ion peak. The contribution due to two deuterium atoms replacing hydrogen atoms would be (0.016 × 0.016)/100 = 0.00000256%, a negligible amount.* To assist in the determination of the ratios of molecular ion, M + 1, and M + 2 peaks, the natural abundances of some common elements and their isotopes are given in Table 6-3.

*The formula for calculating the intensity of the M + 1 peak is:

$$\%\,(M+1) = 100\,\frac{(M+1)}{M} = 1.1 \times \text{number of C atoms} + 0.016 \times \text{number of H atoms} + 0.38 \times \text{number of N atoms} + \text{etc.}$$

The formula for calculating the approximate intensity of the M + 2 peak is:

$$\%\,(M+2) = 100\,\frac{(M+2)}{M} \cong \frac{(1.1 \times \text{number of C atoms})^2}{200} + \frac{(0.016 \times \text{number of H atoms})^2}{200} + 0.20 \times \text{number of O atoms}$$

These formulas are of limited usefulness unless one already is certain of the actual molecular formula. In practice, contributions due to hydrogen isotopes are not included in the calculation. As result, the actual intensities of the M + 1 and M + 2 peaks may be slightly greater than calculated. The exact formula for calculating the intensity of the M + 2 peak is more complex than is shown here (see the reference by Beynon). The formula given above tends to give better agreement with the observed intensity for compounds of high mass.

TABLE 6-3 Natural Abundances of Common Elements and Their Isotopes

Element	Natural Abundance		
Hydrogen	^{1}H 100	^{2}H 0.016	
Carbon	^{12}C 100	^{13}C 1.08	
Nitrogen	^{14}N 100	^{15}N 0.38	
Oxygen	^{16}O 100	^{17}O 0.04	^{18}O 0.20
Fluorine	^{19}F 100		
Silicon	^{28}Si 100	^{29}Si 5.10	^{30}Si 3.35
Phosphorus	^{31}P 100		
Sulfur	^{32}S 100	^{33}S 0.78	^{34}S 4.40
Chlorine	^{35}Cl 100		^{37}Cl 32.5
Bromine	^{79}Br 100		^{81}Br 98.0
Iodine	^{127}I 100		

TABLE 6-4 Isotope Ratios for Propene and Diazomethane

	Relative Intensities		
Compound	M	M + 1	M + 2
C_3H_6	100	3.34	0.05
CH_2N_2	100	1.87	0.01

In order to demonstrate how the intensities of the M + 1 and M + 2 peaks provide a unique value for a given molecular formula, let us consider two molecules of mass 42, propene (C_3H_6) and diazomethane (CH_2N_2). For propene, the intensity of the M + 1 peak should be $(3 \times 1.08) + (6 \times 0.016) = 3.34\%$. The intensity of the M + 2 peak would be 0.05%. The natural abundance of ^{15}N isotopes of nitrogen is 0.38% of the abundance of ^{14}N atoms. In diazomethane, one would expect the relative intensity of the M + 1 peak to be $1.08 + (2 \times 0.016) + (2 \times 0.38)$ or 1.87% of the intensity of the molecular ion peak. The intensity of the M + 2 peak would be 0.01% of the intensity of the molecular ion peak. Table 6–4 summarizes these intensity ratios. As can be seen, the two molecules have the same molecular weight, but the relative intensities of the M + 1 and M + 2 peaks which they yield are quite different. As an additional illustration, Table 6–5 compares the ratios of the molecular ion, M + 1, and M + 2 peaks for three substances of mass = 28, carbon monoxide, nitrogen, and ethene. Again, it should be noticed that the relative intensities of the M + 1 and M + 2 peaks provide a means of distinguishing among these molecules.

TABLE 6-5 Isotope Ratios for CO, N_2 and C_2H_4

	Relative Intensities		
Compound	M	M + 1	M + 2
CO	100	1.12	0.2
N_2	100	0.76	
C_2H_4	100	2.23	0.01

As molecules become larger and more complex, the number of possible combinations which will yield M + 1 and M + 2 peaks becomes greater. For a particular combination of atoms, the intensities of these peaks relative to the intensity of the molecular ion peak are unique. Thus the isotope ratio method can be used to establish the molecular formula of a compound. Tables of possible combinations of carbon, hydrogen, oxygen, and nitrogen and intensity ratios for the M + 1 and M + 2 peaks for each combination have been developed. Appendix 4 and the references by Silverstein, Bassler, and Morrill and by Beynon listed at the end of this chapter contain extensive tables of this sort. For a given molecular weight, one can examine the tables to find which molecular formula corresponds to the isotope ratios observed.

For atoms other than carbon, hydrogen, oxygen, and nitrogen, it is necessary to calculate the intensities of the M + 1 and M + 2 peaks expected for a particular molecular formula. In these cases, one must establish the presence of these other elements by means other than mass spectrometry.

When chlorine or bromine is present, the M + 2 peak becomes very significant. The heavy isotope of each of these elements is two mass units heavier than the lighter isotope. The natural abundance of ^{37}Cl is 32.5% that of ^{35}Cl; the natural abundance of ^{81}Br is 98.0% that of ^{79}Br. When these elements are present, the M + 2 peak becomes quite intense. If a compound contains two chlorine or bromine atoms, a quite distinct M + 4 peak should be observed, as well as an intense M + 2 peak. In these cases, one should exercise caution in identifying the molecular ion peak in a mass spectrum. The mass spectral properties of the organic halogen compounds are discussed in greater detail in Section 6.5–M. Table 6–6 gives the relative intensities of isotope peaks for various combinations of bromine and chlorine atoms. These are also illustrated in Figure 6–7.

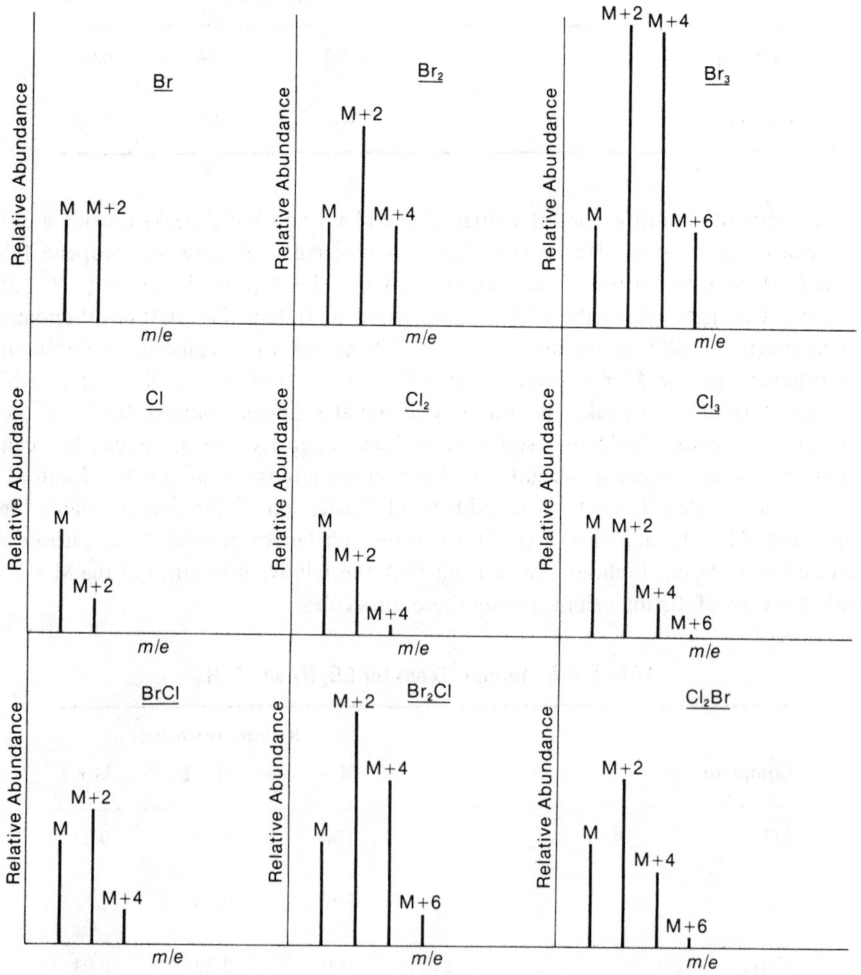

FIGURE 6–7 Mass Spectra Expected for Various Combinations of Bromine and Chlorine.

TABLE 6–6 Relative Intensities of Isotope Peaks for Various Combinations of Bromine and Chlorine

	Relative Intensities			
Halogen	M	M + 2	M + 4	M + 6
Br	100	97.7		
Br_2	100	195.0	95.4	
Br_3	100	293.0	286.0	93.4
Cl	100	32.6		
Cl_2	100	65.3	10.6	
Cl_3	100	97.8	31.9	3.47
BrCl	100	130.0	31.9	
Br_2Cl	100	228.0	159.0	31.2
Cl_2Br	100	163.0	74.4	10.4

6.5 SOME FRAGMENTATION PATTERNS

When a molecule has been bombarded by high-energy electrons in the ionization chamber of a mass spectrometer, besides losing one electron to form an ion, the molecule also absorbs some of the energy transferred in the collision between the molecule and the incident electrons. This extra energy places the molecular ion in an excited vibrational state. The vibrationally excited molecular ion may be unstable, and it may lose some of this extra energy by breaking apart into fragments. If the lifetime of the molecular ion is greater than 10^{-5} second, a peak corresponding to the molecular ion will be observed in the mass spectrum. However, those molecular ions with lifetimes less than 10^{-5} second will break apart into fragments before they are accelerated within the ionization chamber. In such cases, peaks corresponding to the mass-to-charge ratios for these fragments will be observed in the mass spectrum. For a given compound, not all of the molecular ions formed by ionization have precisely the same lifetime. The ions have a distribution of lifetimes; some individual ions may have shorter lifetimes than others. As a result, one usually observes peaks due to both the molecular ion and the fragments in a spectrum.

For most classes of compounds, the mode of fragmentation is somewhat characteristic. It is possible to predict what type of fragmentation a molecule will undergo. In this section, some of the more important modes of fragmentation will be discussed.

Before discussing the fragmentation of each class of compounds, it will be helpful to describe some general principles which govern fragmentation processes. To begin, the ionization of the sample molecule forms a molecular ion which not only carries a positive charge but which also has an unpaired electron. The molecular ion, then, is actually a radical-cation, and it contains an odd number of electrons.

When fragment ions are formed in the mass spectrometer, they are almost always formed by means of unimolecular processes. The pressure of the sample in the ionization chamber is too low to permit a significant number of bimolecular collisions to occur. Those unimolecular processes which are the most favorable energetically will give rise to the most abundant fragment ions.

The fragment ions which are formed are cations. A great deal of the chemistry of these fragment ions can be explained in terms of what is known about carbonium ions in solution. For example, alkyl substitution stabilizes fragment ions (and promotes their formation) in much the same way that it stabilizes carbonium ions. Those fragmentation processes which lead to more stable ions will be favored over processes which lead to the formation of less stable ions.

Often, fragmentation involves the loss of an electrically neutral fragment. This neutral fragment does not appear in the mass spectrum, but its existence can be deduced by noting the difference in masses of the fragment ion and the original molecular ion. Again, processes which lead to the formation of a more stable neutral fragment will be favored over those which lead to the formation of less stable neutral fragments.

The most common mode of fragmentation involves the cleavage of one bond. In this process, the odd-electron molecular ion yields an odd-electron neutral fragment and an even-electron fragment ion. The neutral fragment which is lost is a radical, while the ionic fragment is of the carbonium ion type. Cleavages which lead to the formation of more stable carbonium ions will be favored. Thus, the ease of fragmentation to form ions increases in the order:

$$CH_3^+ < RCH_2^+ < R_2CH^+ < R_3C^+ < CH_2{=}CH{-}CH_2^+ < \phi{-}CH_2^+$$

Examples of fragmentation via the cleavage of one bond are:

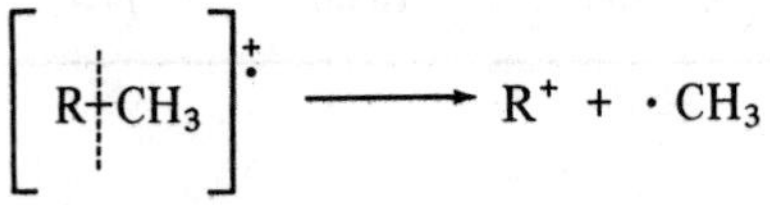

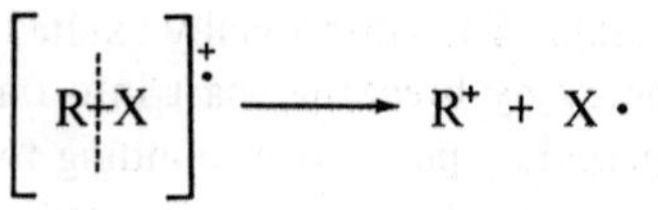

$$\left[R{+}CH_3 \right]^{\ddagger} \longrightarrow R^+ + \cdot CH_3$$

$$\left[\begin{matrix} O \\ \parallel \\ R{+}C{-}R' \end{matrix} \right]^{\ddagger} \longrightarrow \begin{matrix} O \\ \parallel \\ {}^+C{-}R' \end{matrix} + \cdot R$$

$$\left[R{+}X \right]^{\ddagger} \longrightarrow R^+ + X \cdot \quad \begin{matrix} X = \text{halogen, OR, SR, or} \\ NR_2, \text{ where } R = H, \text{ alkyl,} \\ \text{or aryl} \end{matrix}$$

The next most important type of fragmentation involves the cleavage of two bonds. In this process, the odd-electron molecular ion yields an odd-electron fragment ion and an even-electron neutral fragment, usually a small molecule of some type. Examples of this type of cleavage are shown below:

$$\left[\begin{matrix} H \ \ OH \\ | \ \ \ | \\ RCH{-}CHR' \end{matrix} \right]^{\ddagger} \longrightarrow \left[RCH{=}CHR' \right]^{\ddagger} + H_2O$$

$$\left[\begin{matrix} CH_2{-}CH_2 \\ | \ \ \ \ \ | \\ R{-}CH{-}CH_2 \end{matrix} \right]^{\ddagger} \longrightarrow \left[RCH{=}CH_2 \right]^{\ddagger} + CH_2{=}CH_2$$

$$\left[\begin{matrix} \ \ \ \ \ \ \ \ \ \ \ \ \ \ \ O \\ \ \ \ \ \ \ \ \ \ \ \ \ \ \ \ \parallel \\ RCH{-}CH_2{+}O{-}C{-}CH_3 \\ | \\ H \end{matrix} \right]^{\ddagger} \longrightarrow \left[RCH{=}CH_2 \right]^{\ddagger} + \begin{matrix} O \\ \parallel \\ HO{-}C{-}CH_3 \end{matrix}$$

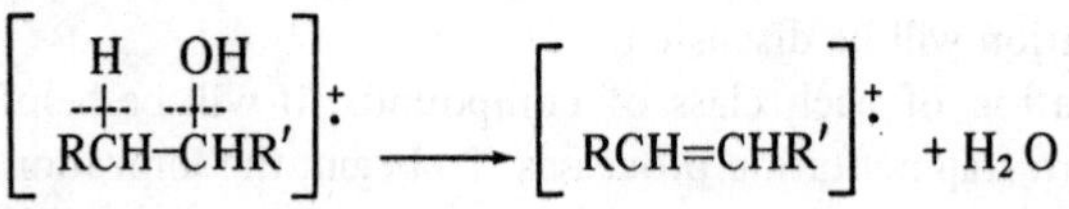

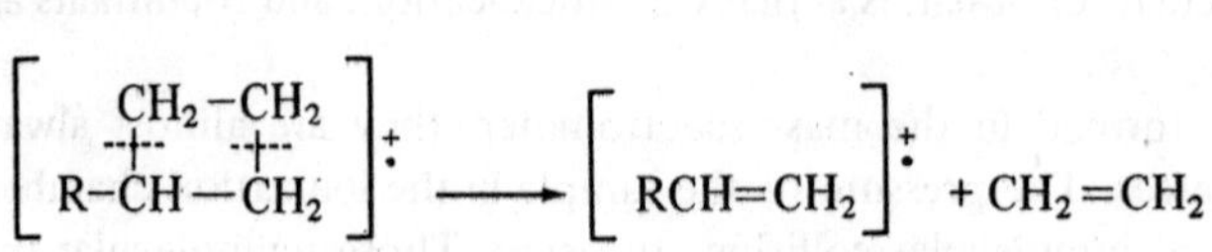

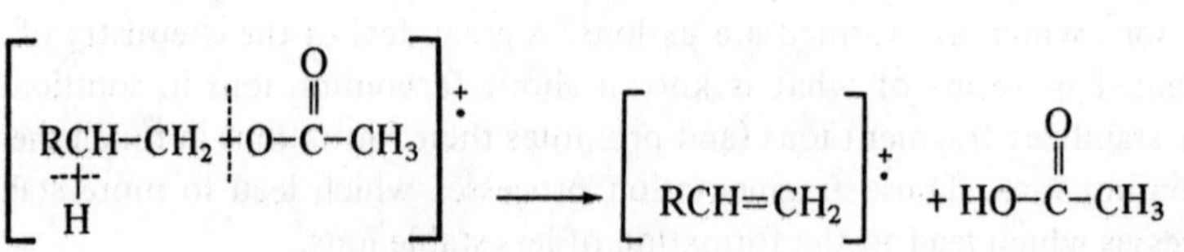

In addition to these processes, fragmentation processes involving rearrangements, migrations of groups, and secondary fragmentations of fragment ions are also possible. These cases occur less often than the two cases described above, and any additional discussion of these modes of fragmentation will be reserved for those compounds in which they are important.

To assist the reader in identifying possible fragment ions, a table listing the molecular formulas for common fragments whose mass is less than 105 is provided in Appendix 5. More complete tables may be found in the reference books by Silverstein, Bassler, and Morrill and by Beynon, listed at the end of this chapter.

A. ALKANES

For saturated hydrocarbons, or organic structures containing a large saturated hydrocarbon skeleton, the methods of fragmentation are quite predictable. What is known about the stabilities of carbonium ions in solution can be used to help us understand the fragmentation patterns of alkanes.

For straight-chain, or "normal," alkanes, a peak corresponding to the molecular ion can be observed, as in the mass spectra of butane (Figure 6–8) and octane (Figure 6–9). As the carbon skeleton becomes more highly branched, the intensity of the molecular ion peak decreases. This effect can be seen easily, when one compares the mass spectrum of butane with that of isobutane (Figure 6–10). In isobutane, the molecular ion peak is much less intense than in butane. A more dramatic illustration of the effect of chain branching on the intensity of the molecular ion peak can be observed by comparing the mass spectra of octane and 2,2,4-trimethylpentane (Figure 6–11). In 2,2,4-trimethylpentane, the molecular ion peak is too weak to be observed, while the molecular ion peak in its straight-chain isomer is quite readily observed. The effect of chain branching on the intensity of the molecular ion peak can be understood by examining the method by which hydrocarbons undergo fragmentation.

Straight-chain hydrocarbons appear to undergo fragmentation by breaking carbon-carbon bonds, resulting in a homologous series of fragmentation products. For example, in the case of butane, cleavage of the carbon-1 to carbon-2 bond results in loss of a methyl radical and in the formation of the propyl carbonium ion (m/e = 43). Cleavage of the carbon-2 to carbon-3 bond results in loss of an ethyl radical and in the formation of the ethyl carbonium ion (m/e = 29). Again, in the case of octane, fragment peaks due to the hexyl ion (m/e = 85), the pentyl ion (m/e = 71), the butyl ion (m/e = 57), the propyl ion (m/e = 43), and the ethyl ion (m/e = 29) are observed. As can be noticed, alkanes fragment to form clusters of peaks which are 14 mass units (corresponding to one CH_2 group) apart from each other. Other fragments within each cluster correspond to additional losses of one or two hydrogen atoms. As can be noticed in the mass spectrum of octane, the three-carbon ions appear to be the most abundant, with the intensities of each cluster uniformly decreasing with increasing fragment weight. Interestingly, for long-chain alkanes, the fragment corresponding to the loss of one carbon atom is generally absent. In the mass spectrum of octane, a seven-carbon fragment would appear at a mass of 99, but it cannot be observed. Straight-chain alkanes have fragments which are always primary carbonium ions. Since these ions are rather unstable, fragmentation is not favored. A significant number of the original molecules survive electron-bombardment without undergoing fragmentation. Consequently, a molecular ion peak of significant intensity is observed.

Cleavage of the carbon-carbon bonds of branched-chain alkanes can lead to secondary or tertiary carbonium ions. These ions are more stable than primary ions, so fragmentation becomes a more favorable process. A greater proportion of the original molecules undergo fragmentation, so the molecular ion peaks of branched-chain alkanes are considerably weaker, or even absent. In isobutane, cleavage of a carbon-carbon bond yields an isopropyl carbonium ion, which is more stable than a normal propyl ion. Isobutane undergoes fragmentation more easily than butane because of the increased stability of the fragmentation products. With 2,2,4-trimethylpentane, the cleavage shown leads to the formation of a *t*-butyl carbonium ion. Since tertiary carbonium ions are the most stable of the saturated alkyl carbonium ions, this cleavage is particularly favorable and accounts for the intense fragment peak at m/e = 57.

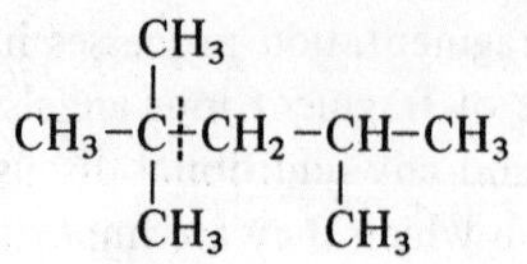

The typical mass spectrum for a cycloalkane shows a relatively intense molecular ion peak. Fragmentation of ring compounds requires the cleavage of two carbon-carbon bonds, which is a more difficult process than cleavage of one such bond. Therefore, a larger proportion of cycloalkane molecules than of acyclic alkane molecules survive electron bombardment without undergoing fragmentation. In the mass spectra of cyclopentane (Figure 6–12) and methylcyclopentane (Figure 6–13), a strong molecular ion peak can be observed.

The fragmentation patterns of cycloalkanes may show mass clusters arranged in a homologous series, as in the alkanes. However, the most significant mode of cleavage of the cycloalkanes involves the loss of a molecule of ethene, either from the parent molecule or from intermediate radical-ions. The peak at m/e = 42 in cyclopentane and the peak at m/e = 56 in methylcyclopentane result from the loss of ethene from the parent molecule. In each case, this fragment peak is the most intense in the mass spectrum.

When the cycloalkane bears a side chain, loss of that side chain is a favorable mode of fragmentation. The fragment peak at m/e = 69 in the mass spectrum of methylcyclopentane is due to the loss of the CH_3 side chain. A secondary carbonium ion results from the loss of the methyl group.

Applying these pieces of information to the mass spectrum of bicyclo[2.2.1]heptane (Figure 6–14), one can identify fragment peaks due to the loss of the side chain (the one-carbon bridge) at m/e = 81 and the loss of ethene at m/e = 68. The fragment ion peak at m/e = 67 is due to the loss of ethene plus an additional hydrogen atom.

Text continued on page 246

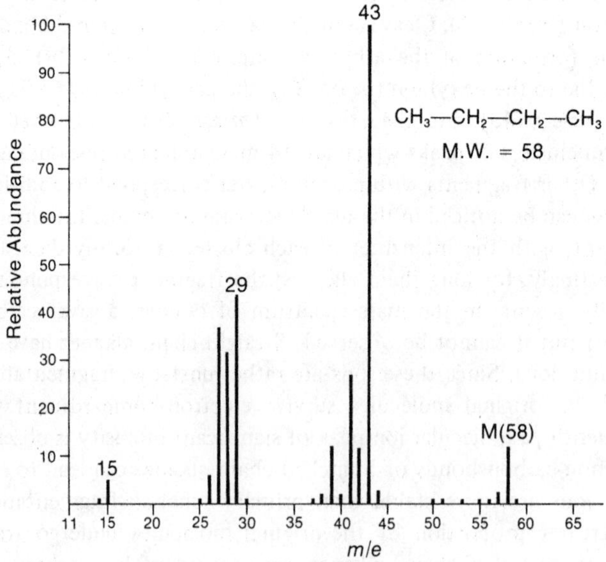

FIGURE 6–8 Mass Spectrum of Butane

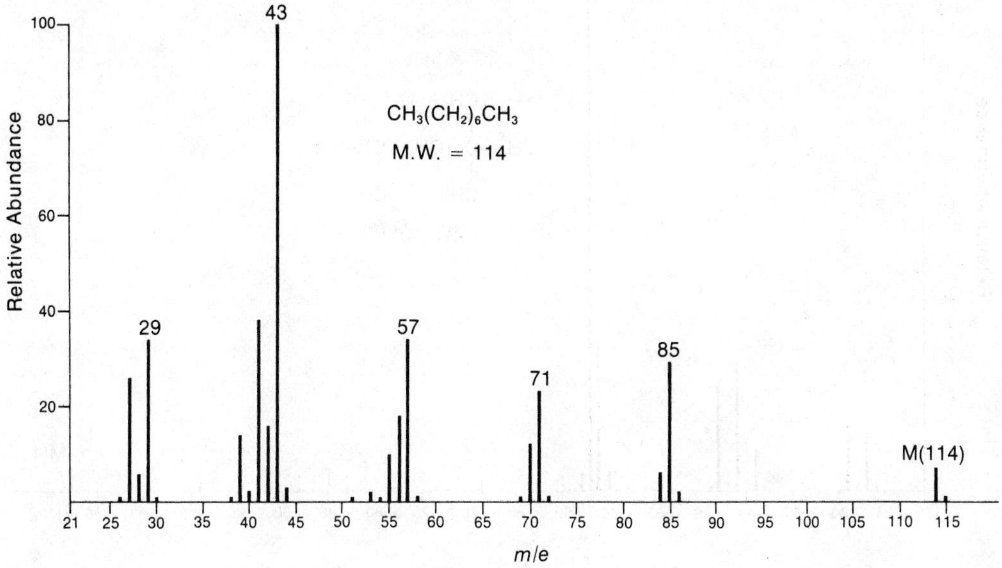

FIGURE 6–9 Mass Spectrum of Octane

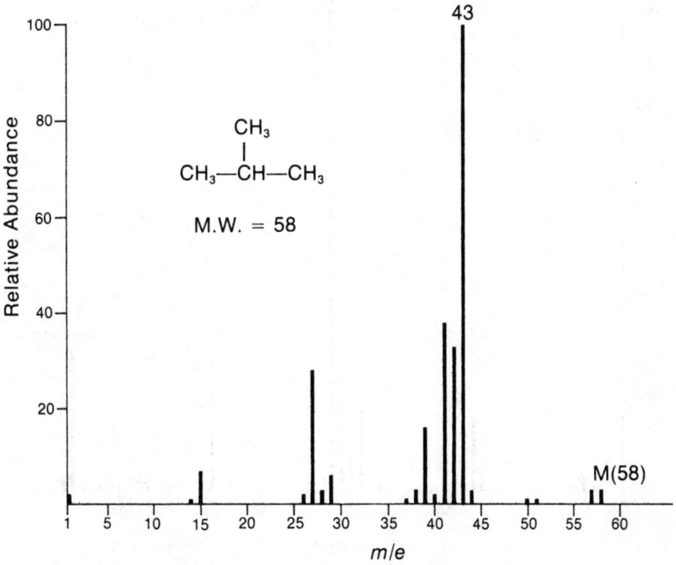

FIGURE 6–10 Mass Spectrum of Isobutane

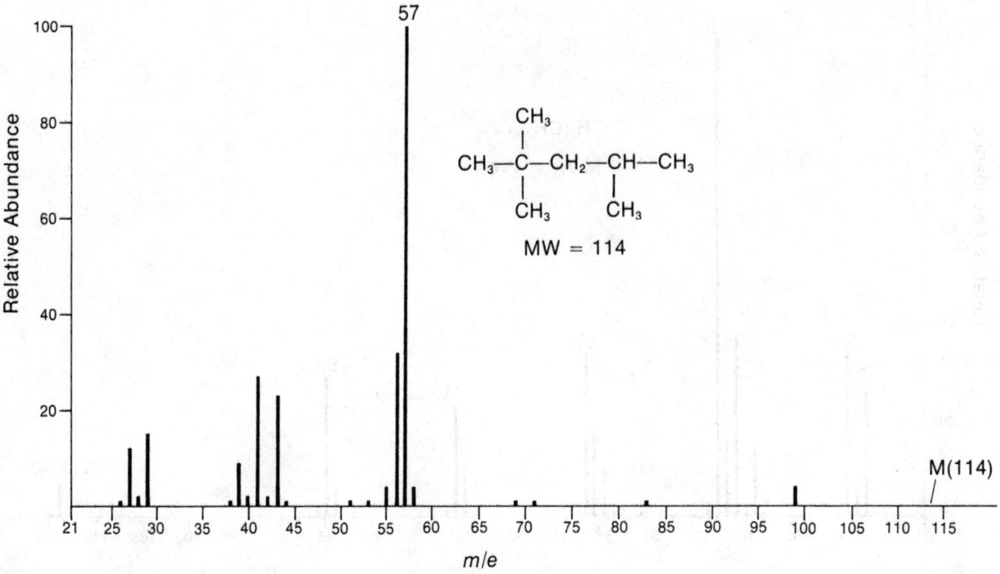

FIGURE 6–11 Mass Spectrum of 2,2,4-Trimethylpentane ("Isooctane")

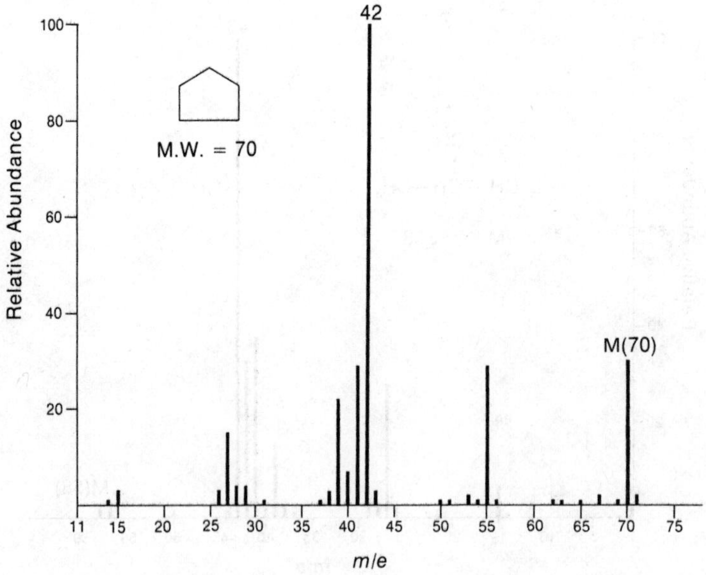

FIGURE 6–12 Mass Spectrum of Cyclopentane

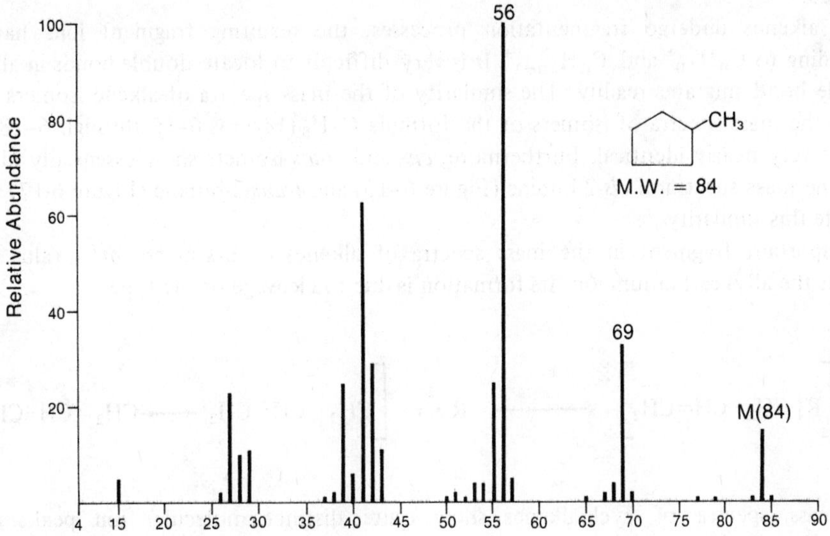

FIGURE 6–13 Mass Spectrum of Methylcyclopentane

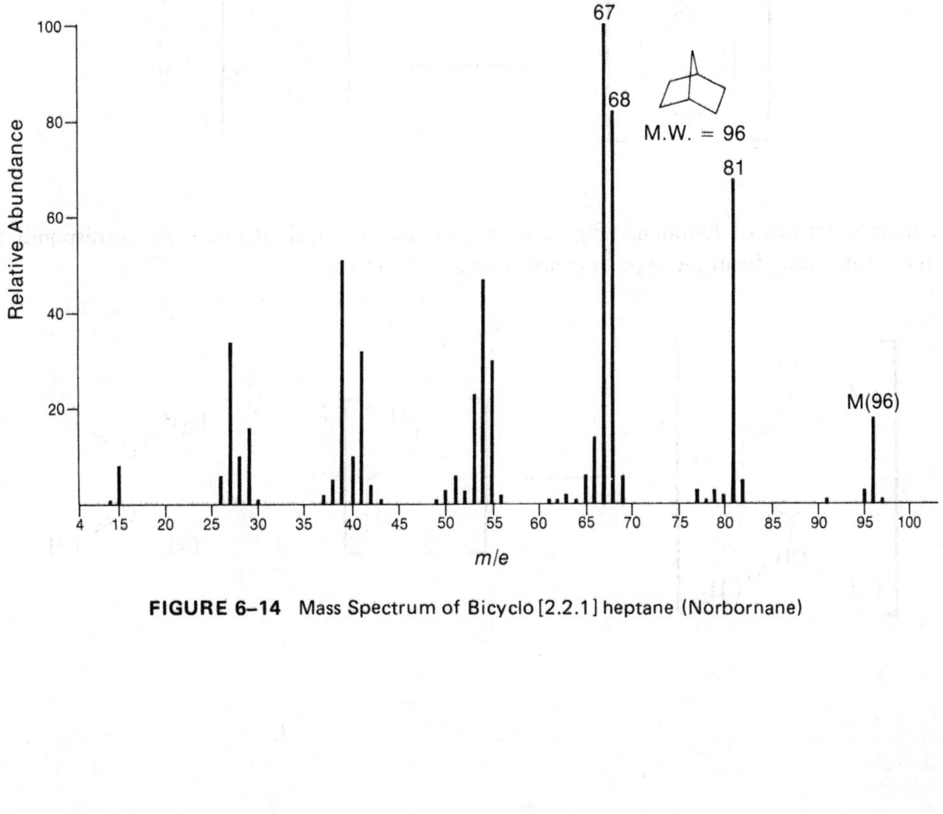

FIGURE 6–14 Mass Spectrum of Bicyclo [2.2.1] heptane (Norbornane)

B. ALKENES

The mass spectra of most alkenes show distinct molecular ion peaks. Apparently, electron bombardment removes one of the electrons in the π-bond, leaving the carbon skeleton relatively undisturbed.

When alkenes undergo fragmentation processes, the resulting fragment ions have formulas corresponding to $C_nH_{2n}^{+}$ and $C_nH_{2n-1}^{+}$. It is very difficult to locate double bonds in alkenes, since the double bond migrates readily. The similarity of the mass spectra of alkene isomers can be seen readily in the mass spectra of isomers of the formula C_4H_8 (Figures 6–15 through 6–18). The mass spectra are very nearly identical. Furthermore, *cis*- and *trans*-isomers show essentially identical mass spectra. The mass spectra of *cis*-2-butene (Figure 6–16) and *trans*-2-butene (Figure 6-17) are included to illustrate this similarity.

An important fragment in the mass spectra of alkenes occurs at an m/e value of 41. This fragment is the allyl carbonium ion. Its formation is due to cleavage of the type:

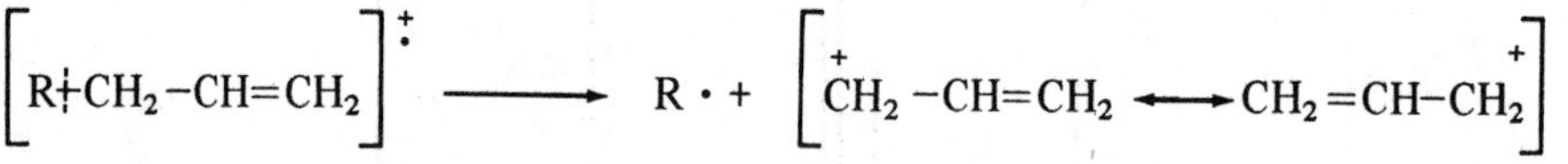

$$\left[R \nmid CH_2{-}CH{=}CH_2\right]^{+\cdot} \longrightarrow R\cdot + \left[\overset{+}{C}H_2{-}CH{=}CH_2 \longleftrightarrow CH_2{=}CH{-}\overset{+}{C}H_2\right]$$

The mass spectra of cycloalkenes show quite distinct molecular ion peaks. For many cycloalkenes, migration of bonds gives mass spectra which are virtually identical. Consequently, it may be impossible to locate the position of the double bond in a cycloalkene, particularly a cyclopentene or a cycloheptene.

Cyclohexenes do have a characteristic fragmentation pattern which corresponds to a reverse Diels-Alder reaction. This cleavage can be illustrated as:

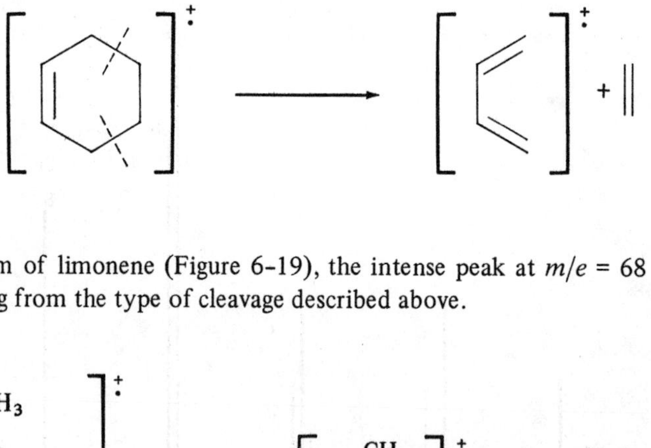

In the mass spectrum of limonene (Figure 6-19), the intense peak at m/e = 68 corresponds to the diene fragment arising from the type of cleavage described above.

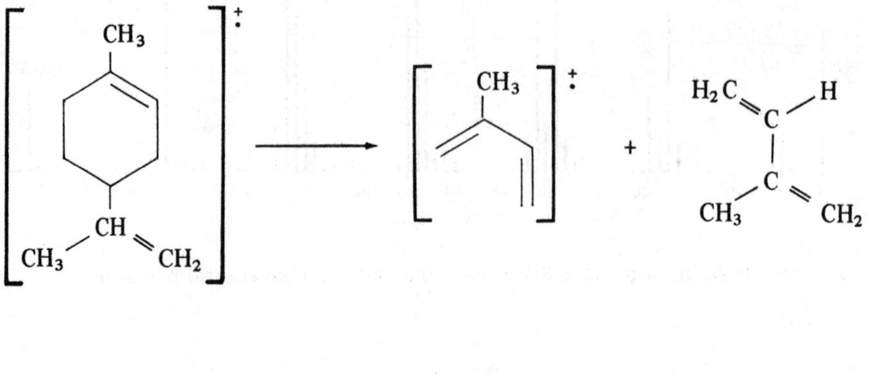

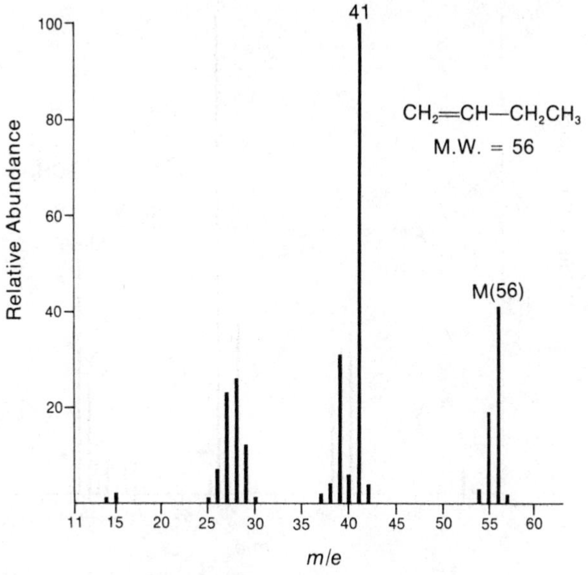

FIGURE 6–15 Mass Spectrum of 1-Butene

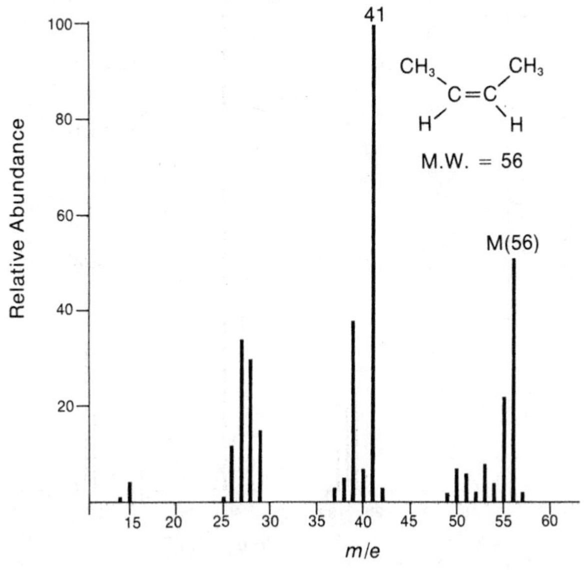

FIGURE 6–16 Mass Spectrum of *cis*-2-Butene

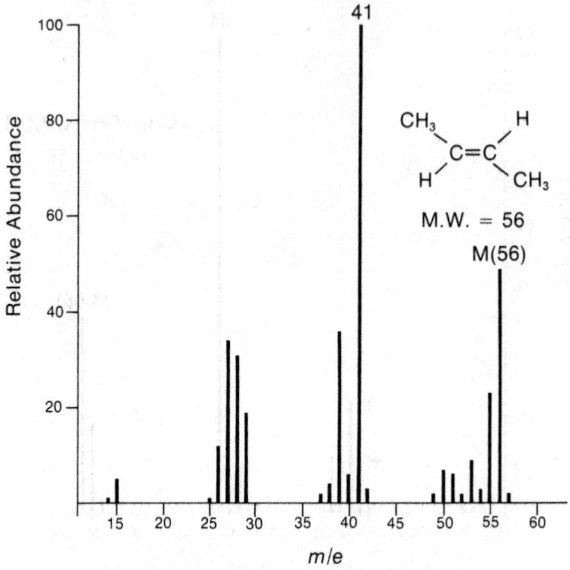

FIGURE 6-17 Mass Spectrum of *trans*-2-Butene

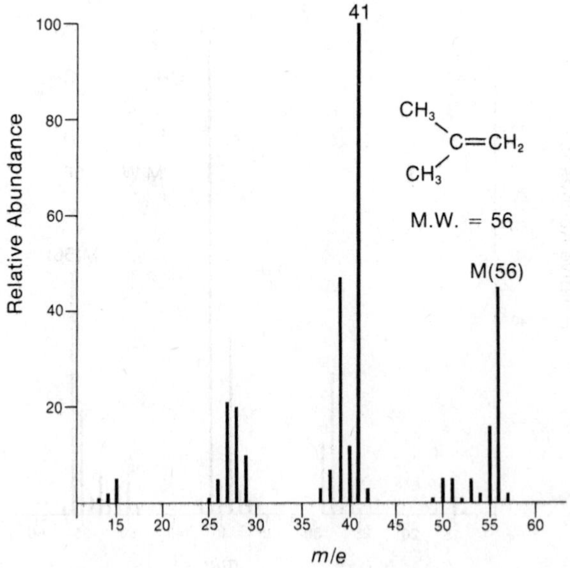

FIGURE 6-18 Mass Spectrum of 2-Methylpropene

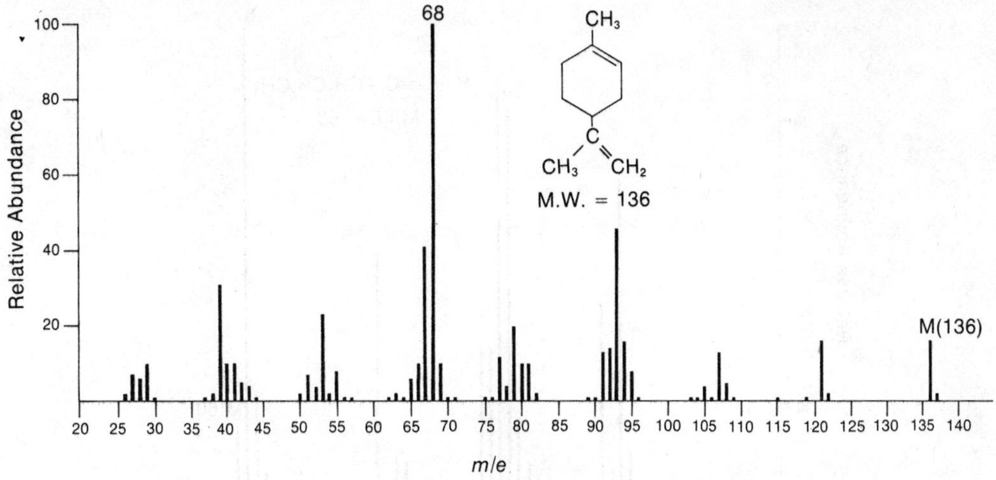

FIGURE 6–19 Mass Spectrum of Limonene

C. ALKYNES

The mass spectra of alkynes are very similar to those of alkenes. The molecular ion peaks tend to be rather intense, and fragmentation patterns parallel those of the alkenes. As can be seen from the mass spectra of propyne (Figure 6–20) and 1-pentyne (Figure 6–21), fragmentation of the type

$$\left[\text{H–C}\equiv\text{C–CH}_2 \dagger \text{R}\right]^{\dot{+}} \longrightarrow \text{R}\cdot + \left[\text{H–C}\equiv\text{C–}\overset{+}{\text{C}}\text{H}_2 \longleftrightarrow \text{H}\overset{+}{\text{C}}\text{=C=CH}_2\right]$$

to yield the propargyl ion (m/e = 39) is important. The formation of the propargyl ion from alkynes is not as important as the formation of the allyl ion from alkenes, since the allyl ion is more stable than the propargyl ion.

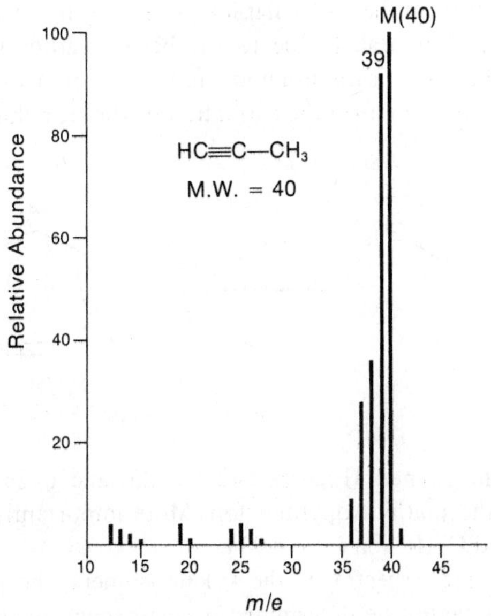

FIGURE 6–20 Mass Spectrum of Propyne

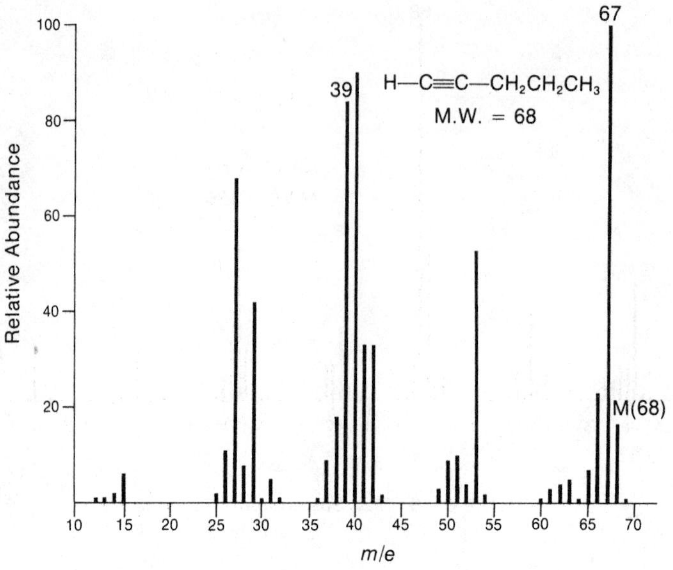

FIGURE 6–21 Mass Spectrum of 1-Pentyne

D. AROMATIC HYDROCARBONS

The mass spectra of most aromatic hydrocarbons show very intense molecular ion peaks. As can be seen from the mass spectrum of benzene (Figure 6–22), fragmentation of the benzene ring requires a great deal of energy. Such fragmentation is not observed to any significant extent.

When an alkyl group is attached to a benzene ring, preferential fragmentation occurs at a benzylic position to form a fragment ion of the formula $C_7H_7^+$ ($m/e = 91$). In the mass spectrum of toluene (Figure 6–23), loss of hydrogen from the molecular ion gives a strong peak at $m/e = 91$. While it may be expected that this fragment ion peak is due to the benzyl carbonium ion, evidence has been amassed which suggests that the benzyl carbonium ion actually rearranges to form the tropylium ion. Isotope labeling experiments tend to confirm the formation of the tropylium ion.

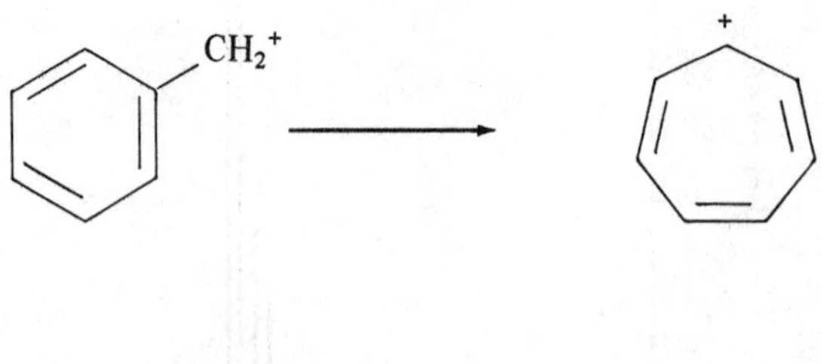

The mass spectra of the xylene isomers (Figures 6-24, 6-25, and 6-26) show a medium peak at $m/e = 105$, which is due to the methyltropylium ion. More importantly, xylene loses one methyl group to form the unsubstituted $C_7H_7^+$ ion ($m/e = 91$).

As can be seen from the mass spectra of the xylene isomers, the position of substitution of polyalkyl substituted benzenes cannot be determined by mass spectrometry. The mass spectra of the three isomers are virtually identical.

The formation of a substituted tropylium ion is typical for alkyl substituted benzenes. In the mass spectrum of isopropylbenzene (Figure 6–27), a strong peak at $m/e = 105$ can be observed. This peak corresponds to loss of a methyl group to form a methyl-substituted tropylium ion:

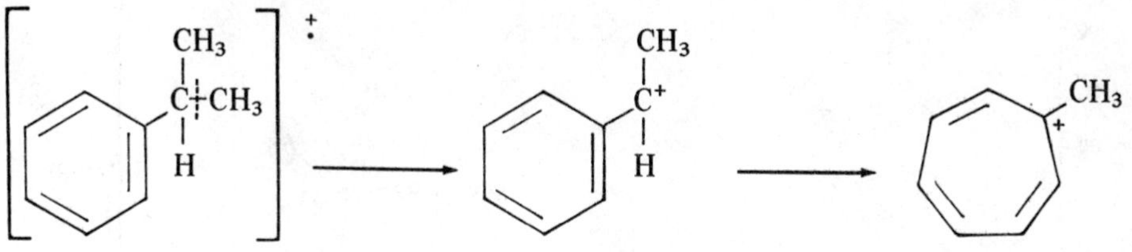

Again in the mass spectrum of propylbenzene (Figure 6–28), a strong peak at $m/e = 91$, due to the tropylium ion, is observed.

When the alkyl group attached to the benzene ring is a propyl group or larger, a type of rearrangement called the *McLafferty rearrangement* occurs. Using butylbenzene as an example, this rearrangement may be depicted:

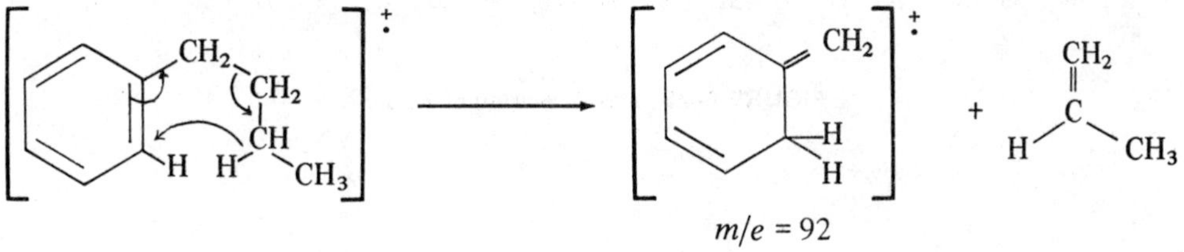

leading to a peak at $m/e = 92$. This peak can also be observed in the mass spectrum of propylbenzene.

Text continued on page 255

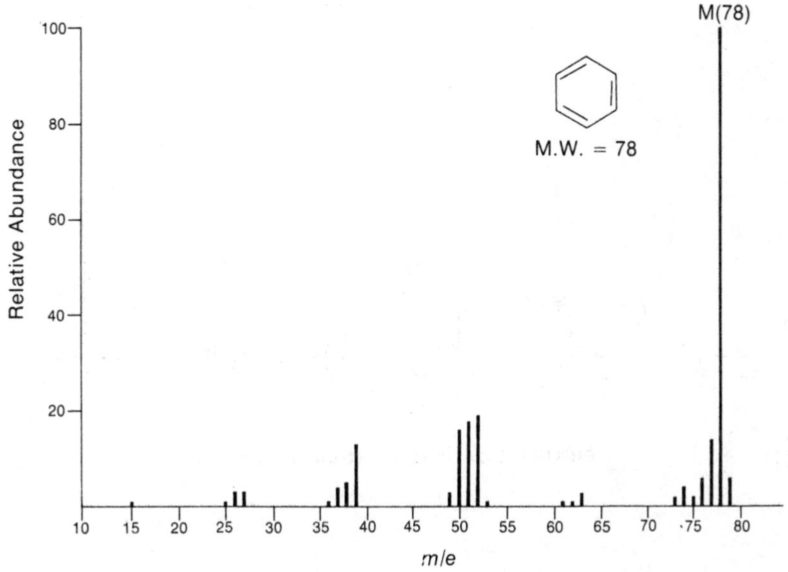

FIGURE 6–22 Mass Spectrum of Benzene

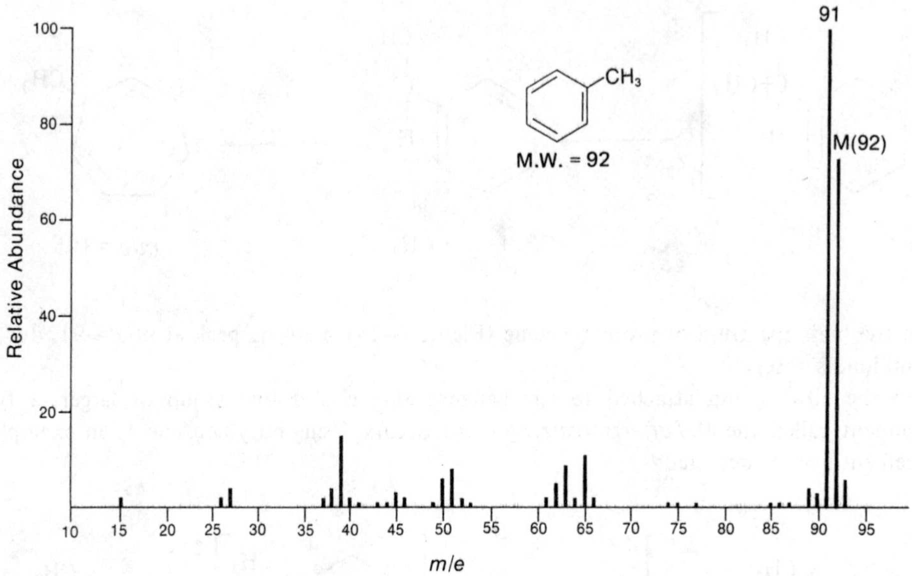

FIGURE 6–23 Mass Spectrum of Toluene

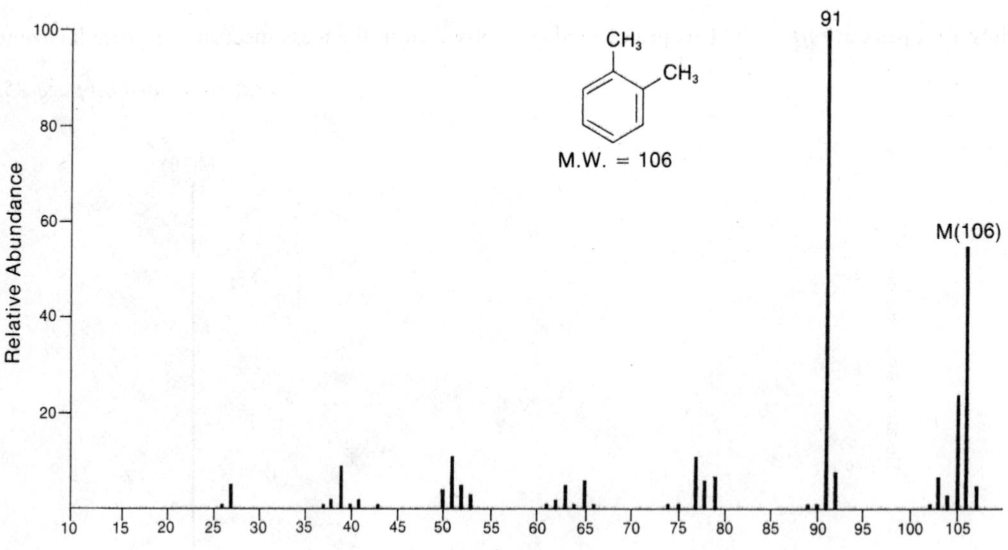

FIGURE 6–24 Mass Spectrum of *o*-Xylene

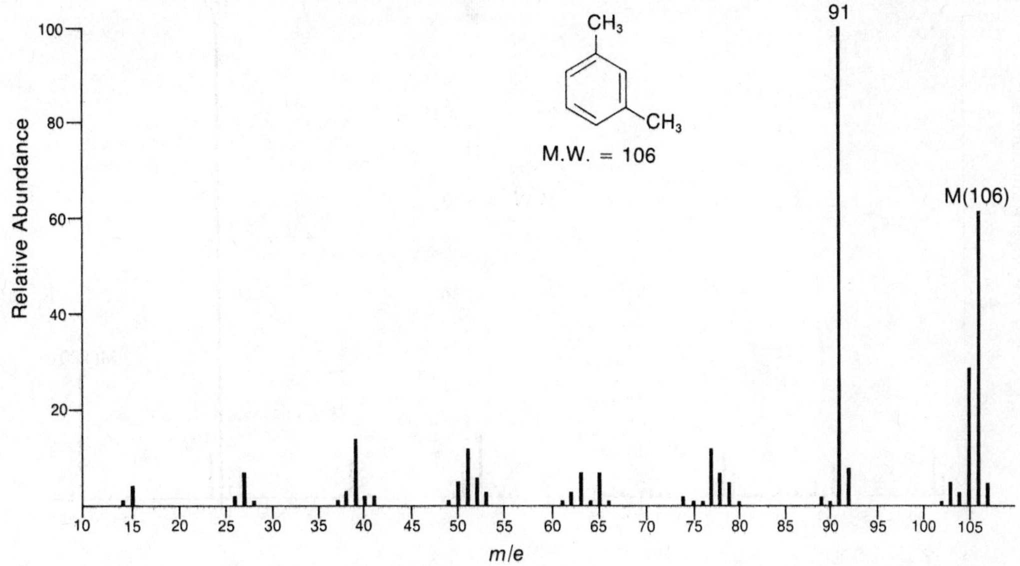

FIGURE 6–25 Mass Spectrum of *m*-Xylene

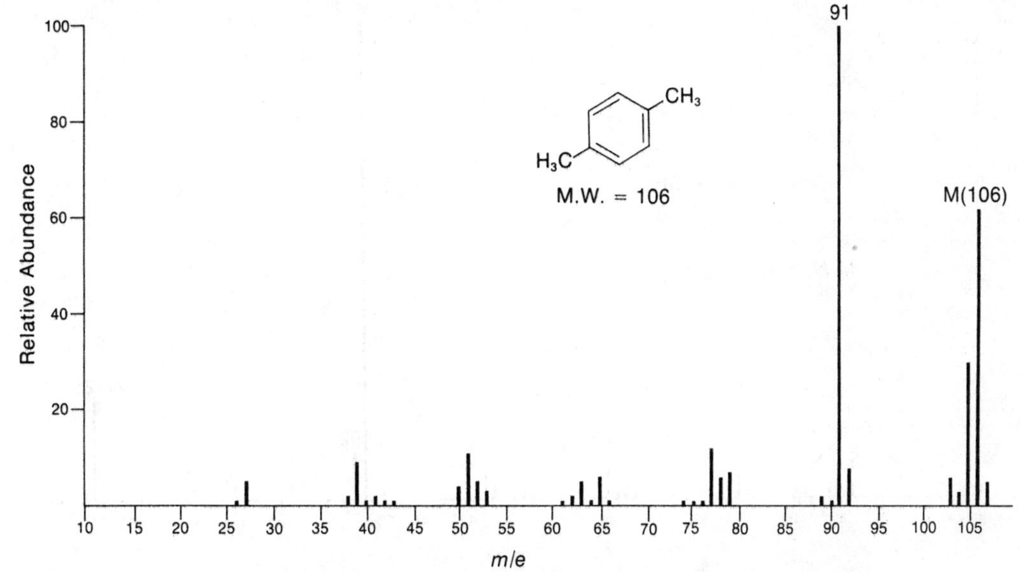

FIGURE 6–26 Mass Spectrum of *p*-Xylene

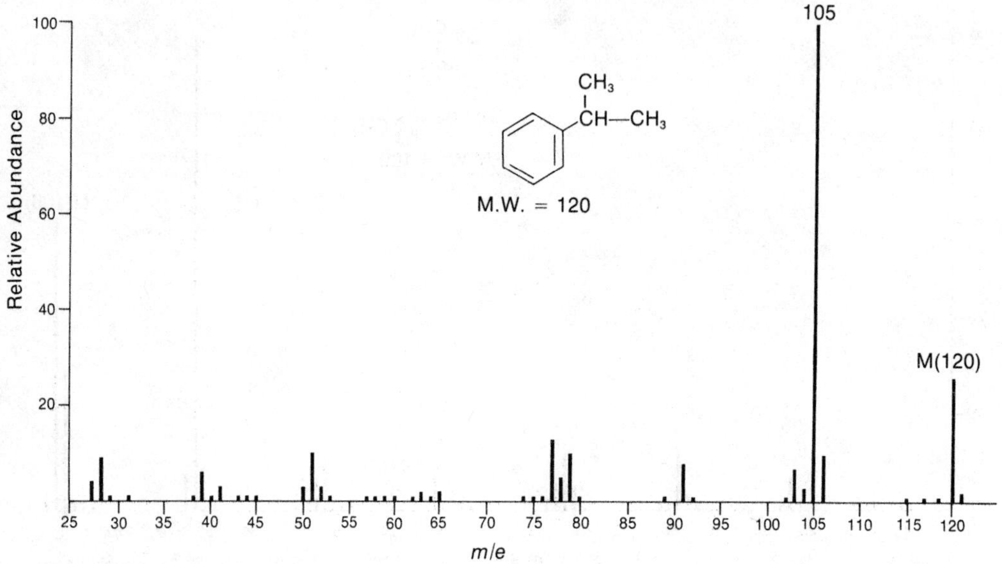

FIGURE 6–27 Mass Spectrum of Isopropylbenzene (Cumene)

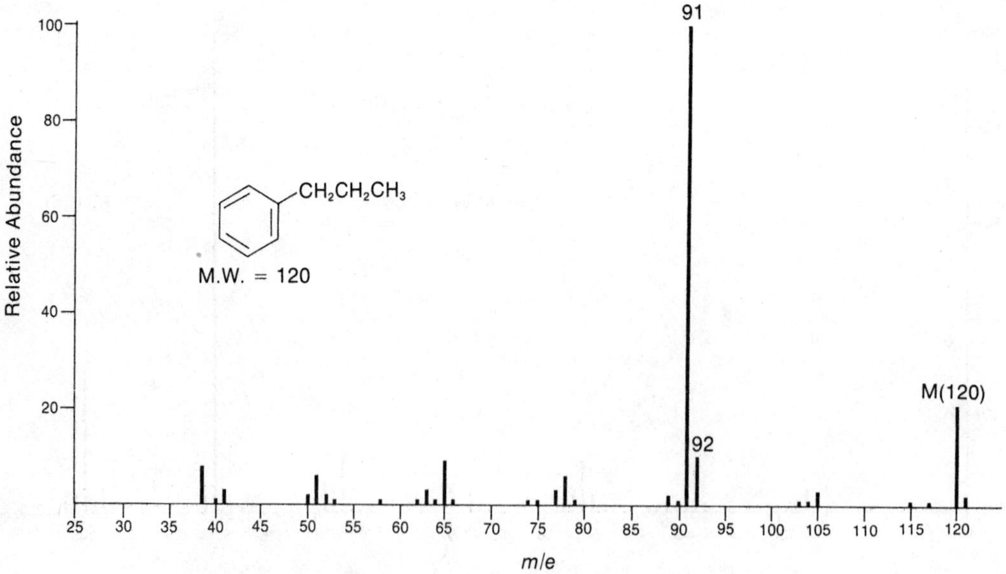

FIGURE 6–28 Mass Spectrum of Propylbenzene

E. ALCOHOLS AND PHENOLS

The intensity of the molecular ion peak in the mass spectrum of a primary or secondary alcohol is usually rather low. The mass spectrum of 1-butanol (Figure 6–29) shows a very weak molecular ion peak at $m/e = 74$, while the mass spectrum of 2-butanol (Figure 6–30) has a molecular ion peak which is too weak to be indicated on the graphical presentation of the spectrum. The molecular ion peak for tertiary alcohols is usually absent, as is seen in the mass spectrum of 2-methyl-2-propanol (Figure 6–31).

The most important fragmentation reaction for alcohols is loss of an alkyl group according to:

$$\left[\begin{array}{c} R' \\ | \\ R{-}\overset{\displaystyle |}{\underset{\displaystyle |}{C}}{-}OH \\ R'' \end{array}\right]^{\stackrel{+}{\cdot}} \longrightarrow R\cdot + \begin{array}{c} R' \\ \diagdown \\ C{=}\overset{+}{O}H \\ \diagup \\ R'' \end{array}$$

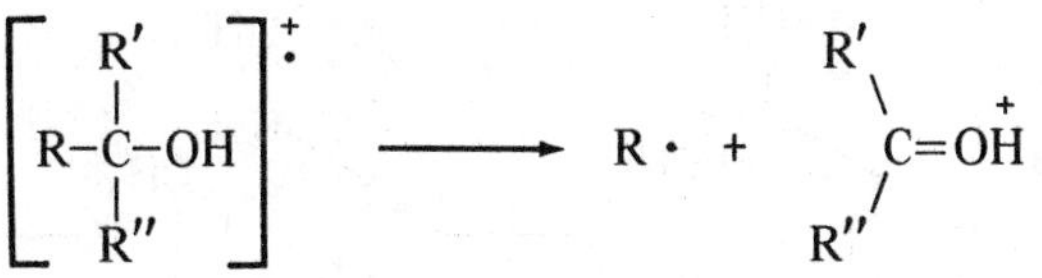

The largest alkyl group is the one which is lost most readily. In the spectrum of 1-butanol (Figure 6–29), the intense peak at $m/e = 31$ is due to the loss of a propyl group to form a $H_2C{=}OH^+$ ion. 2-Butanol (Figure 6–30) loses an ethyl group to form the $CH_3CH{=}OH^+$ fragment at $m/e = 45$. 2-Methyl-2-propanol (Figure 6-31) loses a methyl group to form the $(CH_3)_2C{=}OH^+$ fragment at $m/e = 59$.

A second common mode of fragmentation involves dehydration. The importance of dehydration increases as the chain length of the alcohol increases. While the fragment ion peak resulting from dehydration ($m/e = 56$) is very intense in 1-butanol, it is very weak in the other butanol isomers. However, in the mass spectra of the five-carbon alcohols, this peak due to dehydration of the molecular ion is quite important.

Dehydration may occur by either of two mechanisms. The hot surfaces of the inlet system may stimulate dehydration of the alcohol molecule before the molecule comes in contact with the electrons. In this case, the dehydration is a 1,2-elimination of water. However, the molecular ion, once it is formed, may also lose water. In this case the dehydration is a 1,4-elimination of water *via* a cyclic mechanism:

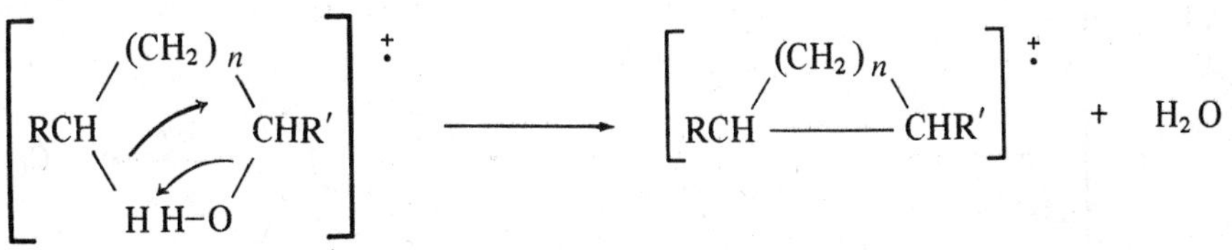

$$\left[\begin{array}{c} \diagup (CH_2)_n \diagdown \\ RCH \qquad\qquad CHR' \\ \diagdown \qquad\qquad \diagup \\ H \ \ H{-}O \end{array}\right]^{\stackrel{+}{\cdot}} \longrightarrow \left[\begin{array}{c} \diagup (CH_2)_n \diagdown \\ RCH {-}\!\!\!-\!\!\!-\!\!\!- CHR' \end{array}\right]^{\stackrel{+}{\cdot}} + H_2O$$

$$n = 1 \text{ or } 2$$

Alcohols containing four or more carbons may undergo the simultaneous loss of both water and ethylene:

$$\left[\begin{array}{c} H \\ | \\ O \\ \diagup \\ CH_2 \qquad H \\ | \qquad | \\ CH_2 \qquad CH{-}R \\ \diagdown \ CH_2 \diagup \end{array}\right]^{\stackrel{+}{\cdot}} \longrightarrow \left[\begin{array}{c} CH{-}R \\ \diagup\!\!\!\diagup \\ CH_2 \end{array}\right]^{\stackrel{+}{\cdot}} + H_2O + CH_2{=}CH_2$$

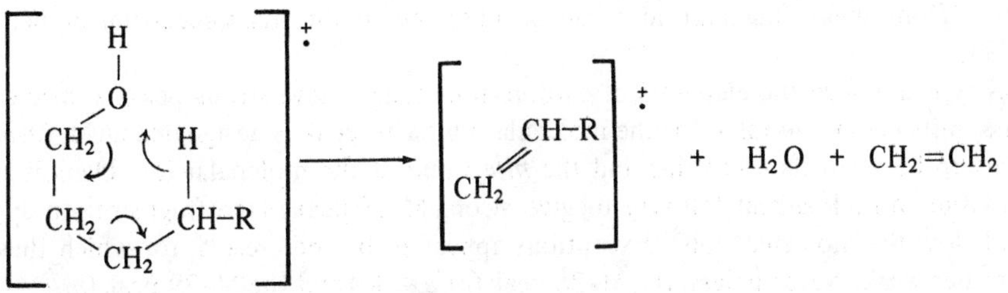

In the case of 1-butanol, the fragment ion is responsible for a rather weak peak at $m/e = 28$. However, with 1-pentanol, the corresponding peak is the most intense in the spectrum.

Cyclic alcohols may undergo fragmentation by at least three different pathways:

1)

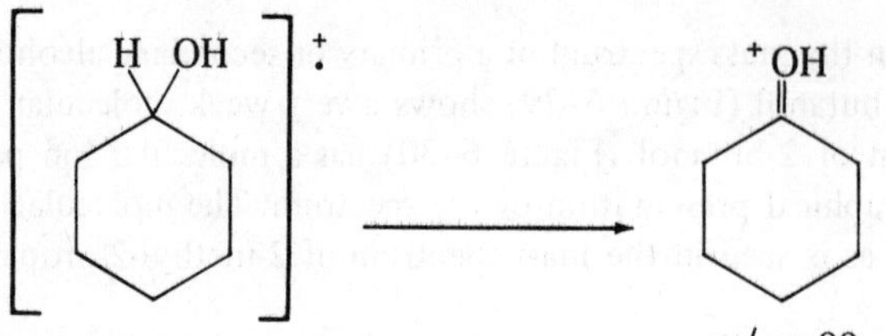

$m/e = 99$

2)

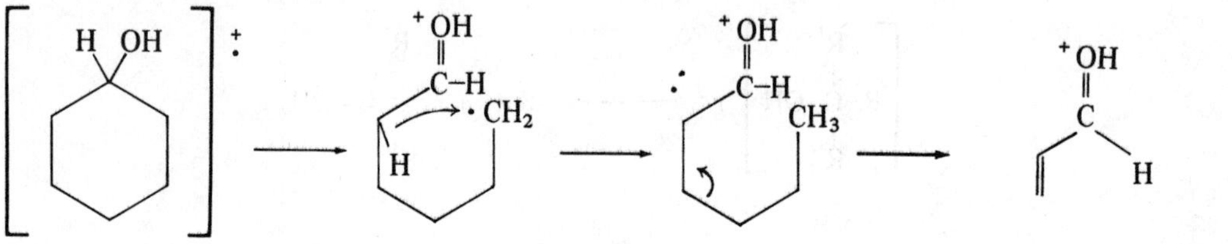

$m/e = 57$ + $C_3H_7\cdot$

3)

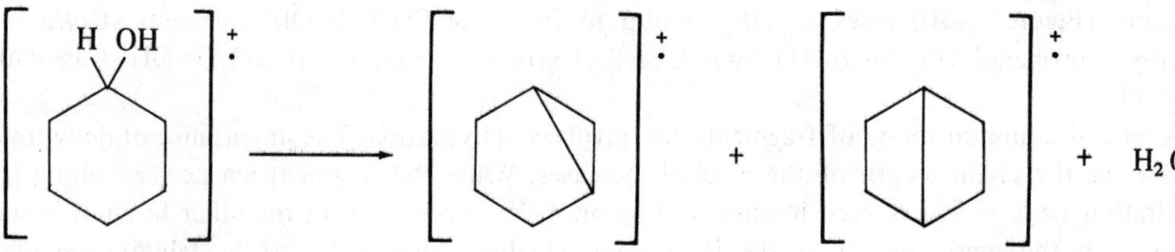

$m/e = 82$

Peaks corresponding to each of these fragment ions may be observed in the mass spectrum of cyclohexanol (Figure 6–32).

Benzylic alcohols exhibit an intense molecular ion peak. Their principal modes of fragmentation are illustrated in the sequence of reactions:

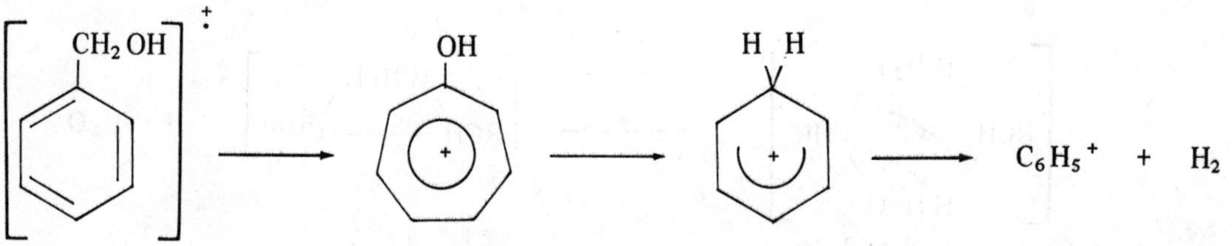

$m/e = 107$ $m/e = 79$ $m/e = 77$

+ H·   + CO

Peaks arising from these fragment ions can be observed in the mass spectrum of benzyl alcohol (Figure 6–33).

Phenols typically lose the elements of carbon monoxide to give strong peaks at m/e values which are 28 mass units below the value for the molecular ion. This peak is designated an M–28 peak, due to the relationship between its m/e value and the m/e value of the molecular ion. Phenols also lose the elements of the formyl radical (HCO·) to give strong M–29 peaks (the fragment ion appears at 29 mass units below the molecular ion). Exceptions appear to be the cresols, for which these fragment peaks are rather weak. Nevertheless, the M–28 peak ($m/e = 80$) and the M–29 peak ($m/e = 79$) may be observed in the mass spectrum of 2-methylphenol (Figure 6–34).

Text continued on page 260

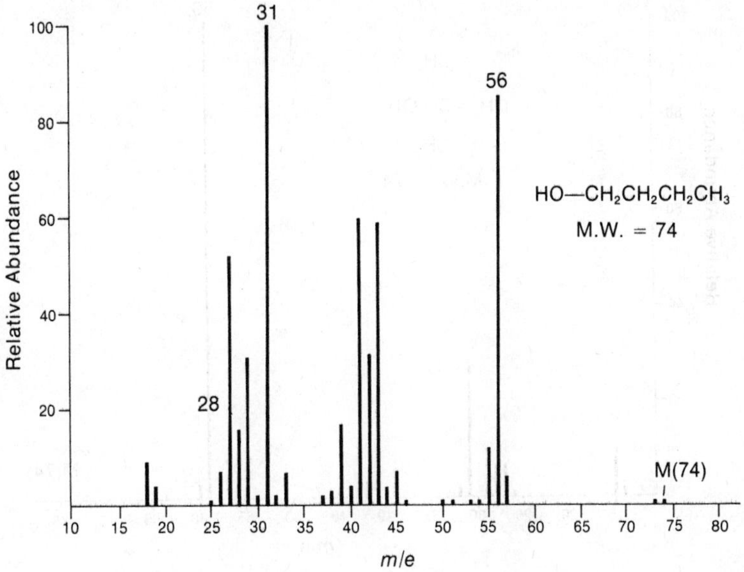

FIGURE 6–29 Mass Spectrum of 1-Butanol

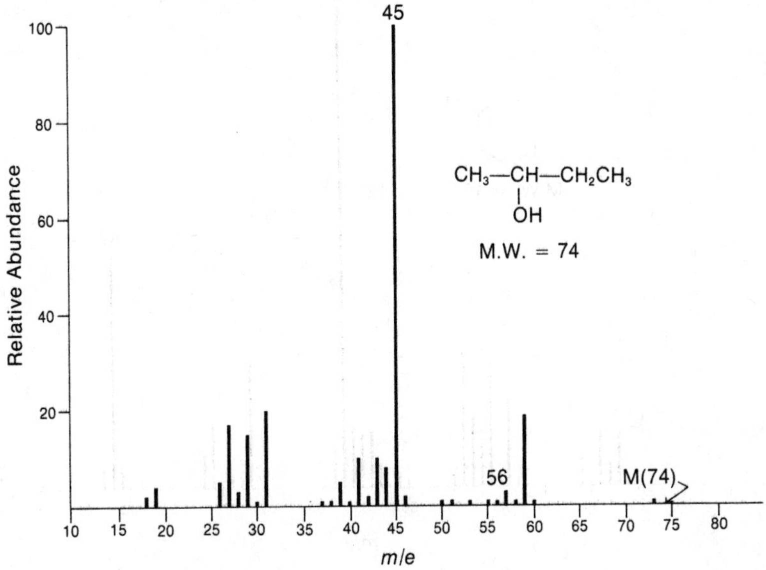

FIGURE 6–30 Mass Spectrum of 2-Butanol

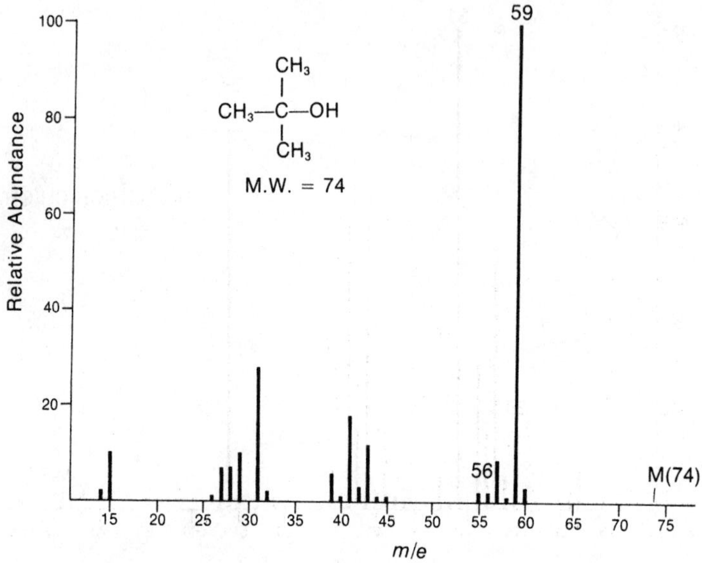

FIGURE 6–31 Mass Spectrum of 2-Methyl-2-propanol

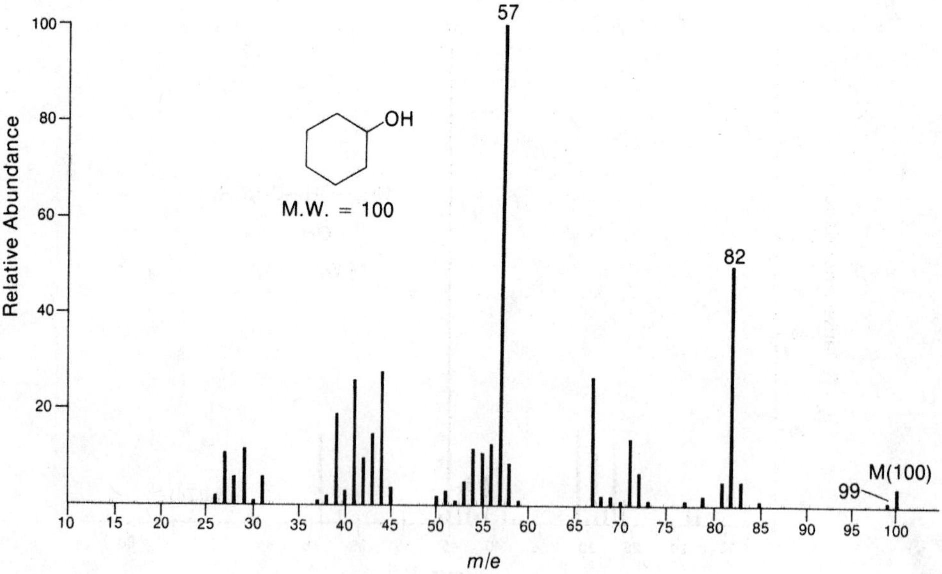

FIGURE 6–32 Mass Spectrum of Cyclohexanol

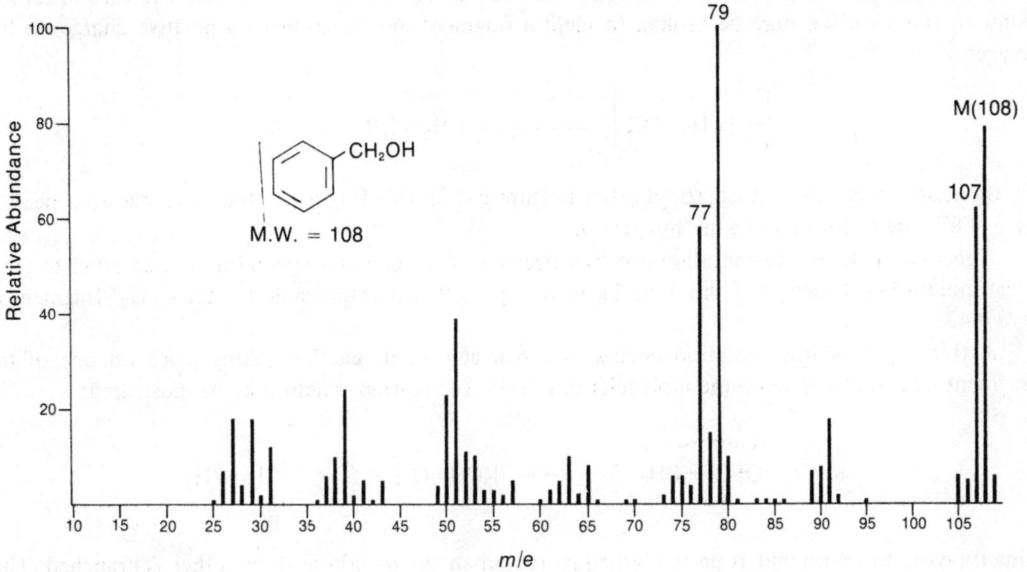

FIGURE 6–33 Mass Spectrum of Benzyl Alcohol

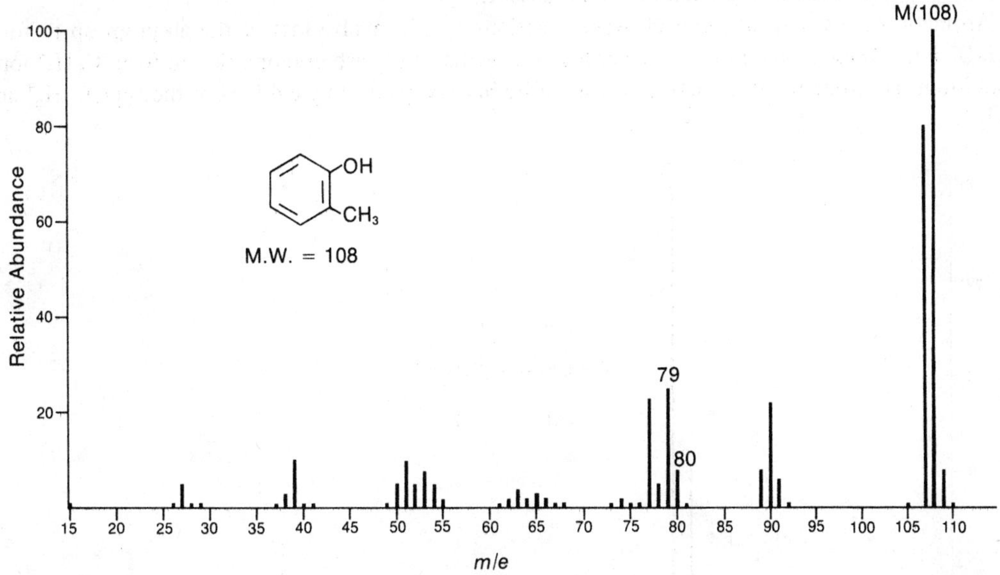

FIGURE 6–34 Mass Spectrum of 2-Methylphenol (o-Cresol)

F. ETHERS

Aliphatic ethers tend to exhibit molecular ion peaks which are stronger than those of alcohols with the same molecular weight. The molecular ion peaks of ethers are, nevertheless, rather weak.

The fragmentation of the ethers is somewhat similar to that of the alcohols. The carbon-carbon bond to the α-carbon may be broken to yield a fragment ion which bears a positive charge on the oxygen.

$$\left[R\!\!+\!\overset{\alpha}{CH_2}\!-\!OR \right]^{\ddagger} \longrightarrow CH_2 = \overset{+}{O}R + R\cdot$$

In the mass spectrum of diisopropyl ether (Figure 6–35), this fragmentation gives rise to a peak at $m/e = 87$, due to the loss of a methyl group.

A second mode of fragmentation involves cleavage of the carbon-oxygen bond of an ether to yield a carbonium ion. Cleavage of this type in diisopropyl ether is responsible for the $C_3H_7{}^+$ fragment at $m/e = 43$.

A third type of fragmentation occurs as a rearrangement reaction taking place on one of the fragment ions, rather than on the molecular ion itself. The rearrangement may be illustrated:

$$R\!-\!CH\!=\!\overset{+}{O}\!\!+\!\!\underset{\underset{R}{|\alpha}}{CH}\!-\!\underset{\beta}{\overset{\overset{H}{|}}{CH_2}} \longrightarrow RCH\!=\!\overset{+}{O}H + \underset{R}{CH\!=\!CH_2}$$

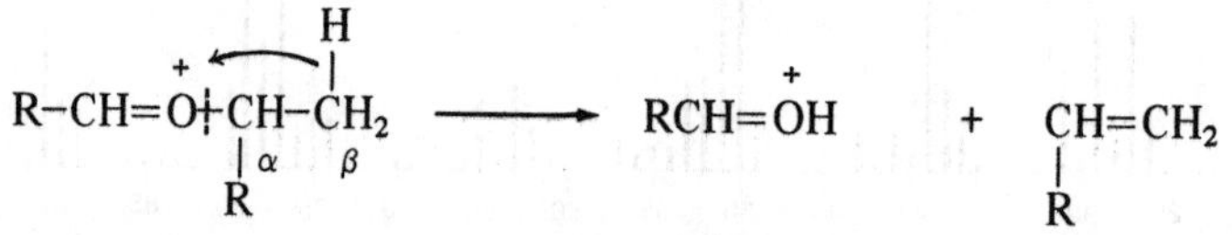

This type of rearrangement is particularly favored when the α-carbon of the ether is branched. This rearrangement in the case of diisopropyl ether gives rise to a $C_2H_5O^+$ fragment ($m/e = 45$).

As a result of these fragmentation processes, the fragment ion peaks of ethers often fall into the series 31, 45, 59, 73, etc. These fragments are of either the RO^+ or the $ROCH_2{}^+$ type.

Acetals and ketals behave in a fashion very similar to that of ethers. However, fragmentation is even more favorable in acetals and ketals than it is in ethers, so the molecular ion peak of an acetal or ketal may be either extremely weak or totally absent.

Aromatic ethers may undergo cleavage reactions which involve loss of the alkyl group to form $C_6H_5O^+$ ions. These fragment ions further lose the elements of carbon monoxide to form $C_5H_5{}^+$ ions. In addition, an aromatic ether may lose the entire alkoxy group to yield ions of the type $C_6H_6{}^+$ and $C_6H_5{}^+$.

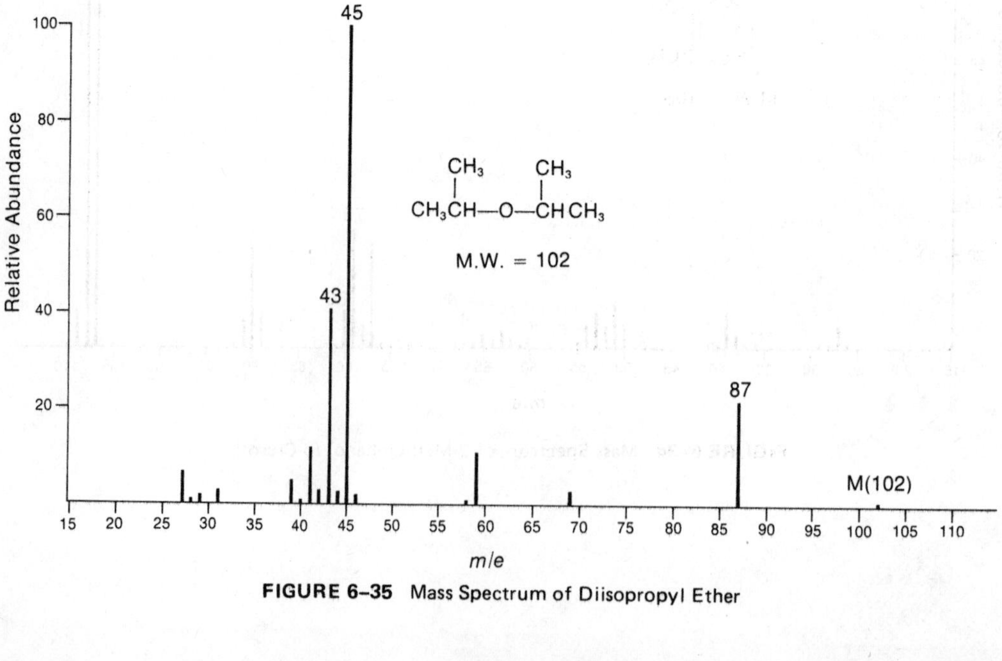

FIGURE 6–35 Mass Spectrum of Diisopropyl Ether

G. ALDEHYDES

The molecular ion peak of an aliphatic aldehyde is usually observable, although at times it may be fairly weak. The most characteristic peaks due to the fragmentation of aldehydes come from three classes of processes.

Cleavage of one of the two bonds to the carbonyl group occurs very commonly. This is sometimes called *α-cleavage,* and it may be outlined as:

$$\left[\mathrm{R{-}CHO}\right]^{+\cdot} \longrightarrow \mathrm{R{-}C{\equiv}\overset{+}{O}} \quad + \quad \mathrm{H}\cdot$$

$$\left[\mathrm{R{-}CHO}\right]^{+\cdot} \longrightarrow \mathrm{H{-}C{\equiv}\overset{+}{O}} \quad + \quad \mathrm{R}\cdot$$

The peak due to the loss of one hydrogen atom is very characteristic of aldehydes. This peak may be observed at *m/e* = 43 in the mass spectrum of acetaldehyde (Figure 6–36) and at *m/e* = 71 in the mass spectrum of butyraldehyde (Figure 6–37). The peak due to the formation of HCO^+ can be observed at *m/e* = 29 in each of the spectra; this is also a very characteristic peak in the mass spectra of aldehydes.

The second important mode of fragmentation for aldehydes is known as *β-cleavage.*

$$\left[\mathrm{R{\mid}CH_2{-}CHO}\right]^{+\cdot} \longrightarrow \mathrm{R^+} \quad + \quad \mathrm{CH_2{=}CH{-}O}\cdot$$

The R^+ fragment ion coincidentally also occurs at *m/e* = 29 in the mass spectrum of butyraldehyde. However, in higher aldehydes it will occur at higher *m/e* values. At any rate, the mass of this peak is 43 mass units less than the mass of the molecular ion.

The third major fragmentation pathway for aldehydes is the *McLafferty rearrangement* (see Section 6.5D for an earlier example).

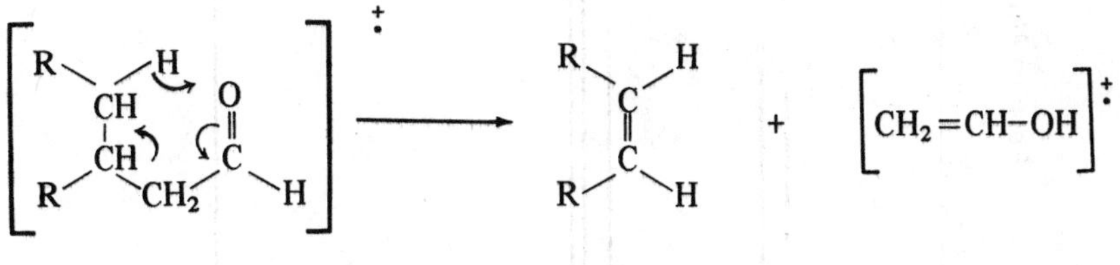

The fragment ion formed in this rearrangement has *m/e* = 44. The presence of a peak at *m/e* = 44 is also considered to be very characteristic of the mass spectra of aldehydes. This peak may be observed in the mass spectrum of butyraldehyde, where it is the most intense peak in the spectrum.

Aromatic aldehydes exhibit a very intense molecular ion peak. The loss of one hydrogen atom *via* α-cleavage is a very favorable process. The resulting M − 1 peak may be more intense than the molecular ion peak. In the mass spectrum of benzaldehyde (Figure 6-38), the M − 1 peak at *m/e* = 105 may be observed. Also observed is a peak at *m/e* = 77, which is due to the loss of the −CHO group to give $C_6H_5^+$.

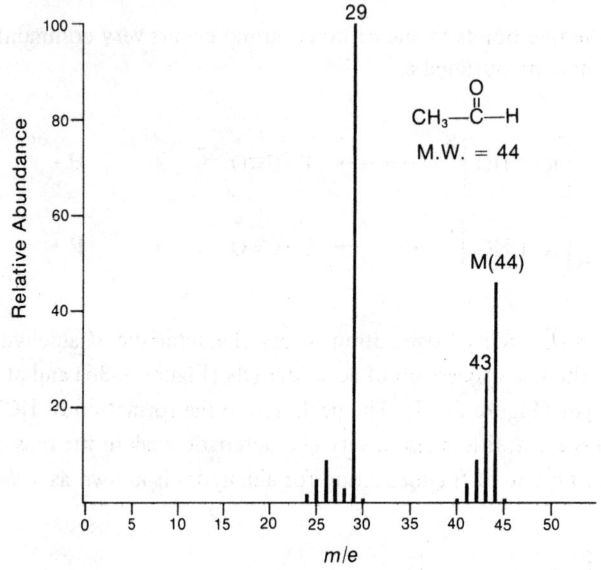

FIGURE 6–36 Mass Spectrum of Acetaldehyde

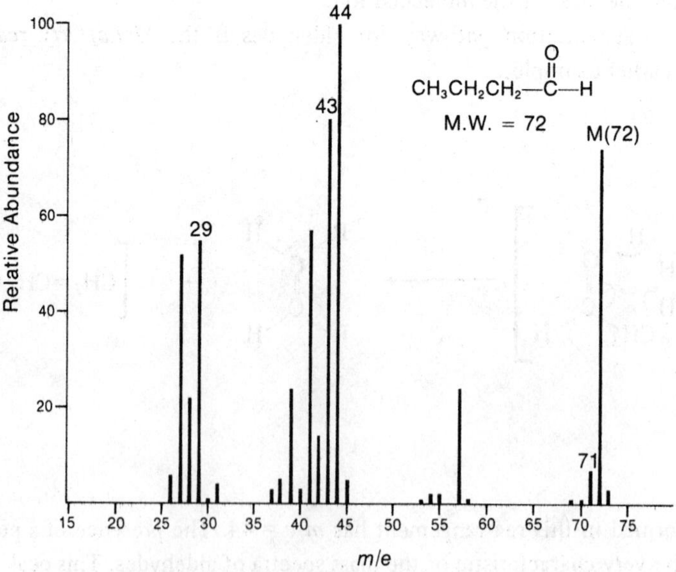

FIGURE 6–37 Mass Spectrum of Butyraldehyde

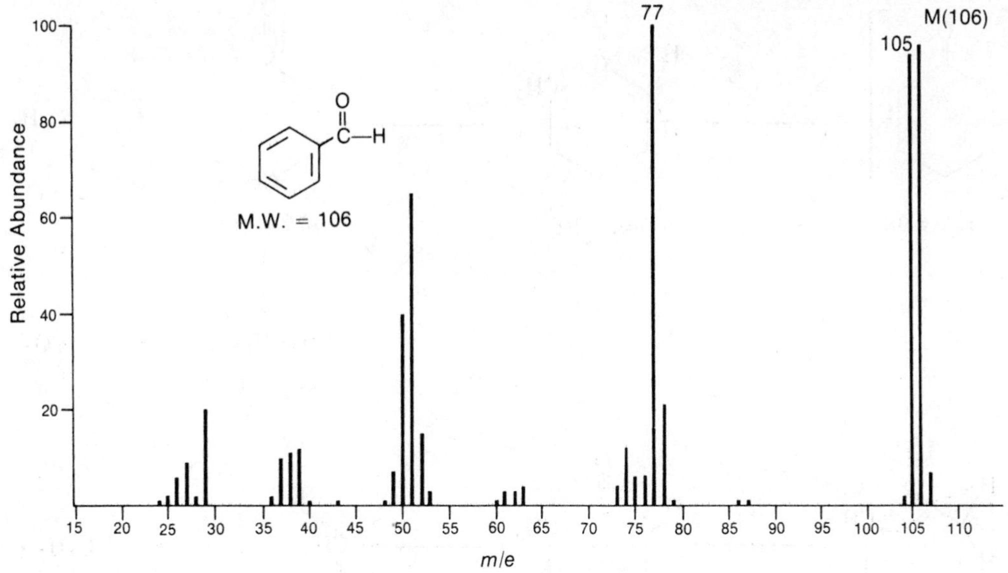

FIGURE 6–38 Mass Spectrum of Benzaldehyde

H. KETONES

The molecular ion peaks of most ketones are quite intense. The pattern of fragmentation for acyclic ketones is very similar to that of aldehydes. Loss of alkyl groups, by means of α-cleavage, is an important mode of fragmentation. The larger of the two alkyl groups attached to the carbonyl group appears to be the group which is more likely to be lost. In the mass spectrum of 2-butanone (Figure 6–39), the peak at $m/e = 43$, due to the loss of the ethyl group, is more intense than the peak at $m/e = 57$, which is due to the loss of the methyl group. Similarly, in the mass spectrum of 2-octanone (Figure 6–40), loss of the hexyl group, giving a peak at $m/e = 43$, is more important than loss of the methyl group, which gives the weak peak at $m/e = 113$.

When the carbonyl group of a ketone has at least one alkyl group attached to it which contains three or more carbon atoms, a McLafferty rearrangement is possible. This rearrangement may be described:

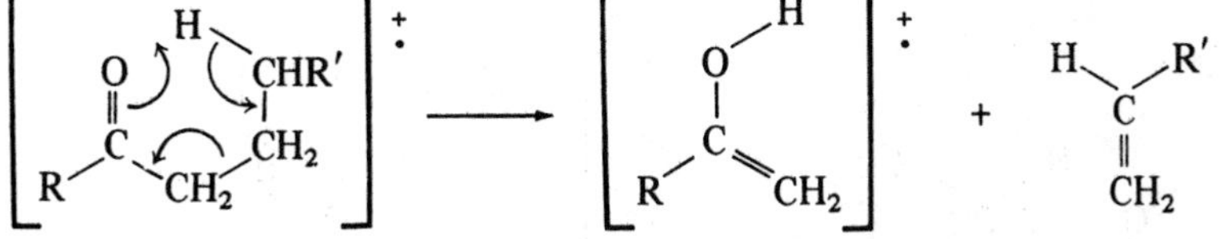

The peak at $m/e = 58$ in the mass spectrum of 2-octanone is due to the fragment ion resulting from this rearrangement.

Cyclic ketones may undergo a variety of fragmentation and rearrangement processes. These processes are outlined below for the case of cyclohexanone. Fragment ion peaks corresponding to each of these processes may be observed in the mass spectrum of cyclohexanone (Figure 6–41).

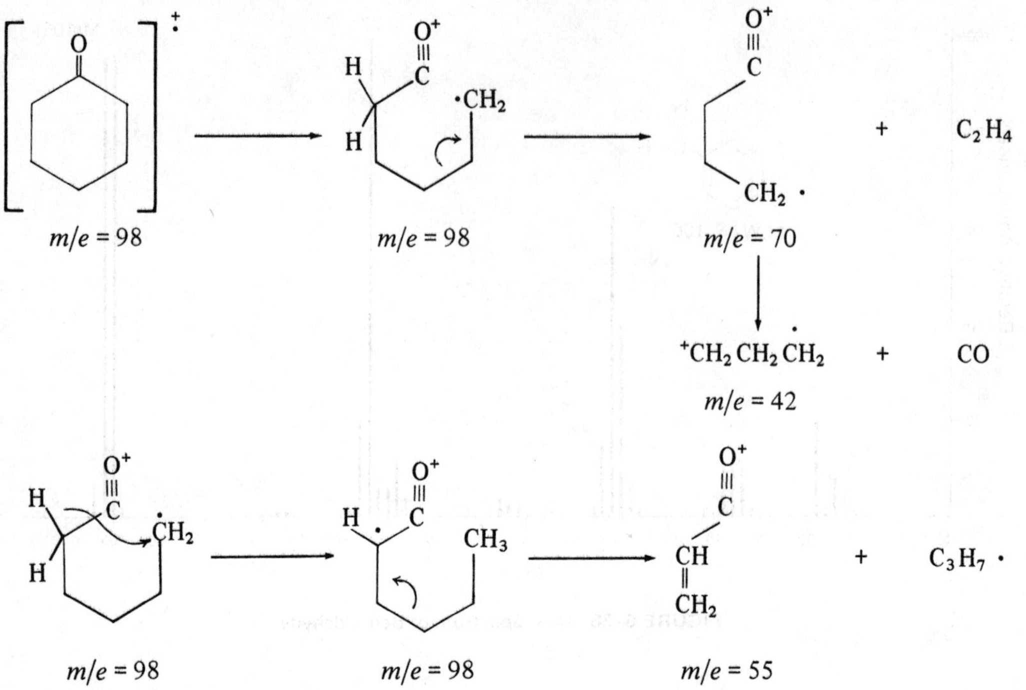

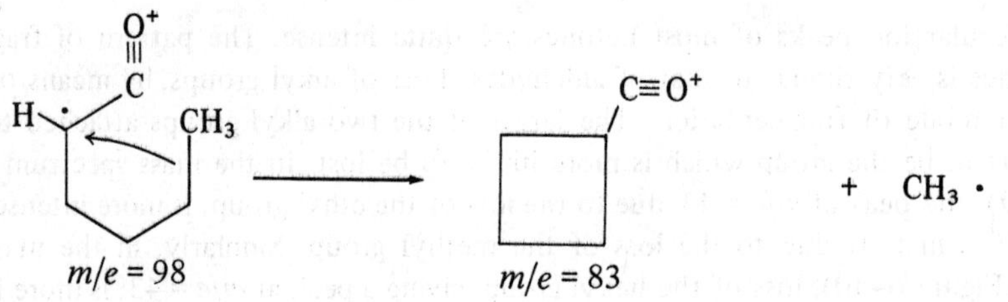

Aromatic ketones undergo α-cleavage to lose the alkyl group and form the $C_6H_5CO^+$ ion (m/e = 105). This ion loses carbon monoxide to form the $C_6H_5^+$ ion (m/e = 77). These peaks appear prominently in the mass spectrum of acetophenone (Figure 6–42). With larger alkyl groups attached to the carbonyl group of an aromatic ketone, rearrangements of the McLafferty type are possible. This rearrangement is illustrated below:

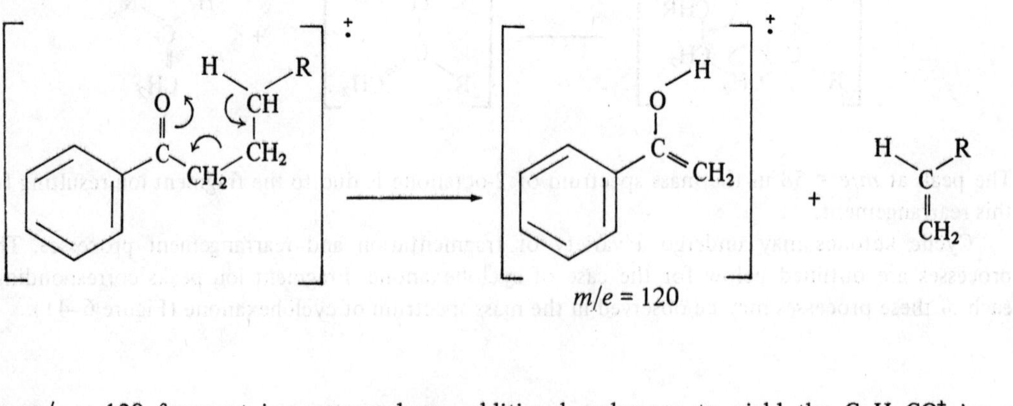

The m/e = 120 fragment ion may undergo additional α-cleavage to yield the $C_6H_5CO^+$ ion at m/e = 105.

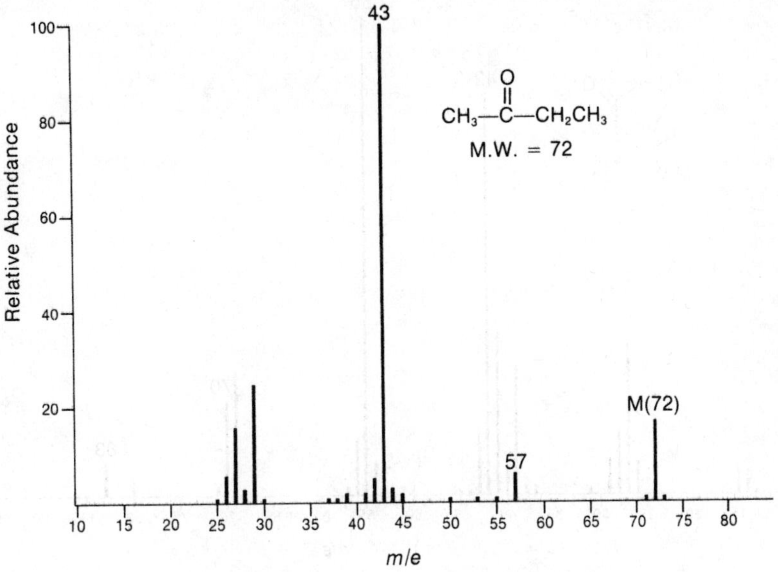

FIGURE 6–39 Mass Spectrum of 2-Butanone

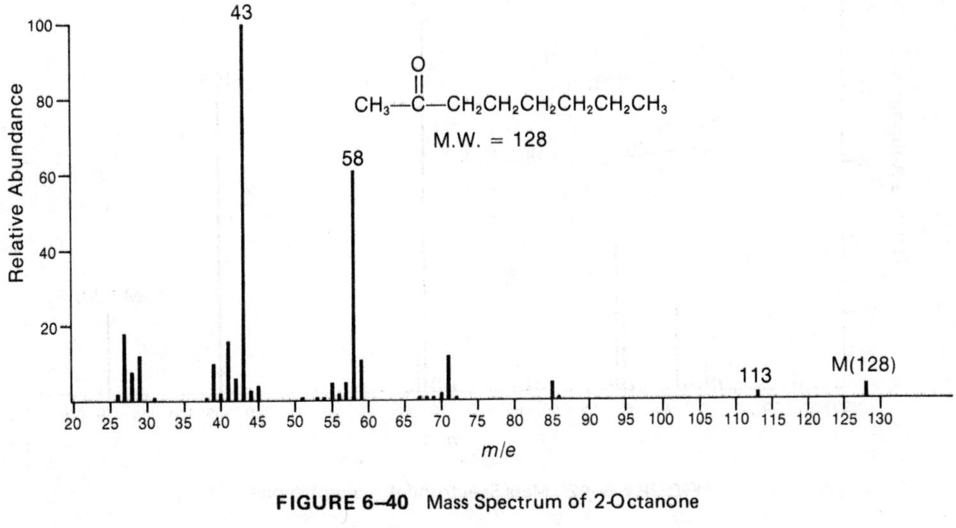

FIGURE 6–40 Mass Spectrum of 2-Octanone

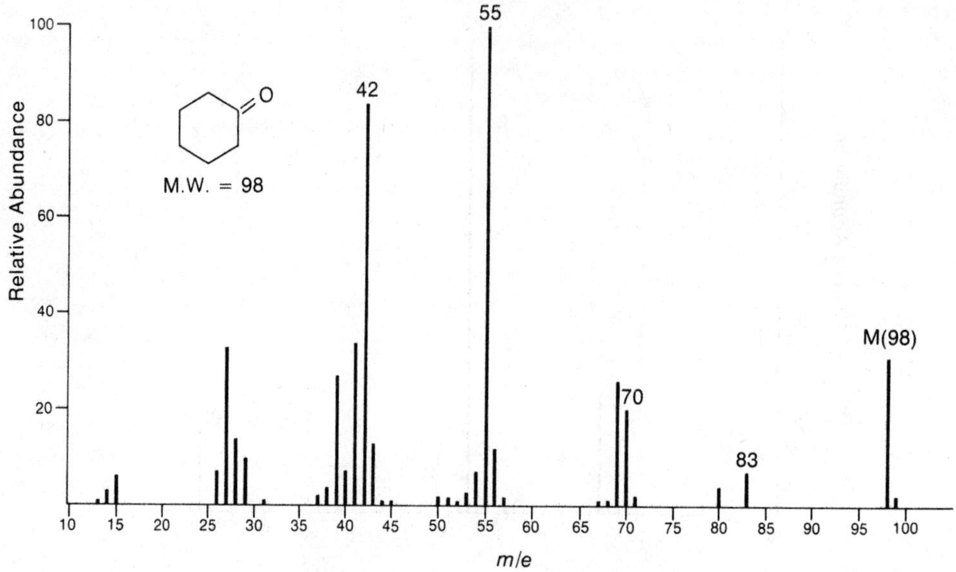

FIGURE 6–41 Mass Spectrum of Cyclohexanone

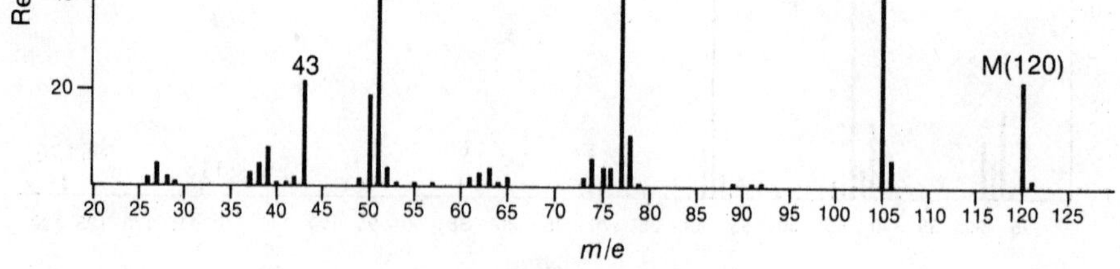

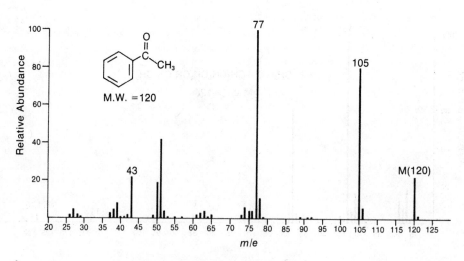

FIGURE 6–42 Mass Spectrum of Acetophenone

I. ESTERS

In the mass spectra of methyl esters, it is usually possible to observe a weak but nevertheless noticeable molecular ion peak. Fragment ions result from both α- and β-cleavage processes.

The most important of the α-cleavage reactions involves the loss of the alkoxy group from an ester to form the corresponding acylium ion. RCO^+. The acylium ion peak may be observed at $m/e = 43$ in the mass spectrum of methyl acetate (Figure 6-43) and at $m/e = 71$ in the mass spectrum of methyl butyrate (Figure 6-44). The acylium ion peak is a useful diagnostic peak when one is examining the mass spectra of esters. A second useful peak results from the loss of the alkyl group from the acyl portion of the ester molecule, leaving a fragment $CH_3{-}O{-}C{=}O^+$, which appears at $m/e = 59$. This peak may also be observed in the mass spectrum of methyl butyrate. Again, this $m/e = 59$ peak, though less intense than the acylium ion peak, is a useful diagnostic peak for methyl esters. Other fragment ion peaks include the $OCH_3{}^+$ fragment and the R^+ fragment from the acyl portion of the ester molecule. These latter ions are much less important than the former two ions.

The most important β-cleavage reaction of methyl esters is the McLafferty rearrangement. The rearrangement may be illustrated:

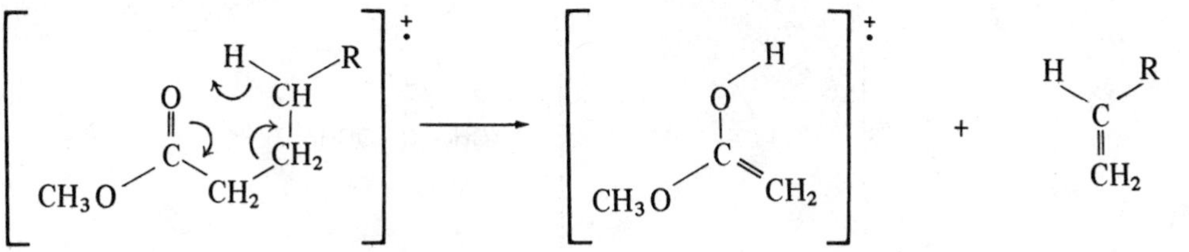

This fragment ion peak is very important in the mass spectra of methyl esters, as may be observed from its prominence in the mass spectrum of methyl butyrate.

The esters of higher alcohols than methanol form much weaker molecular ion peaks. One should contrast the mass spectrum of methyl acetate with that of ethyl acetate (Figure 6-45). In the latter spectrum, the molecular ion peak is much weaker than in the former. Esters of alcohols larger than butanol may form molecular ion peaks which are too weak to be observed.

Ethyl, propyl, butyl, and higher alkyl esters also undergo the α-cleavage and McLafferty rearrangements typical of the methyl esters. However, these esters may also undergo a rearrangement of the alkyl portion of the molecule, in which a hydrogen atom from the alkyl portion is transferred to the carbonyl oxygen of the acyl portion of the ester. This results in fragments of the type:

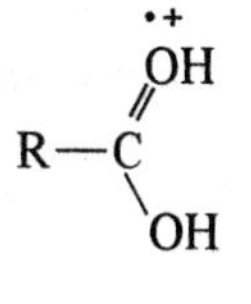

which appear in the series $m/e = 61, 75, 89$, etc. These ions may also appear without the extra hydrogen as $RCOOH^+$ fragment ions.

Benzyl esters undergo rearrangement to eliminate the neutral ketene molecule.

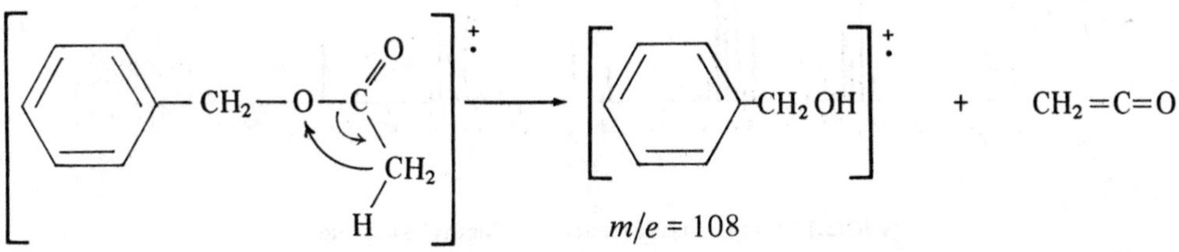

$m/e = 108$

The resulting ion is often the most intense peak in the mass spectrum of such compounds.

Alkyl benzoate esters prefer to lose the alkoxy group to form the $C_6H_5CO^+$ ion (m/e = 105.) This ion may lose carbon monoxide to form the $C_6H_5^+$ ion at m/e = 77. Each of these peaks may be observed in the mass spectrum of methyl benzoate (Figure 6–46).

Alkyl substitution on benzoate esters appears to have little effect on the mass spectral results unless the alkyl group is in the *ortho* position with respect to the ester functional group. In this case, the alkyl group can interact with the ester function, with the elimination of a molecule of alcohol. This reaction is illustrated:

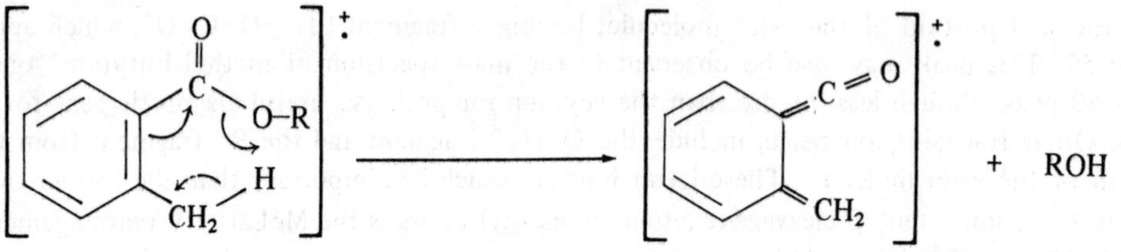

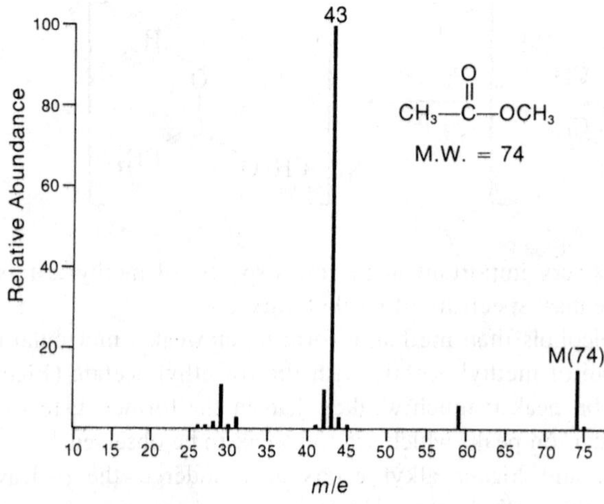

FIGURE 6–43 Mass Spectrum of Methyl Acetate

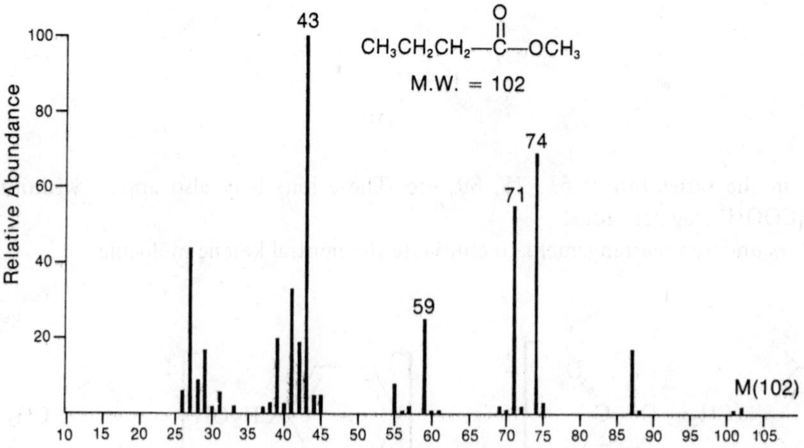

FIGURE 6–44 Mass Spectrum of Methyl Butyrate

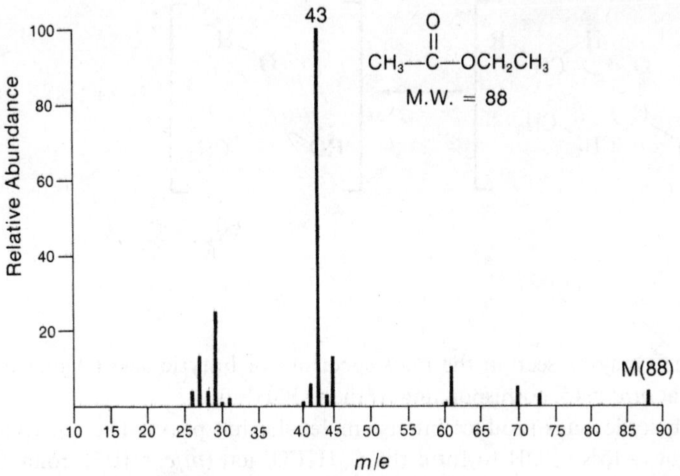

FIGURE 6–45 Mass Spectrum of Ethyl Acetate

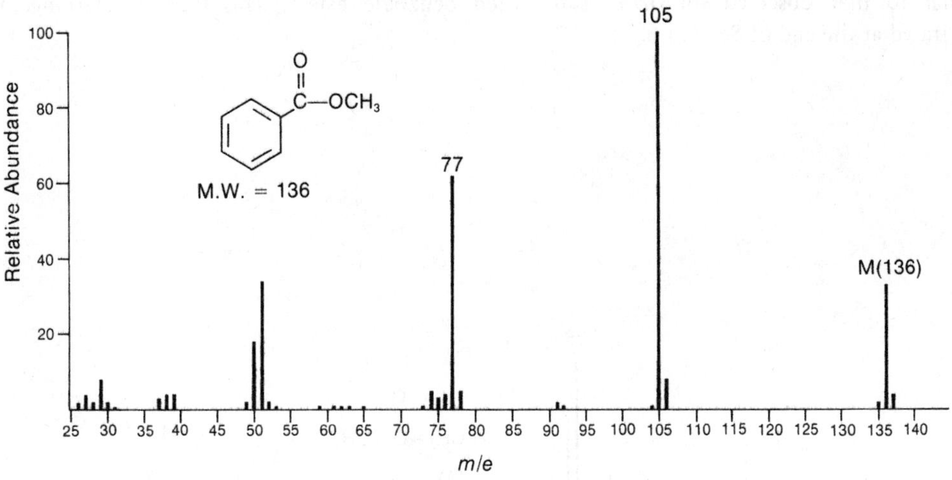

FIGURE 6–46 Mass Spectrum of Methyl Benzoate

J. CARBOXYLIC ACIDS

The mass spectra of aliphatic carboxylic acids show rather weak, but nevertheless observable, molecular ion peaks. Their fragmentation patterns resemble those of the methyl esters.

With short-chain acids, the loss of OH and COOH through α-cleavage on either side of the C=O group may be observed. In the mass spectrum of propionic acid (Figure 6–47), loss of OH gives rise to a peak at m/e = 57. Loss of COOH gives rise to the peak at m/e = 29. The intense peak at m/e = 28 is due to further fragmentation of the alkyl portion of the acid molecule. Loss of the alkyl group as a free radical, leaving the $COOH^+$ ion (m/e = 45), also occurs. This fragment ion peak may also be observed in the mass spectrum. This peak is very characteristic of the mass spectra of carboxylic acids.

With acids containing γ-hydrogens, the principal pathway for fragmentation is the McLafferty rearrangement. In the case of carboxylic acids, this rearrangement produces a prominent peak at m/e = 60.

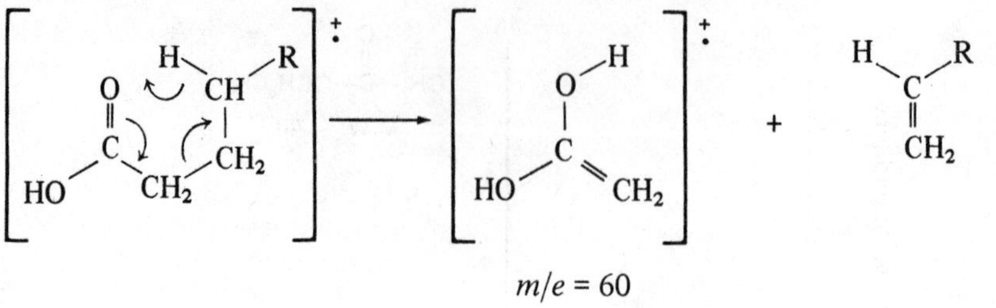

This m/e = 60 peak may be seen in the mass spectrum of butyric acid (Figure 6–48). One may also observe the peak at m/e = 45, corresponding to the $COOH^+$ ion.

Aromatic carboxylic acids produce intense molecular ion peaks. The most important fragmentation pathway involves loss of OH to form the $C_6H_5CO^+$ ion (m/e = 105), followed by loss of CO to form the $C_6H_5^+$ ion. In the mass spectrum of *p*-anisic acid (Figure 6–49), loss of OH gives rise to a peak at m/e = 135. Further loss of CO from this ion gives rise to a peak at m/e = 107.

Benzoic acids bearing *ortho* substituents undergo loss of water through a rearrangement reaction similar to that observed for *ortho* substituted benzoate esters. This type of rearrangement is illustrated at the end of Section 6.5-I.

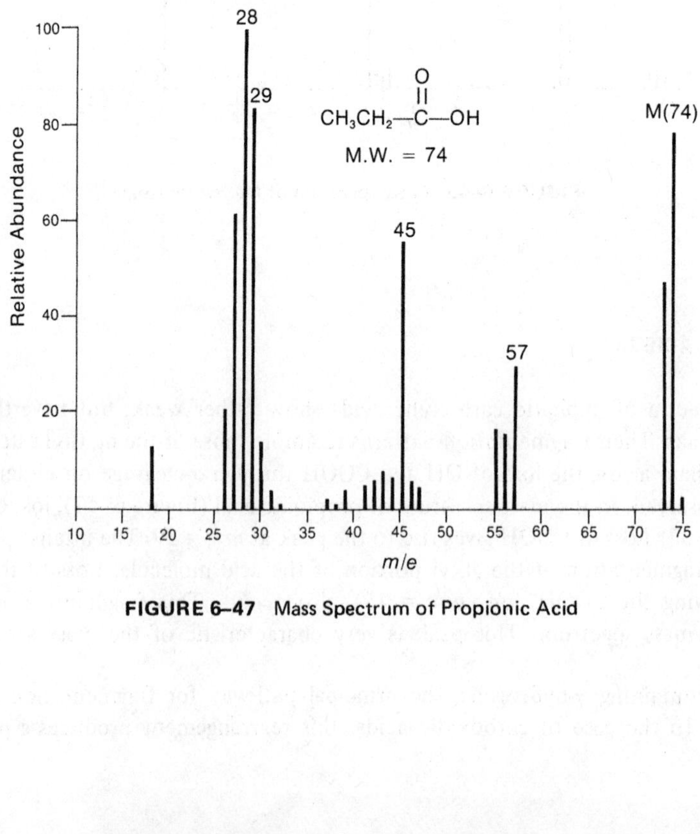

FIGURE 6–47 Mass Spectrum of Propionic Acid

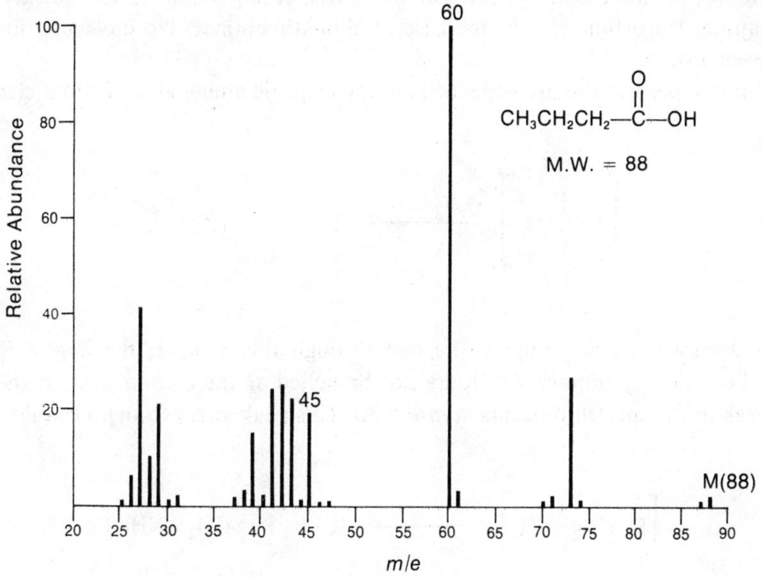

FIGURE 6–48 Mass Spectrum of Butyric Acid

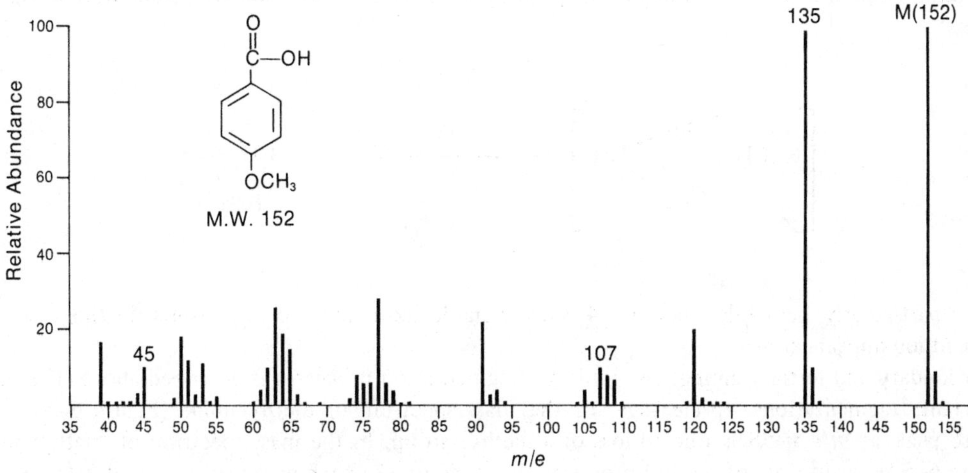

FIGURE 6–49 Mass Spectrum of *p*-Anisic Acid

K. AMINES

The value of the mass of the molecular ion can be of great help in identifying a substances as an amine. As was stated in Section 6.3, compounds which have an odd number of nitrogen atoms must have an odd-numbered molecular weight. On this basis, it is possible to tell quickly if a substance could be an amine. Unfortunately, in the case of aliphatic amines, the molecular ion peak may be very weak or even absent.

The most intense peak in the mass spectrum of an aliphatic amine arises from β-cleavage:

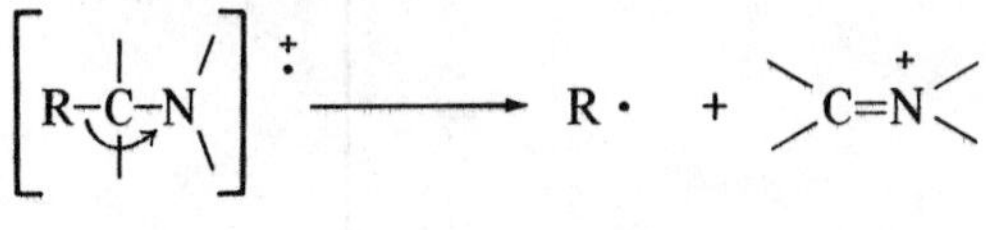

When there is a choice of R groups to be lost through this process, the largest R group is lost preferentially. For primary amines which are not branched at the carbon next to the nitrogen, the most intense peak in the spectrum occurs at $m/e = 30$. This peak arises from β-cleavage:

$$\left[R{-}CH_2{-}NH_2\right]^{+\cdot} \longrightarrow R\cdot + CH_2{=}\overset{+}{N}H_2 \qquad m/e = 30$$

The presence of this peak is good evidence that the test substance is a primary amine, although it is not conclusive. This peak may arise from secondary fragmentation of ions formed from the fragmentation of secondary or tertiary amines as well. In the mass spectrum of ethylamine (Figure 6–50), the $m/e = 30$ peak can be seen clearly.

For long-chain primary amines, the same β-cleavage peak can also occur. Further fragmentation of the R group of the amine leads to clusters of fragments 14 mass units apart, due to sequential loss of CH_2 units from the R group. Long-chain primary amines can also undergo fragmentation via the process:

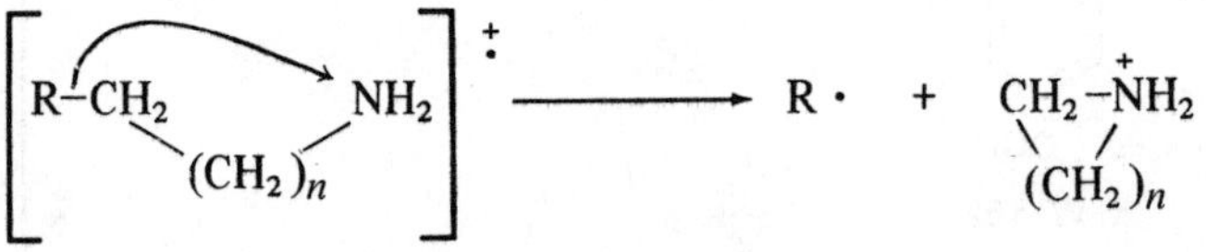

This is particularly favorable when $n = 4$, since a stable six-membered ring results. In this case, the fragment ion appears at $m/e = 86$.

Secondary and tertiary amines also undergo fragmentation processes as described above. The most important fragmentation is β-cleavage. In the mass spectrum of diethylamine (Figure 6–51), the intense peak at $m/e = 58$ is due to loss of a methyl group. In the mass spectrum of triethylamine (Figure 6–52), again, loss of methyl produces the most intense peak in the spectrum, at $m/e = 86$. In each case, further fragmentation of this initially formed fragment ion produces a peak at $m/e = 30$.

Cyclic aliphatic amines usually produce an intense molecular ion peak. Their principal modes of fragmentation are:

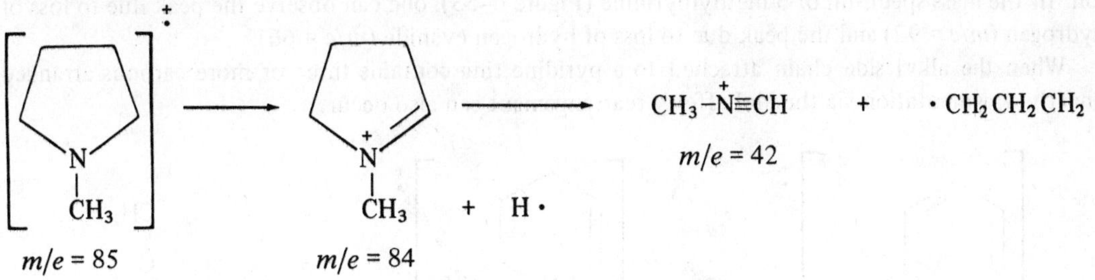

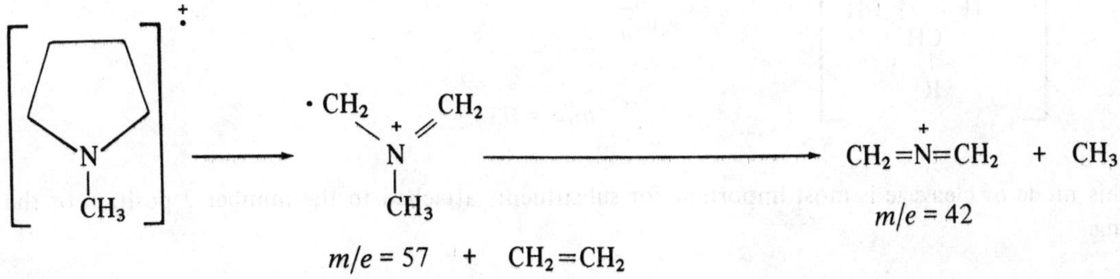

Aromatic amines show an intense molecular ion peak. They may show a moderately intense peak at an m/e value one mass unit less than that of the molecular ion, due to loss of a hydrogen atom. The fragmentation of aromatic amines may be illustrated in the case of aniline:

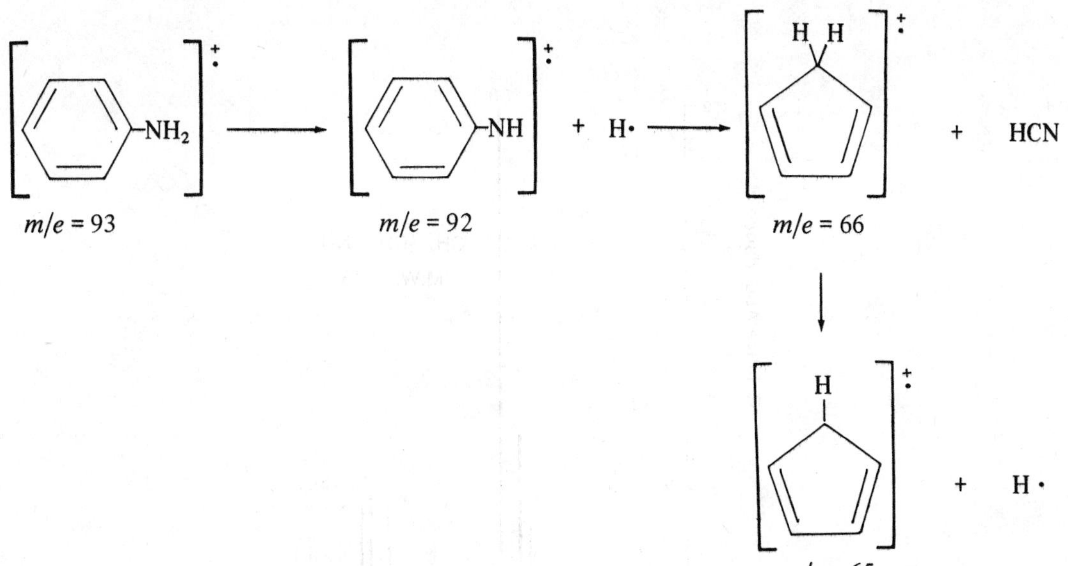

Substituted pyridines are characterized by very intense molecular ion peaks. Frequently, loss of a hydrogen atom to produce a peak at an m/e value which is one mass unit less than the molecular ion is also observed.

The most important fragmentation process for the pyridine ring is loss of the elements of hydrogen cyanide. This produces a fragment ion which is 27 mass units lighter than the molecular ion. In the mass spectrum of 3-methylpyridine (Figure 6–53), one can observe the peak due to loss of hydrogen ($m/e = 92$) and the peak due to loss of hydrogen cyanide ($m/e = 66$).

When the alkyl side chain attached to a pyridine ring contains three or more carbons arranged linearly, fragmentation via the McLafferty rearrangement can also occur.

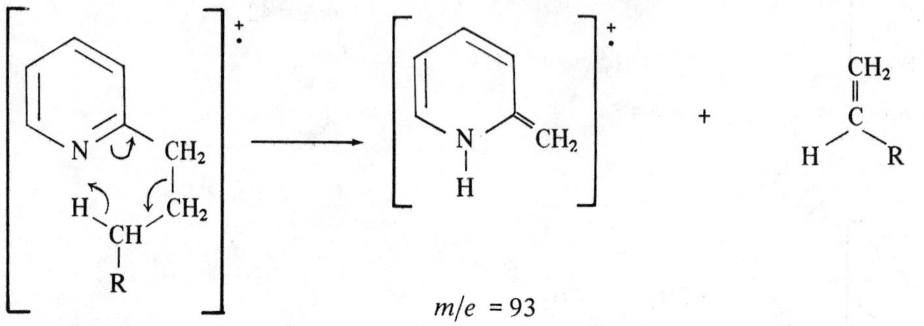

This mode of cleavage is most important for substituents attached to the number 2 position of the ring.

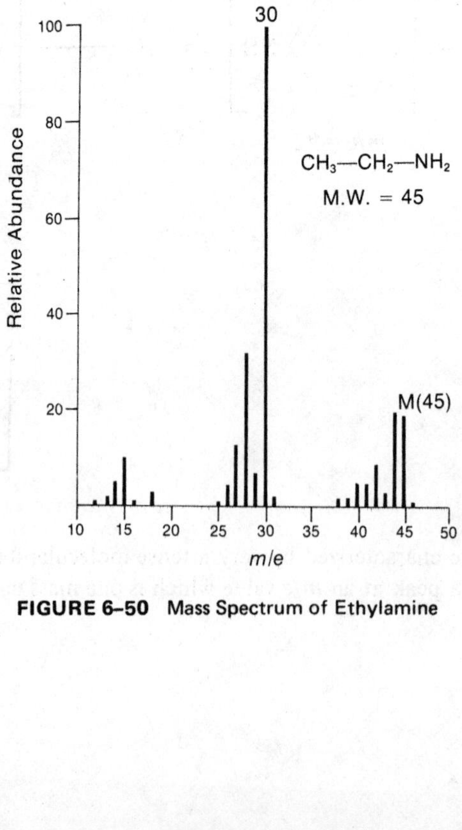

FIGURE 6–50 Mass Spectrum of Ethylamine

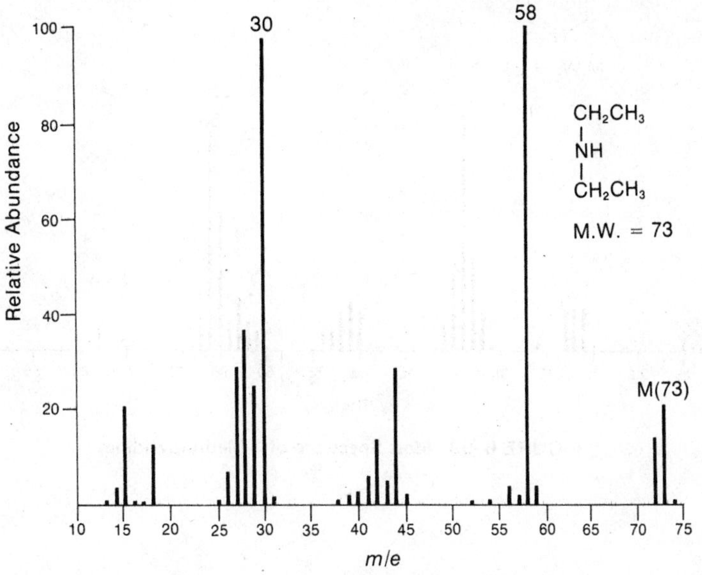

FIGURE 6–51 Mass Spectrum of Diethylamine

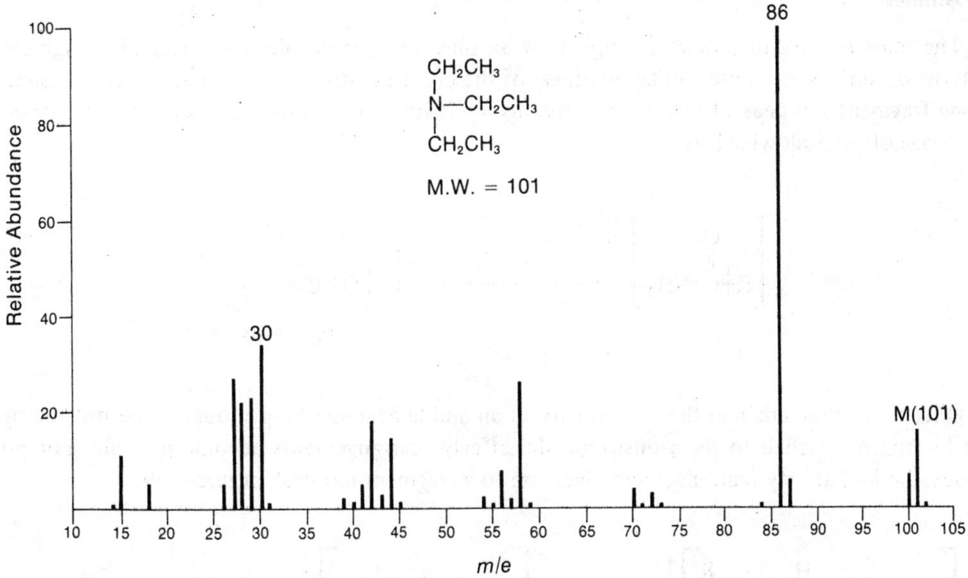

FIGURE 6–52 Mass Spectrum of Triethylamine

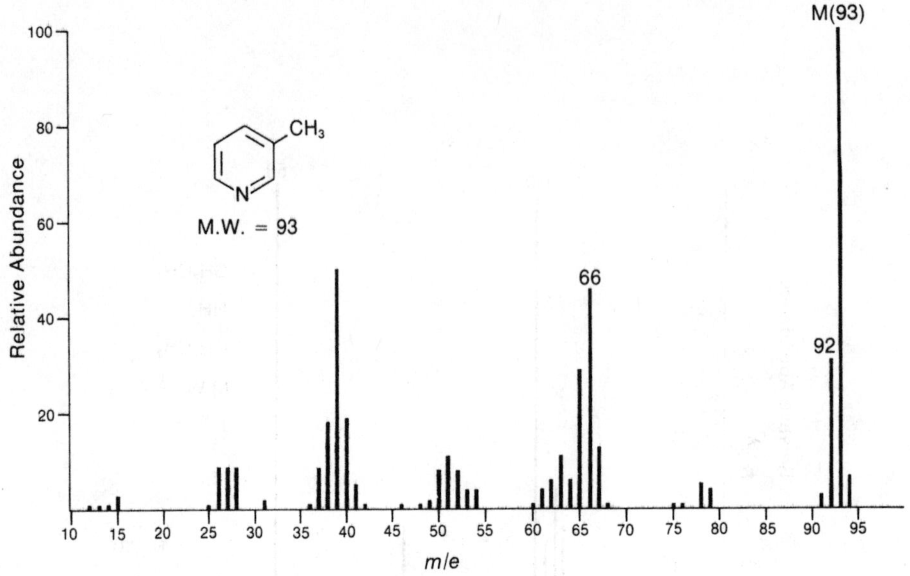

FIGURE 6–53   Mass Spectrum of 3-Methylpyridine

L. SELECTED NITROGEN AND SULFUR COMPOUNDS

As is true of amines, such nitrogen-bearing compounds as amides, nitriles, and nitro compounds must follow the nitrogen rule: If they contain an odd number of nitrogen atoms, they must have an odd-numbered molecular weight. The nitrogen rule is explained more completely in Section 6.3.

Amides

The mass spectra of amides usually show an observable molecular ion peak. The fragmentation patterns of amides are quite similar to those of the corresponding esters and acids. The presence of a strong fragment ion peak at m/e = 44 is usually indicative of a primary amide. This peak arises from α-cleavage of the following sort:

$$\left[\begin{array}{c} O \\ \| \\ R \dotplus C-NH_2 \end{array}\right]^{+\cdot} \longrightarrow R\cdot \; + \; \left[O=C=NH_2\right]^{+}$$
$$m/e = 44$$

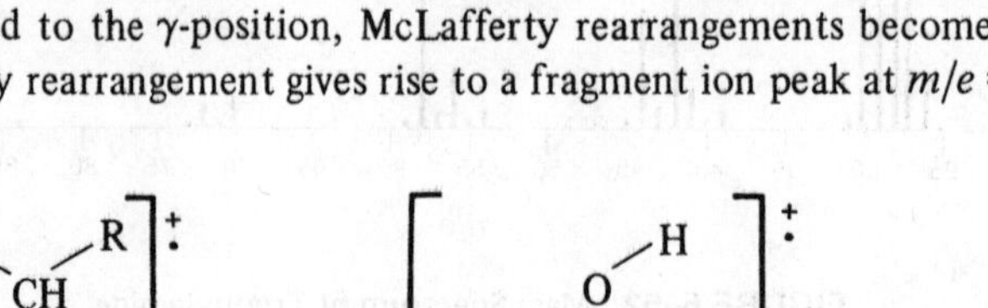

Once the carbon chain in the acyl moiety of an amide becomes long enough to permit the transfer of a hydrogen attached to the γ-position, McLafferty rearrangements become possible. For primary amides, the McLafferty rearrangement gives rise to a fragment ion peak at m/e = 59.

$$\left[\begin{array}{c} H \quad R \\ O \quad CH \\ \| \quad \curvearrowright \\ H_2N \quad C \quad CH_2 \\ \quad CH_2 \end{array}\right]^{+\cdot} \longrightarrow \left[\begin{array}{c} O \quad H \\ \| \\ H_2N \quad C \quad CH_2 \end{array}\right]^{+\cdot} \; + \; \begin{array}{c} H \quad C \quad R \\ \| \\ CH_2 \end{array}$$
$$m/e = 59$$

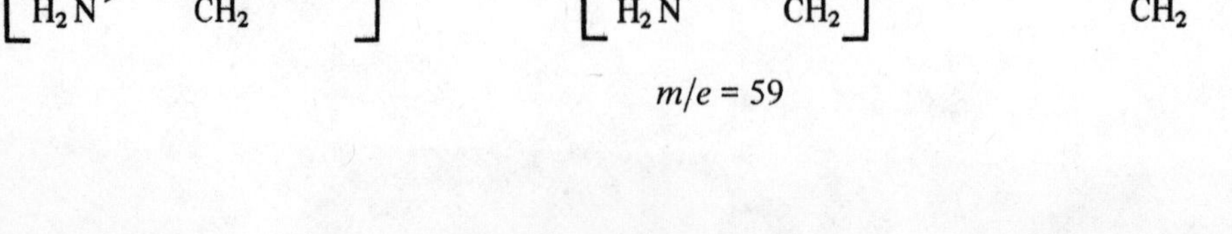

Nitriles

Aliphatic nitriles usually undergo fragmentation so readily that the molecular ion peak is too weak to be observed. However, most nitriles form a peak due to the loss of one hydrogen atom, forming an ion of the type R–CH=C=N⁺. Although this peak may be weak, it is a useful diagnostic peak in characterizing nitriles. In the mass spectrum of hexanonitrile (Figure 6–54), this peak may be observed at $m/e = 96$.

When the alkyl group attached to the nitrile functional group is a propyl group or some longer hydrocarbon group, the most intense peak in the mass spectrum results from a McLafferty rearrangement:

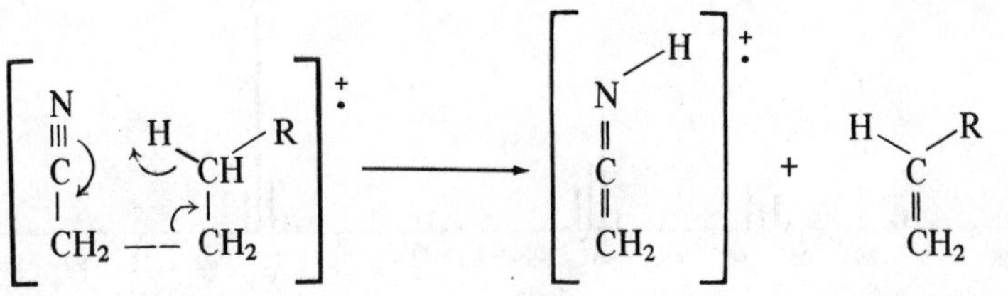

$m/e = 41$

This peak, which may be observed in the mass spectrum of hexanonitrile, can be quite useful in characterizing an aliphatic nitrile. Unfortunately, as the alkyl group of a nitrile becomes longer, the probability of formation of the $C_3H_5^+$ ion, which also appears at $m/e = 41$, increases. With high molecular weight nitriles, most of the fragment ions of mass 41 are $C_3H_5^+$ ions, rather than ions formed as a result of a McLafferty rearrangement.

The strongest peak in the mass spectrum of an aromatic nitrile is the molecular ion peak. Loss of cyanide occurs, giving, in the case of benzonitrile (Figure 6–55), the $C_6H_5^+$ ion at $m/e = 77$. More important fragmentation involves loss of the elements of hydrogen cyanide. In benzonitrile, this gives rise to a peak at $m/e = 76$.

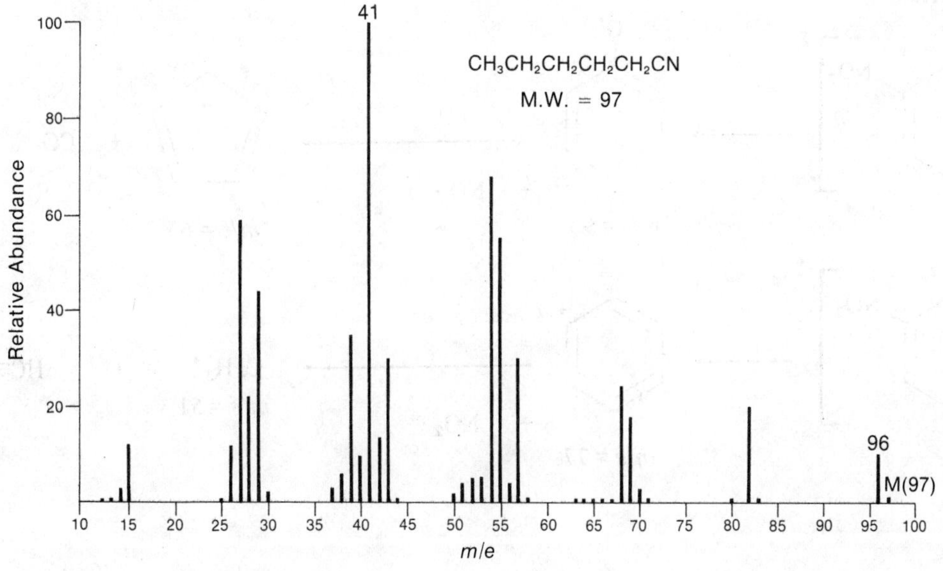

FIGURE 6–54 Mass Spectrum of Hexanonitrile

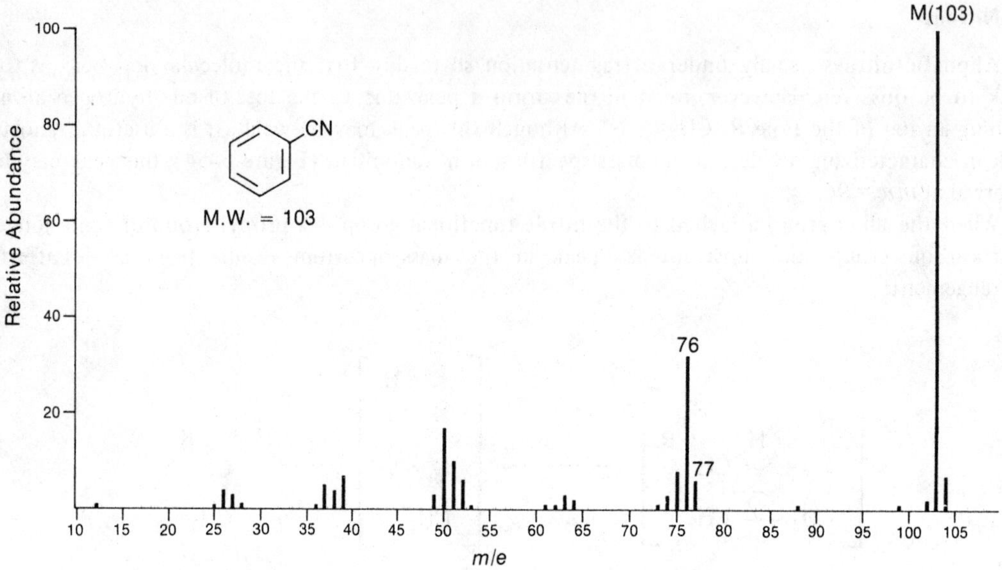

FIGURE 6–55 Mass Spectrum of Benzonitrile

Nitro Compounds

The molecular ion peak for an aliphatic nitro compound is seldom observed. The mass spectrum is the result of fragmentation of the hydrocarbon part of the molecule. However, the mass spectra of nitro compounds may show a moderate peak at m/e = 30, corresponding to the NO^+ ion, and a weaker peak at m/e = 46, corresponding to the NO_2^+ ion. These peaks may be observed in the mass spectrum of 1-nitropropane (Figure 6–56). The intense peak at m/e = 43 is due to the $C_3H_7^+$ ion.

Aromatic nitro compounds show an intense molecular ion peak. The characteristic NO^+ (m/e = 30) and NO_2^+ (m/e = 46) peaks may be found in the mass spectrum. The principal fragmentation pattern, however, involves loss of all or part of the nitro group. Using nitrobenzene (Figure 6-57) as an example, this fragmentation pattern may be described as:

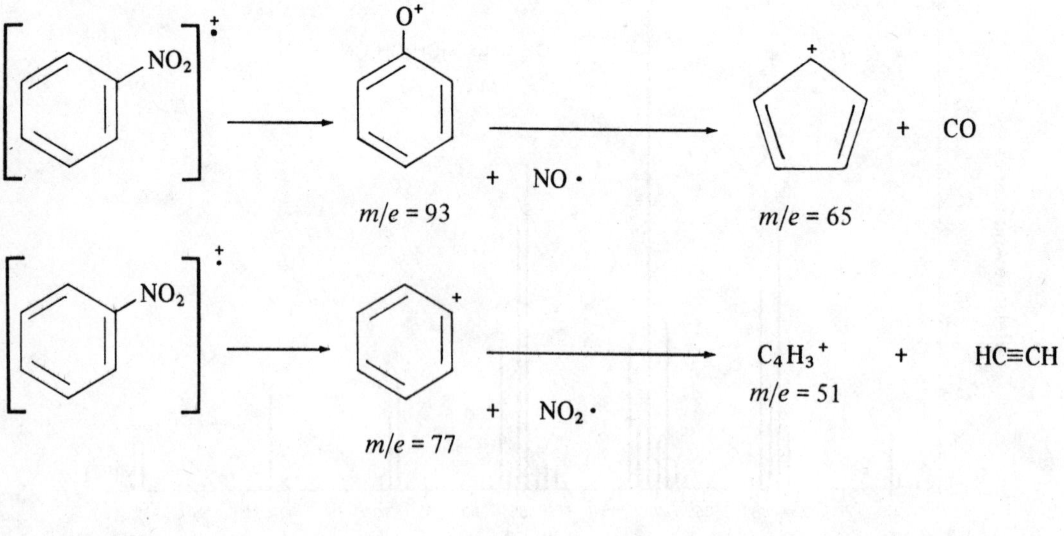

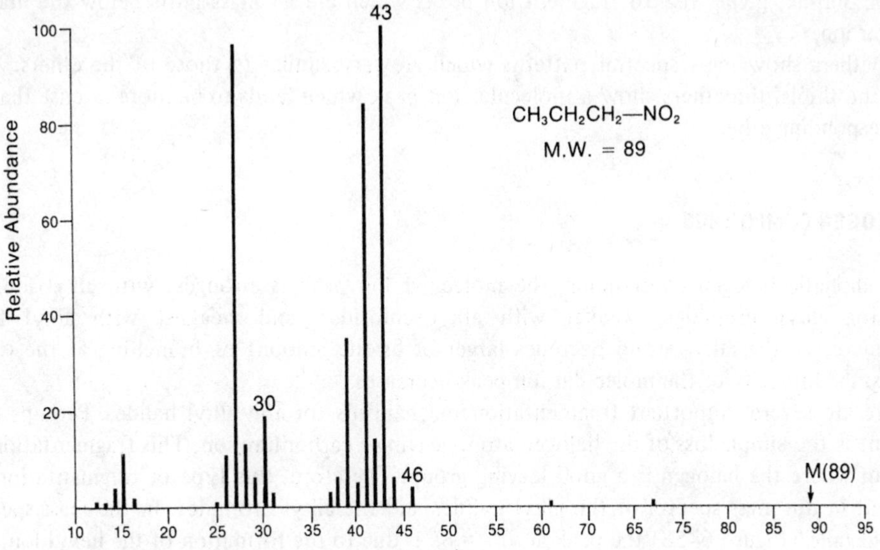

FIGURE 6–56 Mass Spectrum of 1-Nitropropane

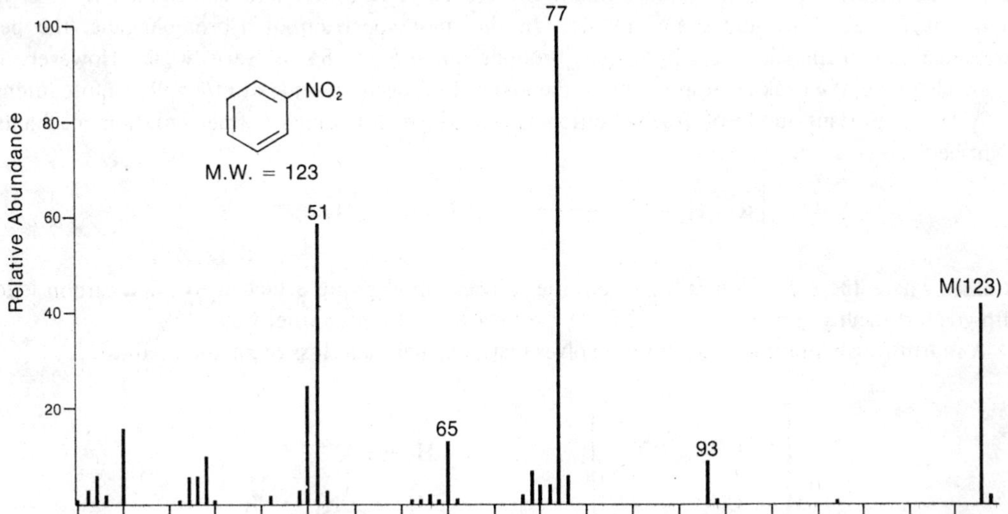

FIGURE 6–57 Mass Spectrum of Nitrobenzene

Thiols and Thioethers

Thiols show molecular ion peaks which are more intense than those shown by the corresponding alcohols. The fragmentation patterns of the thiols are very similar to those of the alcohols. As alcohols tend to undergo dehydration under some conditions, thiols tend to lose the elements of hydrogen sulfide, giving rise to fragment ion peaks which are 34 mass units below the mass of the molecular ion.

Thioethers show mass spectral patterns which are very similar to those of the ethers. As in the case of the thiols, thioethers show a molecular ion peak which tends to be more intense than that of the corresponding ether.

M. HALOGEN COMPOUNDS

For aliphatic halogen compounds, the molecular ion peak is strongest with alkyl iodides, less strong for alkyl bromides, weaker with alkyl chlorides, and weakest with alkyl fluorides. Furthermore, as the alkyl group becomes larger or as the amount of branching at the α-position increases, the intensity of the molecular ion peak decreases.

There are several important fragmentation mechanisms for the alkyl halides. Perhaps the most important is the simple loss of the halogen atom, leaving a carbonium ion. This fragmentation is most important where the halogen is a good leaving group. Therefore, this type of fragmentation is most prominent in the mass spectra of the alkyl iodides and the alkyl bromides. In the mass spectrum of 1-bromohexane (Figure 6–58) the peak at $m/e = 85$ is due to the formation of the hexyl ion. This ion undergoes further fragmentation to form a $C_3H_7^+$ ion at $m/e = 43$. The corresponding heptyl ion peak at $m/e = 99$ in the mass spectrum of 2-chloroheptane (Figure 6–59) is quite weak.

Alkyl halides may also lose a molecule of hydrogen halide according to the following process:

$$\left[R-CH_2CH_2-X\right]^{+\cdot} \longrightarrow \left[RCH=CH_2\right]^{+\cdot} + HX$$

This mode of fragmentation is most important for alkyl fluorides and chlorides, and it is less important for alkyl bromides and iodides. In the mass spectrum of 1-bromohexane, the peak corresponding to the loss of hydrogen bromide at $m/e = 84$ is very weak. However, for 2-chloroheptane, the peak corresponding to the loss of hydrogen chloride at $m/e = 98$ is quite intense.

A less important mode of fragmentation is α-cleavage, for which a fragmentation mechanism might be:

$$\left[R{+}CH_2-X\right]^{+\cdot} \longrightarrow R\cdot + CH_2{=}X^+$$

In cases where the α-position is branched, the heaviest alkyl group attached to the α-carbon is lost with greatest facility. The peaks arising from α-cleavage are usually rather weak.

A fourth fragmentation mechanism involves rearrangement and loss of an alkyl radical:

$$\left[\begin{array}{c} R-CH_2 \rightarrow X \\ | \qquad | \\ CH_2 \quad CH_2 \\ CH_2 \end{array} \right]^{+\cdot} \longrightarrow \begin{array}{c} CH_2 \!-\! X^+ \\ | \qquad | \\ CH_2 \quad CH_2 \\ CH_2 \end{array} + R\cdot$$

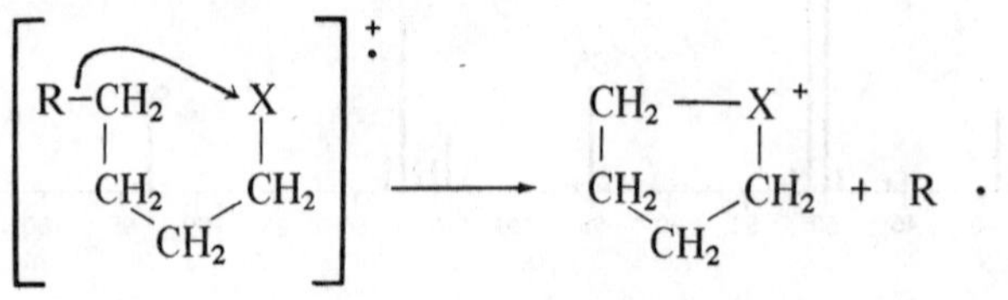

The corresponding cyclic ion can be observed at $m/e = 135$ and 137 in the mass spectrum of 1-bromohexane and at $m/e = 105$ and 107 in the mass spectrum of 2-chloroheptane. Such fragmentation is important only in the mass spectra of long-chain alkyl chlorides or bromides.

The molecular ion peak in the mass spectra of benzyl halides is usually of sufficient intensity to be observed. The most important fragmentation involves loss of halogen to form the $C_7H_7^+$ ion. When the aromatic ring of a benzyl halide carries substituents, a substituted phenyl cation may also be observed.

The molecular ion peak of an aromatic halide is usually quite intense. The most important mode of fragmentation involves loss of halogen to form the $C_6H_5^+$ ion.

Although the fragmentation patterns described above are well characterized, the most interesting feature of the mass spectra of chlorine- and bromine-containing compounds is the fact that these substances show *two* molecular ion peaks. As was indicated in Section 6.4, chlorine occurs naturally in two isotopic forms. Chlorine of mass 37 occurs with a natural abundance which is 32.5% that of chlorine of mass 35. Bromine of mass 81 has a natural abundance which is 98.0% that of ^{79}Br. Therefore, the intensity of the M + 2 peak in a chlorine-containing compound should 32.5% of the intensity of the molecular ion peak. The intensity of the M + 2 peak in a bromine-containing compound should be almost equal to the intensity of the molecular ion peak. These pairs of molecular ion peaks (sometimes called "doublets") can be seen in the mass spectra of ethyl chloride (Figure 6–60) and ethyl bromide (Figure 6–61).

Table 6–6 in Section 6.4 can be used to determine what the ratio of the intensities of the molecular ion and isotopic peaks should be when more than one chlorine or bromine is present in the same molecule. The mass spectra of dichloromethane (Figure 6–62), dibromomethane (Figure 6–63), and 1-bromo-2-chloroethane (Figure 6–64) are included here to illustrate some of the combinations of halogens listed in Table 6–6.

Unfortunately, one cannot always take advantage of these characteristic patterns to identify halogen compounds. Frequently, the molecular ion peaks may be too weak to permit accurate measurement of the ratio of the intensity of the molecular ion peak to isotopic peaks. However, it is often possible to make such a comparison on certain fragment ion peaks in the mass spectrum of a halogen compound. The mass spectrum of 1-bromohexane (Figure 6–58) may be used to illustrate this method. The presence of bromine can be determined using the fragment ion peaks at *m/e* values of 135 and 137.

Since iodine and fluorine exist in nature in the form of only one isotope, their mass spectra do not show isotopic peaks. The presence of halogen must be deduced either by noting the unusually weak M + 1 peak or by observing the mass difference between the fragment ions and the molecular ion.

Text continued on page 285

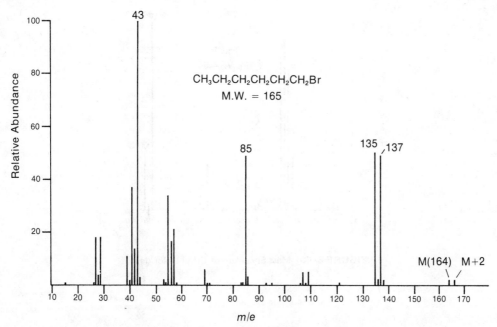

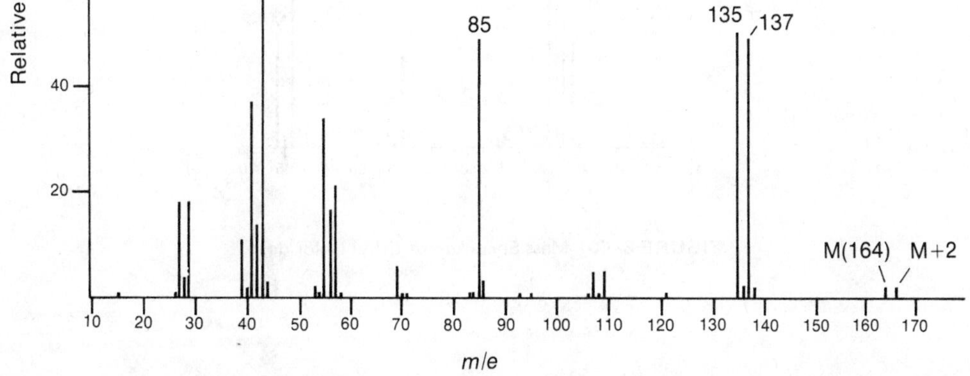

FIGURE 6–58 Mass Spectrum of 1-Bromohexane

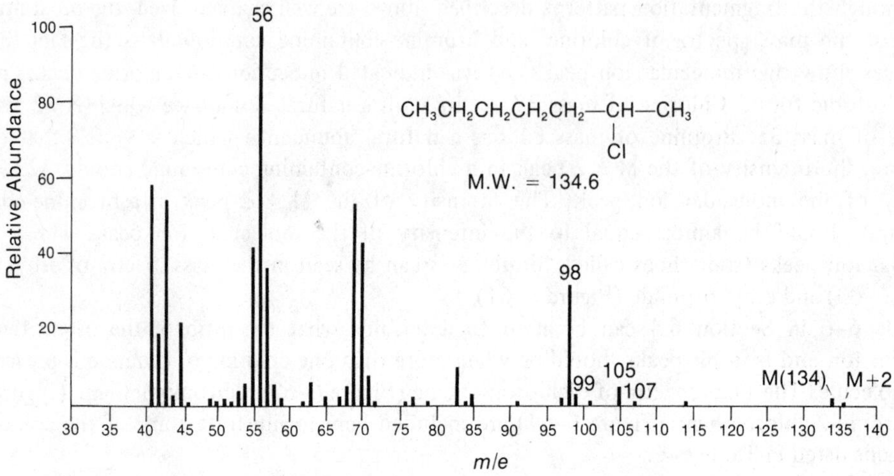

FIGURE 6–59 Mass Spectrum of 2-Chloroheptane

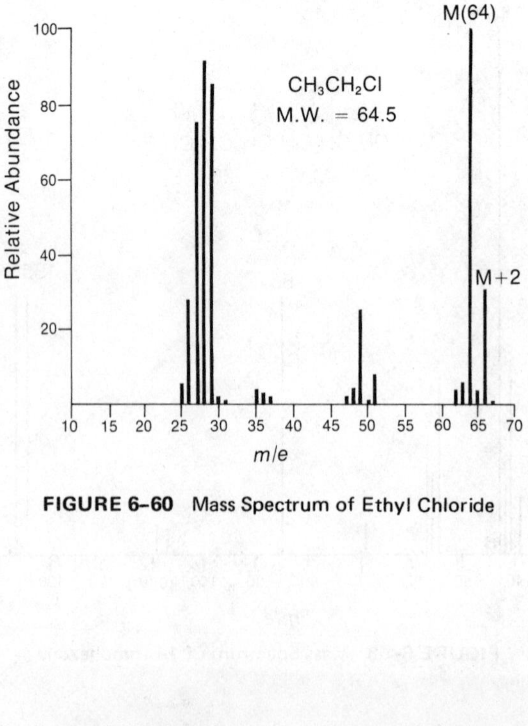

FIGURE 6–60 Mass Spectrum of Ethyl Chloride

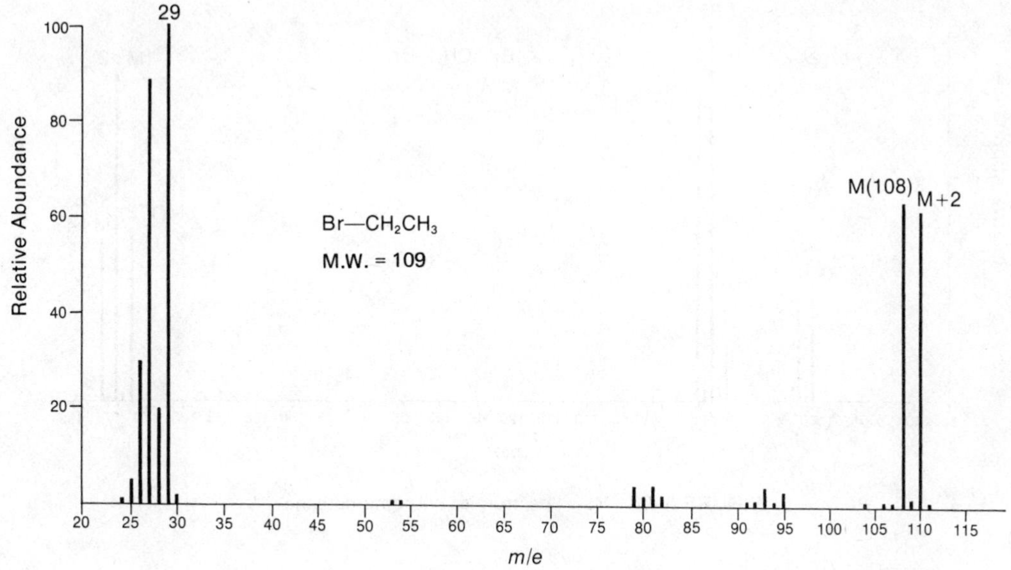

FIGURE 6–61 Mass Spectrum of Ethyl Bromide

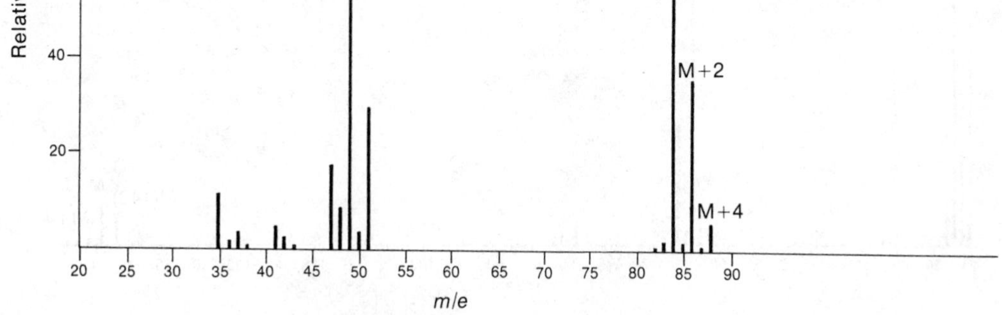

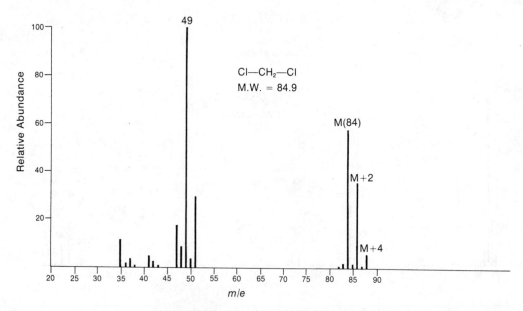

FIGURE 6–62 Mass Spectrum of Dichloromethane

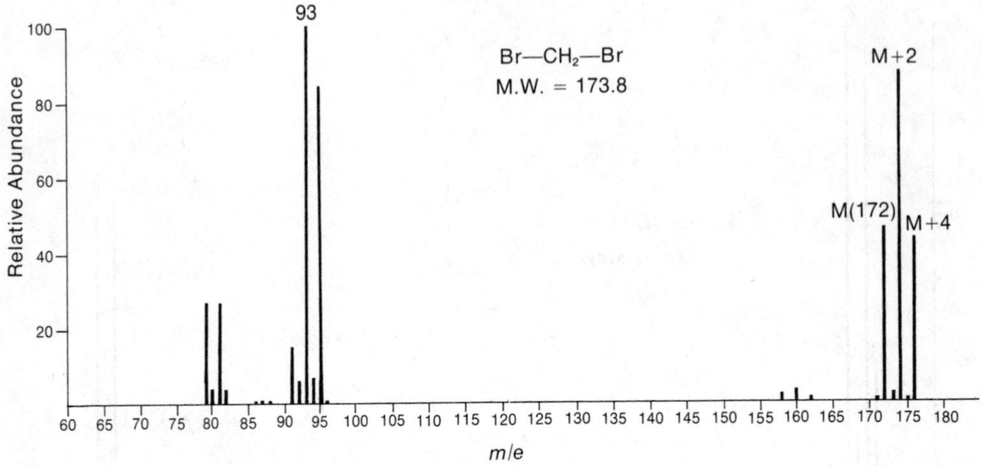

FIGURE 6–63 Mass Spectrum of Dibromomethane

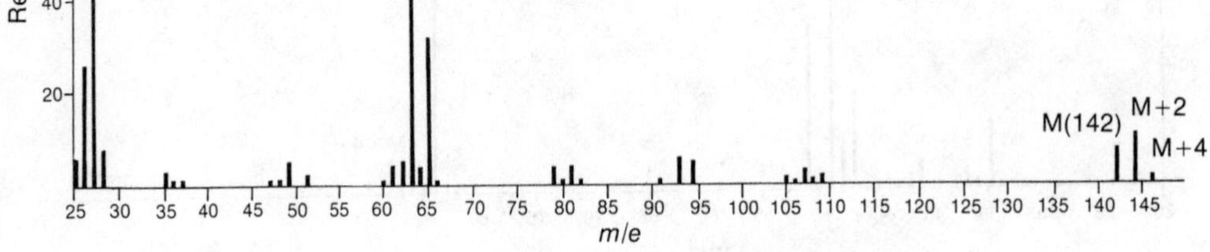

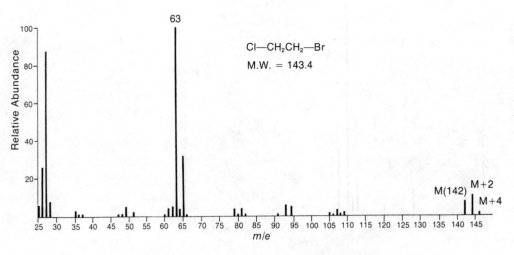

FIGURE 6–64 Mass Spectrum of 1-Bromo-2-chloroethane

6.6 ADDITIONAL TOPICS

As one may have observed during the discussions in Section 6.5, the molecular ions of many classes of compounds are so unstable that they decompose before they can reach the detector of the mass spectrometer. As a result, the molecular ion peaks for these classes of compounds are either very weak or totally absent. In such cases, two recently introduced techniques are useful in studying the molecular ion. These techniques are *field ionization* and *chemical ionization.*

Field ionization involves passing the sample molecules very close to a thin wire which carries a high electrical potential. The strong electric field in the vicinity of this wire ionizes the molecules of sample. Molecular ions formed in this manner do not possess the high degree of vibrational energy found in molecular ions formed by electron impact. Consequently, the ions formed in field ionization are much more stable. The abundance of molecular ions formed in this manner is much higher than in other means of ionization, and the molecular ion peak is usually fairly intense. In some cases, a molecular ion peak may be observed easily by using field ionization, while no peak at all would be observed after electron bombardment.

In chemical ionization, the sample is introduced into the ionization chamber along with 1 or 2 mm Hg of some reagent gas, usually methane. Essentially all of the electrons ionize methane molecules, rather than sample molecules. Once the methane molecules are ionized, a series of ion-molecule collisions yield, among other species, the ions CH_5^+, $C_2H_5^+$, and $C_3H_5^+$. These ions act as strong Lewis acids, and they can react with sample molecules to produce their corresponding conjugate acids. A sample chemical equation illustrates this behavior:

$$CH_5^+ \quad + \quad R\text{–}H \quad \rightarrow \quad RH_2^+ \quad + \quad CH_4$$

These protonated molecules, being cations, are accelerated in the usual way, giving rise to peaks whose mass is one unit higher than the mass of the expected molecular ion. Chemical ionization mass spectra show peaks at one mass unit higher than those expected in electron impact spectra. Occasionally, it is possible for peaks to appear at one mass unit *lower* than expected, due to the loss of hydrogen molecules from the protonated molecular ions. Nevertheless, chemical ionization methods are useful in studying molecular ions, since the stability of the M + 1 peak is usually greater than that of the molecular ion. The M + 1 peak can be observed in cases where the molecular ion peak is either very weak or totally absent.

A special type of mass spectrometer, called a *time-of-flight* mass spectrometer, may be used for certain applications. The time-of-flight instrument does not use a magnetic field to separate ions of varying masses. In this method, ions are produced from sample molecules by means of electron bombardment, as in most mass spectrometric methods. However, rather than using a continuous beam of high-energy electrons, the time-of-flight mass spectrometer passes the electrons through a control grid which has an oscillating positive potential applied across it. The positive potential oscillates at about 10,000 Hz, producing pulses of electrons lasting about 0.25 microsecond. The ions which are produced by these pulses of electrons are accelerated by an electric field which is pulsed at the same frequency as the control grid, but lagging behind the ionization pulse. In this acceleration, all ions will obtain the same kinetic energy, and they will reach a velocity which depends upon their individual mass-to-charge ratios.

Rather than being focused by a magnetic field, the ions travel down a field-free, evacuated drift tube, which is about 1 to 3 meters long. Because the ions are traveling at different velocities, they will arrive at the detector at the end of the drift tube at different times. The output of the detector is amplified and recorded on an oscilloscope, although a chart recorder may also be used.

Since all of the ions which were created by a single electron pulse are collected and recorded on the oscilloscope, it is possible to produce a mass spectrum far more quickly than by conventional methods. This makes the time-of-flight technique particularly useful in the study of short-lived phenomena. The spectrum produced on the oscilloscope is a plot of the number of ions reaching the detector versus mass-to-charge ratio, as usual. However, the mass-to-charge ratio is determined in this case by an accurate electronic measurement of each ion's flight time from the ion source to the detector.

Time-of-flight instruments are not capable of the high resolution which conventional instruments can provide. Resolution of about 200 is possible, with a mass range of from 1 to about 5000. Ions whose value of m/e is greater than 5000 travel so slowly down the drift tube that ions of $m/e = 1$ from the next burst catch up to them before they can reach the detector.

Time-of-flight mass spectrometers are relatively simple, which makes it possible to use them in the field. Because of their value in studying short-lived species, they are particularly useful in kinetic studies, especially with applications to very fast reactions. Such very rapid reactions as combustion and explosions can be investigated with this technique.

A recent innovation in sample introduction systems is the use of a gas chromatograph coupled to a mass spectrometer. This method is known as gas chromatography/mass spectrometry (or gc/ms). In this technique, the gas stream emerging from the gas chromatograph is admitted through a valve into a tube, where the gas stream passes over a molecular leak. Some of the gas stream is thus admitted into the ionization chamber of the mass spectrometer. In this way, it is possible to obtain the mass spectrum of every component in a mixture being injected into the gas chromatograph.

A drawback to this method involves the need for rapid scanning by the mass spectrometer. The mass spectrum of each component in the mixture must be recorded before the next component exits from the gas chromatography column. The mass spectrometer must be capable of recording the spectrum rather quickly in order that one substance is not contaminated by the next fraction before its spectrum has been completely recorded. In some cases, it may be necessary to use a time-of-flight instrument to obtain sufficiently rapid scan speeds.

The mass spectrometer may be disconnected from the gas chromatograph quite easily in a system such as that described here. Therefore, the mass spectrometer may also be used with the types of sample introduction methods described in Section 6.1.

REFERENCES

D. J. Pasto and C. R. Johnson, "Organic Structure Determination," Prentice-Hall, Englewood Cliffs, N.J., 1969. Chapter 8.

S. R. Shrader, "Introductory Mass Spectrometry," Allyn and Bacon, Boston, 1971.

F. W. McLafferty, "Interpretation of Mass Spectra," 2nd Edition, W. A. Benjamin, Inc., Reading, Mass., 1973.

R. M. Silverstein, G. C. Bassler, and T. C. Morrill, "Spectrometric Identification of Organic Compounds," 3rd Edition, John Wiley, New York, 1974. Chapter 2.

H. Budzikiewicz, C. Djerassi, and D. H. Williams, "Mass Spectrometry of Organic Compounds," Holden-Day, San Francisco, 1967.

K. Biemann, "Mass Spectrometry: Organic Chemical Applications," McGraw-Hill, New York, 1962.

J. H. Beynon, "Mass Spectrometry and its Applications to Organic Chemistry," Elsevier, Amsterdam, 1960.

PROBLEMS

1. A low resolution mass spectrum of the alkaloid vobtusine showed the molecular weight to be 718. This molecular weight is correct for the molecular formulas $C_{43}H_{50}N_4O_6$ or $C_{42}H_{46}N_4O_7$. A high resolution mass spectrum provided a molecular weight of 718.3743. Which of the possible formulas is the correct molecular formula for vobtusine?

2. A tetramethyltriacetyl derivative of oregonin, a diarylheptanoid xyloside found in red alder, was found to have a molecular weight of 660 by low resolution mass spectrometry. Possible molecular formulas include $C_{32}H_{36}O_{15}$, $C_{33}H_{40}O_{14}$, $C_{34}H_{44}O_{13}$, $C_{35}H_{48}O_{12}$, $C_{32}H_{52}O_{14}$, and $C_{33}H_{56}O_{13}$. High resolution mass spectrometry indicated that the precise molecular weight was 660.278. What is the correct molecular formula for this derivative of oregonin?

3. An unknown substance shows a molecular ion peak at m/e = 170 with a relative intensity of 100. The M + 1 peak has an intensity of 13.2, and the M + 2 peak has an intensity of 1.00. What is the molecular formula of the unknown?

4. An unknown hydrocarbon has a molecular ion peak at m/e = 84, with a relative intensity of 31.3. The M + 1 peak has a relative intensity of 2.06, and the M + 2 peak has a relative intensity of 0.08. What is the molecular formula for this substance?

5. An unknown substance has a molecular ion peak at m/e = 107, with a relative intensity of 100. The relative intensity of the M + 1 peak is 8.00, and the relative intensity of the M + 2 peak is 0.30. What is the molecular formula for this unknown?

6. The mass spectrum of an unknown liquid shows a molecular ion peak at m/e = 78, with a relative intensity of 23.6. The relative intensities of the isotopic peaks are:

m/e =	Relative Intensity =
79	1.00
80	7.55
81	0.25

What is the molecular formula of this unknown?

7. Predict the mass spectra of the four compounds shown below, and discuss how these spectra might be used to predict and confirm the fragmentation pattern of cyclohexanone.

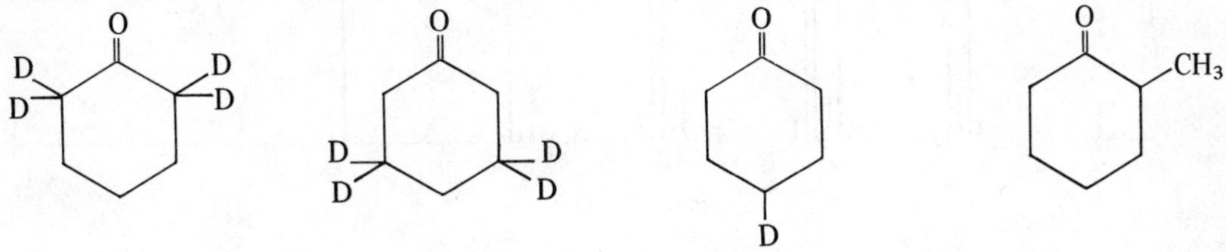

8. Assign a structure which would be expected to give rise to each of the mass spectra shown below. NOTE: There may be more than one reasonable answer to some of these problems.

a)

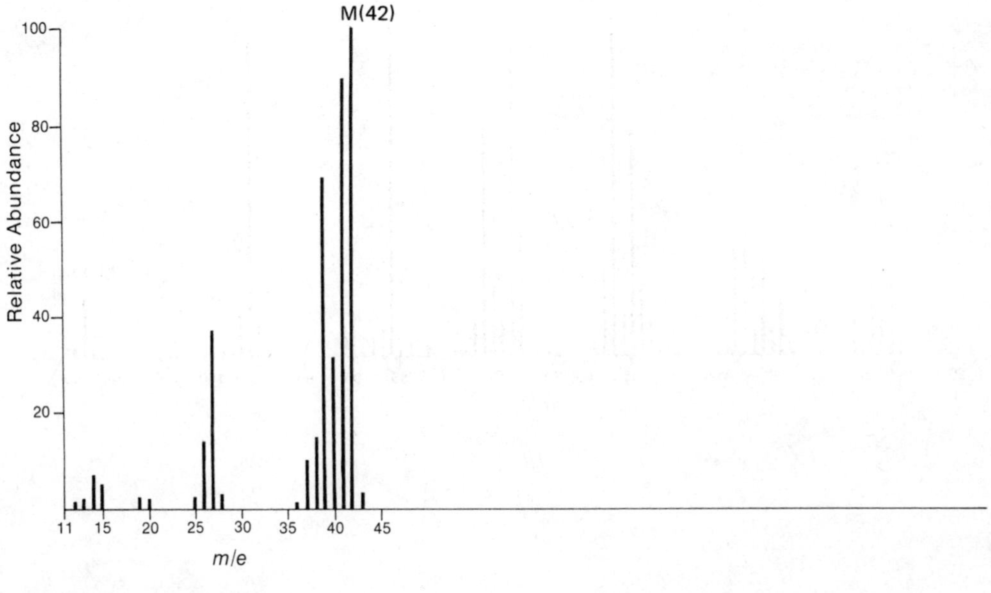

b)

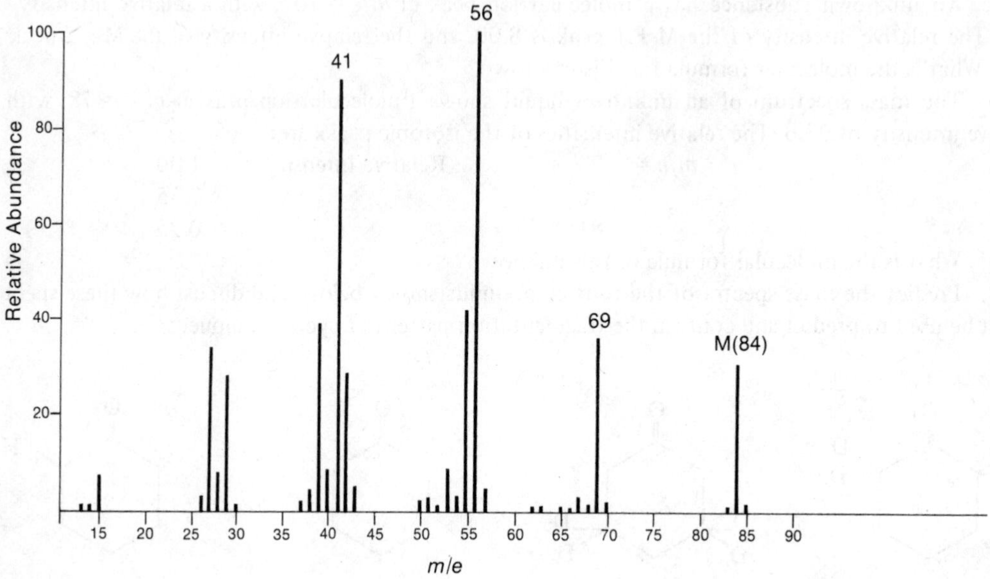
56
41
69
M(84)
Relative Abundance
100
80
60
40
20
15 20 25 30 35 40 45 50 55 60 65 70 75 80 85 90
m/e

c)

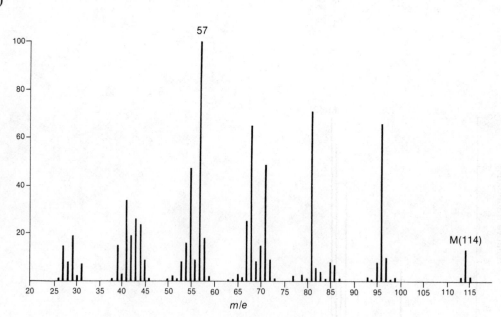

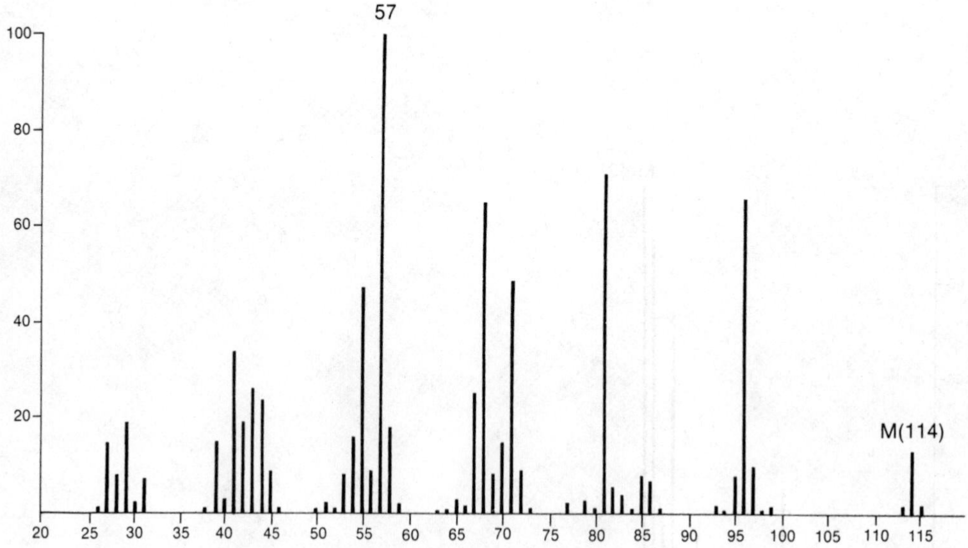
57
M(114)
100
80
60
40
20
20 25 30 35 40 45 50 55 60 65 70 75 80 85 90 95 100 105 110 115
m/e

d)

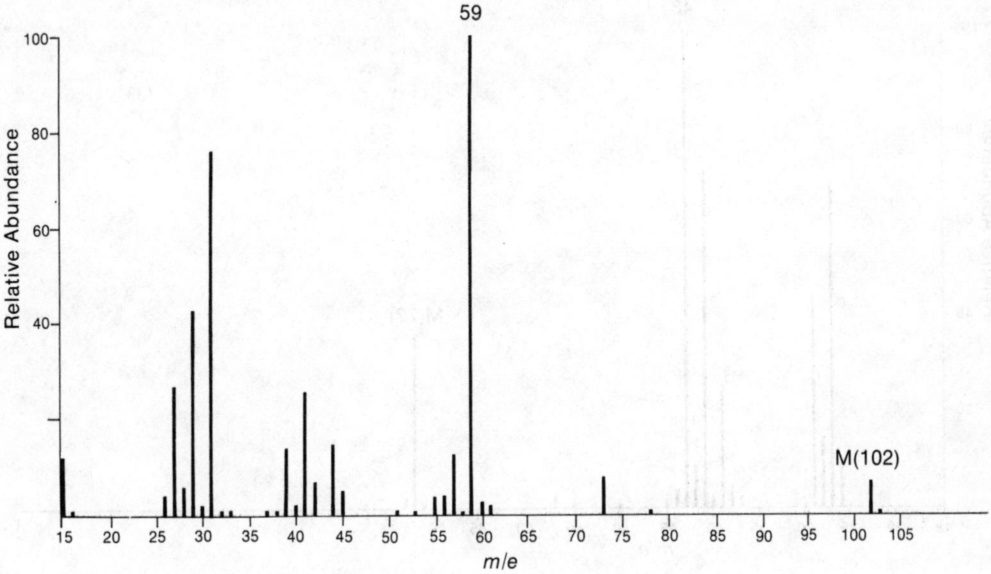
59
M(102)
Relative Abundance
m/e

e)

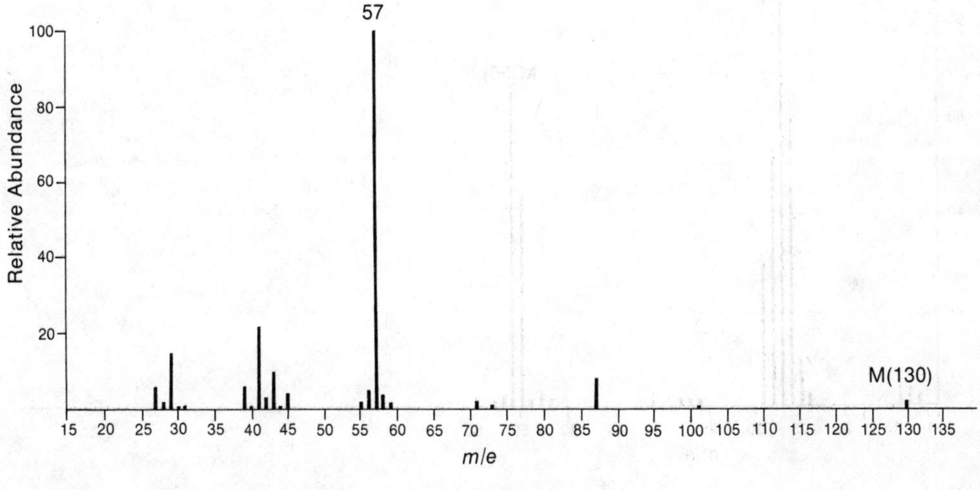
57
M(130)
Relative Abundance
m/e

f)

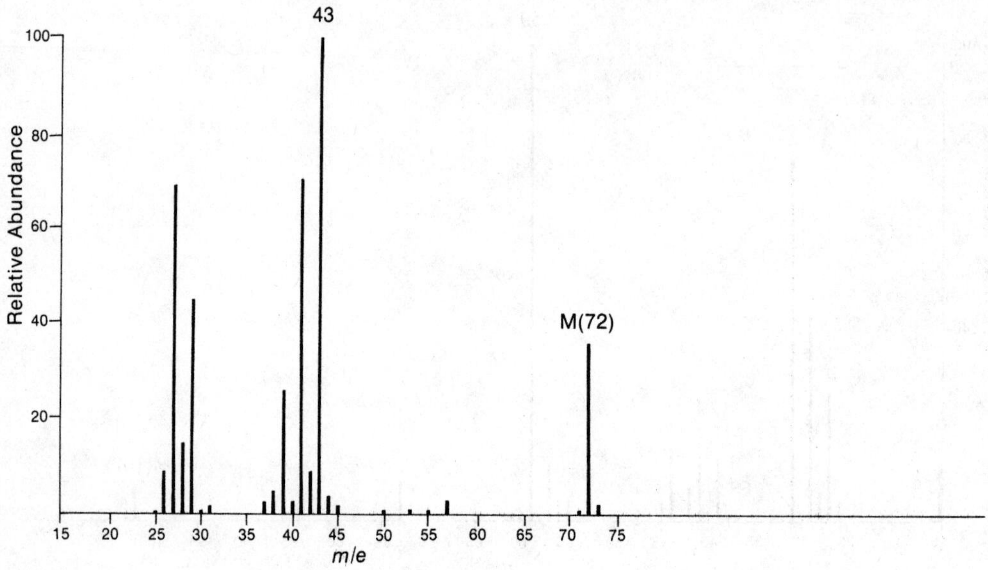
43
M(72)
Relative Abundance
100
80
60
40
20
15
20
25
30
35
40
45
50
55
60
65
70
75
m/e

g)

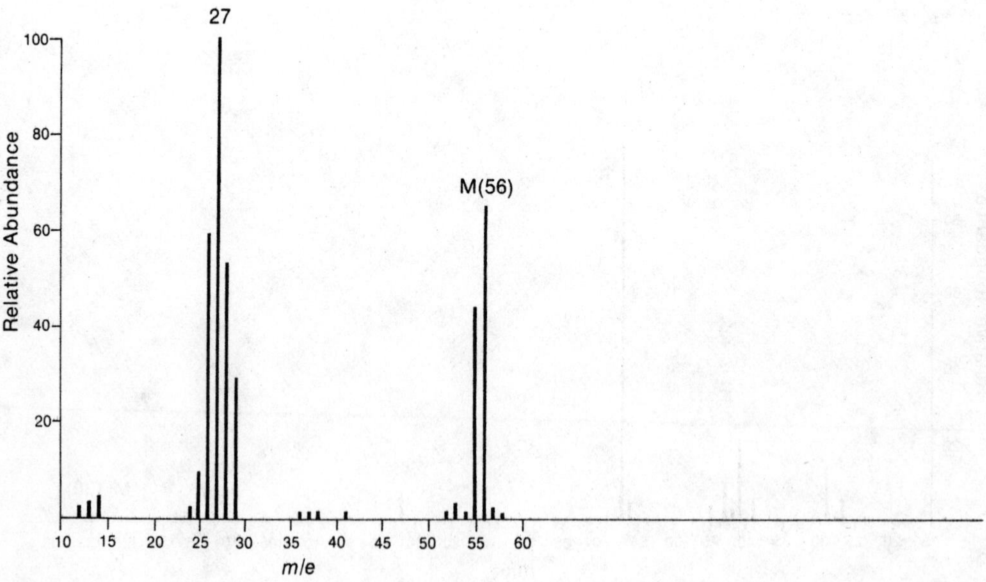
27
M(56)
Relative Abundance
100
80
60
40
20
10
15
20
25
30
35
40
45
50
55
60
m/e

h)

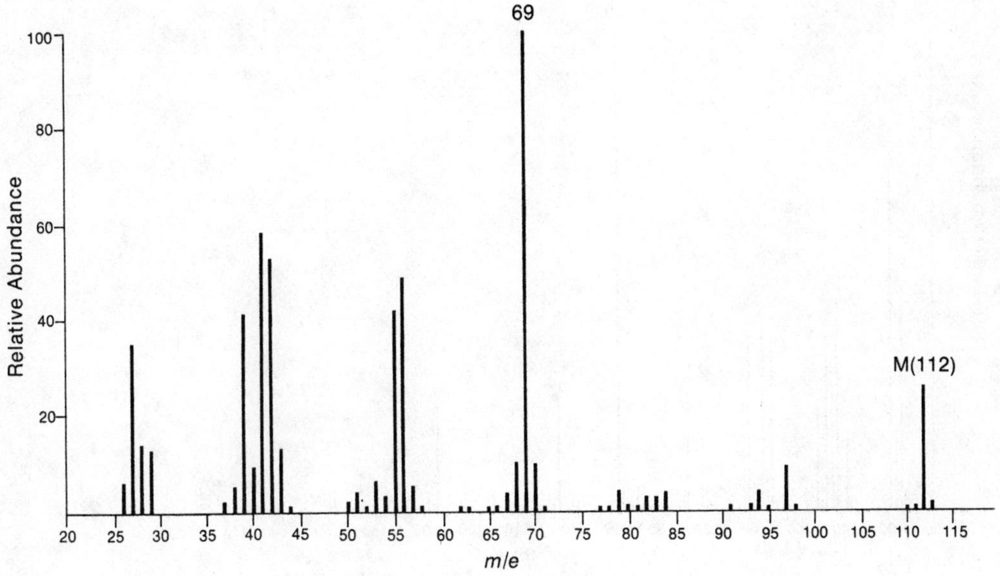
69
100
80
60
40
20
Relative Abundance
M(112)
20 25 30 35 40 45 50 55 60 65 70 75 80 85 90 95 100 105 110 115
m/e

i)

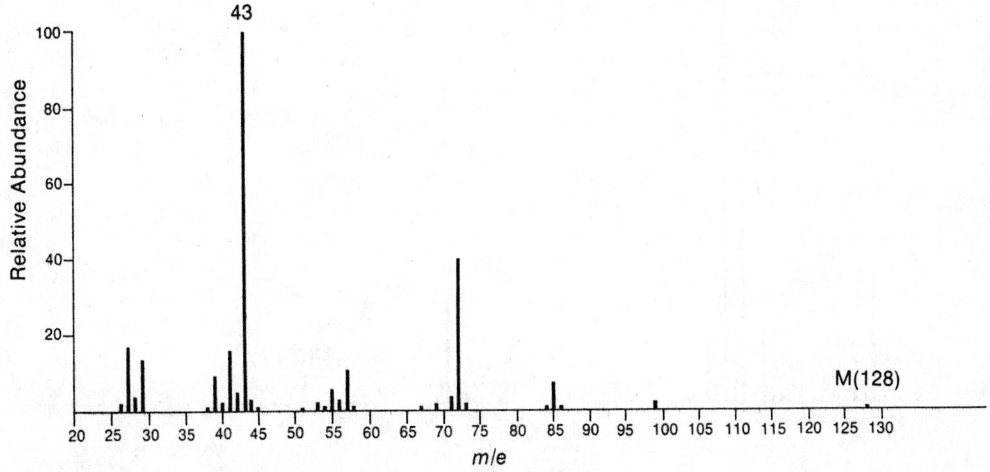
43
100
80
60
40
20
Relative Abundance
M(128)
20 25 30 35 40 45 50 55 60 65 70 75 80 85 90 95 100 105 110 115 120 125 130
m/e

j)

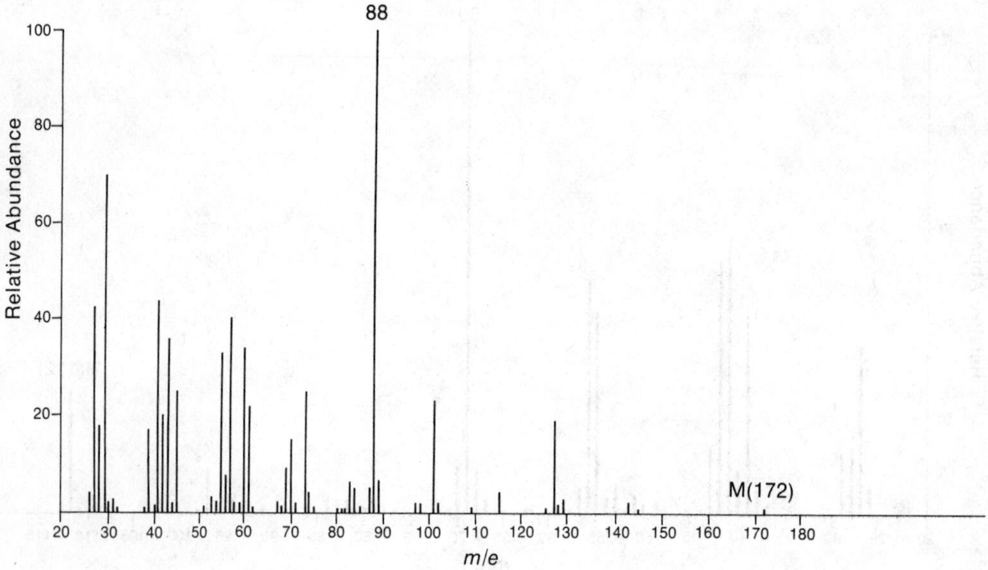
88
M(172)
Relative Abundance
m/e
100
80
60
40
20
20
30
40
50
60
70
80
90
100
110
120
130
140
150
160
170
180

k)

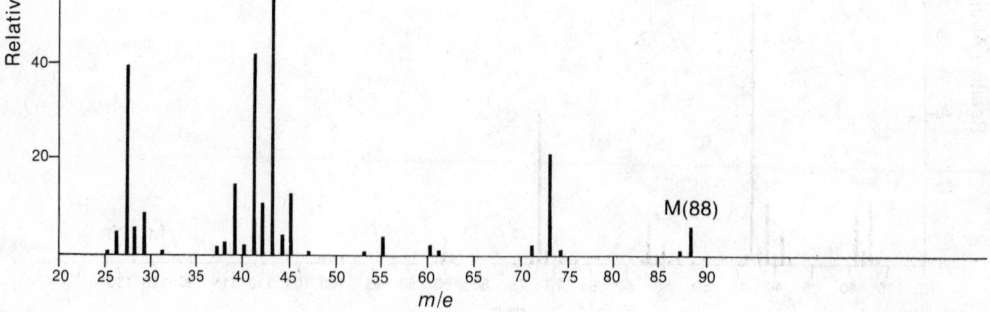
43
M(88)
Relative Abundance
m/e
100
80
60
40
20
20
25
30
35
40
45
50
55
60
65
70
75
80
85
90

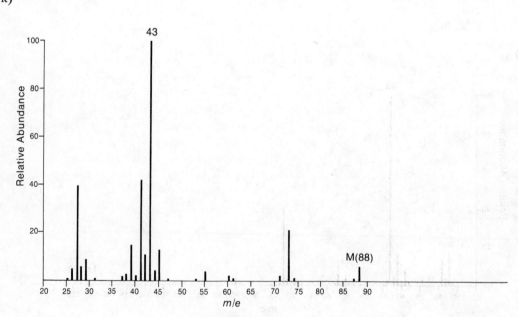

l)

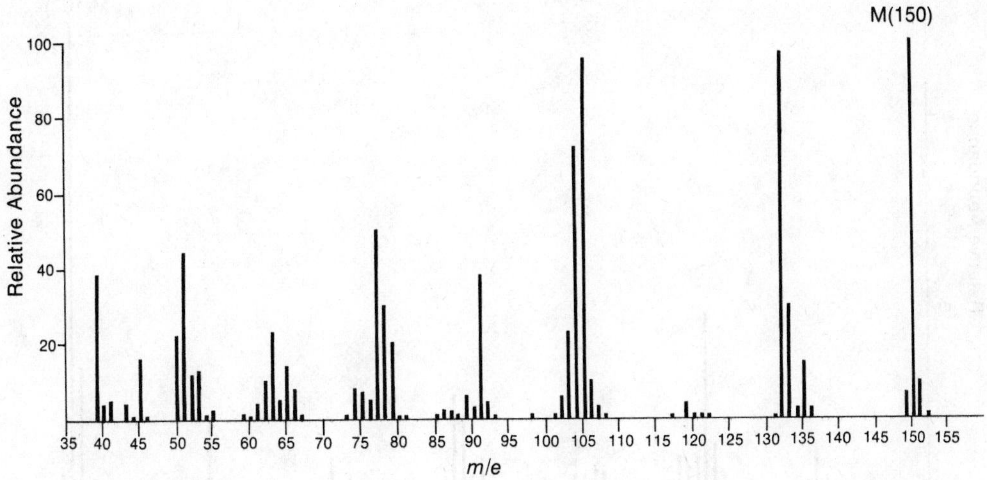
M(150)
100
80
60
40
20
Relative Abundance
35 40 45 50 55 60 65 70 75 80 85 90 95 100 105 110 115 120 125 130 135 140 145 150 155
m/e

m)

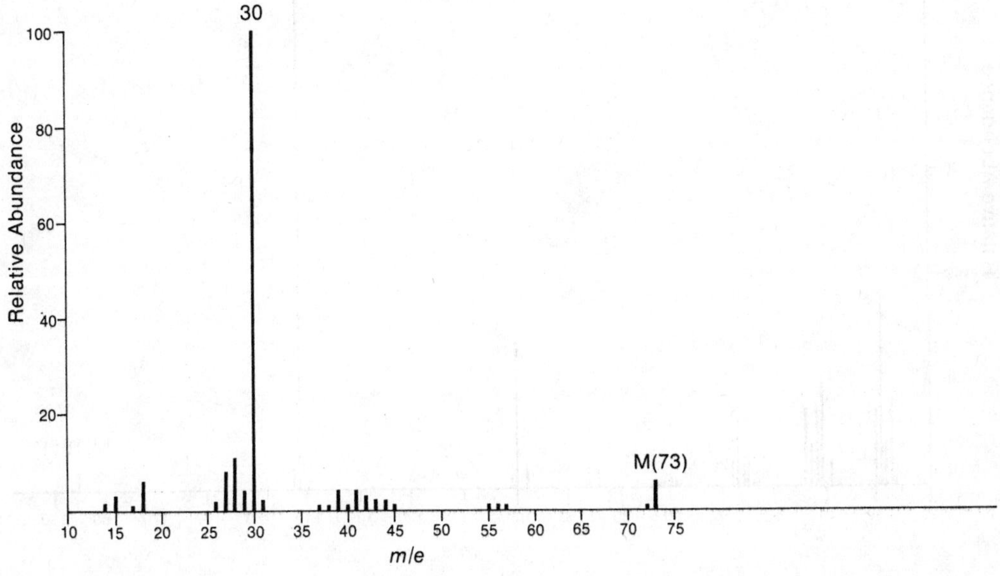
30
M(73)
100
80
60
40
20
Relative Abundance
10 15 20 25 30 35 40 45 50 55 60 65 70 75
m/e

n)

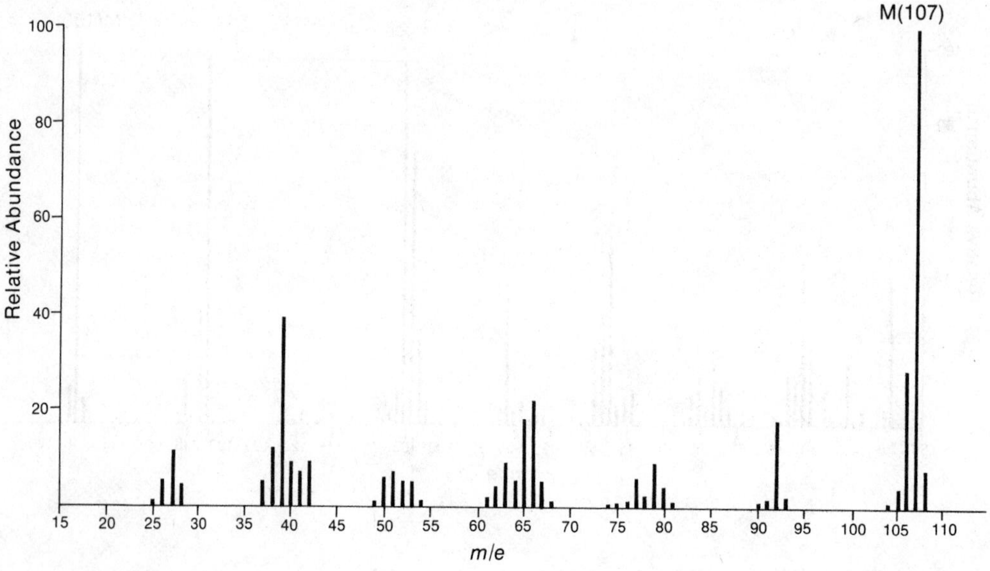
M(107)
100
80
60
40
20
Relative Abundance
15
20
25
30
35
40
45
50
55
60
65
70
75
80
85
90
95
100
105
110
m/e

o)

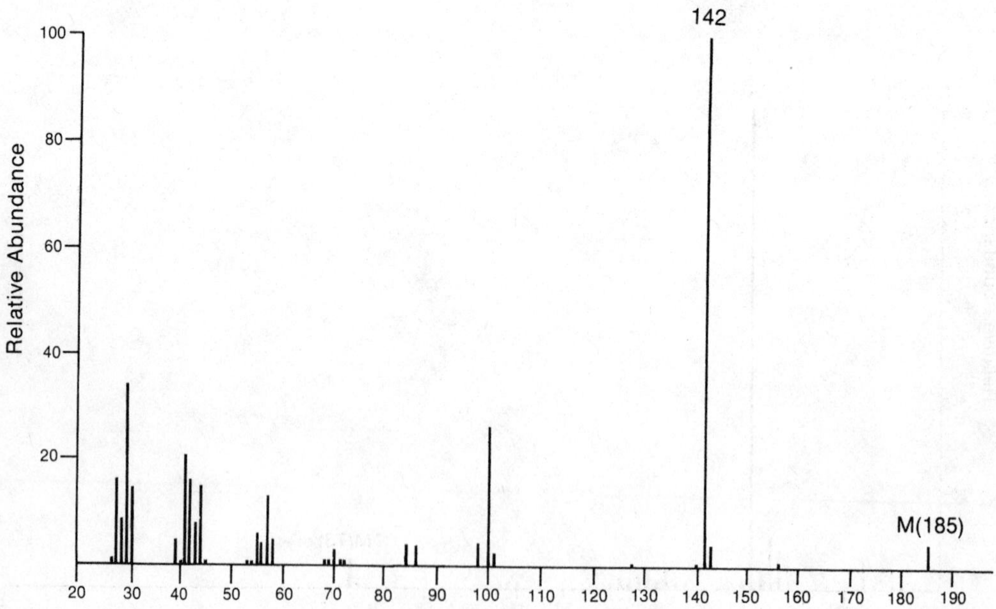
142
M(185)
100
80
60
40
20
Relative Abundance
20
30
40
50
60
70
80
90
100
110
120
130
140
150
160
170
180
190
m/e

p)

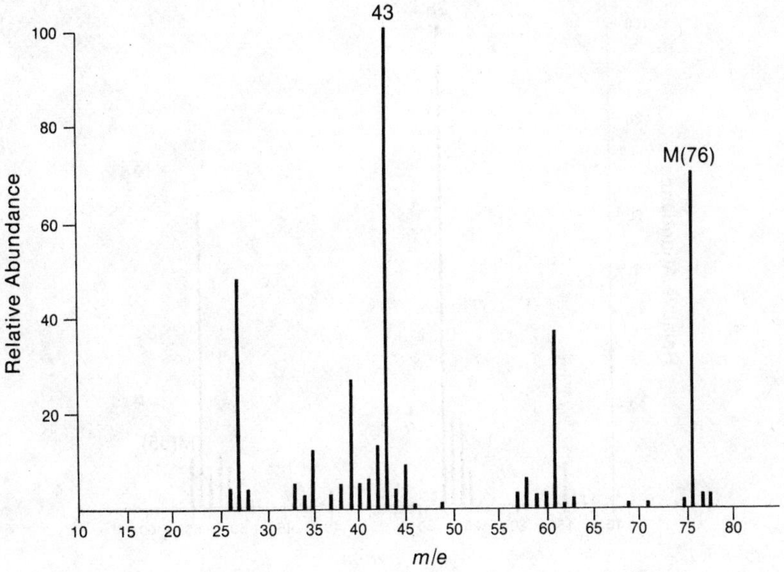
43
M(76)
Relative Abundance
100
80
60
40
20
10
15
20
25
30
35
40
45
50
55
60
65
70
75
80
m/e

q)

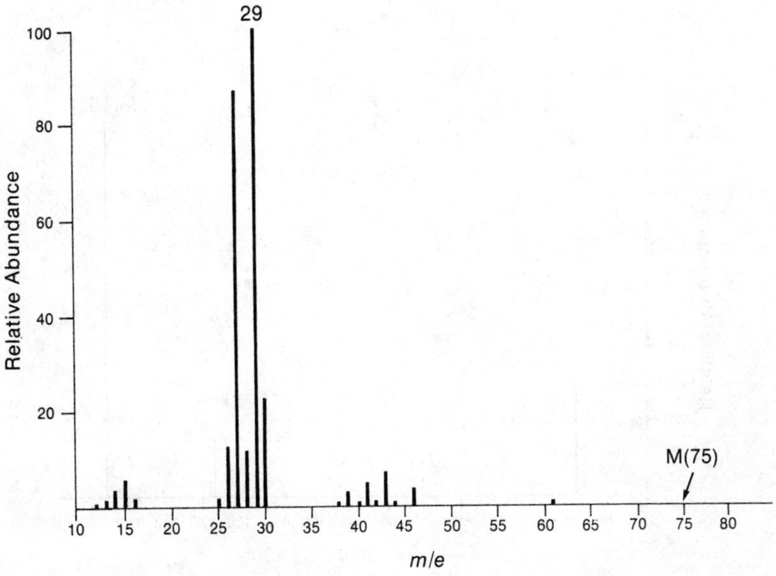
29
M(75)
Relative Abundance
100
80
60
40
20
10
15
20
25
30
35
40
45
50
55
60
65
70
75
80
m/e

r)

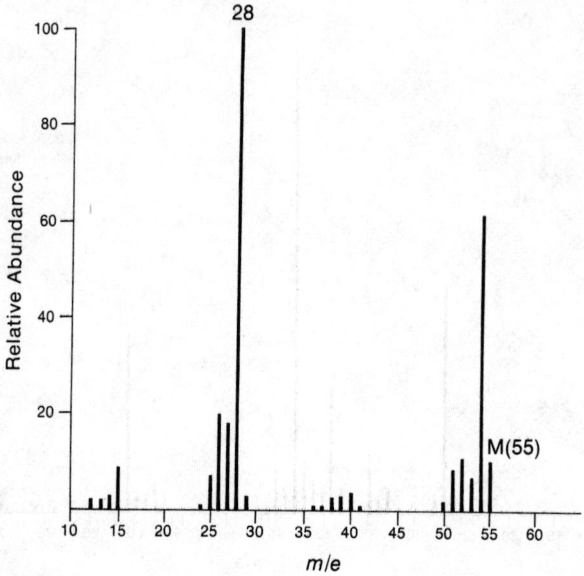
28
M(55)
Relative Abundance
100
80
60
40
20
10 15 20 25 30 35 40 45 50 55 60
m/e

s)

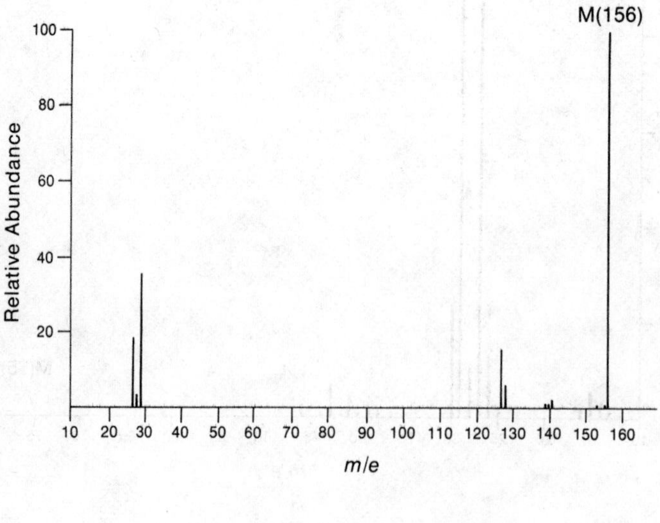
M(156)
Relative Abundance
100
80
60
40
20
10 20 30 40 50 60 70 80 90 100 110 120 130 140 150 160
m/e

t)

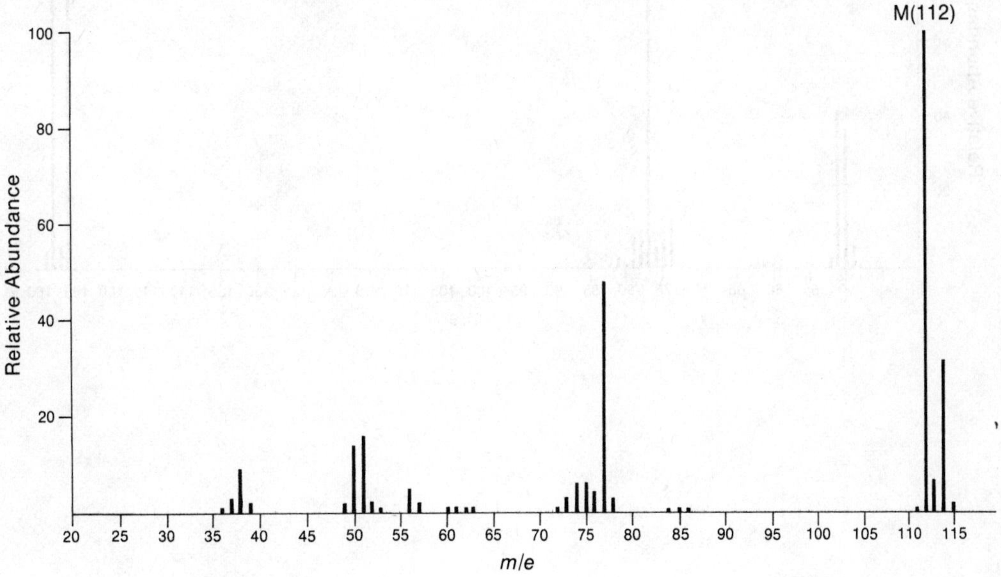
M(112)
100
80
60
40
20
Relative Abundance
20 25 30 35 40 45 50 55 60 65 70 75 80 85 90 95 100 105 110 115
m/e

u)

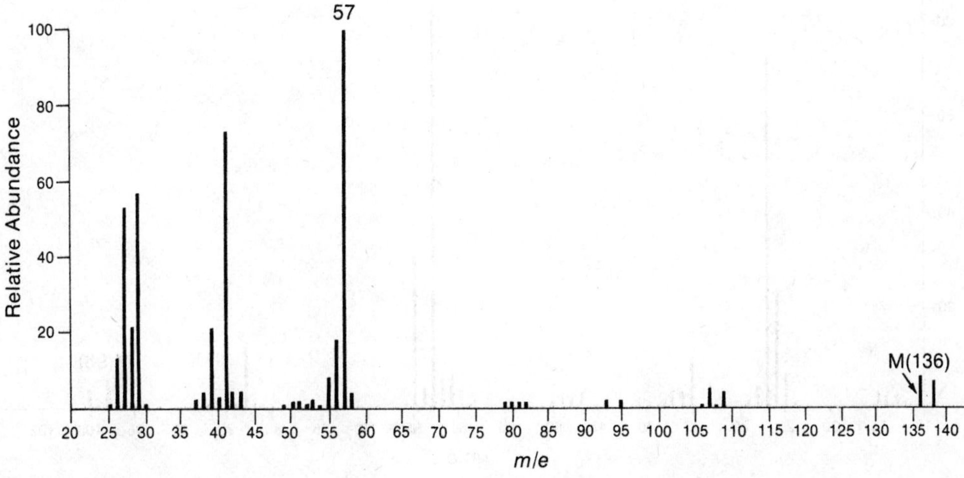
57
M(136)
100
80
60
40
20
Relative Abundance
20 25 30 35 40 45 50 55 60 65 70 75 80 85 90 95 100 105 110 115 120 125 130 135 140
m/e

v)

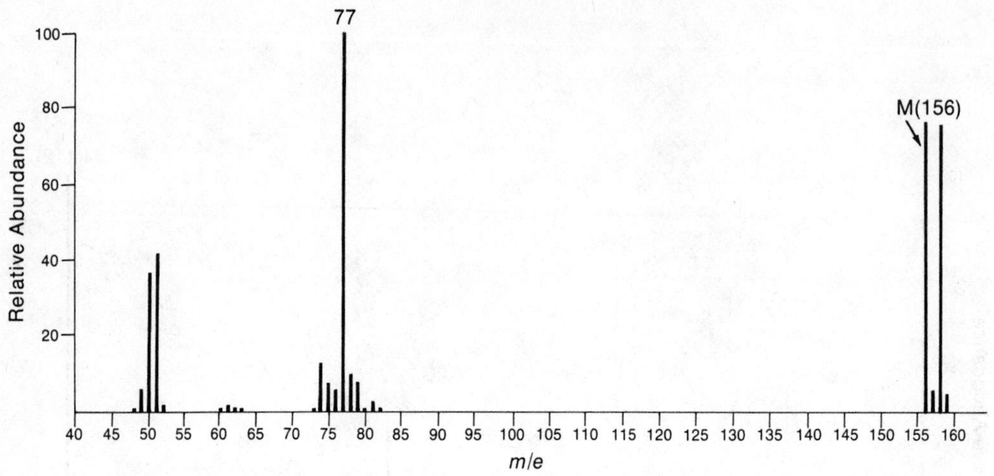
77
M(156)
Relative Abundance
100
80
60
40
20
40 45 50 55 60 65 70 75 80 85 90 95 100 105 110 115 120 125 130 135 140 145 150 155 160
m/e

w)

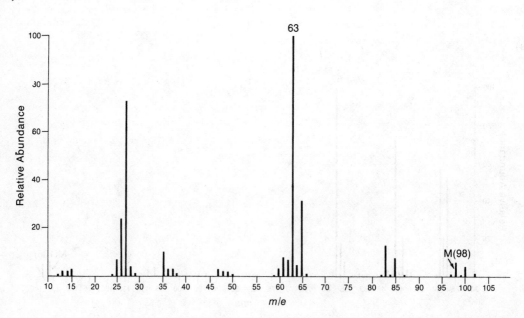

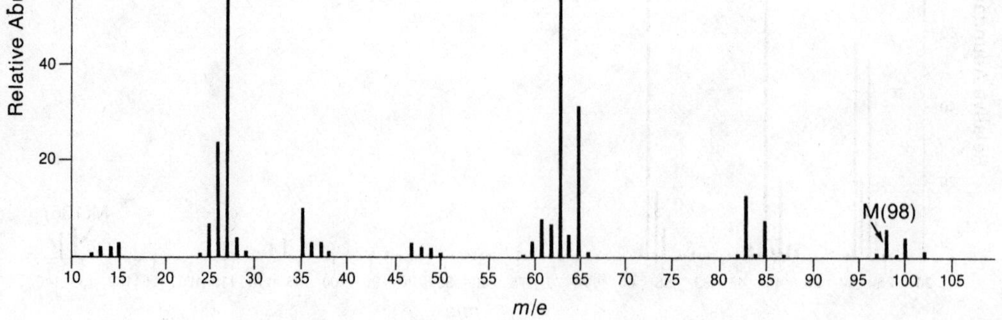
63
M(98)
Relative Abundance
100
30
60
40
20
10 15 20 25 30 35 40 45 50 55 60 65 70 75 80 85 90 95 100 105
m/e

x)

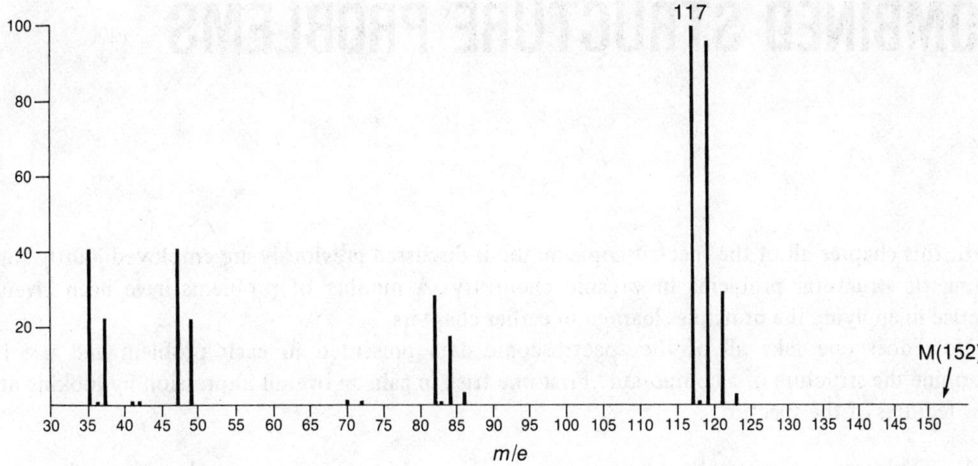
117
M(152)
Relative Abundance
100
80
60
40
20
30 35 40 45 50 55 60 65 70 75 80 85 90 95 100 105 110 115 120 125 130 135 140 145 150
m/e

CHAPTER 7

COMBINED STRUCTURE PROBLEMS

In this chapter all of the spectroscopic methods discussed previously are employed jointly for the solving of structural problems in organic chemistry. A number of problems have been given for practice in applying the principles learned in earlier chapters.

How does one take all of the spectroscopic data presented in each problem and use it to determine the structure of a compound? First one tries to gain an overall impression by looking at the gross features of the spectra:

1. The mass spectrum should give a molecular weight and may provide information on the number and type of halogen atoms present.

2. The infrared spectrum provides some idea of the functional groups that may be present or not present.

3. The nuclear magnetic resonance spectrum gives the structure of the attached carbon-hydrogen containing groups.

4. The ultraviolet spectrum indicates something about the unsaturation in the carbon skeleton.

After having obtained this first impression, then one may look at each spectrum in more detail so that a final structure may be determined.

Several comments should be made about the problems which follow the examples:

1. The lowest whole number integral ratios have been indicated on most of the nmr spectra. These values may not correspond to the molecular ratios. A few nmr spectra have integrals traced on them in the style of Figure 3–11.

2. Each mass spectrum is labeled to show the masses of some fragment ions. In most cases, the molecular ion is identified.

3. The compounds may contain the following elements: C, H, O, N, S, F, Cl, Br, I.

4. Most of the problems include data from all four of the spectral methods, but in a few problems less information is provided. The problems near the end of the chapter may be somewhat more difficult to solve than the earlier ones.

5. One should obtain a unique answer for each problem in nearly every case. If one experiences difficulty in solving the problem, the molecular formulas are listed at the end of this chapter.

6. Additional problems of the type given here are available in a number of excellent books. They are listed in the References section at the end of this chapter.

Four examples of solved problems follow. It should be noted that several alternate approaches may be taken to the solution of these example problems.

EXAMPLE ONE

THE PROBLEM

The uv spectrum of this compound shows strong **end absorption** (strong off-scale absorption below 200 nm) in 95% ethanol, with shoulders at λ_{max} 252 nm (log ϵ 2.4) and 280 nm (log ϵ 2.2). The 252 nm band shows fine structure. The ir spectrum is obtained on a **neat** liquid sample (no solvent).

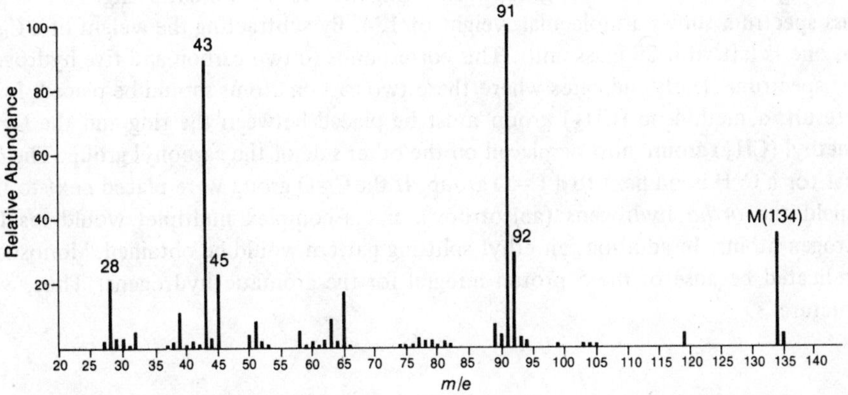

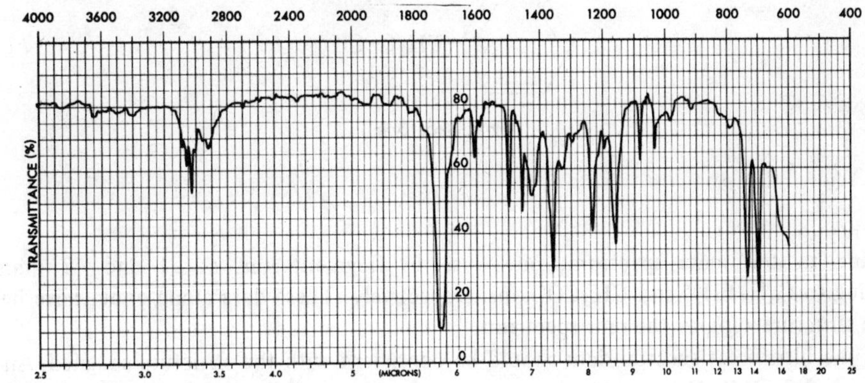

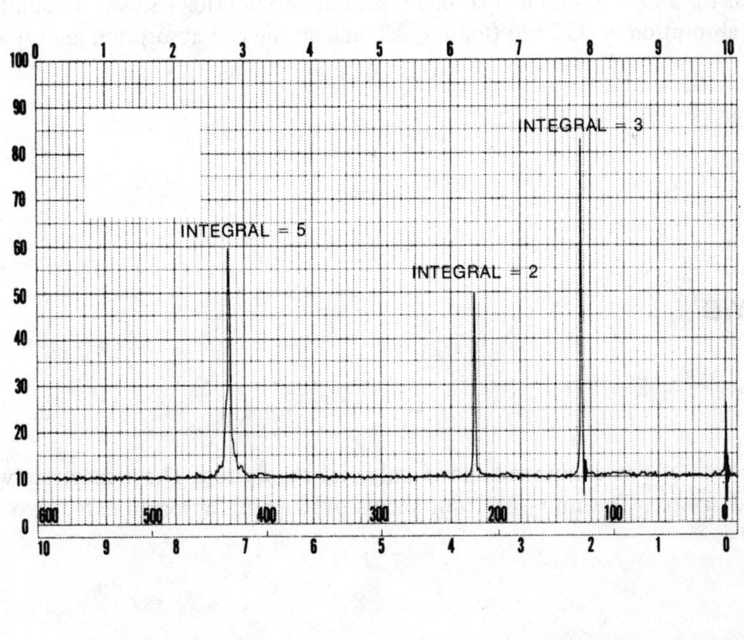

THE SOLUTION

The infrared spectrum shows aromatic C–H absorptions to the left of 3000 cm^{-1} and aliphatic hydrogens to the right of this value. In addition, one observes a C=O group at 1720 cm^{-1}, benzene ring absorption between 1600 and 1460 cm^{-1} and a monosubstituted aromatic ring pattern at 740 and 700 cm^{-1}. Since no bands appear at 2850 and 2750 cm^{-1}, an aldehyde is excluded. In addition, no strong C–O absorption occurs in the region from 1300 to 1000 cm^{-1}, which excludes an ester. Since other carbonyl-containing groups may also be eliminated, one is left with a ketone as the functional group responsible for the C=O absorption at 1720 cm^{-1}. This value is typical for an unconjugated ketone, and excludes the group from being next to the aromatic ring.

The mass spectrum shows a molecular weight of 134. By subtracting the weight of a C_6H_5 and a C=O group, one is left with 29 mass units. This corresponds to two carbon and five hydrogen atoms.

The nmr spectrum clearly indicates where these two carbon atoms should be placed. In order for singlets to result, a methylene (CH_2) group must be placed between the ring and the C=O group, while the methyl (CH_3) group must be placed on the other side of the carbonyl group. The δ value of 2.1 is typical for a C–H bond next to a C=O group. If the C=O group were placed next to the ring, it would deshield the *ortho* hydrogens (anisotropy), and a complex multiplet would result for the phenyl hydrogen atoms. In addition, an ethyl splitting pattern would be obtained. Monosubstitution is clearly indicated because of the 5 proton integral for the aromatic hydrogens. Thus, we are left with the structure:

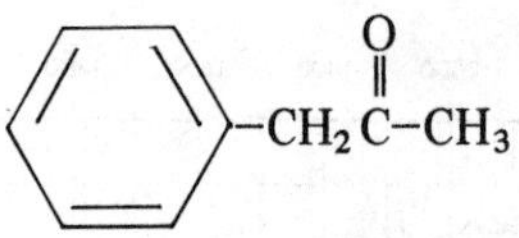

Phenylacetone

Returning to the mass spectrum, one observes fragment ions at 91 and 43 mass units, corresponding to $C_6H_5CH_2^+$ and CH_3CO^+ ions, respectively. These data alone could have been used to solve the structure *without* the nmr spectrum.

Finally, the ultraviolet spectral data should be consistent with the structure assigned. An $n \rightarrow \pi^*$ transition typical for a C=O group in a ketone is seen at 280 nm (log ϵ small). In addition, secondary (fine structure) absorption at 252 nm (log ϵ small) and strong end absorption are typical of aromatic compounds.

EXAMPLE TWO

THE PROBLEM

The uv spectrum of this compound shows only end absorption. The ir spectrum was determined on a neat liquid sample. The nmr spectrum shows overlapping absorption for two similar groups centering on 1.2 δ.

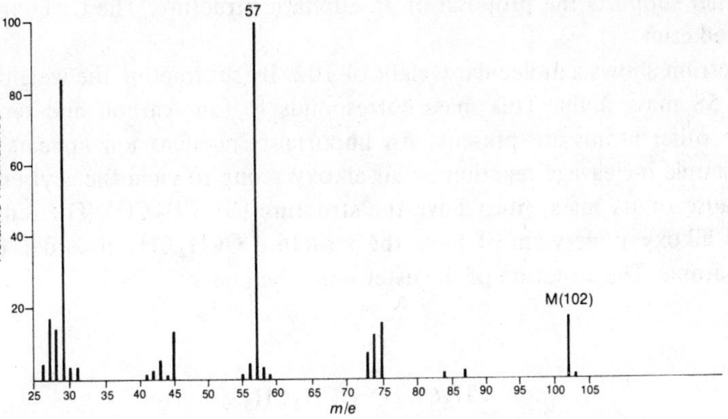

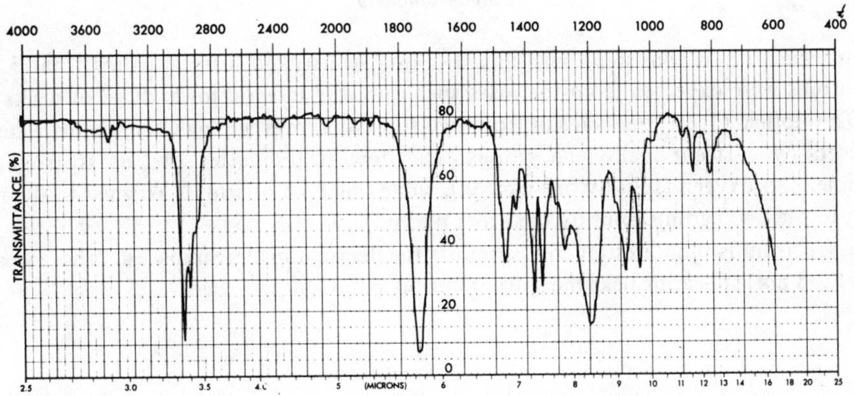

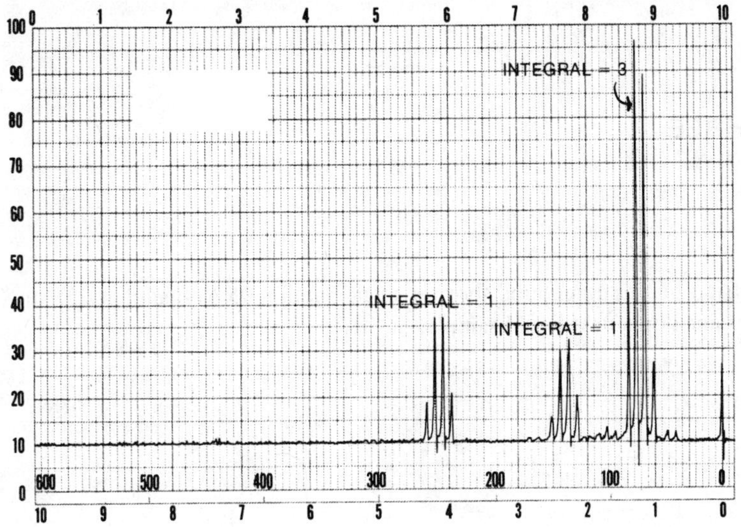

THE SOLUTION

The infrared spectrum shows aliphatic C–H absorption to the right of 3000 cm^{-1}, a C=O group at 1740 cm^{-1}, and C–O absorption (strong and broad) at 1200 cm^{-1}. This evidence suggests an aliphatic ester. Notice that no vinyl or phenyl C= C absorptions appear between 1600 and 1480 cm^{-1}, which further supports the proposal of an aliphatic structure. The C=O absorption is normal for an unconjugated ester.

The mass spectrum shows a molecular weight of 102. By subtracting the weight of a −COO−unit, one is left with 58 mass units. This mass corresponds to four carbon and ten hydrogen atoms, assuming that no other atoms are present. An important fragment ion appears at 57 mass units, resulting from a simple α-cleavage reaction of an alkoxy group to yield the acylium ion (RCO^+). This acylium ion, because of its mass, must have the structure $CH_3CH_2CO^+$. The remaining part of the ester, namely the alkoxy moiety, must have the structure OCH_2CH_3 in order to account for the remaining carbon atoms. The structure of the ester must then be:

$$CH_3CH_2\overset{\overset{\large O}{\|}}{C}{-}OCH_2CH_3$$

Ethyl Propionate

The nmr spectrum is consistent with this structure. First, the integral ratios total 5. Since 10 hydrogen atoms were expected from the mass spectrum, the integral for each set of peaks should be doubled. The upfield (1.2 δ) peaks result from two somewhat overlapping CH_3 triplets. The splittings are produced by adjacent methylene groups. The downfield quartet (4.1 δ) is assigned to the methylene next to oxygen (split by one methyl), while the 2.3 δ quartet is assigned to the methylene group next to the C=O group (split by the other methyl).

The uv spectrum is uninteresting, but supports the decision made. Simple esters have weak $n \rightarrow \pi^*$ transitions (205 nm) appearing near the solvent cut-off point.

EXAMPLE THREE

THE PROBLEM

The uv spectrum of this compound shows a single band in 95% ethanol at 258 nm (log ε 2.6). The ir spectrum is obtained on a neat liquid sample.

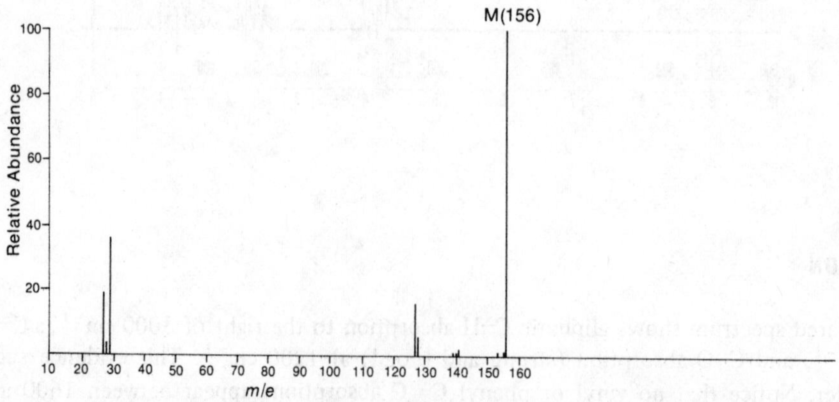

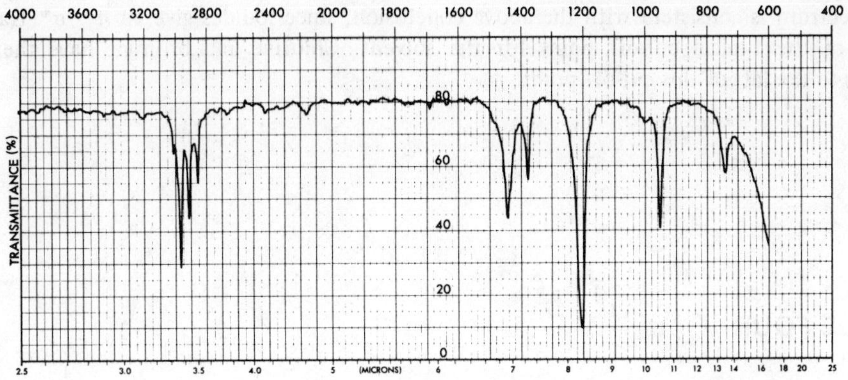

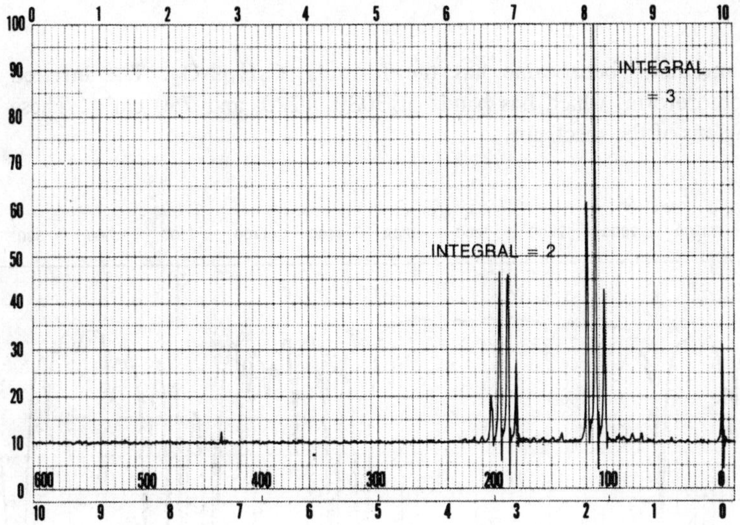

THE SOLUTION

The infrared spectrum indicates that the compound is extremely simple in its structure. An alkane might first come to mind, but the strong band at 1200 cm^{-1} would not be found in this class of compounds. Ethers absorb near this value, but usually appear near 1120 cm^{-1} for simple dialkyl substituted compounds. An alkyl halide is a strong possibility because the CH_2 wagging vibration for this class of compounds appears here.

The nmr spectrum obviously indicates an ethyl pattern, but an ether is excluded because the CH_2 group would appear closer to 3.6 δ in that case. This leaves the only possibility, an alkyl halide. The chemical shift value for the CH_2 groups would be consistent with this conclusion.

The mass spectrum provides the best clue to the nature of the halogen atom. The molecular weight of the compound is 156. If one subtracts 29 mass units for the ethyl group, the remainder is 127. Iodine has a mass of 127. Looking at the spectrum somewhat differently, one observes that the molecular ion peak is not accompanied by any other peaks (M + 2, M + 4, etc.). This excludes all of the halogens except fluorine and iodine, which are the only monoisotopic halogens. Since fluorine has a low mass, iodine remains as the only possibility. The compound is:

$$CH_3CH_2{-}I$$

Ethyl Iodide

The uv spectrum is consistent with the above conclusion, since iodides give an $n \rightarrow \sigma^*$ transition to the right of the typical cut-off point for the solvent. Chlorine and bromine have their $n \rightarrow \sigma^*$ transitions to the left of this cut-off point.

EXAMPLE FOUR

THE PROBLEM

The local anesthetic benzocaine has the formula $C_9H_{11}NO_2$. The infrared spectrum was determined in chloroform, which has bands at 3030, 1220, and 750 cm^{-1}. These bands partially obscure these regions of the spectrum.

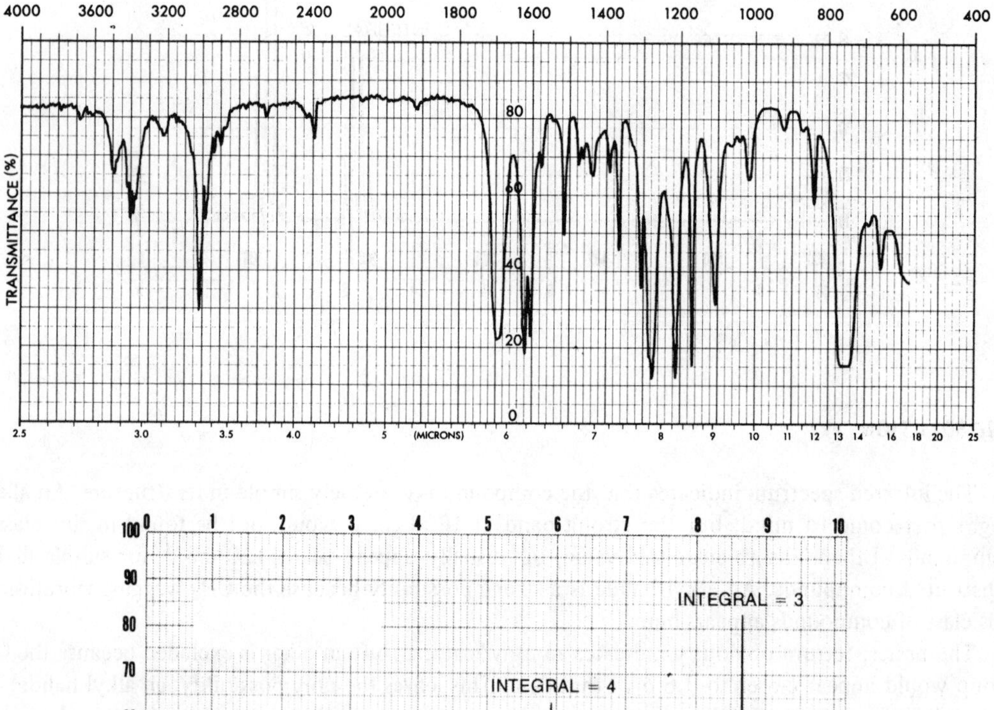

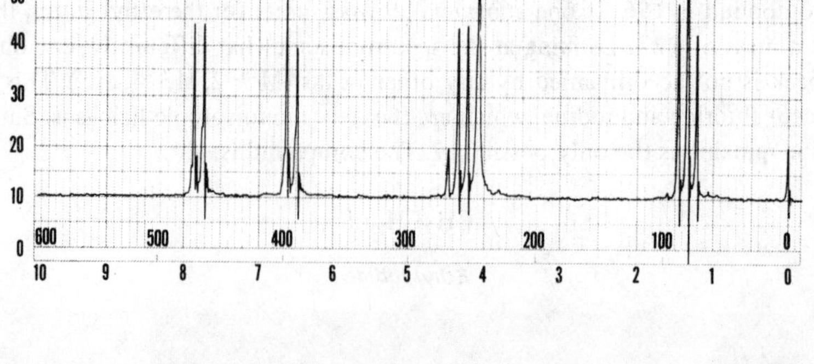

THE SOLUTION

The infrared spectrum shows the typical double peak patterns for a primary amine at 3500 and 3400 cm^{-1}. The N–H bending band overlaps the C=C aromatic ring absorptions between 1640 and 1450 cm^{-1}. In addition, the appearance of a C=O band at 1690 cm^{-1} together with several bands appearing in the range from 1300 to 1050 cm^{-1} suggest the presence of a conjugated ester. It should also be mentioned that the C–N stretch occurs in this range, as well. The substitution pattern on the ring is difficult to establish because the solvent peak at 750 cm^{-1} partially obscures the C–H oop region of the spectrum.

The nmr spectrum indicates a *para* substitution pattern (doublets at 6.6 and 7.8 δ integral = 4). The broad nmr peak at 4.1 δ is assigned to the NH_2 group. Assuming that the NH_2 peak obscures one leg of a quartet at 4.3 δ, one is left with an ethyl pattern. The compound must have the structure:

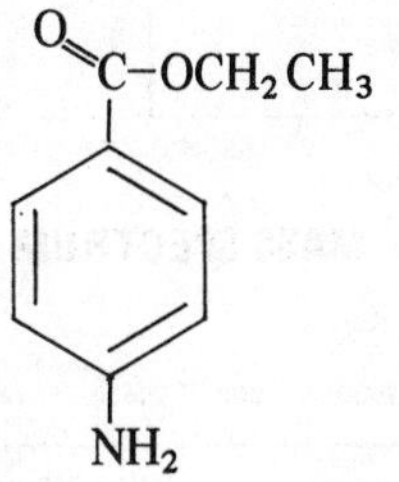

Although the mass spectrum was not given, one would have expected an *odd* mass for this compound because of the presence of a single nitrogen atom.

PROBLEM 1

The uv spectrum of this compound is determined in 95% ethanol: λ_{max} 290 nm (log ϵ 1.3). The ir spectrum is obtained on a neat liquid sample.

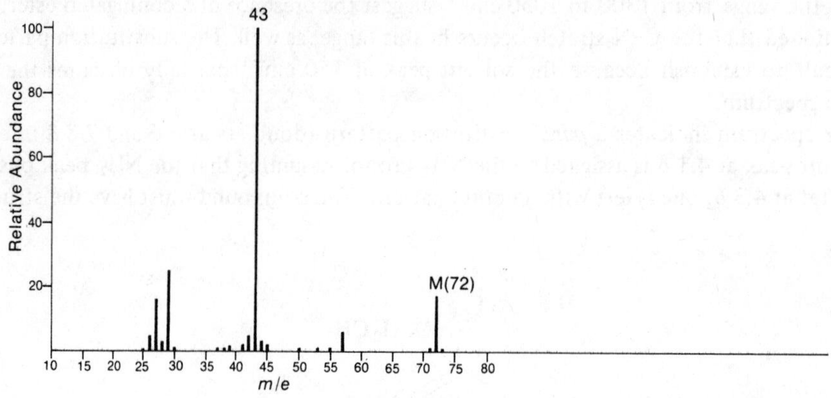

MASS SPECTRUM

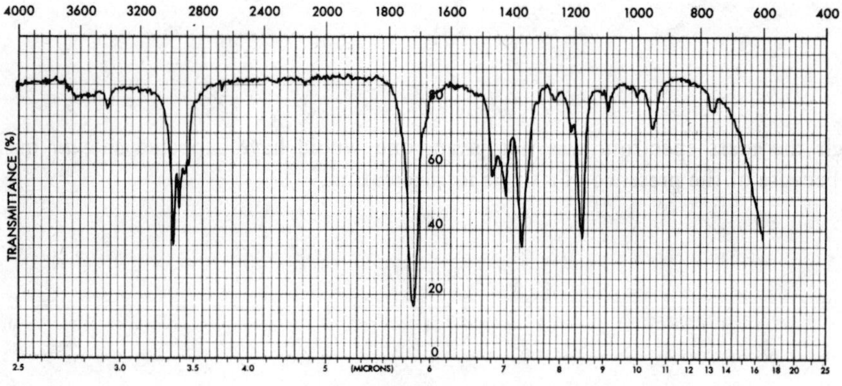

INFRARED SPECTRUM

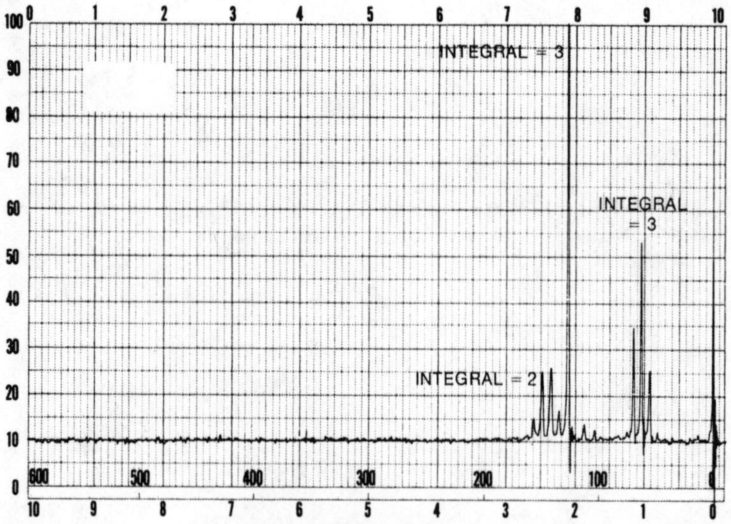

NMR SPECTRUM

PROBLEM 2

The uv spectrum of this compound shows no maximum above 205 nm. The ir spectrum is obtained on a neat liquid sample.

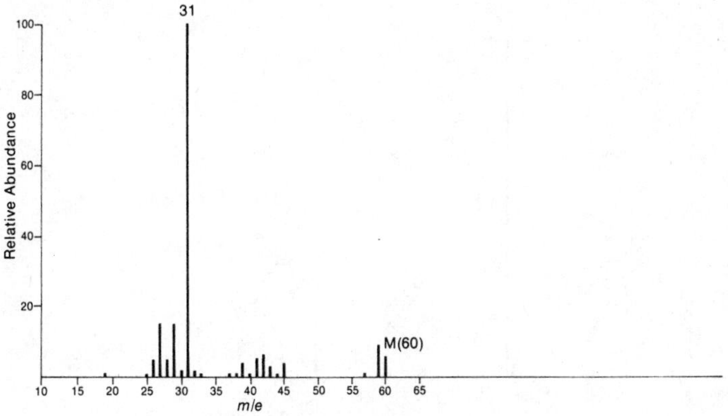

MASS SPECTRUM

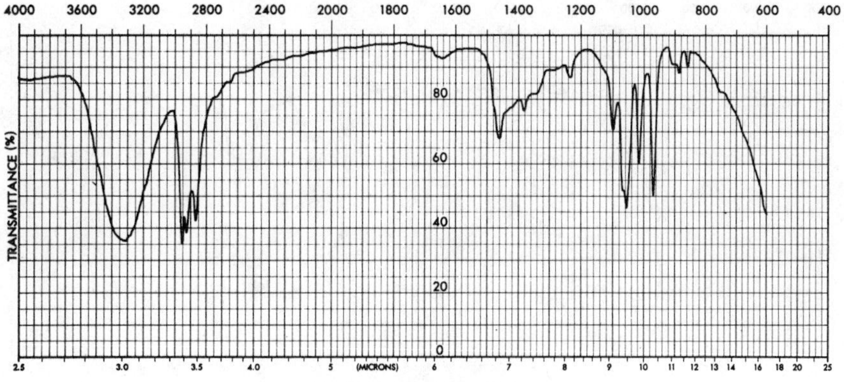

INFRARED SPECTRUM

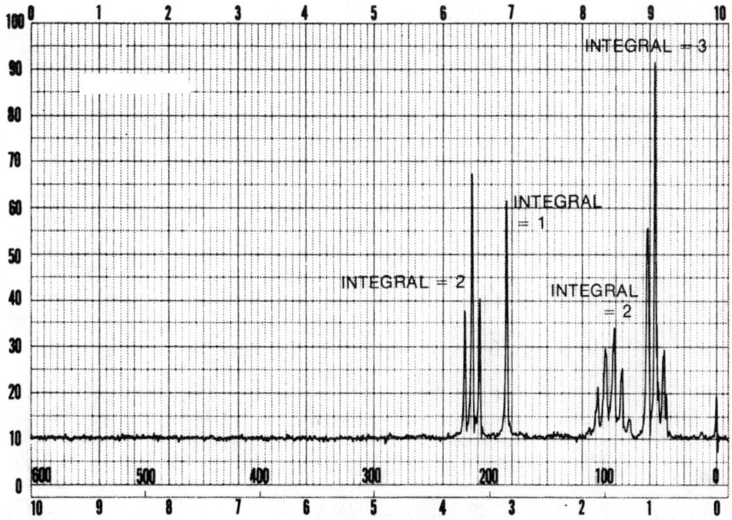

NMR SPECTRUM

PROBLEM 3

The uv spectrum of this compound is determined in 95% ethanol: λ_{max} 280 nm (log ϵ 1.3). The ir spectrum is obtained on a neat liquid sample.

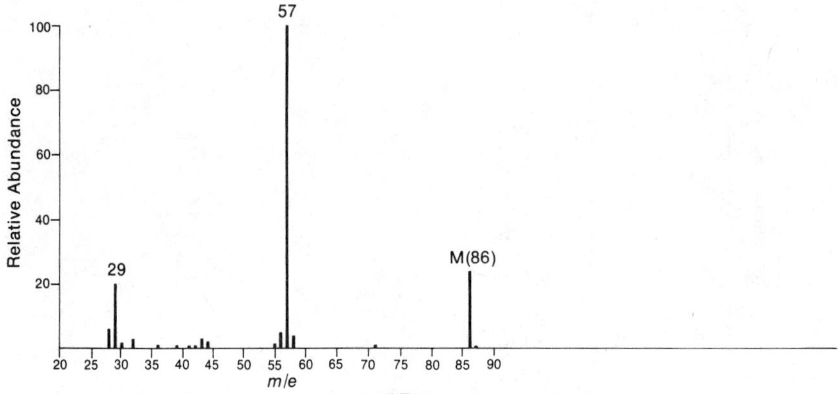

MASS SPECTRUM

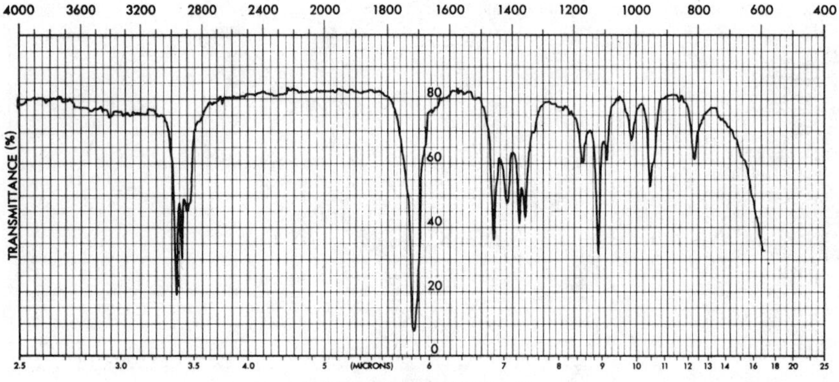

INFRARED SPECTRUM

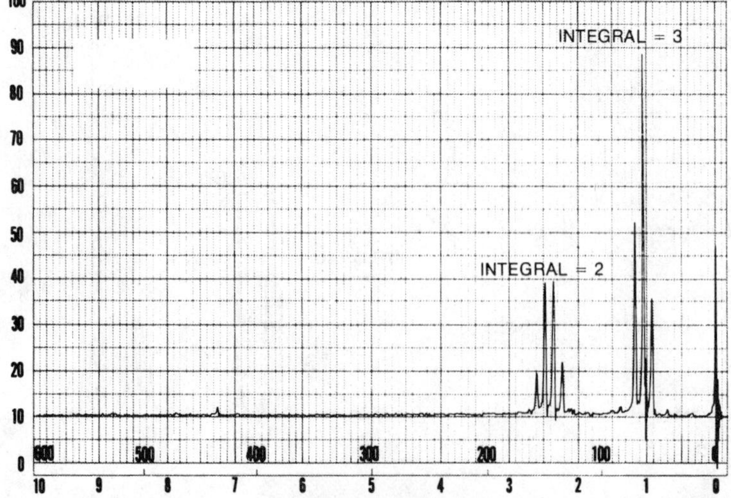

NMR SPECTRUM

PROBLEM 4

The formula for this compound is $C_6H_{12}O_2$. Solve the structure of this compound with only the ir and nmr spectra. The ir spectrum is obtained on a neat liquid sample.

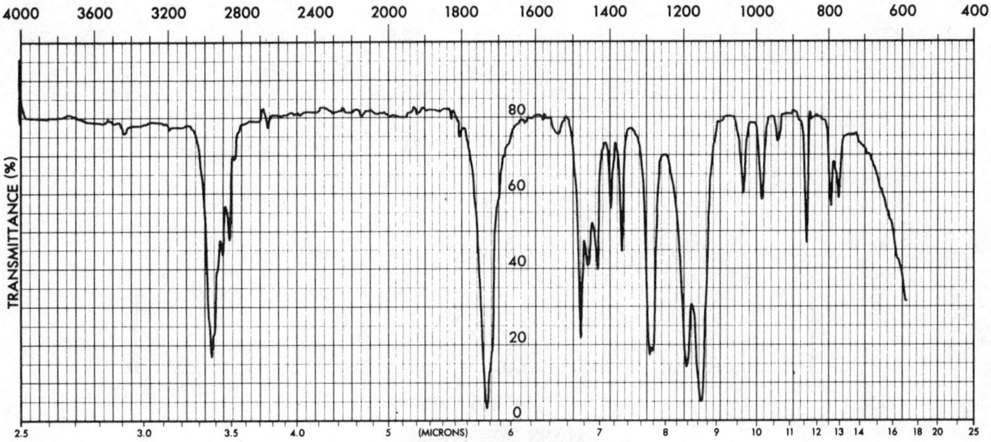

INFRARED SPECTRUM

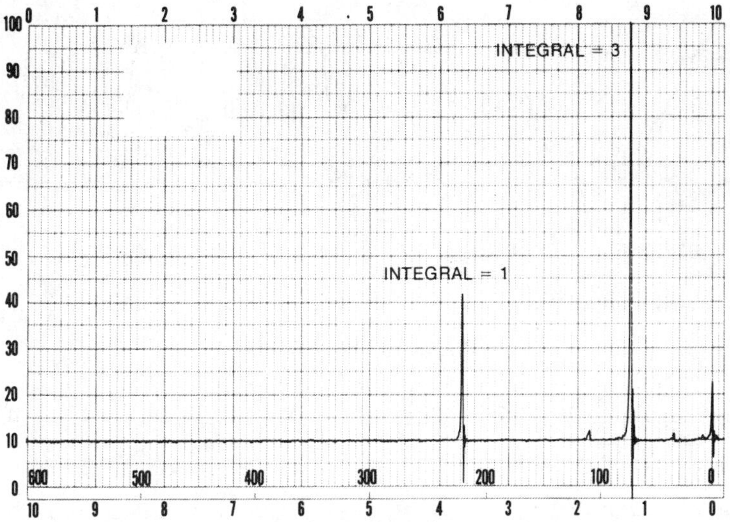

NMR SPECTRUM

PROBLEM 5

The uv spectrum of this compound is determined in 95% ethanol: strong end absorption and a band with fine structure appearing at λ_{max} 257 nm (log ϵ 2.4). The ir spectrum is determined in solution (CCl_4).

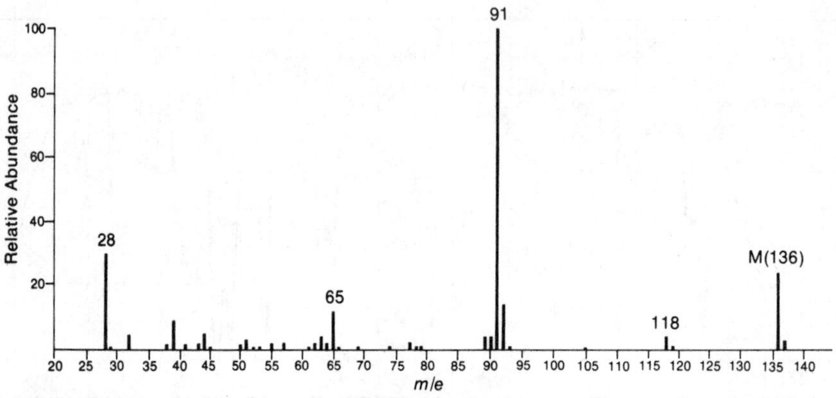

MASS SPECTRUM

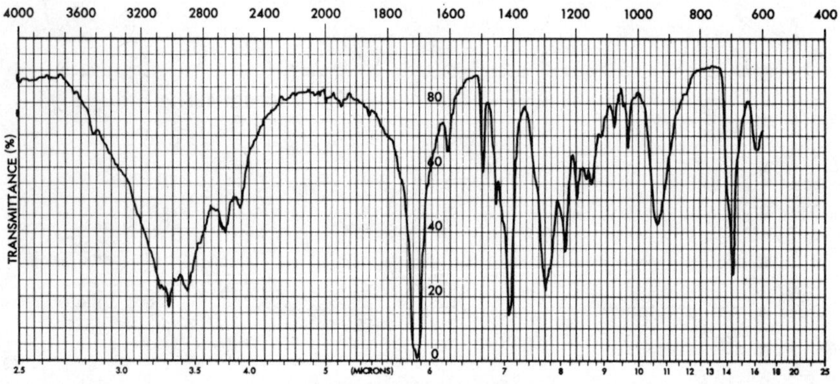

INFRARED SPECTRUM

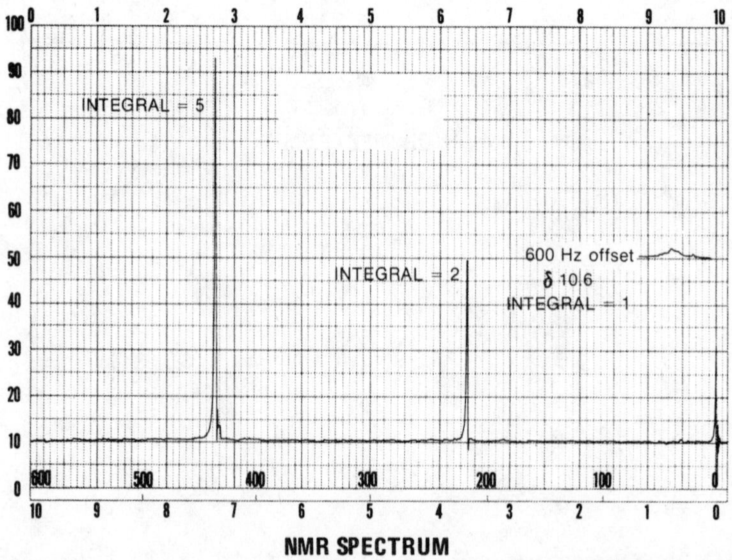

NMR SPECTRUM

PROBLEM 6

The mass spectrum of this compound shows an intense molecular ion at 172 mass units and an M + 2 peak of approximately the same size. The largest fragment ion appears at 65 mass units. The infrared spectrum of this solid compound was obtained by casting a film on the salt plates from a carbon tetracholoride solution.

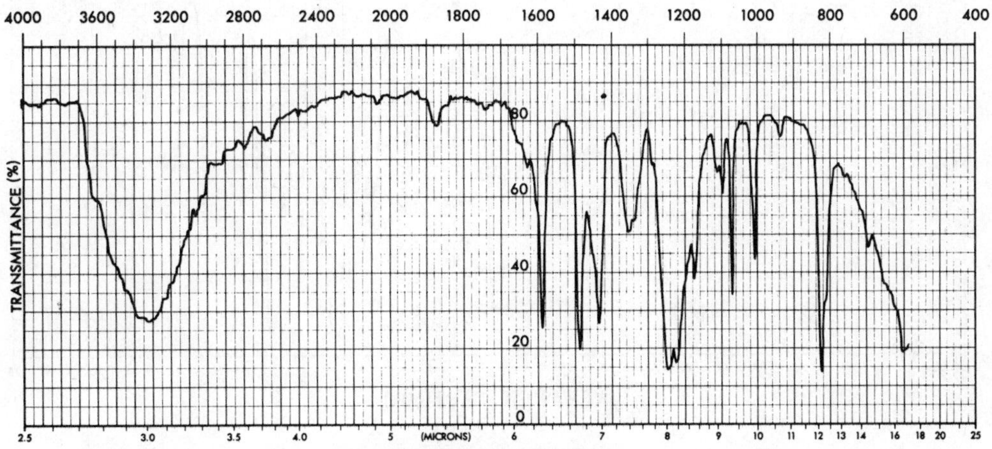

INFRARED SPECTRUM

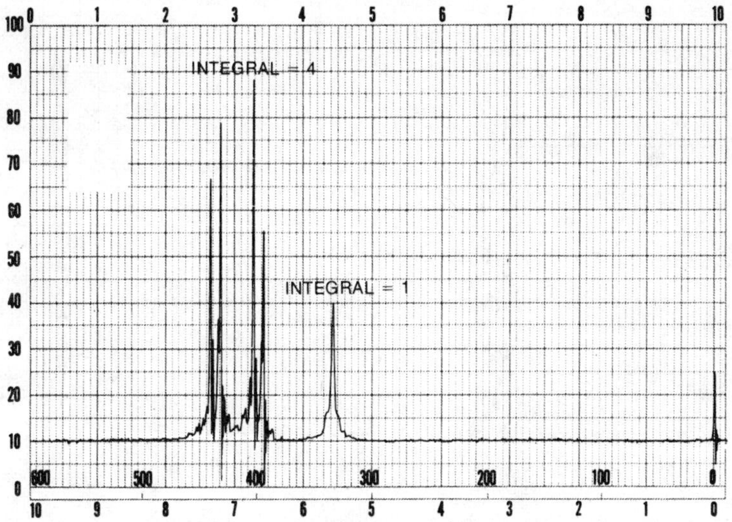

NMR SPECTRUM

PROBLEM 7

The formula for this compound is $C_{10}H_{12}O_2$. Solve the structure of this compound with only the ir and nmr spectra. The ir spectrum is determined on a neat liquid sample.

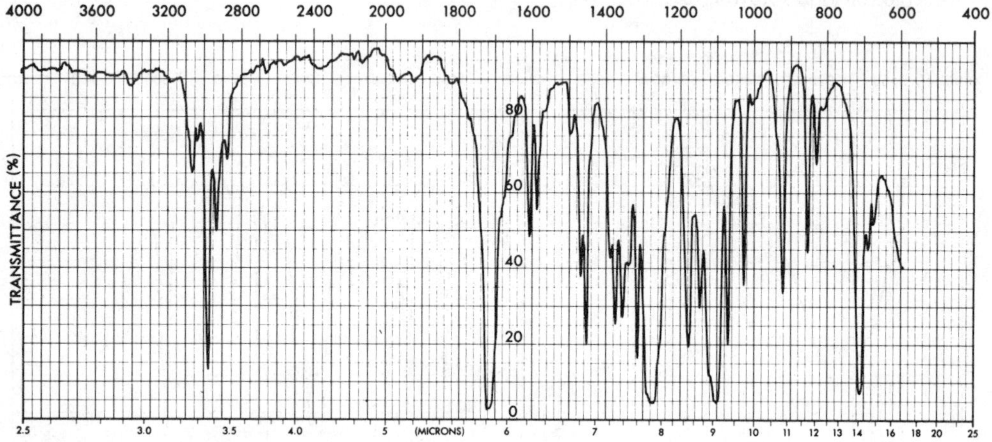

INFRARED SPECTRUM

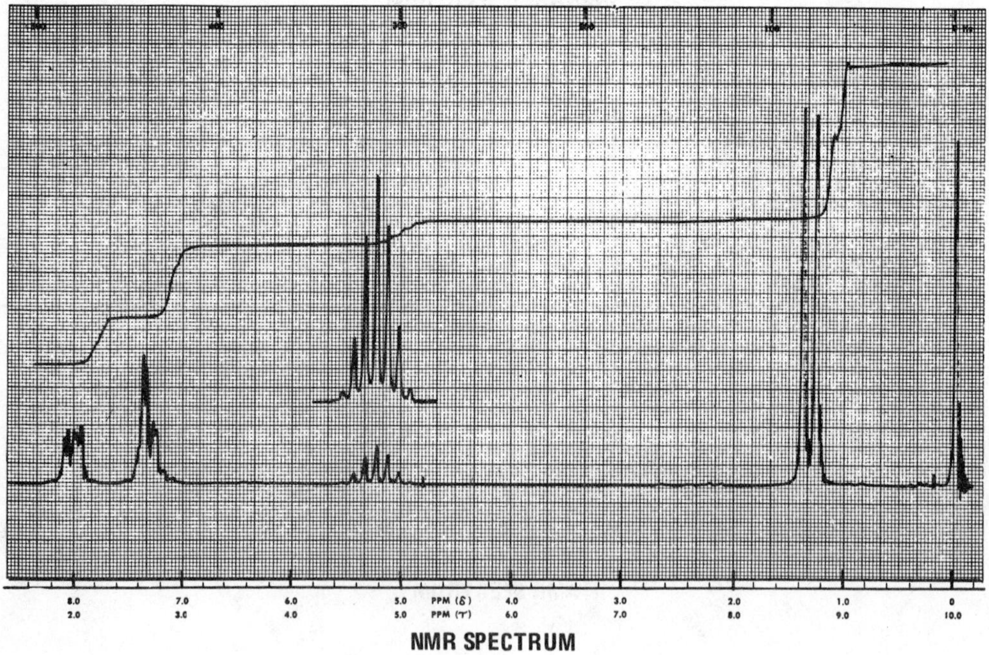

NMR SPECTRUM

PROBLEM 8

The uv spectrum of this compound is determined in 95% ethanol: λ_{max} 220 nm (log ϵ 4.0) and λ_{max} 275nm (log ϵ 3.1). The ir spectrum is obtained on a neat liquid sample.

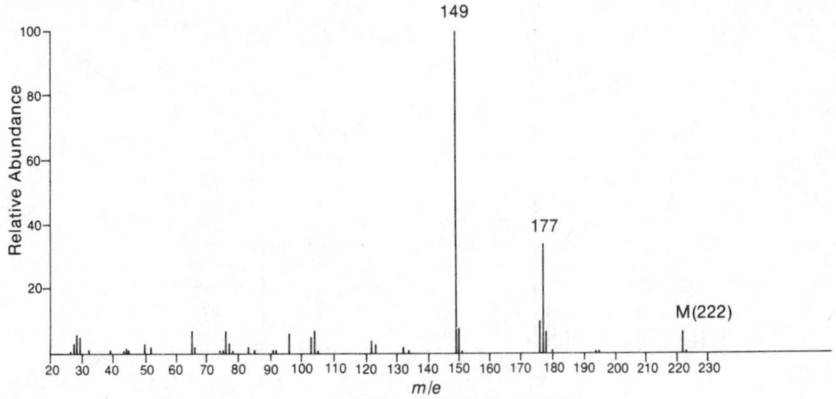

MASS SPECTRUM

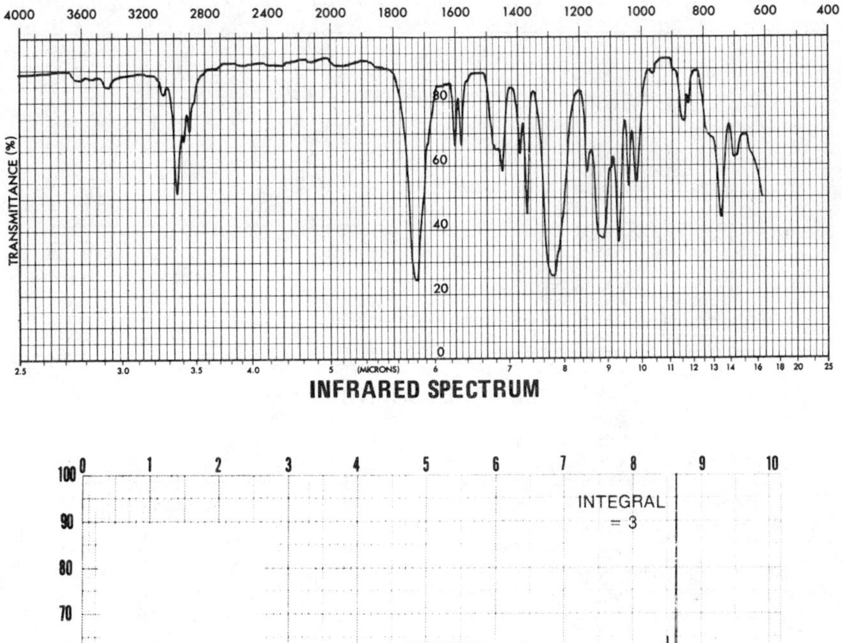

INFRARED SPECTRUM

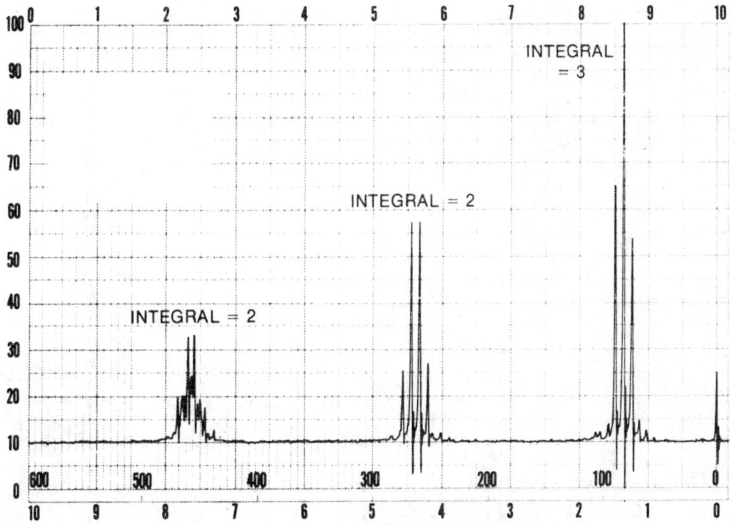

NMR SPECTRUM

PROBLEM 9

The uv spectrum of this compound shows no maximum above 205 nm. The ir spectrum is obtained on a neat liquid sample.

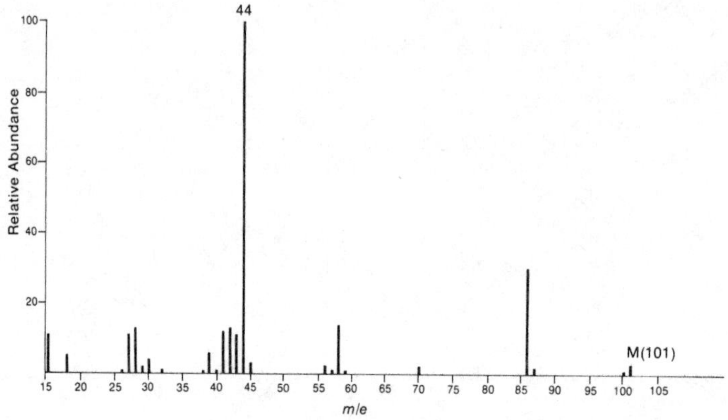

MASS SPECTRUM

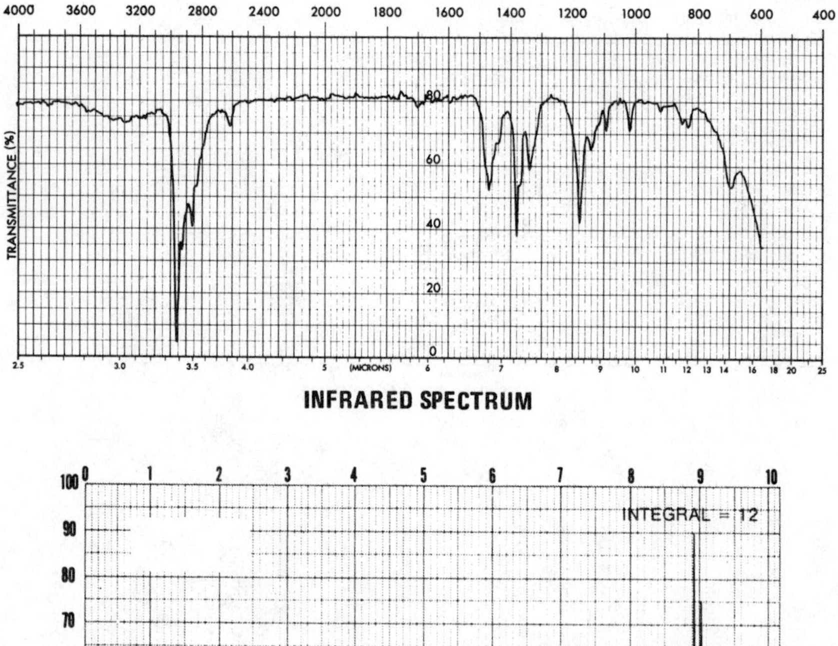

INFRARED SPECTRUM

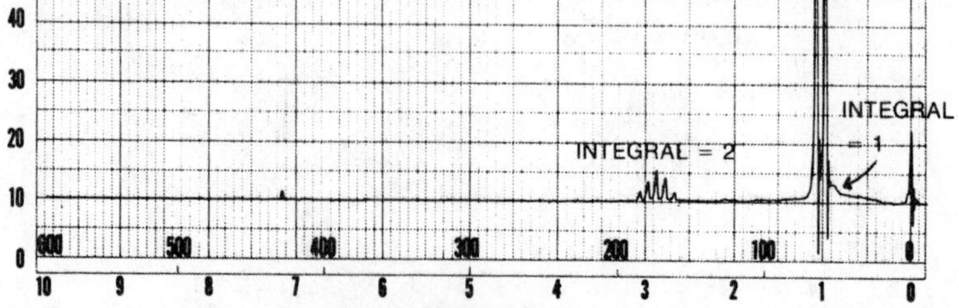

NMR SPECTRUM

PROBLEM 10

The uv spectrum of this compound shows no maximum above 205 nm. The ir spectrum is obtained on a neat liquid sample.

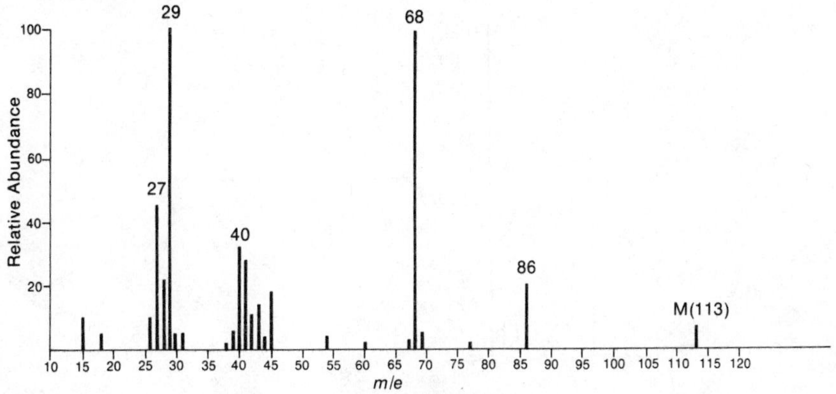

MASS SPECTRUM

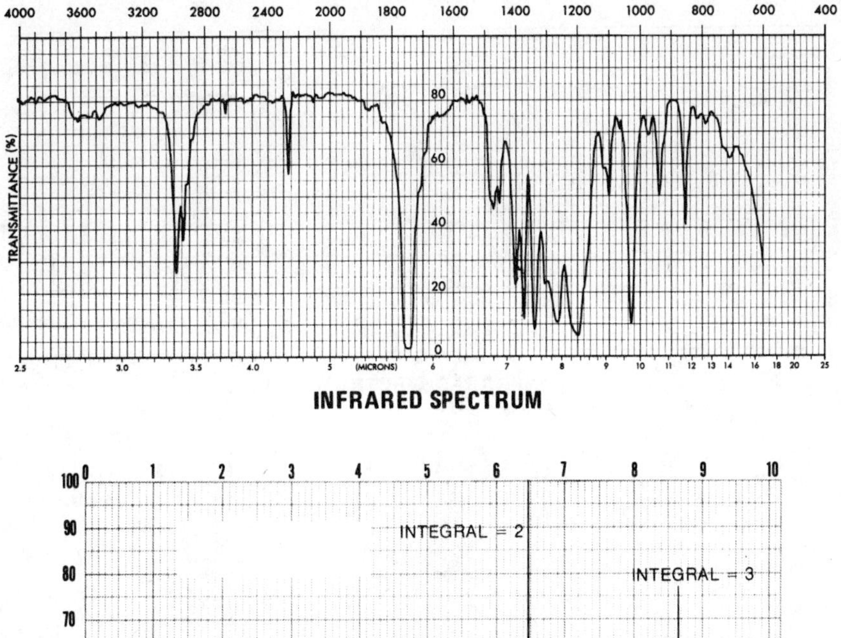

INFRARED SPECTRUM

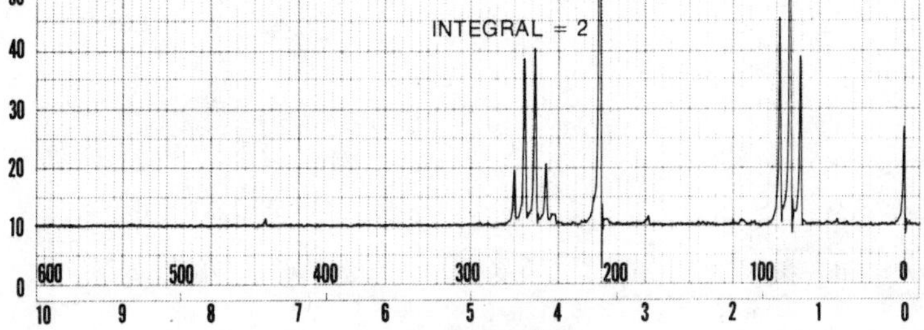

NMR SPECTRUM

PROBLEM 11

The uv spectrum of this compound is determined in 95% ethanol: λ_{max} 280 nm (log ϵ 1.3). The ir spectrum is obtained on a neat liquid sample.

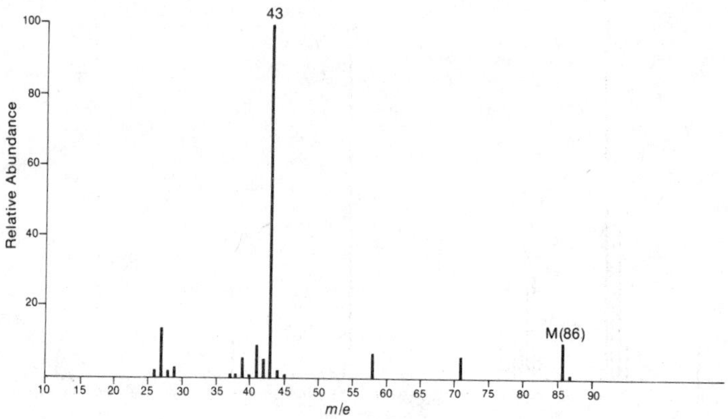

MASS SPECTRUM

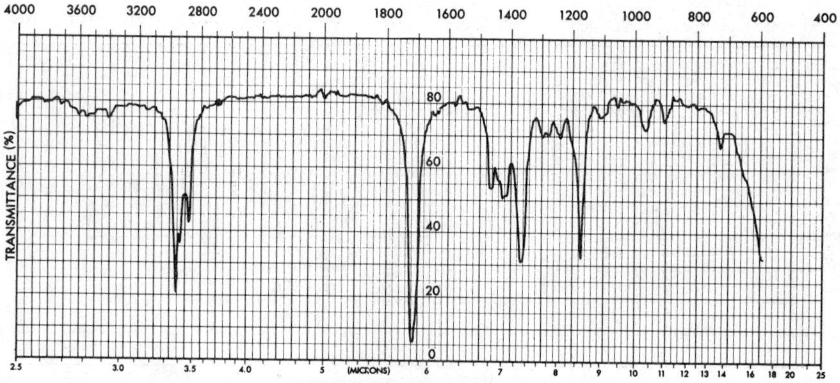

INFRARED SPECTRUM

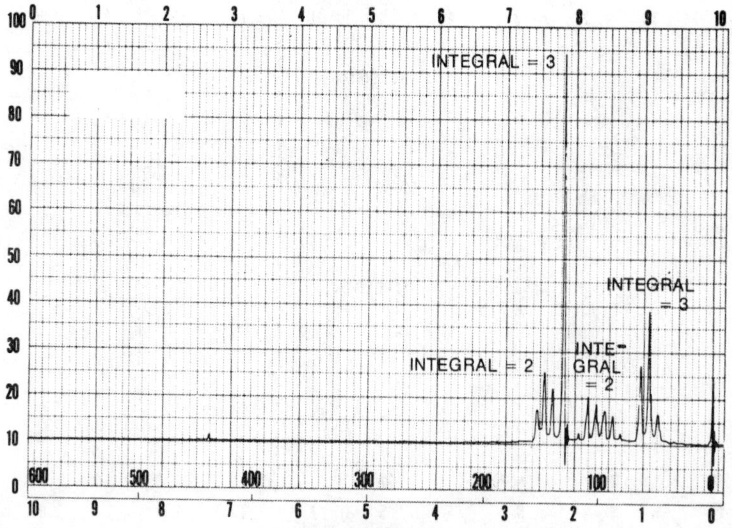

NMR SPECTRUM

PROBLEM 12

The uv spectrum of this compound shows no maximum above 205 nm. The ir spectrum is obtained on a neat liquid sample.

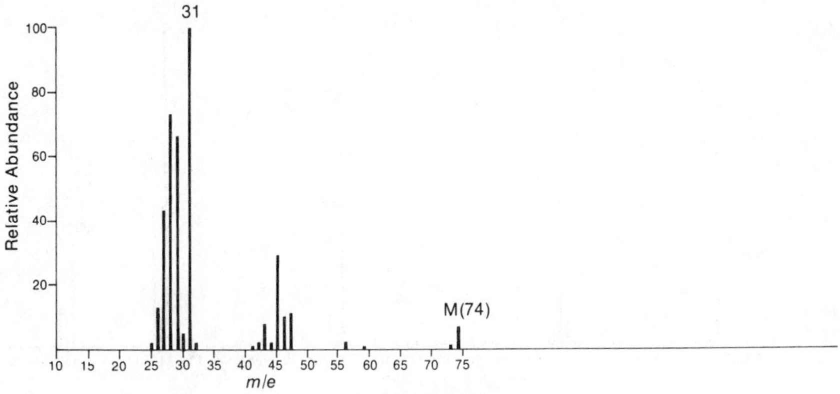

MASS SPECTRUM

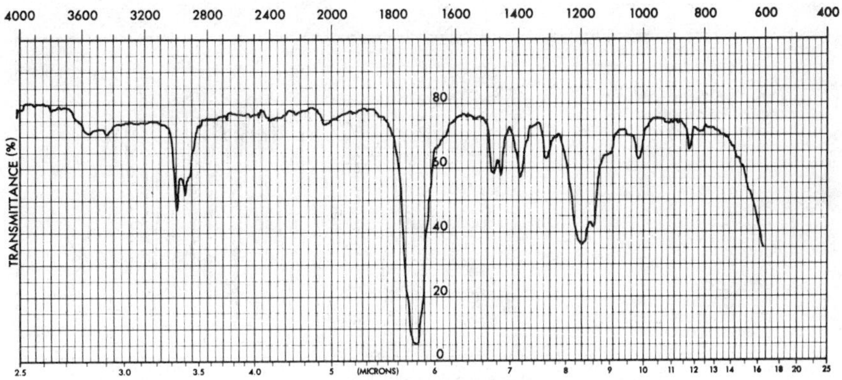

INFRARED SPECTRUM

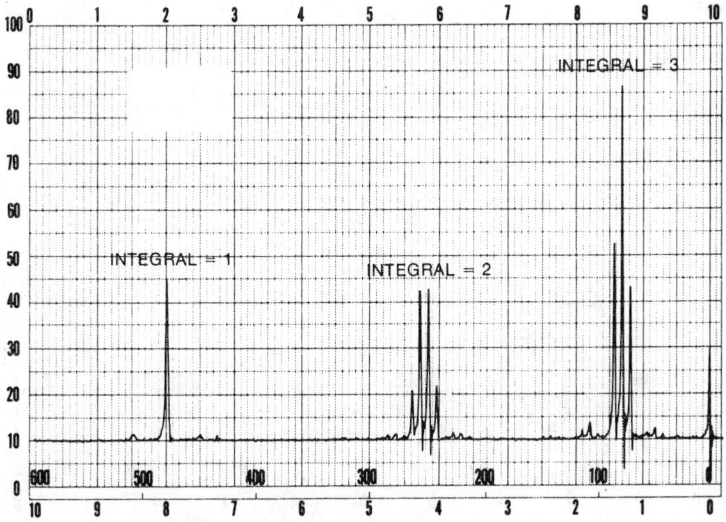

NMR SPECTRUM

PROBLEM 13

The uv spectrum of this compound is determined in 95% ethanol: λ_{max} 250 nm (log ϵ 3.2). The ir spectrum is obtained on a neat liquid sample.

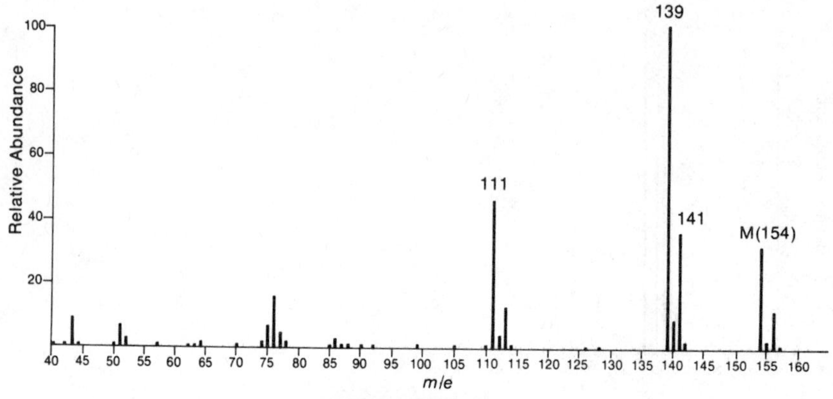

MASS SPECTRUM

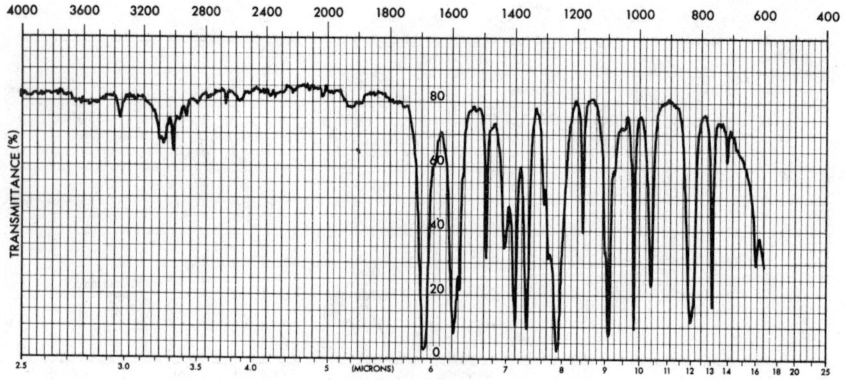

INFRARED SPECTRUM

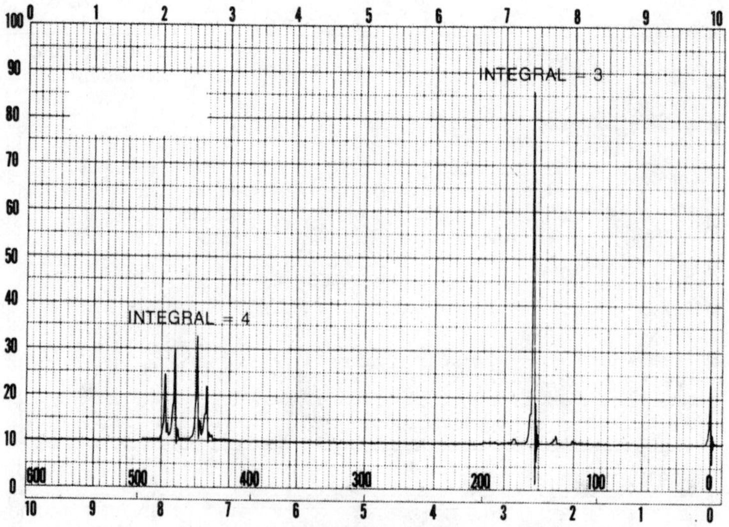

NMR SPECTRUM

PROBLEM 14

The uv spectrum of this compound is determined in 95% ethanol: λ_{max} 290 nm (log ϵ 1.4). The ir spectrum is obtained on a neat liquid sample.

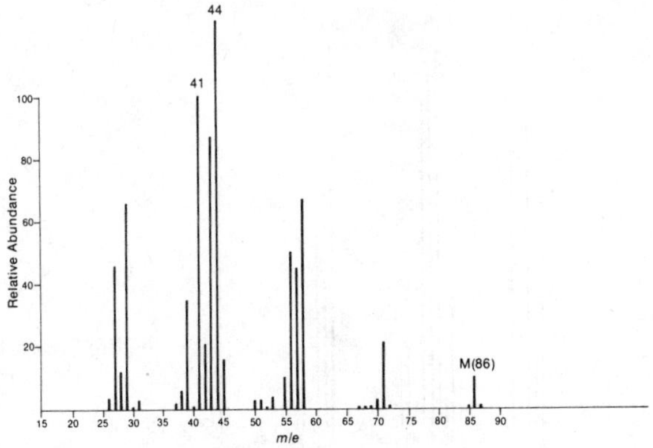

MASS SPECTRUM

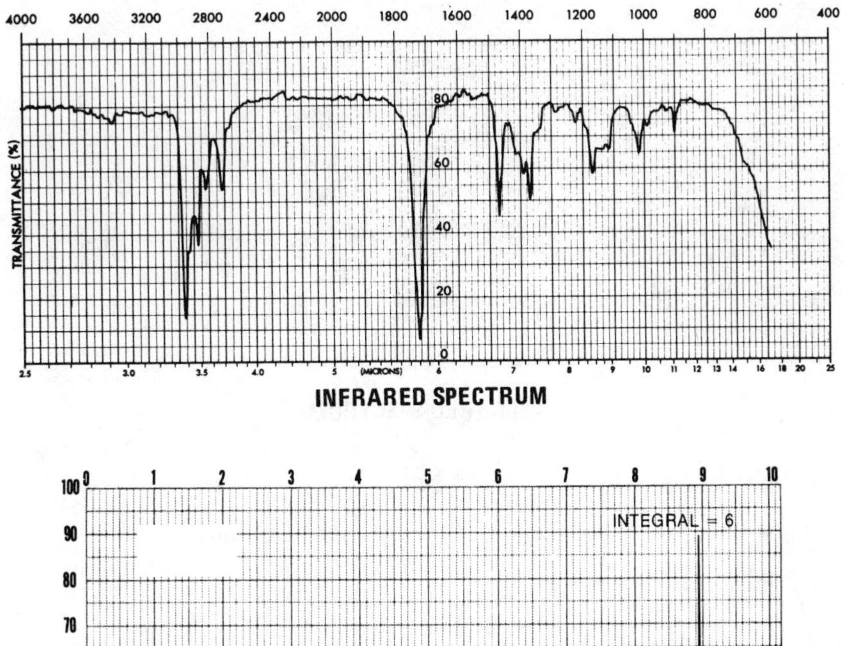

INFRARED SPECTRUM

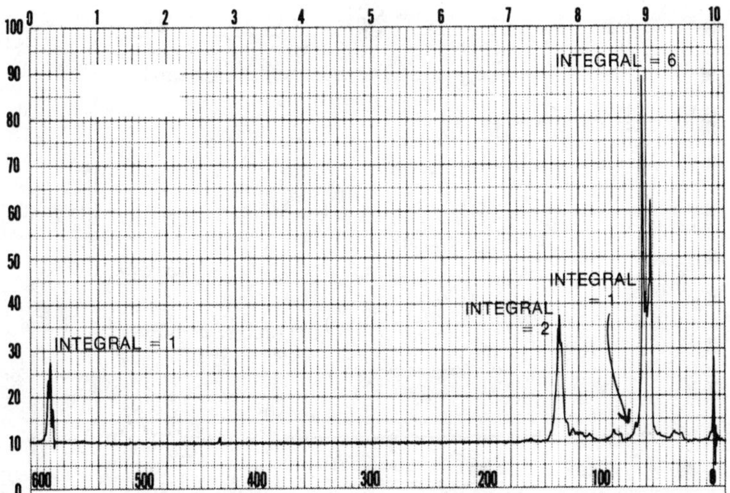

NMR SPECTRUM

PROBLEM 15

The uv spectrum of this compound shows no maximum above 205 nm. The molecular ion does not appear in the mass spectrum; it is readily dehydrated to a fragment ion of lower mass. The ir spectrum is obtained on a neat liquid sample.

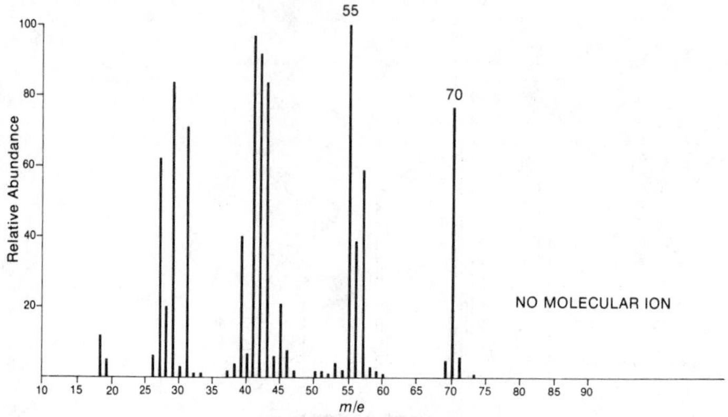

MASS SPECTRUM

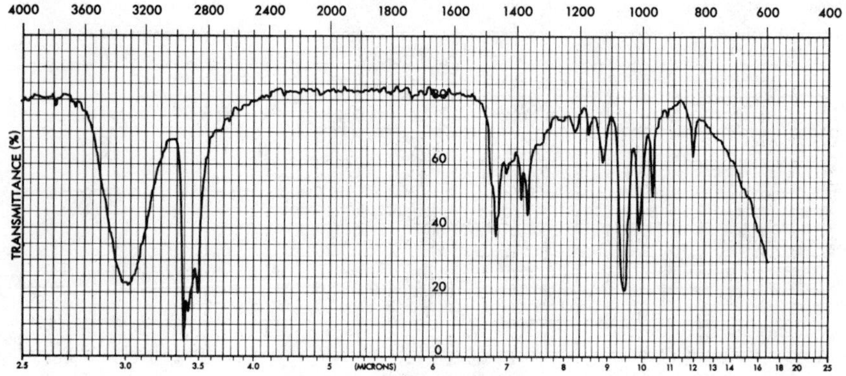

INFRARED SPECTRUM

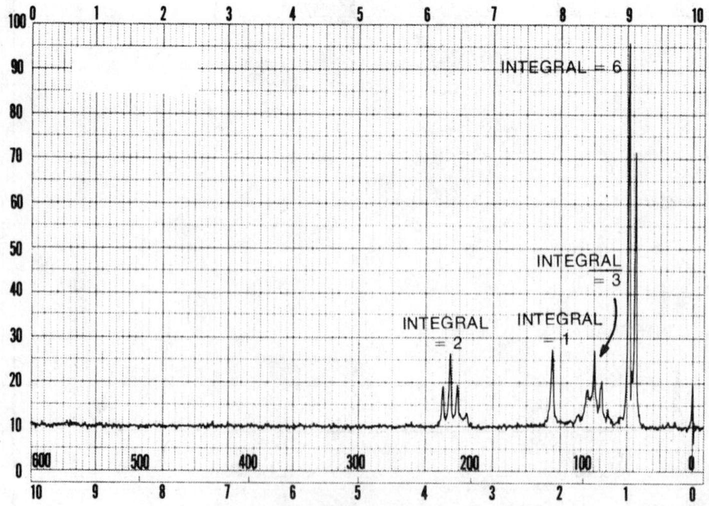

NMR SPECTRUM

PROBLEM 16

The uv spectrum of this compound shows no maximum above 205 nm. The molecular ion does not appear in the mass spectrum; it readily loses one mass unit to give a fragment ion of lower mass. The ir spectrum is obtained on a neat liquid sample. In the nmr spectrum, the 1.65 δ peak is concentration dependent. This compound is an isomer of the structure derived in Problem 15.

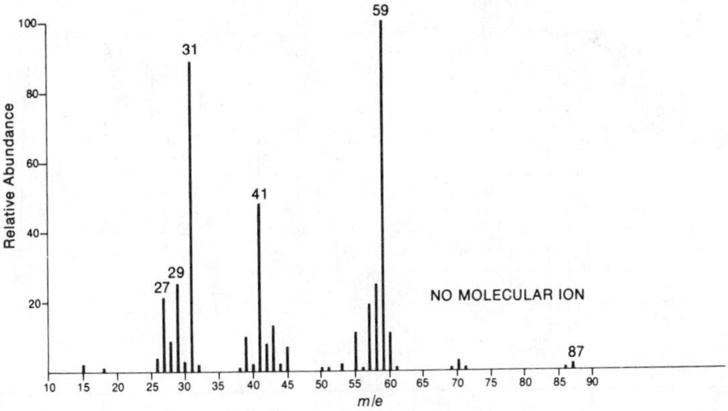

MASS SPECTRUM

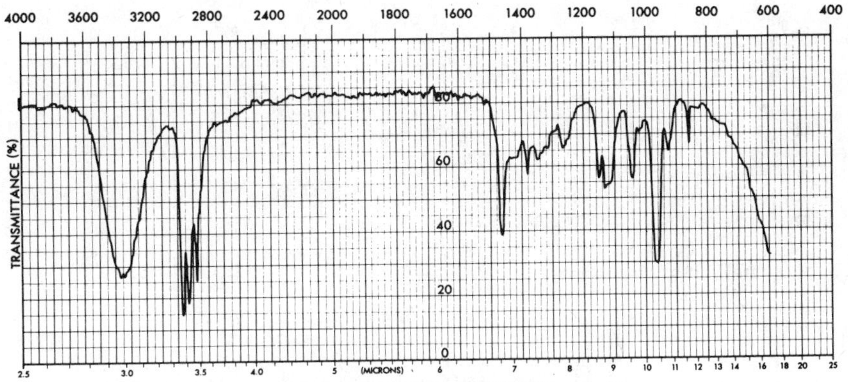

INFRARED SPECTRUM

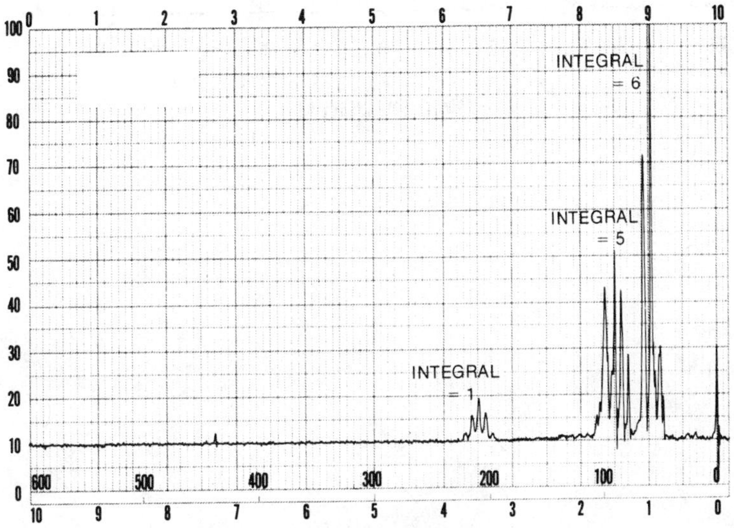

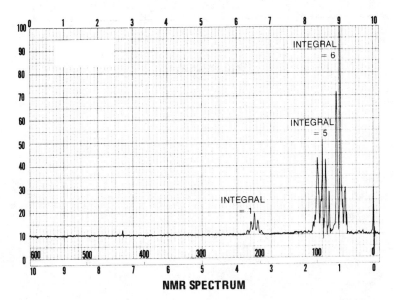

NMR SPECTRUM

PROBLEM 17

The formula for this compound is $C_7H_6O_2$. Solve the structure of this compound with only the ir and nmr spectra. The ir spectrum is obtained on a neat liquid sample.

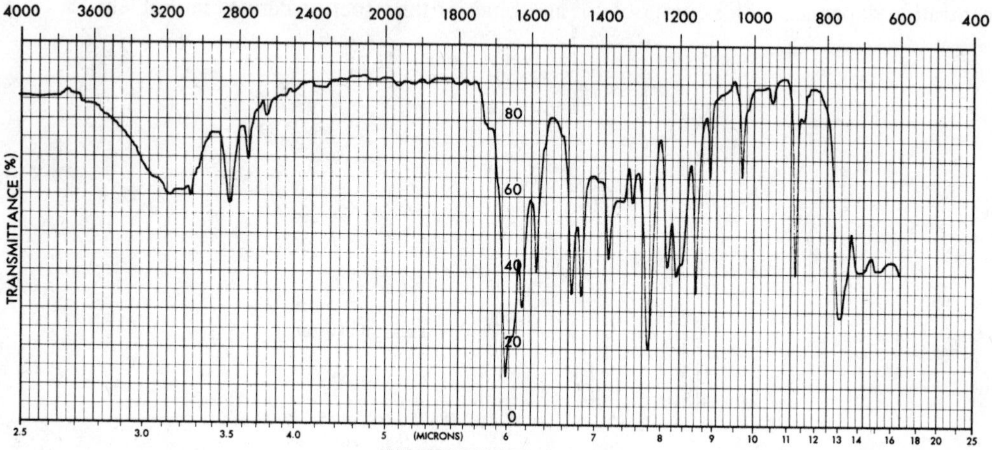

INFRARED SPECTRUM

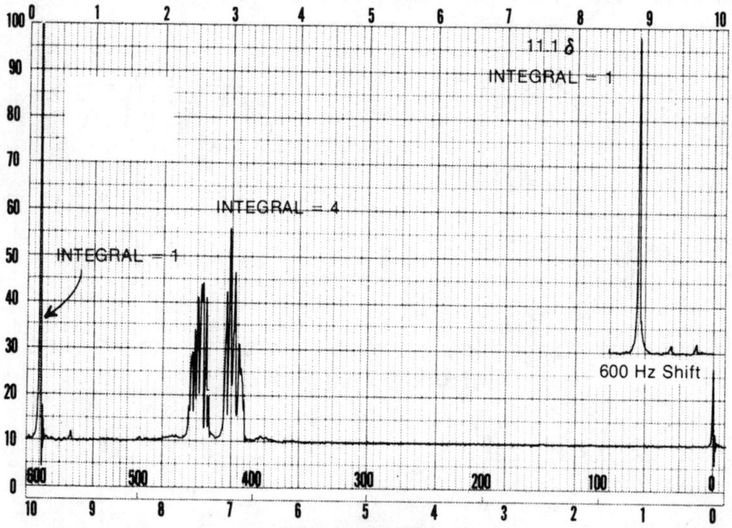

NMR SPECTRUM

PROBLEM 18

The uv spectrum of this compound is determined in 95% ethanol: λ_{max} 237 nm (log ϵ 4.1) and 310 nm (log ϵ 1.8). The ir spectrum is obtained on a neat liquid sample.

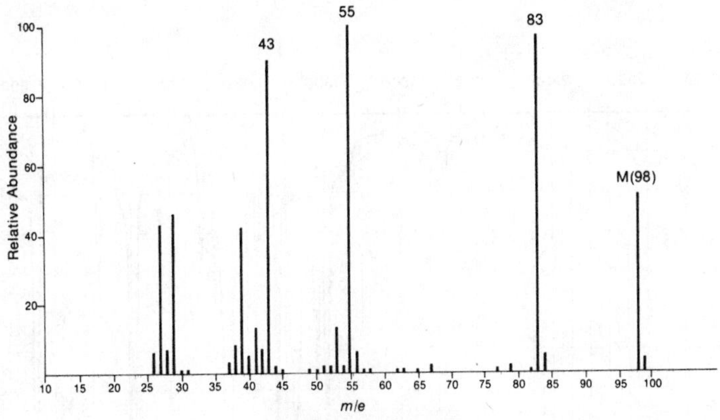

MASS SPECTRUM

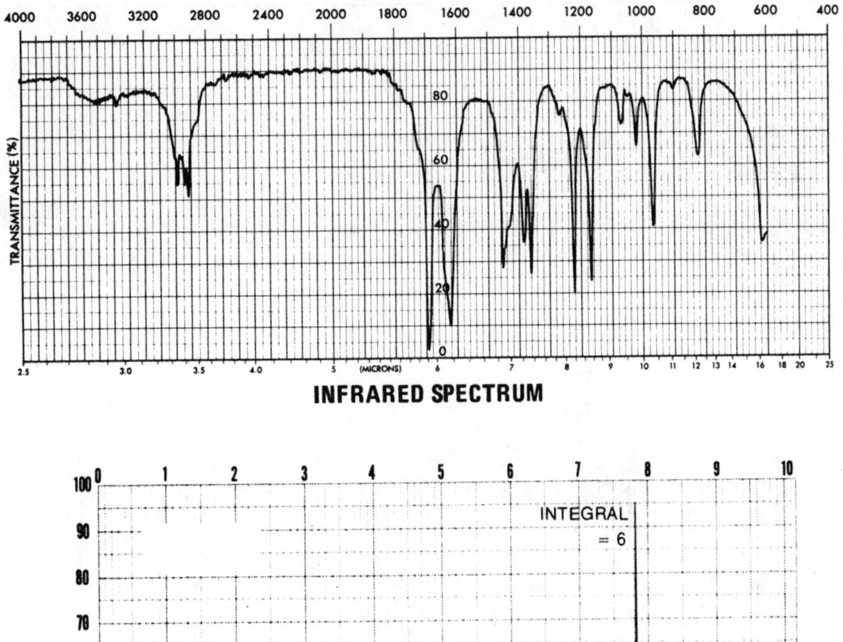

INFRARED SPECTRUM

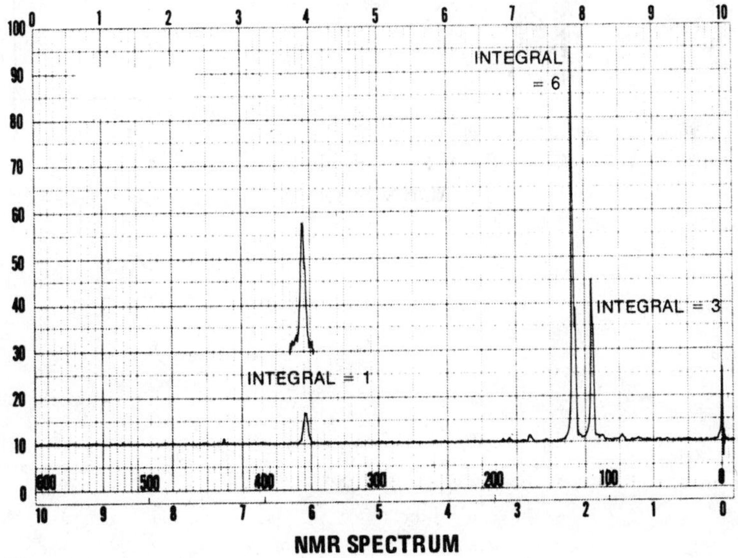

NMR SPECTRUM

PROBLEM 19

The mass spectrum of this compound shows a molecular ion peak (157 mass units) and an M + 2 peak in a ratio of 3:1. Several large fragment ions appear at 127 and 111 mass units, with each of these accompanied by M + 2 peaks in a ratio of 3:1. The ir spectrum is determined in solution (CCl_4).

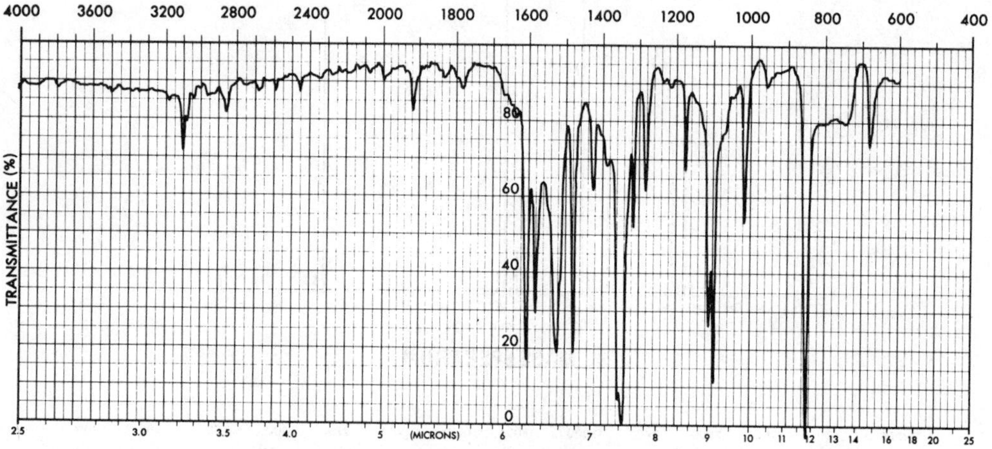

INFRARED SPECTRUM

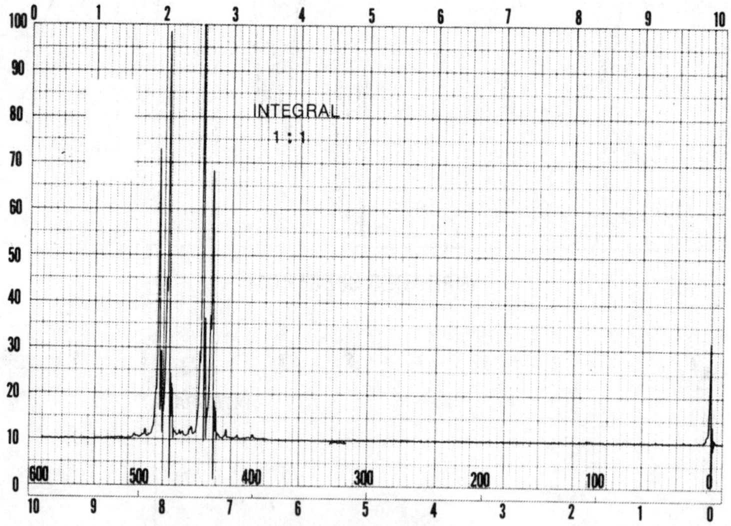

NMR SPECTRUM

PROBLEM 20

The uv spectrum of this compound shows no maximum above 205 nm. The ir spectrum is obtained on a neat liquid sample.

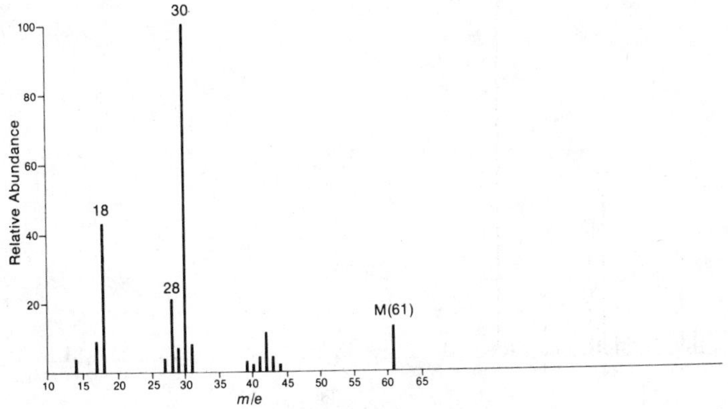

MASS SPECTRUM

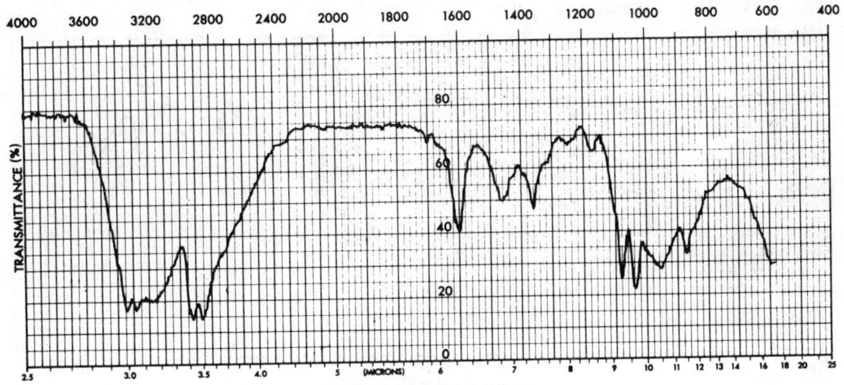

INFRARED SPECTRUM

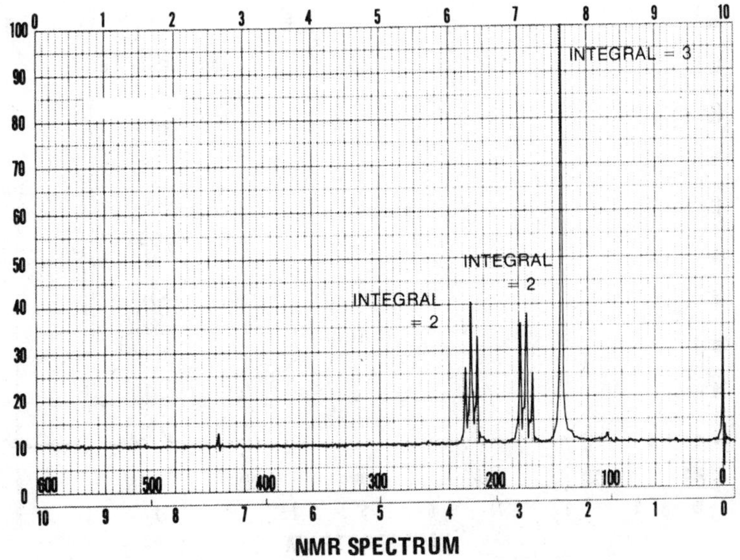

NMR SPECTRUM

PROBLEM 21

The uv spectrum of this compound shows no maximum above 205 nm. The ir spectrum is obtained on a neat liquid sample.

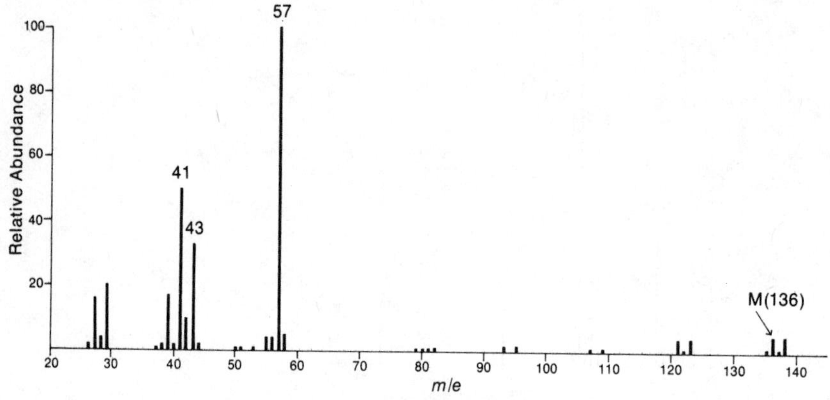

MASS SPECTRUM

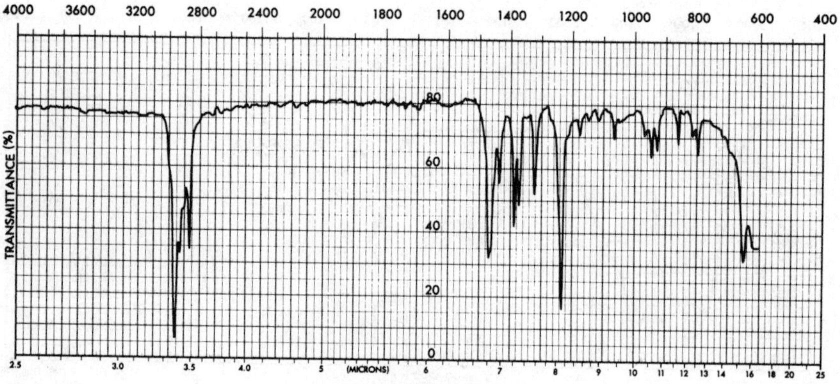

INFRARED SPECTRUM

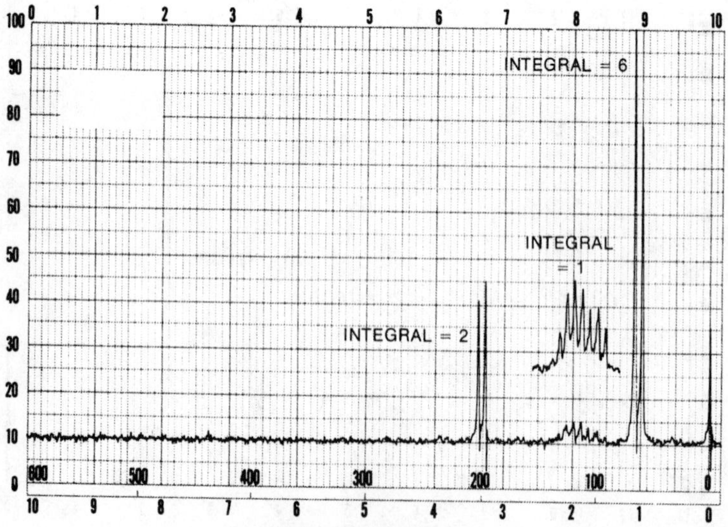

NMR SPECTRUM

PROBLEM 22

The formula for this compound is $C_{11}H_{12}O_2$. Solve the structure, including stereochemistry, using only the ir and nmr spectra. The ir spectrum is obtained on a neat liquid sample.

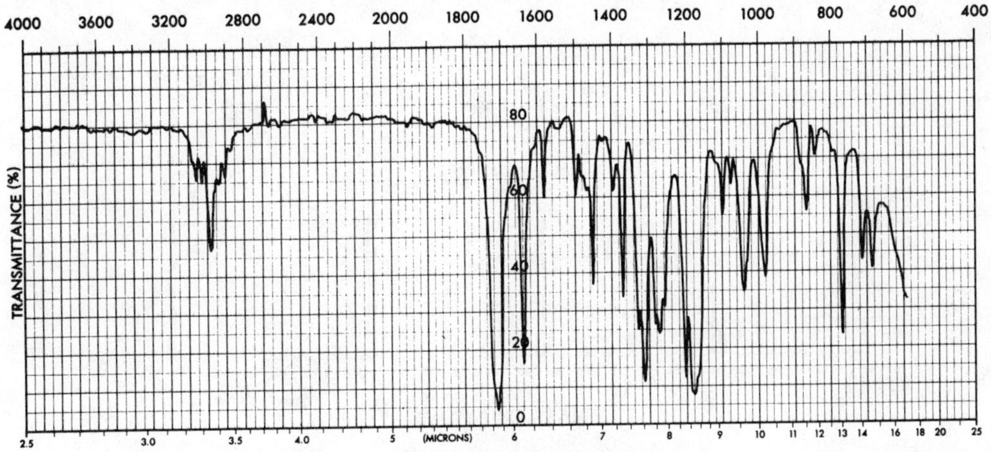

INFRARED SPECTRUM

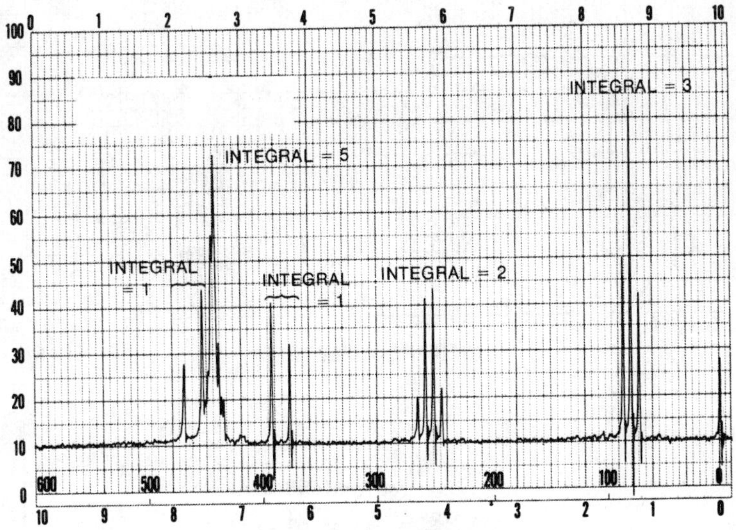

NMR SPECTRUM

PROBLEM 23

The uv spectrum of this compound shows an intense primary band, λ_{max} 251 nm (log $\epsilon > 4$). The ir spectrum is obtained on a neat liquid sample.

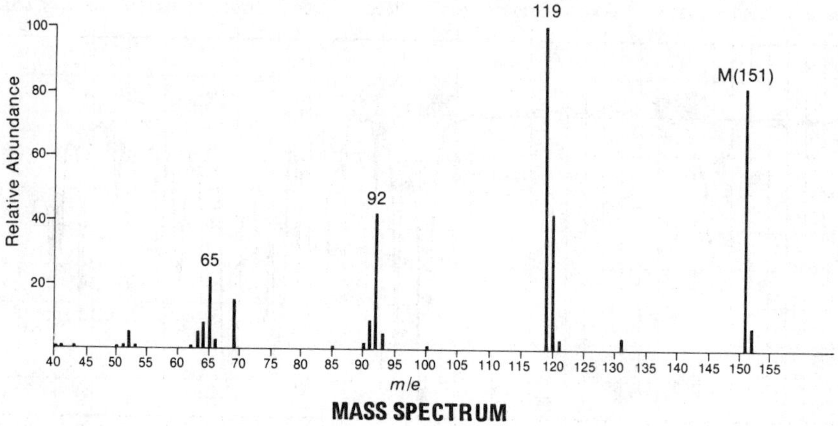

MASS SPECTRUM

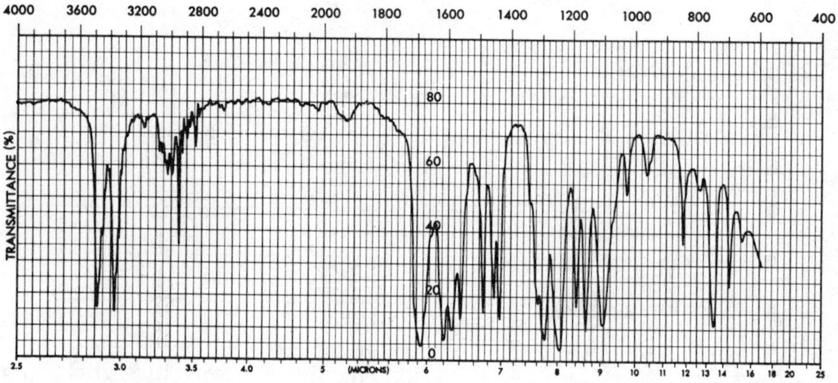

INFRARED SPECTRUM

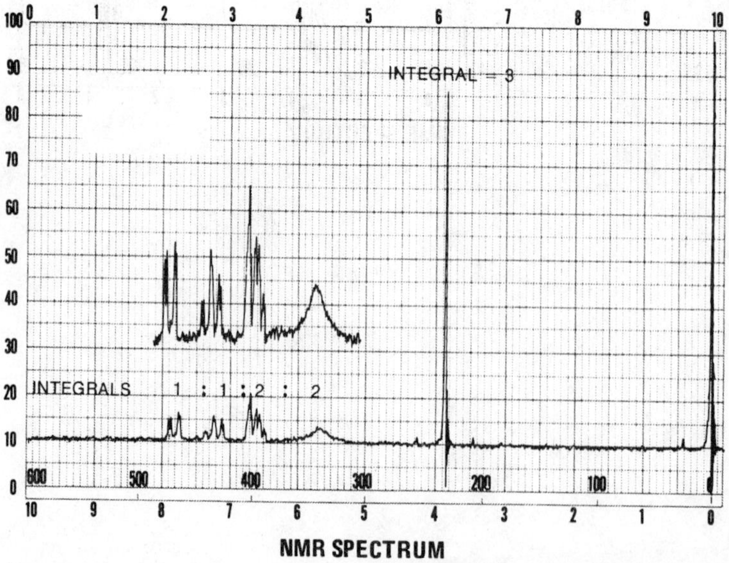

NMR SPECTRUM

PROBLEM 24

The uv spectrum of this compound shows no maximum above 205 nm. In the mass spectrum, notice that the patterns for the M, M + 2, and M + 4 peaks have a ratio of 1:2:1. What are the structures of the mass 135 and 137 peaks? The ir spectrum is obtained on a neat liquid sample.

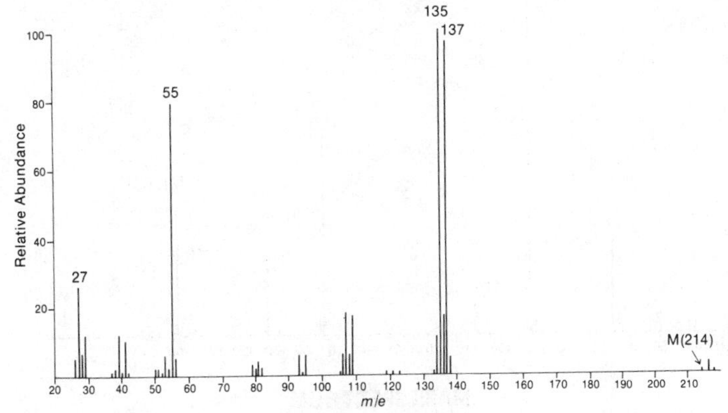

MASS SPECTRUM

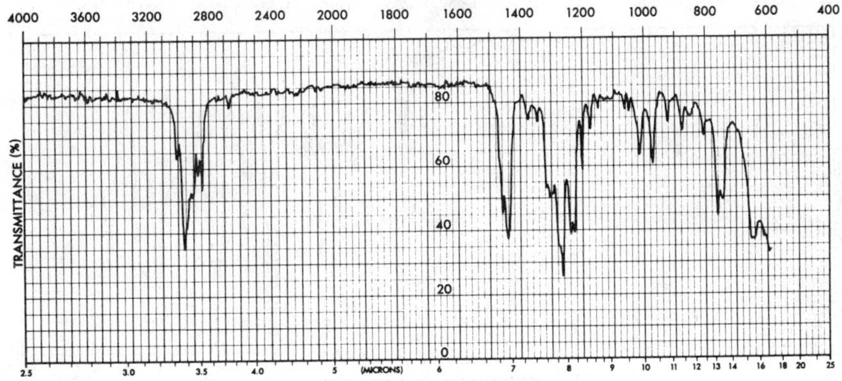

INFRARED SPECTRUM

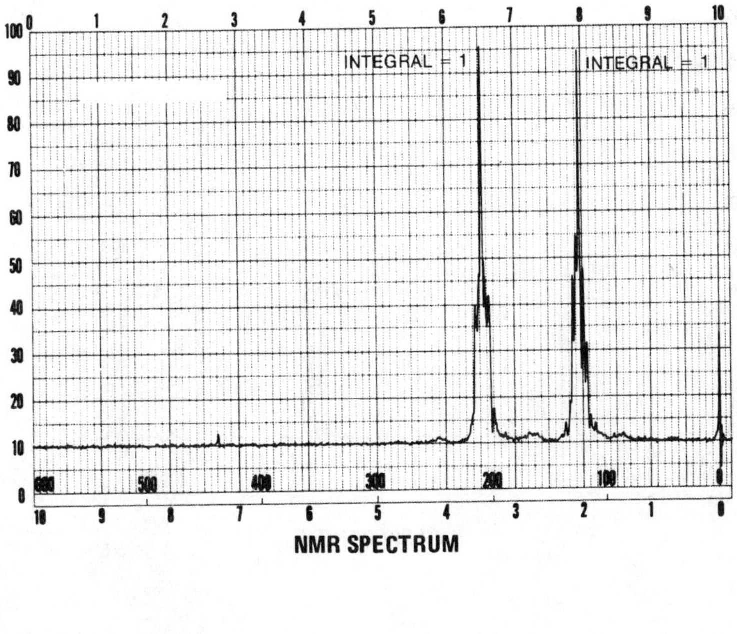

NMR SPECTRUM

PROBLEM 25

The uv spectrum of this compound is determined in 95% ethanol: λ_{max} 225 nm (log ϵ 4.0) and 270 nm (log ϵ 2.8). The ir spectrum is obtained on a neat liquid sample.

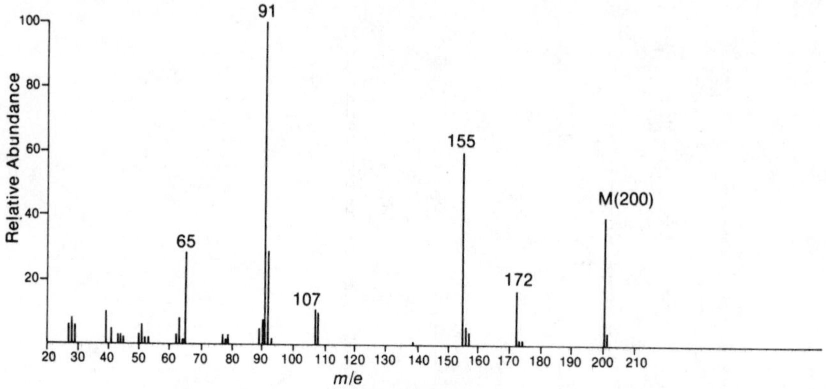

MASS SPECTRUM

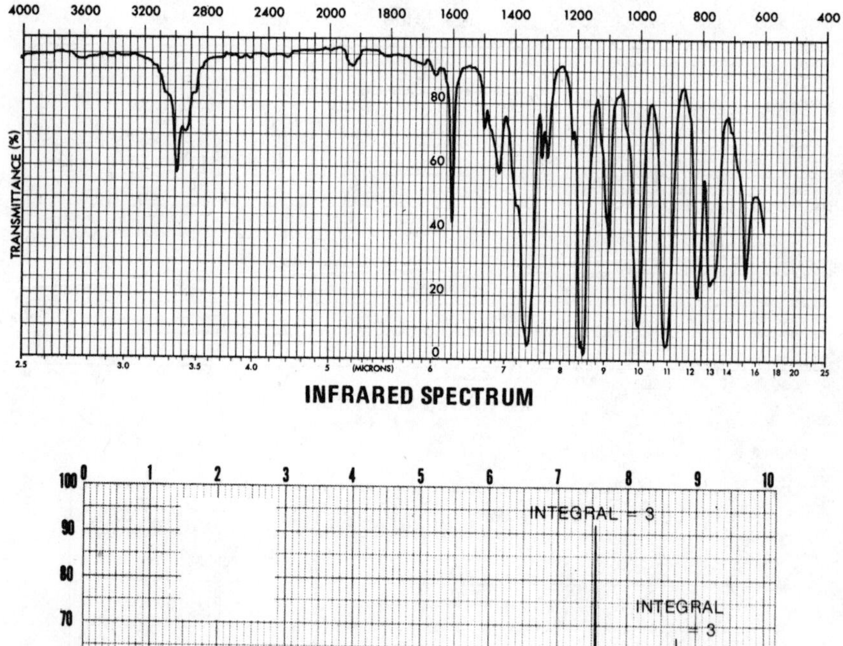

INFRARED SPECTRUM

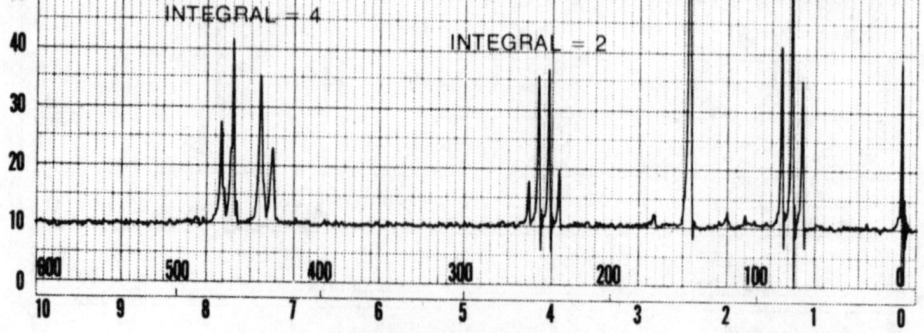

NMR SPECTRUM

PROBLEM 26

The uv spectrum of this compound shows no maximum above 205 nm. The molecular ion does not appear in the mass spectrum; it readily loses a methyl group to give a fragment ion of lower mass. The ir spectrum is obtained on a neat liquid sample.

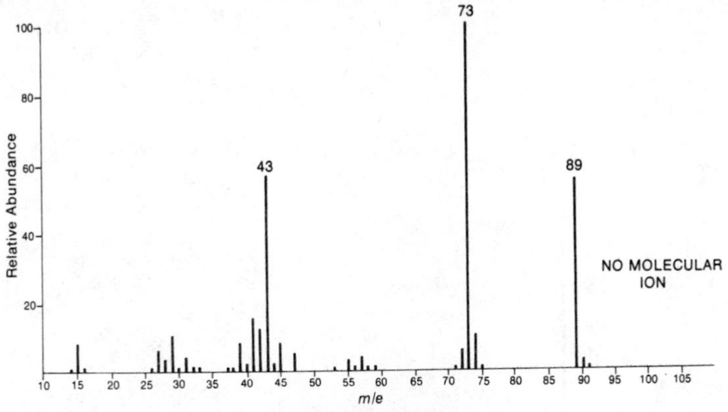

MASS SPECTRUM

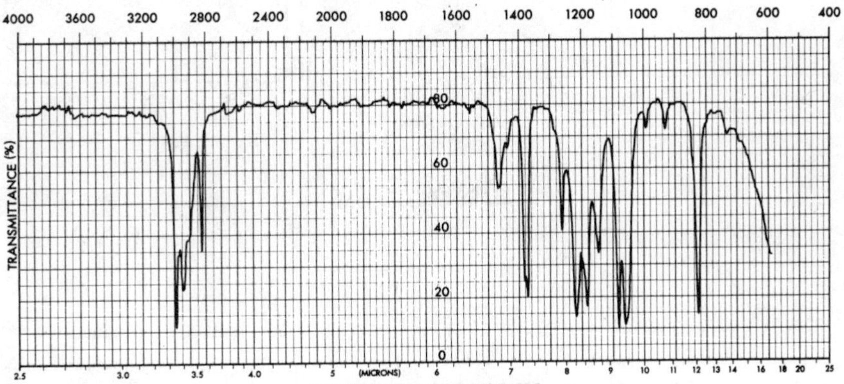

INFRARED SPECTRUM

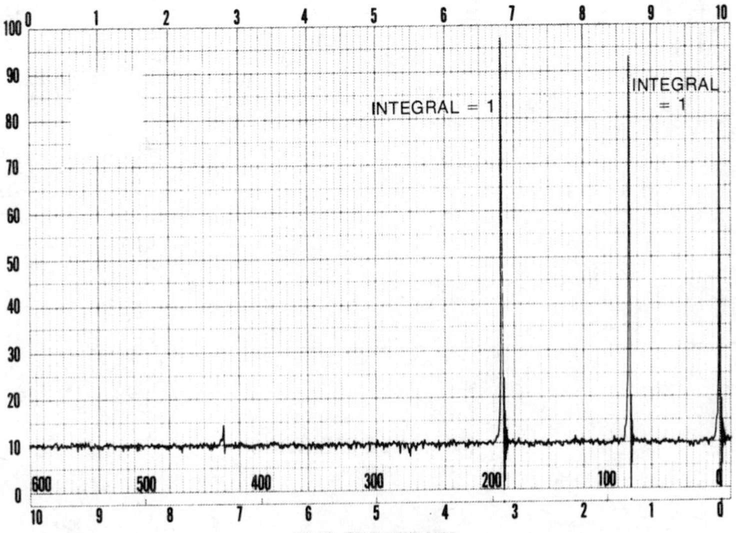

NMR SPECTRUM

PROBLEM 27

The uv spectrum of this compound is determined in 95% ethanol: λ_{max} 245 nm (log ϵ 2.8). The ir spectrum is obtained on a neat liquid sample. This compound exists in two tautomeric forms, with one tautomer predominating in the equilibrum mixture. The tiny nmr peaks at 5.0 δ, 2.0 δ, and an offset proton (not shown) at 12.0 δ, are assigned to the minor tautomer. Give the structures of both tautomers.

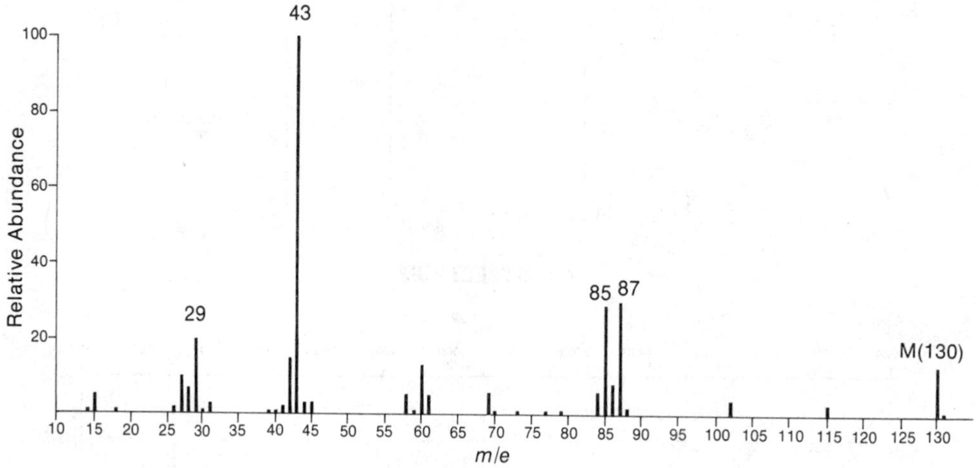

MASS SPECTRUM

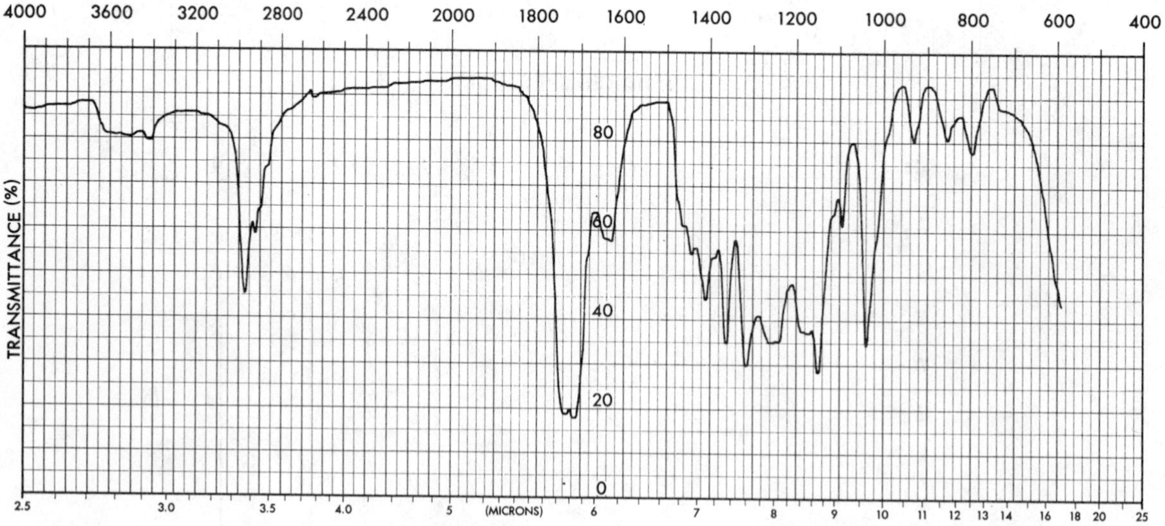

INFRARED SPECTRUM

(See nmr spectrum on p. 335)

PROBLEM 27 (Continued)

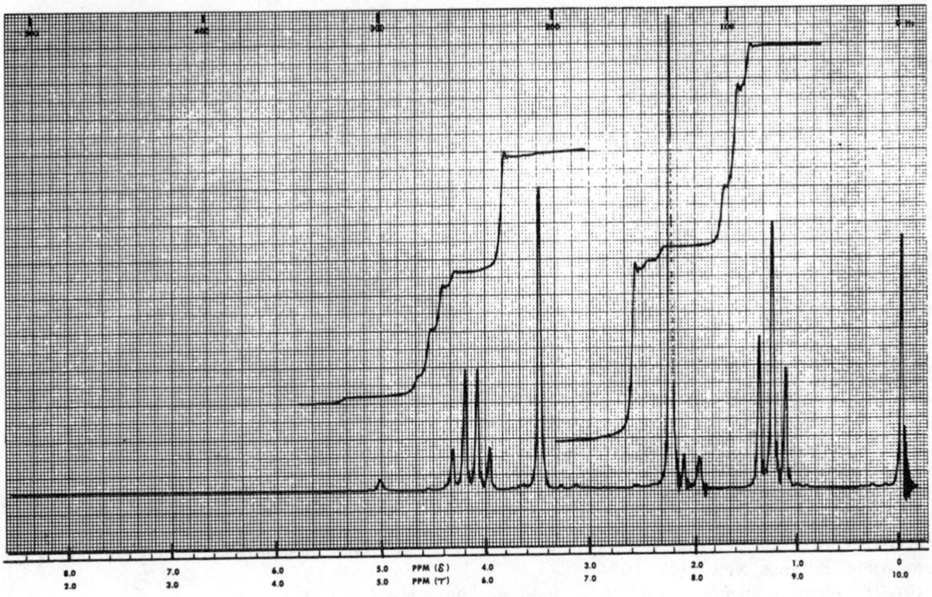

NMR SPECTRUM

PROBLEM 28

The anesthetic procaine (Novocaine) has the formula $C_{13}H_{20}N_2O_2$ and gives the nmr spectrum shown below. The "quintet" at about 2.7 δ actually arises from the overlap of a upfield quartet (4H) and a downfield triplet (2H). An impurity ($CHCl_3$) appears at about 7.25 δ. The infrared spectrum shows a double absorption at 3350 cm^{-1} (2.98 μ), and other principal bands at about 3000 cm^{-1} (3.33 μ), 1700 cm^{-1} (5.88 μ), 1600 cm^{-1} (6.25 μ), 1450 cm^{-1} (6.90 μ), and 1280 cm^{-1} (7.81 μ, strong and broad). What is the structure of procaine?

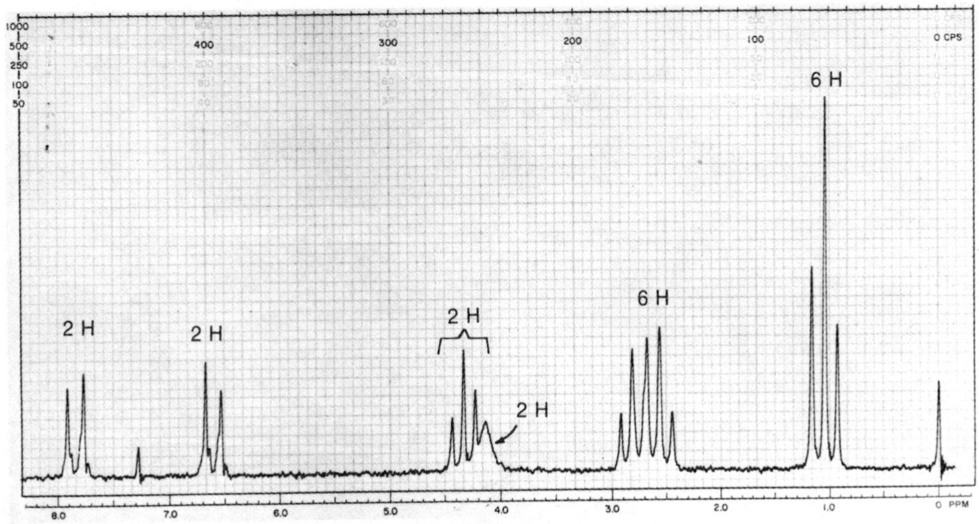

NMR SPECTRUM

MOLECULAR FORMULAS FOR PROBLEMS

1. C_4H_8O
2. C_3H_8O
3. $C_5H_{10}O$
4. –
5. $C_8H_8O_2$
6. C_6H_5OBr
7. –
8. $C_{12}H_{14}O_4$
9. $C_6H_{15}N$
10. $C_5H_7NO_2$
11. $C_5H_{10}O$
12. $C_3H_6O_2$
13. C_8H_7ClO
14. $C_5H_{10}O$
15. $C_5H_{12}O$
16. $C_5H_{12}O$
17. –
18. $C_6H_{10}O$
19. $C_6H_4NO_2Cl$
20. C_2H_7NO
21. C_4H_9Br
22. –
23. $C_8H_9NO_2$
24. $C_4H_8Br_2$
25. $C_9H_{12}SO_3$
26. $C_5H_{12}O_2$
27. $C_6H_{10}O_3$
28. –

REFERENCES

A. Ault, "Problems in Organic Structure Determination," McGraw-Hill Book Company, New York, 1967.
J. R. Dyer, "Organic Spectral Problems," Prentice-Hall, Inc., Englewood Cliffs, N.J., 1972.
R. H. Shapiro and C. H. DePuy, "Exercises in Organic Spectroscopy," 2nd ed., Holt, Rinehart and Winston, New York, 1977.
R. M. Silverstein, G. C. Bassler, and T. C. Morrill, "Spectrometric Identification of Organic Compounds," 3rd ed., John Wiley and Sons, New York, 1974.
B. M. Trost, "Problems in Spectroscopy," W. A. Benjamin, Inc., New York, 1967.
D. H. Williams and I. Fleming, "Spectroscopic Problems in Organic Chemistry, "McGraw-Hill, Ltd., London, 1967.

ANSWERS TO PROBLEMS

CHAPTER 1

1. a) 90.50% carbon; 9.50% hydrogen b) C_4H_5
2. 32.0% carbon; 5.4% hydrogen; 62.8% chlorine; $C_3H_6Cl_2$
3. $C_2H_5NO_2$
4. 181.2 = molecular weight. Molecular formula is $C_9H_8O_4$.
5. Equivalent weight = 52.3
6. a) 6, b) 1, c) 3, d) 6, e) 12
7. The index of hydrogen deficiency = 1. There cannot be a triple bond.
8. a) 59.96% carbon; 5.75% hydrogen; 34.29% oxygen. b) $C_7H_8O_3$ c) $C_{21}H_{24}O_9$
   d) A maximum of two aromatic rings

CHAPTER 2

1. a) 3-Methyl-1-butanol b) 3-Bromopropyne c) 3-Penten-2-one (*trans*)
   d) 4-Isopropyltoluene e) 2-Aminotoluene f) 3-Hydroxytoluene
   g) 1-Phenyl-1-propanone h) 1-Phenyl-2-propanone i) Butanedioic anhydride (succinic anhydride)
   j) *N*-Ethylaniline k) 2-Chloropropanoic acid
   l) 3-Dimethylaminopropanonitrile m) Ethyl 3-phenylpropenoate (ethyl cinnamate)
2. Citronellal (3,7-Dimethyl-6-octenal)
3. 4-Aminobenzoic acid
4. 3-Phenylpropenal (cinnamaldehyde)
5. Polymethyl methacrylate
   Polyethylene
   Nylon
   Polystyrene

CHAPTER 3

1. a) −1, 0, +1; b) −1/2, +1/2; c) −5/2, −3/2, −1/2, +1/2, +3/2, +5/2;
   d) −1/2, +1/2
2. 2.13 δ
3. a) 180 Hz; b) 1.50 δ
4. See Figures 3–15 and 3–16. The methyl hydrogens are in the shielding region. Acetonitrile shows similar anisotropic behavior to acetylene.

5. *o*-Hydroxyacetophenone is intramolecularly hydrogen bonded. The proton is deshielded (12.05 δ). Changing concentration does not alter the extent of hydrogen bonding. Phenol is intermolecularly hydrogen bonded. The extent of hydrogen bonding depends upon concentration.
6. The methyl groups are in the shielding region of the double bonds. See Figure 3–16.
7. The carbonyl group deshields the *ortho* hydrogens owing to its anisotropy.
8. The methyl groups are in the shielding region of the double bonded system. See Figure 3–17.
9. The spectrum will be similar to Figure 3–18, with some differences in chemical shifts. Spin arrangements: For H_A, identical to Figure 3–25; for H_B, it will see two spin states (+½ and −½) and will appear as a doublet.
10. The isopropyl group will appear as a septet for the α-H. From Pascal's triangle, the intensities are 1:6:15:20:15:6:1. The CH_3's will be a doublet.
11. Downfield doublet, area = 2, for C–1 and C–3 hydrogens; upfield triplet, area = 1, for C–2 hydrogen.
12. $X{-}CH_2CH_2{-}Y$, where $X \neq Y$.
13. Upfield triplet for C–3 hydrogens, area = 3; intermediate sextet for C–2 hydrogens, area = 2; and downfield triplet for C–1 hydrogens, area = 2.
14. Ethyl ethanoate
15. Isopropylbenzene
16. 1-Nitropropane
17. 2-Bromobutanoic acid
18. a) Methyl propanoate b) 1-Phenyl-1-propanone c) *t*-Butylbenzene d) 2-Phenylethyl ethanoate e) Phenylethanonitrile f) Ethyl dichloroethanoate g) 4-Heptanone h) Isopropyl ethanoate i) *t*-Butyl ethanoate j) Benzyl ethanoate k) Ethyl phenylethanoate l) 1-Phenylethyl ethanoate m) 3-Hydroxy-2-butanone n) 1-Phenyl-2-butanone o) 1,1-Dibromoethane p) 1,3-Dibromopropane q) 1-Bromo-1-phenylethane r) 1,2-Diphenylethane s) *N,N*-Diethylbenzylamine t) 2-Chloropropanoic acid u) 3-Chloropropanoic acid v) 4-Isopropyltoluene w) Diethyl propanedioate x) 2-Indanone y) Indane z) 2-Oxetanone (β-Propiolactone) aa) Ethyl 3-bromopropanoate bb) Ethyl 2-Bromopropanoate cc) Ethyl *N,N*-dimethylaminoethanoate

CHAPTER 4

1. Downfield doublet for CH_2; upfield singlet for CH_3
2. Triplet (equal intensity for each peak) for CH_2; singlet for CH_3
3. Downfield singlet for CH_2; upfield singlet for CH_3
4. $J_{HD} = J_{HH}/3.25 = 1.5$ Hz
5. a) $J_{ab} = 0$ Hz b) $J_{ab} \cong 10$ Hz c) $J_{ab} = 0$ Hz d) $J_{ab} \cong 3$ Hz e) $J_{ab} = 0$ Hz f) $J_{ab} \cong 10$ Hz g) $J_{ab} = 0$ Hz h) $J_{ab} = 0$ Hz i) $J_{ab} \cong 10$ Hz; $J_{ac} \cong 16$ Hz; $J_{bc} \cong 3$ Hz
6. $H_a = 2.25\ \delta$; $H_b = 4.95\ \delta$; $H_c = 5.18\ \delta$; $H_d = 6.43\ \delta$
   $J_{bd} = 17$ Hz; $J_{cd} = 10$ Hz; $J_{bc} = 0$ Hz

7.

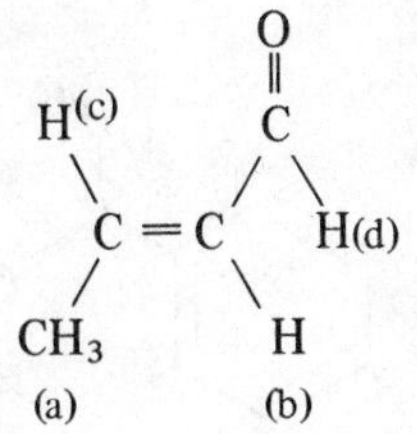

$H_a = 2.03\ \delta$; $H_b = 6.13\ \delta$; $H_c = 6.87\ \delta$; $H_d = 9.48\ \delta$

H_a is split into a doublet by H_c, and again split by H_b (allylic coupling – small J).

H_b is split by H_c into a doublet, and again split by H_d into doublets. Each peak is again split slightly by H_a (allylic coupling).

H_c is split by H_b into a doublet, and is again split into *two* quartets by H_a.

8.

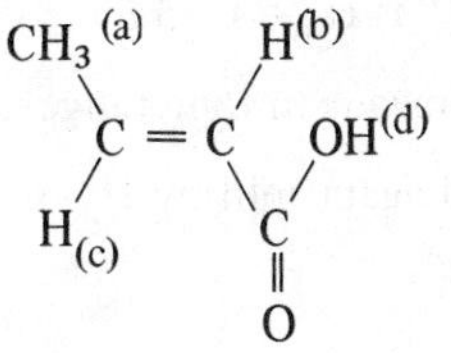

$H_a = 1.90\ \delta$; $H_b = 5.83\ \delta$; $H_c = 7.10\ \delta$; $H_d = 12.18\ \delta$

H_a is split into a doublet by H_c, and again split by H_b (allylic coupling) into two doublets.

H_b is split by H_c into a doublet, and split by H_a (allylic coupling) into quartets.

H_c is split by H_b into a doublet, and then each doublet is split further into quartets by H_a.

9. Methylene cyclohexane

10. Diethyl fumarate. Diethyl maleate would also give a similar spectrum.

$CH_3 = 1.32\ \delta$; $CH_2 = 4.27\ \delta$; vinyl-H = $6.83\ \delta$

11. Spectrum A: 4-Chlorostyrene

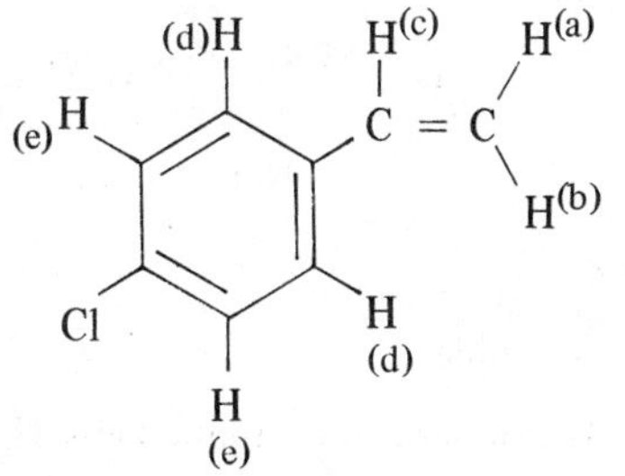

$H_a = 5.28\ \delta$; $H_b = 5.73\ \delta$; $H_c = 6.69\ \delta$; H_d and $H_e = 7.32\ \delta$

H_a is split by H_c into a doublet and again split into doublets by H_b.

H_b is split by H_c into a doublet and again split into doublets by H_a.

H_c is split by H_b into a doublet and again split by H_a into doublets.

Spectrum B: β-Chlorostyrene

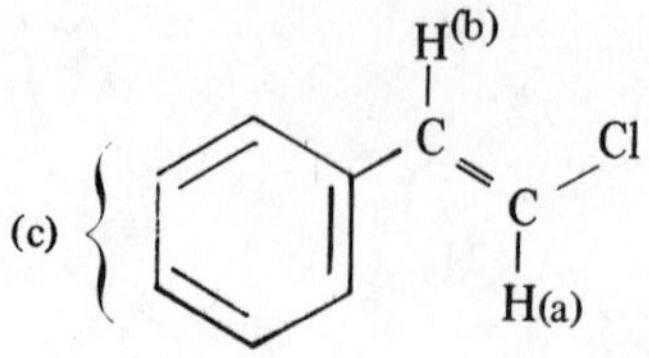

$H_a = 6.75\ \delta$; $H_b = 7.10\ \delta$; $H_c = 7.29\ \delta$

H_a is split by H_b into a doublet.

H_b is split by H_a into a doublet.

$J_{ab} = 15$ Hz (*trans*)

12. Structure A would show allylic coupling.
13. $H_a = 3.08\ \delta$; $H_b = 3.80\ \delta$; $H_c = 4.52\ \delta$; $H_d = 6.35\ \delta$

    H_a is split into a doublet by H_c by means of long range coupling.

    H_c is split into a doublet by H_d and again split by H_a through long range coupling.

    H_d is split into a doublet by H_c.
14. Alanine. The peak at 4.9 δ is from the solvent, which is D_2O. Deuterium exchange between hydroxyl and amino protons occurs, and the peak arises from HOD present in the solvent.
15. *N,N*-Dimethylethanamide. Restricted rotation makes the methyl groups non-equivalent.
16. 4-Methoxybenzaldehyde
17. 4-Ethoxyacetanilide
18. 4-Propenylanisole
19. 2,4-Dinitrophenol

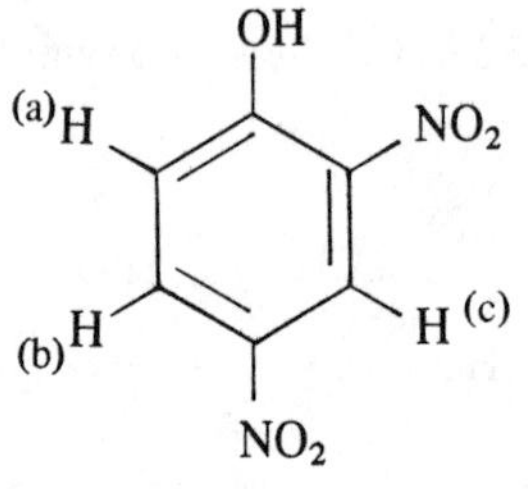

$H_a = 7.3\ \delta$; $H_b = 8.3\ \delta$; $H_c = 8.7\ \delta$

H_a is split by H_b (*ortho*) into a doublet.

H_b is split into a doublet by H_a and again into doublets by H_c (*meta*).

H_c is split into a doublet by H_b (*meta*)

20. Rapid equilibration at room temperature between chair conformations leads to one peak. As one lowers the temperature, the interconversion is slowed down until, at temperatures below −66.7°, peaks due to the axial and equatorial hydrogens are observed. Axial and equatorial hydrogens have different chemical shifts under these conditions.
21. The *t*-butyl-substituted rings are conformationally locked. The hydrogen at C–4 has different chemical shifts, depending upon whether it is axial or equatorial. 4-Bromocyclohexanes are conformationally mobile. No difference between axial and equatorial hydrogens is observed until the rate of chair-chair interconversion is decreased by lowering the temperature.

22. J_{HF} = 8 Hz. F acts like H in coupling, since it also has spin = ½. The CH_2 peak will be split by the three adjacent fluorines into a quartet. The fluorine resonance cannot be seen at 60 MHz. One needs to observe this resonance at lower field strengths.

23. γ-Valerolactone

24. A.

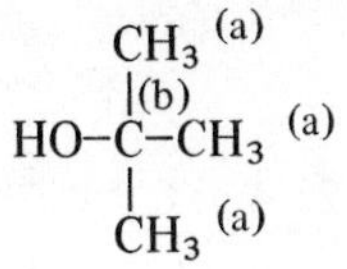

$C_a = 31.2\ \delta$; $C_b = 68.9\ \delta$

B.

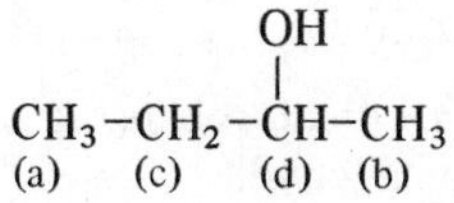

$C_a = 10.0\ \delta$; $C_b = 22.7\ \delta$; $C_c = 32.0\ \delta$; $C_d = 69.2\ \delta$

C.

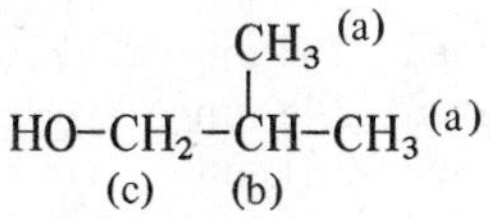

$C_a = 18.9\ \delta$; $C_b = 30.8\ \delta$; $C_c = 69.4\ \delta$

25. D.

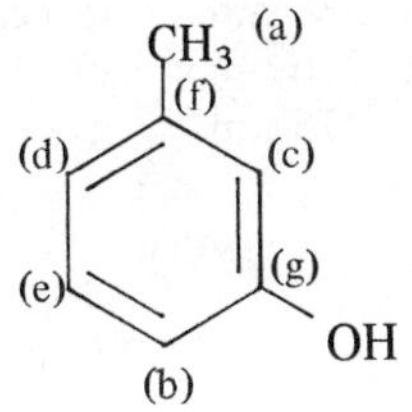

C_a = 21.1 δ; C_b = 112.5 δ; C_c = 116.2 δ; C_d = 121.8 δ; C_e = 129. 4 δ; C_f = 139.8 δ; C_g = 155.0 δ

E.

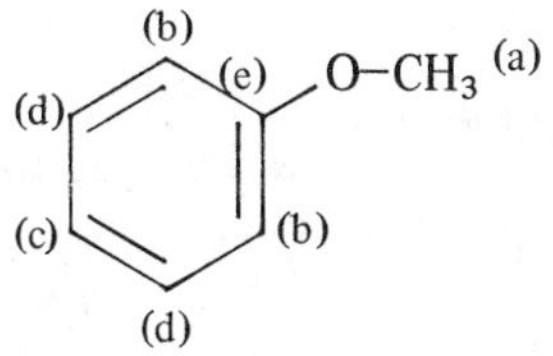

$C_a = 54.7\ \delta$; $C_b = 114.1\ \delta$; $C_c = 120.7\ \delta$; $C_d = 129.5\ \delta$; $C_e = 160.2\ \delta$

26. F.

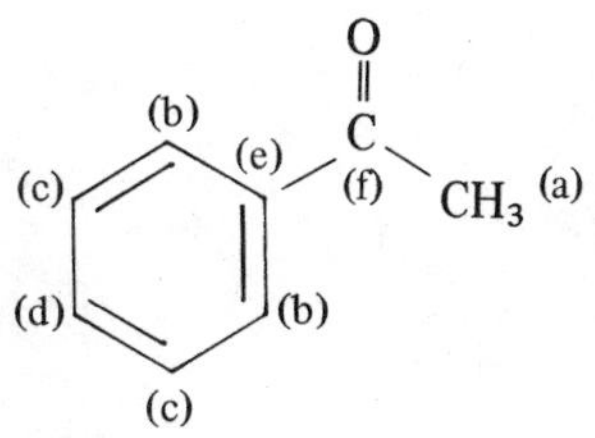

$C_a = 26.3\ \delta$; $C_b = 128.2\ \delta$; $C_c = 128.4\ \delta$; $C_d = 132.9\ \delta$; $C_e = 137.1\ \delta$; $C_f = 197.6\ \delta$

G.

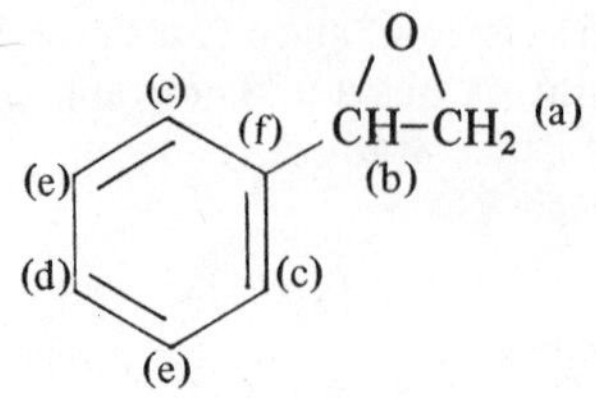

C_a = 50.8 δ; C_b = 52.1 δ; C_c = 125.4 δ; C_d = 128.0 δ; C_e = 128.4 δ; C_f = 137.7 δ

CHAPTER 5

1. a) $\epsilon = 13{,}000$ b) $I_0/I = 1.26$

2. a) 2,4-Dichlorobenzoic acid or 3,4-dichlorobenzoic acid b) 4,5-Dimethyl-4-hexen-3-one
c) 2-Methyl-1-cyclohexenecarboxaldehyde

3. a) calculated: 215 nm observed: 213 nm b) calculated: 225 nm observed: 220 nm
c) calculated: 227 nm observed: 224 nm d) calculated: 249 nm observed: 249 nm
e) calculated: 214 nm observed: 218 nm f) calculated: 254 nm observed: 256 nm
g) calculated: 280 nm observed: 284 nm h) calculated: 317 nm observed: 315 nm
i) calculated: 351 nm observed: 348 nm j) calculated: 234 nm observed: 236 nm
k) calculated: 229 nm observed: 231 nm l) calculated: 273 nm observed: 265 nm
m) calculated: 244 nm observed: 245 nm n) calculated: 283 nm observed: 282 nm
o) calculated: 313 nm observed: 315 nm p) calculated: 303 nm observed: 306 nm
q) calculated: 359 nm observed: 361 nm r) calculated: 249 nm observed: 245 nm
s) calculated: 281 nm observed: 288 nm t) calculated: 292 nm observed: 295 nm
u) calculated: 274 nm observed: 276 nm v) calculated: 281 nm observed: 278 nm
w) calculated: 269 nm observed: 275 nm x) calculated: 275 nm observed: 274 nm

4. 166 nm: $\pi \rightarrow \pi^*$
189 nm: $n \rightarrow \sigma^*$
279 nm: $n \rightarrow \pi^*$

5. Each transition is due to $n \rightarrow \sigma^*$ transitions. As one goes from the chloro to the bromo to the iodo group, the electronegativity of the halogens decreases. The orbitals interact to different degrees, and the energies of the *n* and the σ^* states differ.

6. It would be shifted by 11 nm to *lower* wavelength

7. The band was a $\pi \rightarrow \pi^*$ band.

8. a) $\sigma \rightarrow \sigma^*$, $\sigma \rightarrow \pi^*$, $\pi \rightarrow \pi^*$, and $\pi \rightarrow \sigma^*$
b) $\sigma \rightarrow \sigma^*$, $\sigma \rightarrow \pi^*$, $\pi \rightarrow \pi^*$, $\pi \rightarrow \sigma^*$, $n \rightarrow \sigma^*$, and $n \rightarrow \pi^*$
c) $\sigma \rightarrow \sigma^*$ and $n \rightarrow \sigma^*$ d) $\sigma \rightarrow \sigma^*$, $\sigma \rightarrow \pi^*$, $\pi \rightarrow \pi^*$, $\pi \rightarrow \sigma^*$, $n \rightarrow \sigma^*$, and $n \rightarrow \pi^*$
e) $\sigma \rightarrow \sigma^*$ and $n \rightarrow \sigma^*$ f) $\sigma \rightarrow \sigma^*$

CHAPTER 6

1. $C_{43}H_{50}N_4O_6$

2. $C_{34}H_{44}O_{13}$

3. $C_{12}H_{10}O$

4. C_6H_{12}

5. C_7H_9N

6. C_3H_7Cl

7. See Section 6.5H. Each of the compounds can be used to confirm at least one of the steps shown in the mechanism of fragmentation of cyclohexanone.

8. a) Cyclopropane
b) 2-Methyl-1-pentene
c) 2-Methylcyclohexanol
d) Ethyl isobutyl ether
e) Diisobutyl ether
f) 2-Methylpropanal
g) 2-Propenal
h) 3-Methylcyclohexanone
i) 3-Methyl-2-heptanone
j) Ethyl octanoate
k) 2-Methylpropanoic acid
l) 2,3-Dimethylbenzoic acid
m) Butylamine
n) 2,6-Dimethylpyridine
o) Tributylamine
p) 2-Propanethiol
q) Nitroethane
r) Propanonitrile
s) Iodoethane
t) Chlorobenzene
u) 1-Bromobutane
v) Bromobenzene
w) 1,1-Dichloroethane
x) Carbon tetrachloride

CHAPTER 7

1. 2-Butanone
2. 1-Propanol
3. 3-Pentanone
4. Methyl 2,2-dimethylpropanoate (Methyl pivalate)
5. Phenylacetic acid
6. 4-Bromophenol
7. Isopropyl benzoate
8. Diethyl phthlate
9. Diisopropylamine
10. Ethyl cyanoacetate
11. 2-Pentanone
12. Ethyl formate
13. 4-Chloroacetophenone
14. 3-Methylbutanal (Isovaleraldehyde)
15. 3-Methyl-1-butanol
16. 3-Pentanol
17. Salicylaldehyde
18. 4-Methyl-3-penten-2-one (Mesityl oxide)
19. 1-Chloro-4-nitrobenzene
20. 2-Aminoethanol
21. 1-Bromo-2-methylpropane
22. Ethyl 3-phenylpropenoate (Ethyl cinnamate)
23. Methyl anthranilate
24. 1,4-Dibromobutane
25. Ethyl *p*-toluenesulfonate
26. 2,2-Dimethoxypropane
27. Ethyl acetoacetate
28. $4\text{-}H_2NC_6H_4\text{—}COOCH_2CH_2N(CH_2CH_3)_2$

APPENDICES

APPENDIX ONE

INFRARED ABSORPTION FREQUENCIES OF COMMON FUNCTIONAL GROUPS

(N.B. Colthup; Courtesy of American Cyanamide Company, Stamford, Conn. Used with permission of the Journal of the Optical Society of America.)

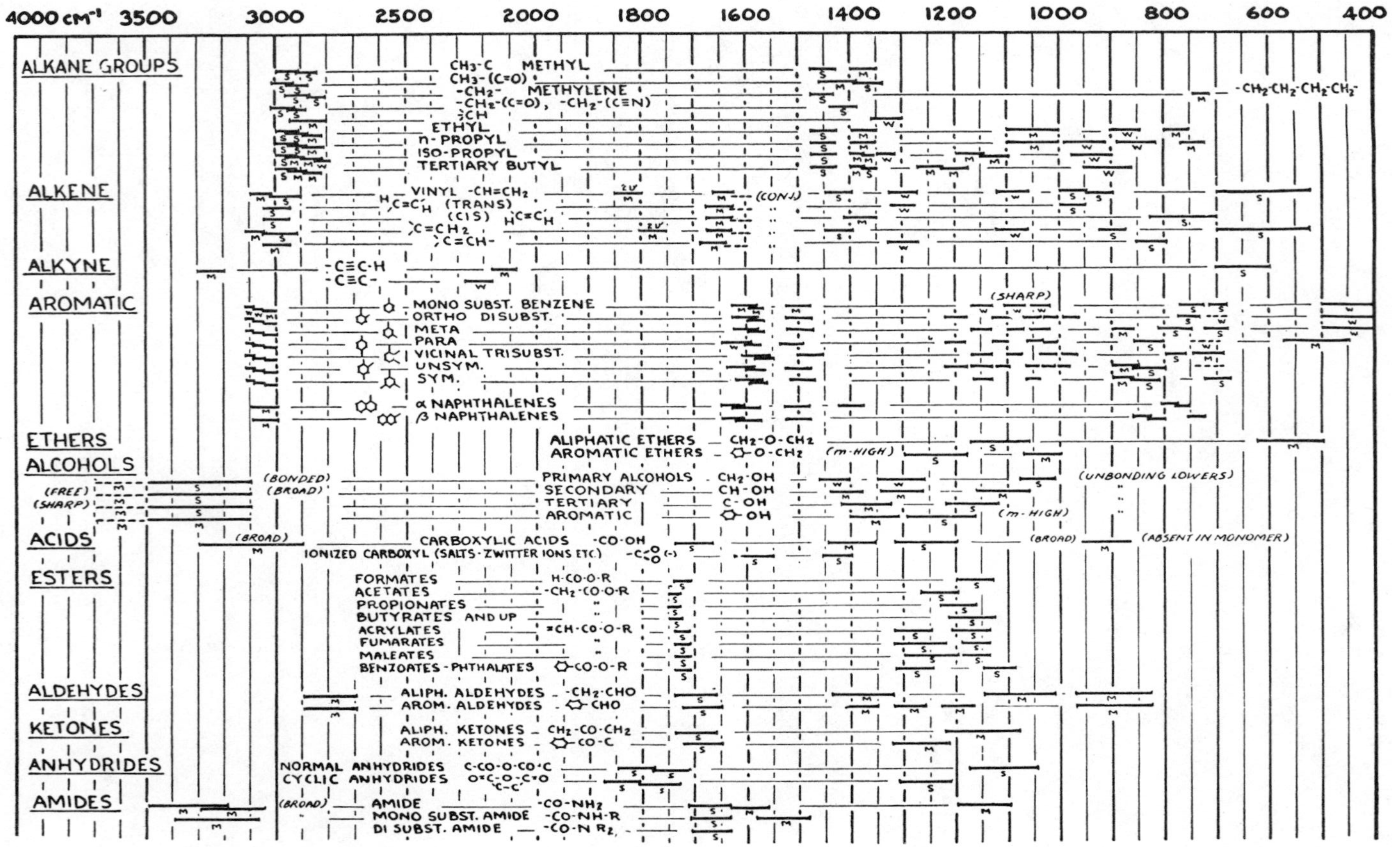

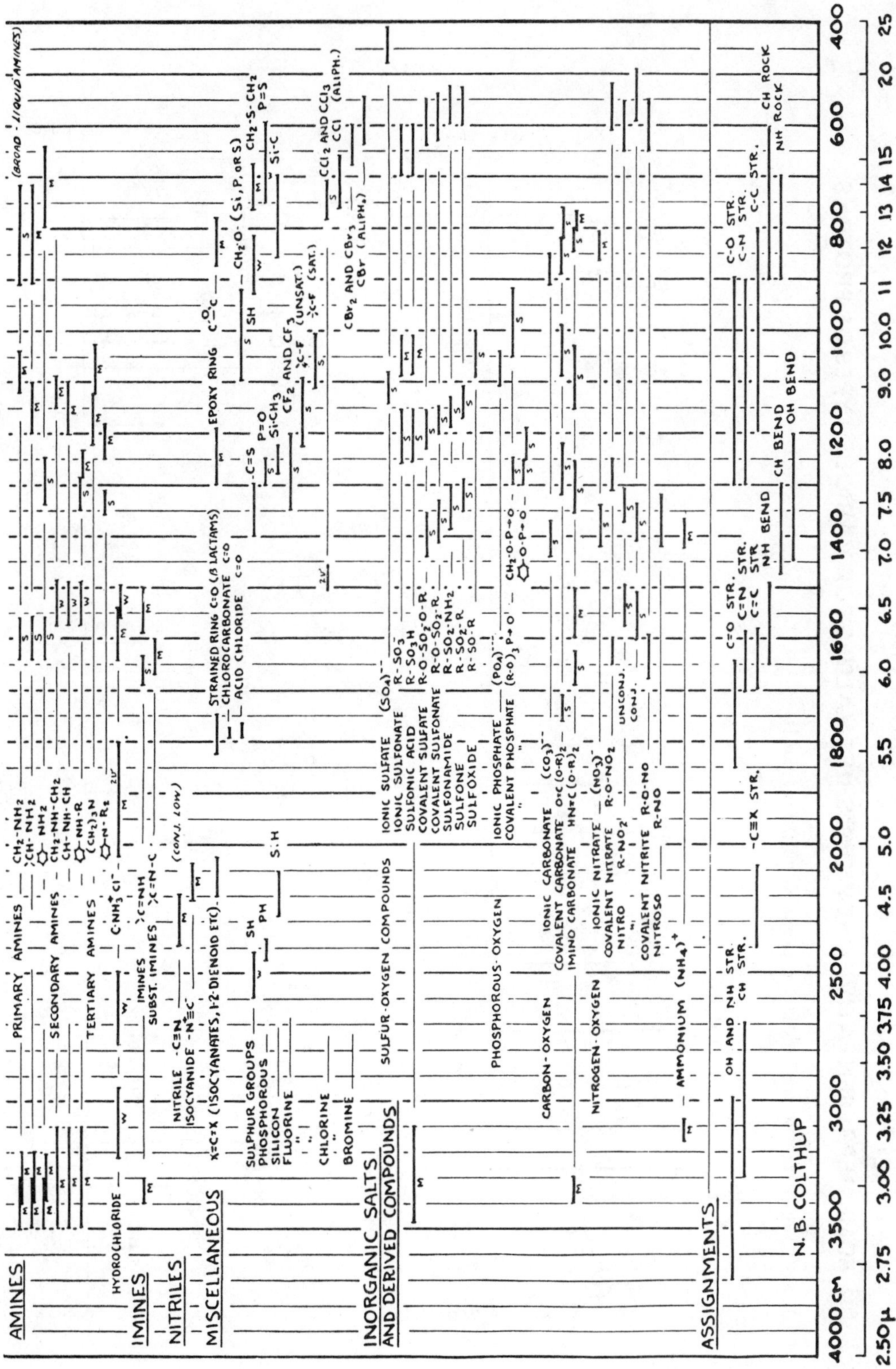
AMINES
IMINES
NITRILES
MISCELLANEOUS
INORGANIC SALTS AND DERIVED COMPOUNDS
ASSIGNMENTS
N. B. COLTHUP
4000 CM 3500 3000 2500 2000 1800 1600 1400 1200 1000 800 600 400
2.50µ 2.75 3.00 3.25 3.50 3.75 4.00 4.5 5.0 5.5 6.0 6.5 7.0 7.5 8.0 9.0 10.0 11 12 13 14 15 20 25

APPENDIX TWO

SOME REPRESENTATIVE CHEMICAL SHIFT VALUES FOR VARIOUS TYPES OF PROTONS*

(Chemical shift values refer to the boldface protons **H**, not to regular H)

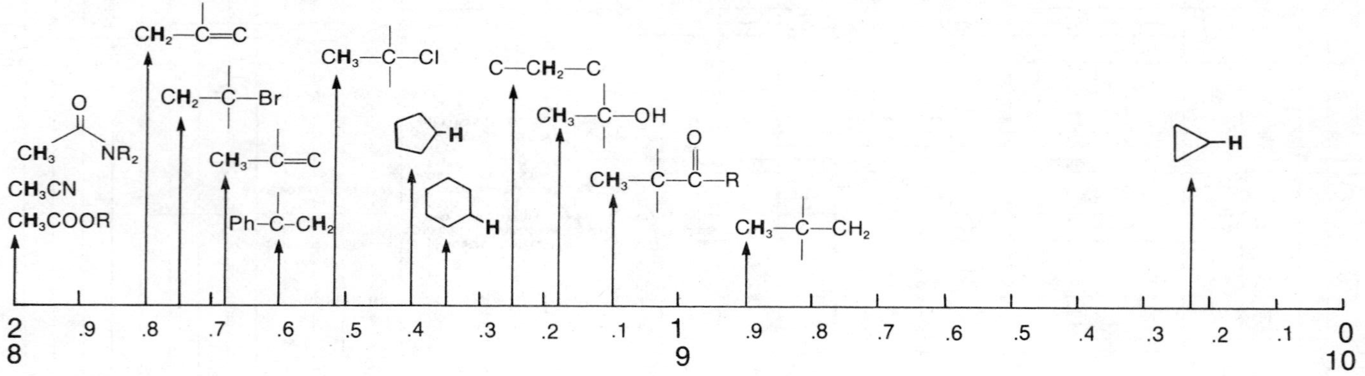

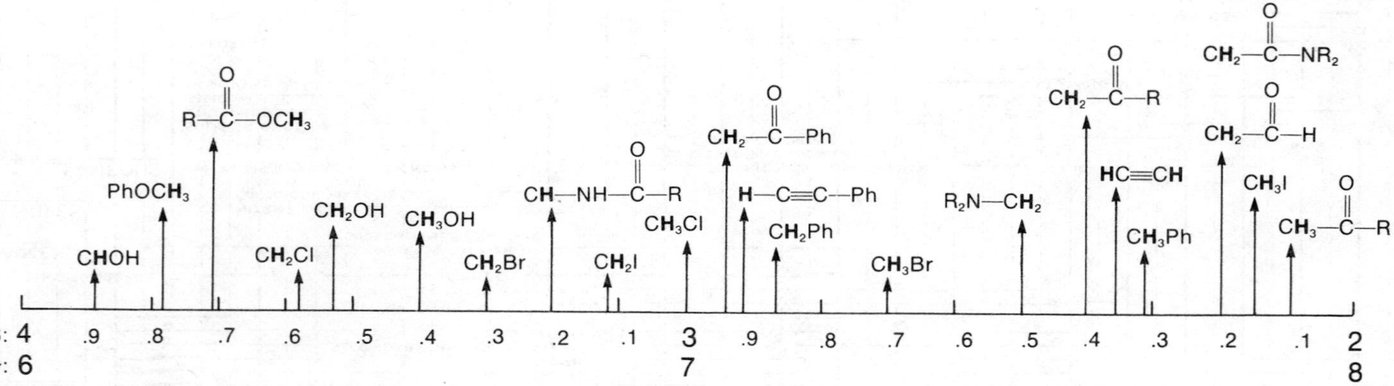

*Adapted with permission from J. A. Landgrebe, "Theory and Practice in the Organic Laboratory," Second Edition, D. C. Heath, Lexington, Mass., 1977.

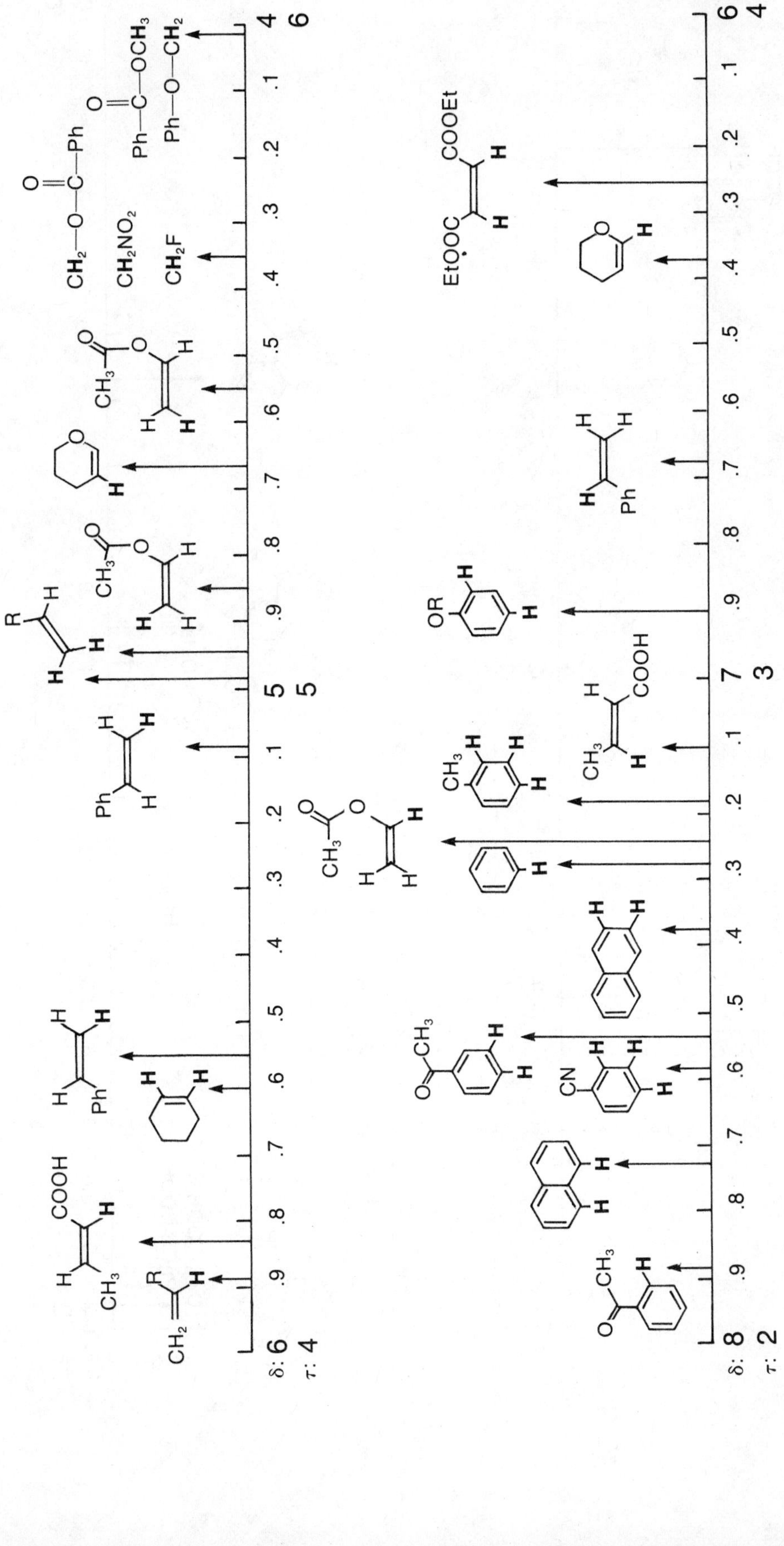
δ: 6
τ: 4
δ: 8
τ: 2

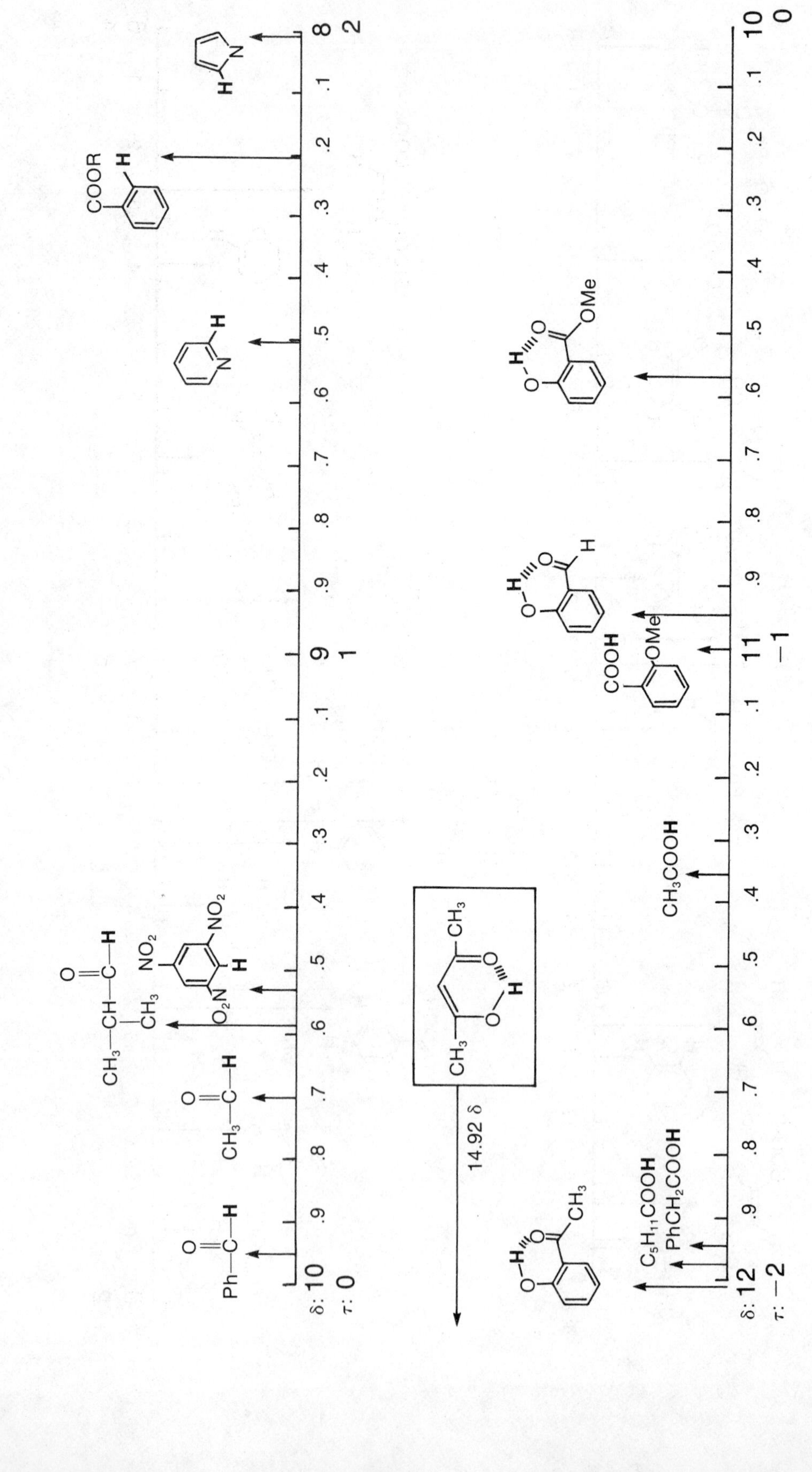
δ: 10
τ: 0
δ: 12
τ: −2
14.92 δ
CH3COOH
C5H11COOH
PhCH2COOH

APPENDIX THREE

TYPICAL PROTON-PROTON COUPLING CONSTANTS

Alkanes

		Typical Value	*Range*	
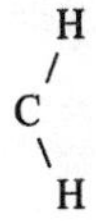	geminal	12 Hz	12–15 Hz	(depends on HCH angle)
	vicinal	7	6–8	(depends on HCCH dihedral angle)
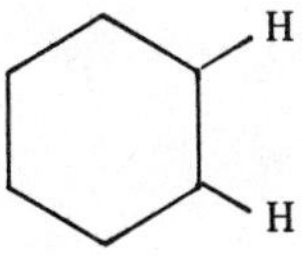	a, a a, e e, e	10 5 3	8–14 0–7 0–5	in conformationally rigid systems (in systems undergoing inversion, all $J \approx$ 7–8 Hz)
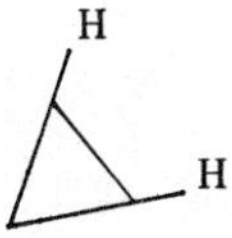	*cis* *trans*	9 6	6–12 4–8	
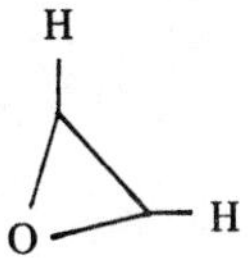	*cis* *trans*	3 4	2–4 2–5	
		0	0–7	(W config. obligatory – strained systems have the larger values)

Alkenes/Alkynes

		Typical Value	*Range*
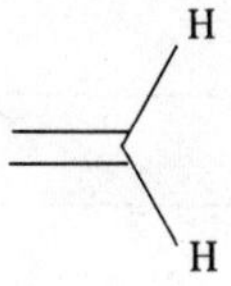	gem	0 Hz	0–5 Hz
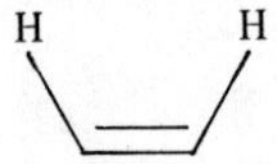	*cis*	10	6–15
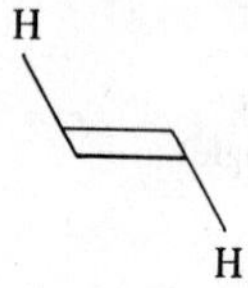	*trans*	16	11–18
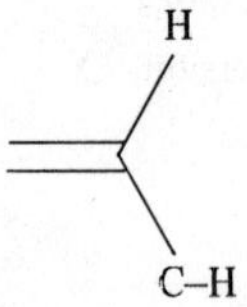	vic	5	4–10
H–C=C–CH Allylic	(*cis* or *trans*)	1	0–3
HC–C=C–CH Homoallylic		0	0–1.5
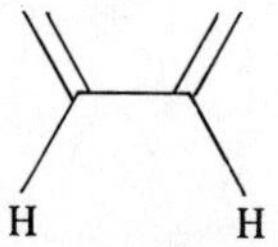		10	9–13

	Typical Value	*Range*
H–C≡C–CH Allylic	2 Hz	2-3 Hz
HC–C≡C–CH Homoallylic	2	2–3
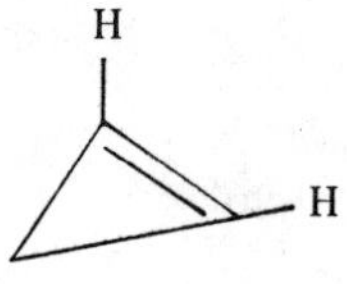	2	0–2
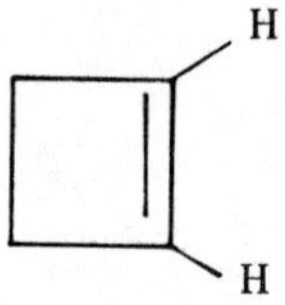	4	2–4
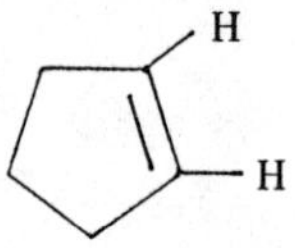	6	5–7
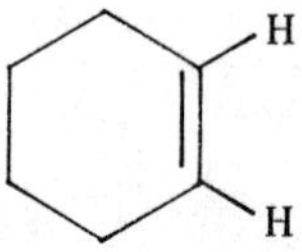	10	8–11

Aromatics/Heterocycles

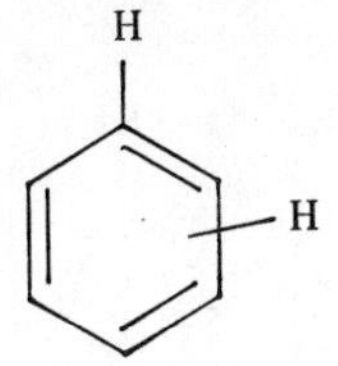

	Typical Value	Range
ortho	8 Hz	6–10 Hz
meta	3	1–4
para	1	0–2

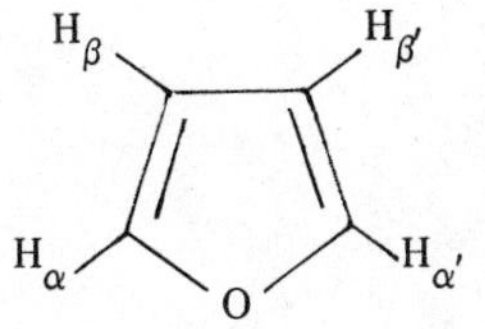

	Range
$\alpha\beta$	1.6–2.0
$\alpha\beta'$	0.6–1.0
$\alpha\alpha'$	1.3–1.8
$\beta\beta'$	3.2–3.8

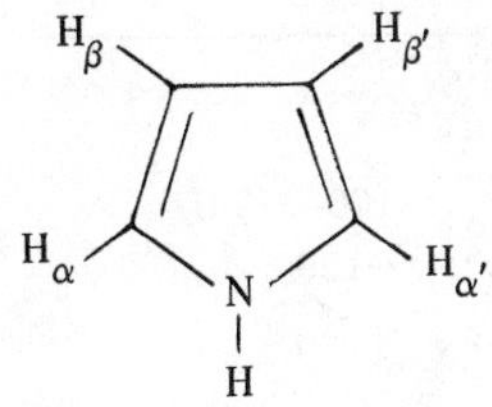

	Range
$\alpha\beta'$	2.0–2.6
$\alpha\beta'$	1.5–2.2
$\alpha\alpha'$	1.8–2.3
$\beta\beta'$	2.8–4.0

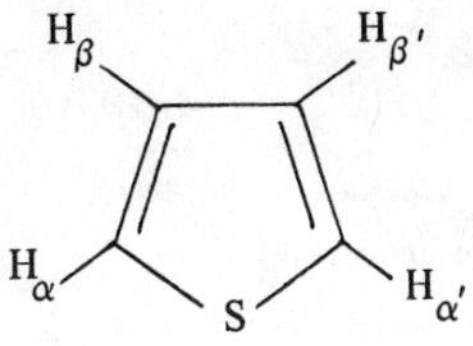

	Range
$\alpha\beta$	4.6–5.8 Hz
$\alpha\beta'$	1.0–1.8
$\alpha\alpha'$	2.1–3.3
$\beta\beta'$	3.0–4.2

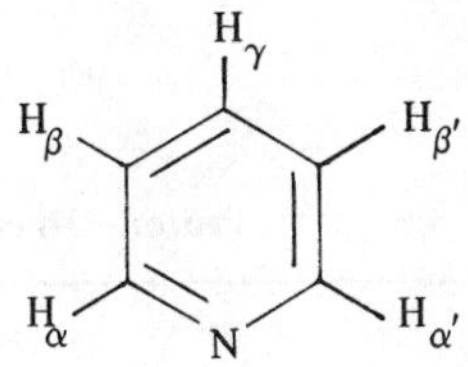

	Range
$\alpha\beta$	4.9–5.7
$\alpha\gamma$	1.6–2.6
$\alpha\beta'$	0.7–1.1
$\alpha\alpha'$	0.2–0.5
$\beta\gamma$	7.2–8.5
$\beta\beta'$	1.4–1.9

Alcohols

	Typical Value	*Range*
H–C(–)–OH (no exchange occurring)	5 Hz	4–10 Hz

Aldehydes

	Typical Value	*Range*
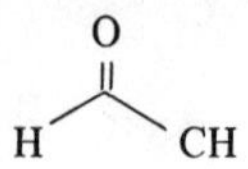	2 Hz	1–3 Hz
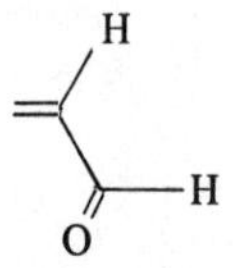	6	5–8

Proton–Other Nucleus Coupling Constants

(Typical Values)

H–C–F (geminal)	~60 Hz	N–H	~52 Hz	^{13}C–H	100–250 Hz
				sp^3	~120
				sp^2	170
H–C–C–F	~20 Hz	H–C–N–H	0 Hz	*sp*	250
				(satellite peaks– *J* depends on hybridization)	
D–C–H	~2 Hz	P–H	~34 Hz		
H–C–C–D	<1 Hz (leads only to peak broadening)	^{13}C–C–H	~34 Hz		

APPENDIX FOUR

Tables of Masses and Isotopic Abundance Ratios for Molecular Ions Under Mass 100 Containing Carbon, Hydrogen, Nitrogen, and Oxygen*

	M + 1	M + 2
16		
CH_4	1.15	
17		
NH_3	0.43	
18		
H_2O	0.07	0.20
26		
C_2H_2	2.19	0.01
27		
CHN	1.48	
28		
N_2	0.76	
CO	1.12	
C_2H_4	2.23	0.01
29		
CH_3N	1.51	
30		
CH_2O	1.15	0.20
C_2H_6	2.26	0.01
31		
CH_5N	1.54	
32		
O_2	0.08	0.40
N_2H_4	0.83	
41		
C_2H_3N	2.59	0.02
42		
CH_2N_2	1.88	0.01
C_2H_2O	2.23	0.21
C_3H_6	3.34	0.04
43		
C_2H_5N	2.62	0.02
44		
N_2O	0.80	0.20
CO_2	1.16	0.40
C_2H_4O	2.26	0.21
C_3H_8	3.37	0.04
45		
CH_3NO	1.55	0.21
C_2H_7N	2.66	0.02

	M + 1	M + 2
46		
NO_2	0.46	0.40
CH_2O_2	1.19	0.40
C_2H_6O	2.30	0.22
47		
CH_5NO	1.58	0.21
48		
CH_4O_2	1.22	0.40
52		
C_4H_4	4.39	0.07
53		
C_3H_3N	3.67	0.05
54		
C_4H_6	4.42	0.07
55		
C_3H_5N	3.70	0.05
56		
$C_2H_4N_2$	2.99	0.03
C_3H_4O	3.35	0.24
C_4H_8	4.45	0.08
57		
C_3H_7N	3.74	0.05
58		
$C_2H_2O_2$	2.27	0.42
$C_2H_6N_2$	3.02	0.03
C_3H_6O	3.38	0.24
C_4H_{10}	4.48	0.08
59		
C_2H_5NO	2.66	0.22
C_3H_9N	3.77	0.05
60		
$C_2H_4O_2$	2.30	0.04
$C_2H_8N_2$	3.05	0.03
C_3H_8O	3.41	0.24
61		
CH_3NO_2	1.59	0.41
C_2H_7NO	2.69	0.22
62		
CH_2O_3	1.23	0.60
$C_2H_6O_2$	2.34	0.42

*Adapted with permission from J. H. Beynon, "Mass Spectrometry and Its Application to Organic Chemistry," Elsevier, Amsterdam, 1960.

	M + 1	M + 2
64		
CH_4O_3	1.26	0.60
66		
C_5H_6	5.50	0.12
67		
C_4H_5N	4.78	0.09
68		
C_4H_4O	4.43	0.28
C_5H_8	5.53	0.12
69		
C_4H_7N	4.82	0.09
70		
$C_3H_2O_2$	3.35	0.44
$C_3H_6N_2$	4.10	0.07
C_4H_6O	4.46	0.28
C_5H_{10}	5.56	0.13
71		
C_4H_9N	4.85	0.09
72		
$C_3H_4O_2$	3.38	0.44
$C_3H_8N_2$	4.13	0.07
C_4H_8O	4.49	0.28
C_5H_{12}	5.60	0.13
73		
C_3H_7NO	3.77	0.25
$C_4H_{11}N$	4.88	0.10
74		
$C_3H_6O_2$	3.42	0.44
$C_3H_{10}N_2$	4.17	0.07
$C_4H_{10}O$	4.52	0.28
75		
$C_2H_5NO_2$	2.70	0.43
C_3H_9NO	3.81	0.25
76		
$C_2H_4O_3$	2.34	0.62
$C_2H_8N_2O$	3.09	0.24
$C_3H_8O_2$	3.45	0.44
77		
$C_2H_7NO_2$	2.73	0.43
78		
$C_2H_6O_3$	2.38	0.62
C_6H_6	6.58	0.18
79		
C_5H_5N	5.87	0.14
80		
C_6H_8	6.61	0.18
81		
C_5H_7N	5.90	0.14

	M + 1	M + 2
82		
C_5H_6O	5.54	0.32
C_6H_{10}	6.64	0.19
83		
C_5H_9N	5.93	0.15
84		
$C_4H_4O_2$	4.47	0.48
$C_4H_8N_2$	5.21	0.11
C_5H_8O	5.57	0.33
C_6H_{12}	6.68	0.19
85		
C_4H_7NO	4.86	0.29
$C_5H_{11}N$	5.96	0.15
86		
$C_4H_6O_2$	4.50	0.48
$C_4H_{10}N_2$	5.25	0.11
$C_5H_{10}O$	5.60	0.33
C_6H_{14}	6.71	0.19
87		
$C_3H_5NO_2$	3.78	0.45
C_4H_9NO	4.89	0.30
$C_5H_{13}N$	5.99	0.15
88		
$C_4H_8O_2$	4.53	0.48
$C_4H_{12}N_2$	5.28	0.11
$C_5H_{12}O$	5.63	0.33
89		
$C_3H_7NO_2$	3.81	0.46
$C_4H_{11}NO$	4.92	0.30
90		
$C_3H_6O_3$	3.46	0.64
$C_3H_{10}N_2O$	4.20	0.27
$C_4H_{10}O_2$	4.56	0.48
91		
$C_2H_5NO_3$	2.74	0.63
$C_2H_9N_3O$	3.49	0.25
$C_3H_9NO_2$	3.85	0.46
92		
$C_3H_8O_3$	3.49	0.64
C_7H_8	7.69	0.26
93		
$C_2H_7NO_3$	2.77	0.63
C_6H_7N	6.98	0.21
94		
$C_2H_6O_4$	2.41	0.82
$C_5H_6N_2$	6.26	0.17
C_6H_6O	6.62	0.38
C_7H_{10}	7.72	0.26

	M + 1	M + 2
95		
$C_4H_5N_3$	5.55	0.13
C_5H_5NO	5.90	0.34
C_6H_9N	7.01	0.21
96		
$C_5H_8N_2$	6.29	0.17
C_6H_8O	6.65	0.39
C_7H_{12}	7.76	0.26
97		
C_5H_7NO	5.94	0.35
$C_6H_{11}N$	7.04	0.21
98		
$C_5H_6O_2$	5.58	0.53
$C_5H_{10}N_2$	6.33	0.17
$C_6H_{10}O$	6.68	0.39
C_7H_{14}	7.79	0.26

	M + 1	M + 2
99		
$C_4H_5NO_2$	4.86	0.50
$C_4H_9N_3$	5.61	0.13
C_5H_9NO	5.97	0.35
$C_6H_{13}N$	7.07	0.21
100		
$C_4H_8N_2O$	5.25	0.31
$C_5H_8O_2$	5.61	0.53
$C_5H_{12}N_2$	6.36	0.17
$C_6H_{12}O$	6.72	0.39
C_7H_{16}	7.82	0.26

APPENDIX FIVE

Common Fragment Ions Under Mass 105*

m/e	Ions
14	CH_2
15	CH_3
16	O
17	OH
18	H_2O
	NH_4
19	F
	H_3O
26	$C{\equiv}N$
27	C_2H_3
28	C_2H_4
	CO
	N_2 (air)
	$CH{=}NH$
29	C_2H_5
	CHO
30	CH_2NH_2
	NO
31	CH_2OH
	OCH_3
32	O_2 (air)
33	SH
	CH_2F
34	H_2S
35	Cl
36	HCl
39	C_3H_3
40	$CH_2C{=}N$
41	C_3H_5
	$CH_2C{=}N$ + H
	C_2H_2NH
42	C_3H_6
43	C_3H_7
	$CH_3C{=}O$
	C_2H_5N
44	$CH_2CH{=}O$ + H
	CH_3CHNH_2
	CO_2
	$NH_2C{=}O$
	$(CH_3)_2N$
45	CH_3CHOH
	CH_2CH_2OH
	CH_2OCH_3
	$\overset{\overset{O}{\|}}{C}{-}OH$
	$CH_3CH{-}O$ + H
46	NO_2
47	CH_2SH
	CH_3S
48	CH_3S + H
49	CH_2Cl
51	CHF_2
53	C_4H_5
54	$CH_2CH_2C{\equiv}N$
55	C_4H_7
	$CH_2{=}CHC{=}O$
56	C_4H_8
57	C_4H_9
	$C_2H_5C{=}O$
58	$CH_3{-}\underset{\underset{CH_2}{\vert}}{C}{=}O$ + H
	$C_2H_5CHNH_2$
	$(CH_3)_2NHCH_2$
	$C_2H_5NHCH_2$
	C_2H_2S
59	$(CH_3)_2COH$
	$CH_2OC_2H_5$
	$\overset{\overset{O}{\|}}{C}{-}OCH_3$
	$NH_2\underset{\underset{CH_2}{\vert}}{C}{=}O$ + H
	CH_3OCHCH_3
	CH_3CHCH_2OH
60	$CH_2\underset{\underset{OH}{\vert}}{C}{=}O$ + H
	CH_2ONO
61	$\overset{\overset{O}{\|}}{C}{-}OCH_3$ + 2H
	CH_2CH_2SH
	CH_2SCH_3

*Adapted with permission from R. M. Silverstein, G. C. Bassler, and T. C. Morrill, "Spectrometric Identification of Organic Compounds," Third Edition, John Wiley and Sons, Inc., New York, 1974.

m/e	Ions
65	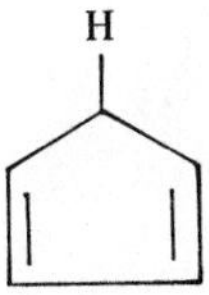 (or C_5H_5)
66	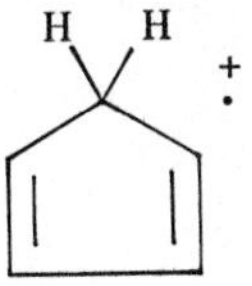 (or C_5H_6)
67	C_5H_7
68	$CH_2CH_2CH_2C{\equiv}N$
69	C_5H_9
	CF_3
	$CH_3CH{=}CHC{=}O$
	$CH_2{=}C(CH_3)C{=}O$
70	C_5H_{10}
71	C_5H_{11}
	$C_3H_7C{=}O$
72	$C_2H_5\overset{O}{\overset{\|}{C}}{-}CH_2$
	$C_3H_7CHNH_2$
	$(CH_3)_2N{=}C{=}O$
	$C_2H_5NHCHCH_3$ and isomers
73	Homologs of 59
74	$CH_2{-}\overset{O}{\overset{\|}{C}}{-}OCH_3$ + H
75	$\overset{O}{\overset{\|}{C}}{-}OC_2H_5$ + 2H
	$CH_2SC_2H_5$
	$(CH_3)_2CSH$
	$(CH_3O)_2CH$
77	C_6H_5
78	C_6H_5 + H
79	C_6H_5 + 2H
	Br
80	CH_3SS + H
81	C_6H_9
	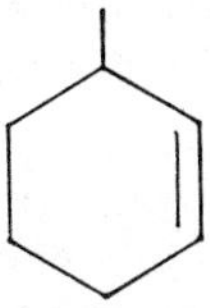
82	$CH_2CH_2CH_2CH_2C{\equiv}N$
	CCl_2
	C_6H_{10}
83	C_6H_{11}
	$CHCl_2$
85	C_6H_{13}
	$C_4H_9C{=}O$
	$CClF_2$
86	$C_3H_7\overset{O}{\overset{\|}{C}}{-}CH_2$ + H
	$C_4H_9CHNH_2$ and isomers
87	$C_3H_7\overset{O}{\overset{\|}{C}}O$
	Homologs of 73
	$CH_2CH_2\underset{O}{\underset{\|}{C}}OCH_3$
88	$CH_2{-}\overset{O}{\overset{\|}{C}}{-}OC_2H_5$ + H
89	$\overset{O}{\overset{\|}{C}}{-}OC_3H_7$ + 2H
	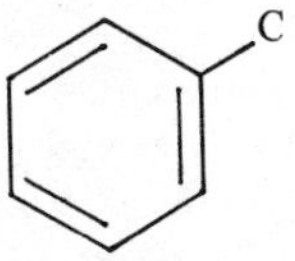
90	CH_3CHONO_2
	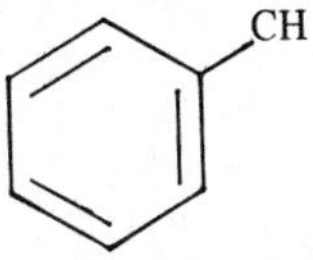
91	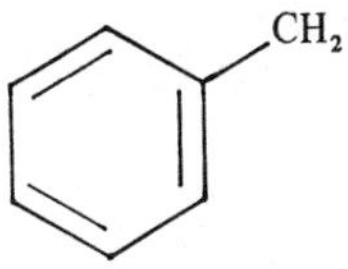 or 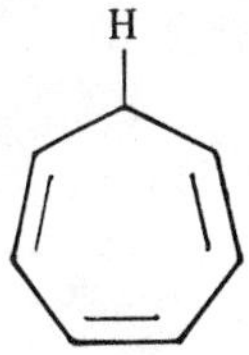
	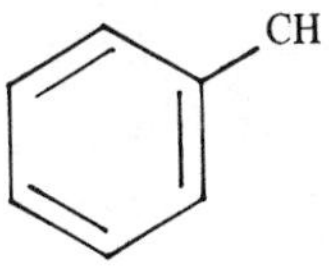 + H
	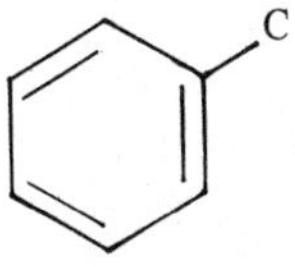 + 2H
	$(CH_2)_4Cl$
	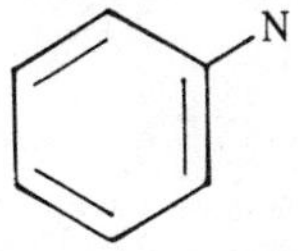

m/e	Ions
92	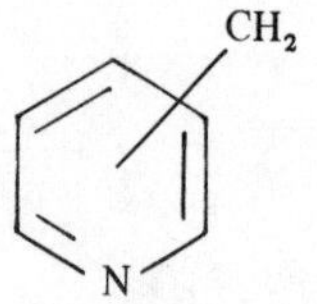
	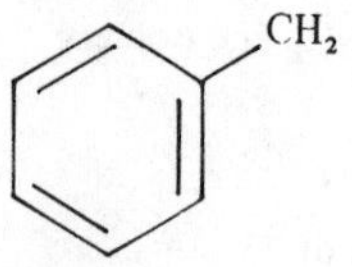 + H
93	CH_2Br
	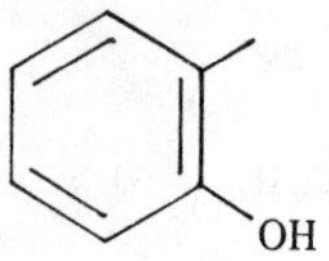
	C_7H_9
	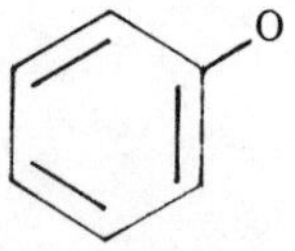
94	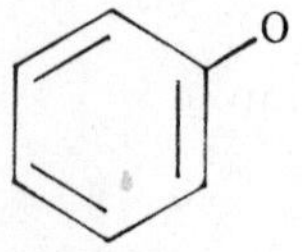 + H
96	$CH_2CH_2CH_2CH_2CH_2C{\equiv}N$
97	C_7H_{13}
99	C_7H_{15}
	$C_6H_{11}O$
100	$C_4H_9C(=O){-}CH_2$ + H
	$C_5H_{11}CHNH_2$
101	$C(=O){-}OC_4H_9$
102	$CH_2C(=O){-}OC_3H_7$ + H
103	$C(=O){-}OC_4H_9$ + 2H
	$C_5H_{11}S$
	$CH(OCH_2CH_3)_2$
104	$C_2H_5CHONO_2$
105	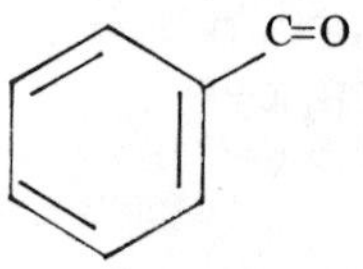
	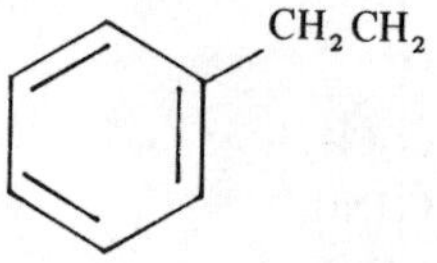
	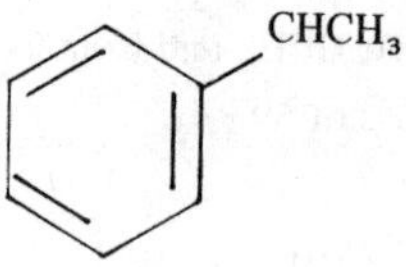

APPENDIX SIX
INDEX OF SPECTRA

INFRARED SPECTRA:

MASS SPECTRA:

INDEX

Entries are alphabetized by disregarding prefixes.